POLLUTION,
POLITICS,
and POWER

Thomas O. McGarity

POLLUTION, POLITICS, and POWER

The Struggle for Sustainable Electricity

Harvard University Press

Cambridge, Massachusetts, and London, England   2019

First printing

Library of Congress Cataloging-in-Publication Data

Names: McGarity, Thomas O., author.

Title: Pollution, politics, and power : the struggle for sustainable
  electricity / Thomas O. McGarity.

Description: Cambridge, Massachusetts : Harvard University Press,
  2019. | Includes bibliographical references and index.

Identifiers: LCCN 2019006756 | ISBN 9780674545434 (hardcover : alk.
  paper)

Subjects: LCSH: Electric utilities—United States—History. | Electric
  power—United States—History. | Environmental policy—United
  States—History. | Renewable energy sources—United States—
  History. | Coal trade—United States—History. | Coal-fired power
  plants—United States—History.

Classification: LCC HD9685.U5 M3333 2019 | DDC 333.793/20973—
  dc23 LC record available at https://lccn.loc.gov/2019006756

This book is dedicated to my grandsons, Brendan, William, and Ryan

Contents

POLLUTION,
POLITICS,
and POWER

Introduction

The Kentucky Coal Mining Museum is housed in a former commissary next to the City Hall of the tiny town of Benham in Harlan County, Kentucky. Owned and operated by the Southeast Kentucky Community and Technical College, the museum is a tribute to the state's most prominent natural resource, the mining industry that dominates the state's economy and politics, and the miners, past and present, who bored deep into the Kentucky mountains to extract what President Donald J. Trump has called an "indestructible" mineral. The museum features dozens of exhibits depicting work in the mines, life in the small mining towns spread throughout the Appalachian Mountains, and models of mines and the massive power plants that burn the coal to light the homes and businesses of residents throughout the Eastern Seaboard and the Midwest. To gain a better sense of what it is like to work in the confined space of a coal mine, visitors may don hard hats and tour a former working mine.

The lights, air conditioners, computers, and other electronic equipment in the museum are powered not by coal but by the sun. In April 2017, the museum installed rooftop solar panels to supply most of its electricity and send some back to the grid to partially offset the power it draws from the grid after the sun sets. Unlike President Carter, who had solar panels installed on the White House roof as a symbolic measure to encourage greater reliance on renewable resources (only to have them removed by President Reagan, who considered them a "joke"), the college installed the panels because the electricity they provided was much cheaper than the power it could purchase from coal-fired plants. Throughout the country, thousands of commercial establishments and millions of homeowners have installed solar panels for the same reason. In August 2018, electric power giant Alliant Energy announced that it would join several other electric power companies in eliminating coal use altogether by 2050.[1]

While coal may or may not be "indestructible," coal mining companies are fragile. They teeter on the edge of bankruptcy as electric power companies throughout the country shutter coal-fired power plants and replace them with turbines fired by natural gas, wind farms, solar arrays, and "negawatts" of electricity produced by their customers as they employ "smart meters" to reduce energy consumption during times when demand is high or supplies are low. Lower demand for coal for power plants has caused mines to close throughout Appalachia with a corresponding loss of jobs. Communities throughout the region have suffered from high unemployment, reduced resources, and a spiraling opioid epidemic.

One of the reasons that coal-fired power is more expensive than renewable power in many parts of the country is that over the past forty years, plant owners have been forced to install expensive technologies to remove pollutants that cause disease and death, turn pristine lakes into acid, and contribute a large proportion of the greenhouse gases that contribute to climate change. The Environmental Protection Agency (EPA), the regulatory agency that is responsible for protecting the nation's air and water, and its counterparts in the states have insisted that new power plants install the best available technology to control air pollution and that existing power plants do their part to ensure that their emissions do not violate EPA-promulgated air quality standards and do not result in significant deterioration of clean air areas. EPA has also promulgated regulations to reduce the environmental risks caused by massive amounts of solid wastes that result from burning coal and from the devices that operators install to remove air pollutants.

The Kentucky Coal Mining Museum's move to solar power is emblematic of a remarkable transformation of the electric power industry that has taken place during the last four decades. Electric power companies have morphed from highly polluting regulated monopolies into cleaner deregulated generators, transmitters, and distributors of electrical power in a far more competitive economic environment. They are investing heavily in natural gas and renewable resources, and they are not building any new coal-fired plants. The distribution side of the business has become a facilitator of end-use efficiency and managed demand and a purchaser of excess electricity produced by rooftop solar panels and backyard wind turbines.

This book tells the story of this transformation and explores the roles that environmental regulation, increased competition due to price deregulation, and market forces have played in setting us on a path toward sustainable generation, transmission, distribution, and use of the electrical power that has become an essential component of a modern postindustrial economy. It highlights the political, institutional, and technical struggles over environmental regulation that accompanied the transformation, and it focuses on the ongoing "war on fossil fuels" that environmental groups and some state attorneys general have waged against power plants burning coal and natural gas in a variety of forums, including Congress, state legislatures, EPA, the Federal Energy Regulatory Commission, state environmental agencies, state public utility commissions, local city councils, and the courts.

Chapter 1 begins with the stories of eight of the largest and most heavily polluting power plants and their evolution into much cleaner plants or early retirements. They were among the two hundred or so "big dirties" that accounted for the lion's share of emissions of pollutants that contributed to acid rain, mercury-contaminated fish, urban smog, and global warming in the late 1970s. Their histories provide a useful introduction to the struggles that have taken place over power plants fired by fossil fuels in the intervening four decades. The chapter also provides an overview of the issues that arose during those struggles and of the strategies that environmental groups employed in their war on fossil fuels.

Chapter 2 provides a description, for readers who need it, of the electric power industry, how power plants operate and how they pollute, the traditional model of economic regulation of public utilities, the advent of economic deregulation, the modern grid, environmental regulation, and how power plant pollution can be reduced to acceptable levels.

Chapters 3 through 9 then describe the technical, legal, and political struggles over power plant pollution under the statutes that Congress enacted in the 1970s as well as struggles over attempts to change that legislation to make environmental regulation more or less comprehensive and stringent. The chapters also trace the history of economic deregulation and the advent of inexpensive natural gas due to a technology called hydraulic fracturing. Electric power companies have responded to the sometimes conflicting, sometimes complementary forces of environmental regulation, economic regulation, and technology-driven changes in markets for fuels, renewable resources, and electricity in complex ways that have varied over time. Because EPA's leadership changed, sometimes radically, with changing presidents, each chapter focuses on a single presidential administration, three of which lasted for eight years. The electric power industry's evolution in response to the interplay of environmental regulation and market forces and the role of politics in shaping these forces are major themes of these chapters and Chapter 11, which analyzes the causes underlying the industry's transformation.

The focus turns in Chapter 10 to the environmental groups' war on fossil fuels as they first focused on preventing new plants from being built during the rush of new coal-fired plant construction toward the end of the George W. Bush administration and then turned their attention to forcing existing plants to retire at the outset of the Obama administration. The natural gas industry supported the environmental groups, sometimes with cash, during these salvos, but did an abrupt about-face when some of the groups began to object to new gas-fired plants and new natural gas pipelines.

Chapter 11 describes the transformation of the electric power industry as it stood prior to the 2016 elections and highlights the many reasons for that transformation, including environmental regulation, determined environmental group opposition, deregulation, and market forces brought on by dramatic decreases in the cost of natural gas and renewable power. Chapter 12 assesses the impact of the Trump administration on the transition and concludes that its impact so far has been modest, but this assessment could change if resistance by environmental groups, sympathetic states, and companies that have largely adapted to prior environmental regulations declines. Chapter 13 offers many suggestions for attaining a sustainable grid while at the same time alleviating the suffering of the miners and electric power workers who have lost their jobs and the communities that have lost a primary source of income. Finally, the Conclusion provides some thoughts on the future of the electric power industry, offering some reasons for pessimism and some reasons for optimism about the prospect of achieving a sustainable grid that provides a reliable source of power with minimal adverse effects on our health and our shared environment.

CHAPTER 1

The Big Dirties

When the Ohio Power Company, an electric utility company serving Columbus and surrounding towns, assembled Ohio's largest coal-fired power plant along the Ohio River next to the old town of Cheshire in 1974–1975, it named the plant after James Maurice "Jumpin' Jim" Gavin, the commanding general of the Eighty-Second Airborne Division during World War II. Because the plant burned local coal with a high sulfur content, it was the largest emitter in the Western Hemisphere of sulfur dioxide (SO_2), a corrosive pollutant that forms when sulfur burns. Ohio Power built a giant 1,100-foot-tall smokestack to propel the SO_2 emissions high into the air and far over the rooftops of the town nestled below. When completed, the plant's two electric generating units (EGUs), each of which was large enough to hold the Statue of Liberty, had a capacity of 2,600 megawatts (enough to supply roughly 2.6 million homes) and consumed six million tons of coal per year. Railroads brought coal from nearby mines in southern Ohio to be transported over long conveyer belts to the boilers that turned water into the steam that drove two enormous turbines that in turn produced electricity. Large pumps pulled water from the river to fill condensers that turned the steam back into water, which was cycled back to the boilers to produce more steam. Ohio Power later built two huge cooling towers to conserve water and to keep fish and other aquatic organisms from becoming crushed in the plant's cooling water intake structures.[1]

The plant's SO_2 emissions mixed with emissions from many other power plants in the Midwest and returned to earth as acid rain in the Adirondack Mountains of New York, rendering the sensitive lakes there incapable of sustaining aquatic life. The acid rain problem became a hot political topic during the George H. W. Bush administration, prompting Congress to create an acid rain "cap-and-trade" program under which power plants in the Midwest could only emit a total of nine million tons per year of SO_2. The Gavin plant, which was then owned by the American Electric Power Company (AEP), elected to comply by installing "scrubbers" capable of reducing SO_2 emissions by 95 percent at a cost of $616 million. When AEP sought permission from the Ohio Public Utility Commission to charge its customers for the cost of the scrubbers, several large commercial ratepayers and the Sierra Club objected on the ground that AEP was really insisting that its customers subsidize the local coal companies whose high-sulfur coal would otherwise not be marketable after EPA's acid rain program went into effect. The Ohio Supreme Court, however, affirmed the commission's approval of the rate hike. When the scrubbers began operating in December 1994, the Gavin plant instantly disappeared from the list of the nation's heaviest SO_2 emitters. AEP also got crosswise with its customers when it used some of the allowances produced by installing the scrubbers to meet the SO_2 reduction requirements of plants in other states whose ratepayers did not have to pay for scrubbers. The utility commissions of the affected states ultimately came up with an equitable arrangement that was approved by the Federal Energy Regulatory Commission, the federal agency that regulates wholesale electric rates.

In 1999, the Gavin plant's emissions of oxides of nitrogen (NO_x), which contributed to the smog that blanketed the Eastern Seaboard, became one of the targets of a major lawsuit filed by New York attorney general Eliot Spitzer and several environmental groups alleging that the plant had undertaken improvements that increased NO_x emissions without installing the best available technology as required by the Clean Air Act's new source review (NSR) program. The lawsuit was an early manifestation of the war on coal that some state attorneys general and environmental groups have waged from the late 1990s through the present day. EPA followed with its own NSR lawsuit later that year. As we shall see, this decade-long initiative was the largest enforcement action that EPA had ever launched, and it ultimately brought about dramatic reductions in emissions from the power plants that it targeted.

AEP attempted to solve the NO_x problem in 2001 by installing state-of-the-art "selective catalytic reduction" (SCR) technology at a cost of $195 million. In a bizarre twist, however, the technology interacted with the sulfur dioxide in the plant's exhaust to form the reactive compound sulfur trioxide (SO_3), which occasionally sent a dark blue plume across the town, causing the town's residents to suffer blistered lips and sore throats. The plant solved the problem of complaining neighbors by paying $20 million for ninety residences, the bait shop, the post office, the service station, and the one-hundred-year-old church in return for the inhabitants' agreeing not to sue the company. It was the first time a company had solved a pollution problem by purchasing and dissolving an entire town. The industry ultimately solved the blue plume problem by injecting sorbents into the exhaust after it exited the SCR unit. In February 2017, AEP sold the Gavin plant to ArcLight Capital Partners, a private equity consortium that planned to keep the plant running as a nonutility "merchant" plant and sell its output on the competitive wholesale electricity market.

The Gavin plant is just one of the hundreds of coal-fired power plants that investor-owned utility companies, municipally owned companies, rural cooperatives, and the Tennessee Valley Authority (TVA) built from the 1940s through the late 1970s to meet the dramatic increase in demand for electricity during the economic expansion following World War II. These enormous power plants were, by today's standards, quite inefficient. They were not subject to any federal pollution control requirements until the early 1970s. To avoid complaints from neighbors, they constructed tall stacks to disperse the pollutants they emitted as widely as possible. As time passed and state and federal pollution control programs went into effect, the plants were only thinly regulated or exempted entirely from the requirements that were applicable to newly constructed plants. By the turn of the twenty-first century, about two hundred of these aging plants, dubbed the "big dirties" by environmental groups, accounted for the lion's share of emissions of pollutants that contributed to acid rain, mercury-contaminated fish, urban smog, and global warming.

This chapter will tell the stories of a representative sample of these power plants as an introduction to the description and analysis in the following chapters of the attempts by EPA to write regulations to protect human health and the environment from power plant pollutants, the efforts by environmental groups and some state attorneys general to force the retirement of old power plants fired by fossil fuels and prevent the construction of new ones, and the remarkable transformation of the electric power industry caused by environmental regulations, public concern over the damage done by coal-fired power plants, deregulation of wholesale and many retail electricity markets, and market forces driven by plentiful natural gas due to new hydraulic fracturing technology, stunning gains in energy efficiency resulting from digital technologies, and the increasing availability of renewable power from wind and solar energy.

SOME OF THE BIGGEST DIRTIES

When Congress enacted the Clean Air Act in 1970 it provided for very little direct regulation of existing power plants on the assumption that companies would soon be replacing plants that were built between the 1940s and 1960s with state-of-the-art facilities that complied with standards that the statute empowered EPA to write for new power plants. That assumption proved incorrect. As states proved reluctant to impose controls on the older plants, they obtained a competitive advantage over new plants, and they continued to emit a disproportionate share of pollutants without controls until well into the twenty-first century. Plants built before 1978 provided 45 percent of the nation's fossil fuel–fired electrical power in 2010, but accounted for 75 percent of the SO_2 emissions from power plants, 64 percent of their NO_x emissions, and 54 percent of their carbon dioxide (CO_2) emissions. Consequently, the electric power industry continued to be a major contributor to the nation's air pollution. The greatest percentage of emissions came from power plants in the Southeast, South Central, and Great Lakes regions, but emissions from individual coal-fired power plants in the West were the primary contributors to visibility impairment in national parks and wilderness areas in that region.[2] The following descriptions of a representative sample of the big dirties provides a bird's-eye view of the many struggles that have taken place through the years over power plant emissions and their effects on human health and the environment.

Beckjord

One of the oldest and largest of the big dirties was Duke Energy's 1,400-megawatt W. C. Beckjord plant on the Ohio River near Cincinnati.[3] Its six coal-fired units were built between 1948 and 1969 and housed within a huge brick building that loomed over the river like a high red cliff. The plant installed electrostatic precipitators in the 1970s to reduce the particulate matter (PM) that settled onto nearby buildings and residences but made no effort to capture the thousands of tons of SO_2 and NO_x and even larger amounts of CO_2 that it forced into the air through five small smokestacks every year. The solid coal combustion residuals, or "coal ash," that the plant produced on a continuing basis stacked up in giant impoundments from which toxic metals could leach into groundwater and ultimately reach the adjacent river. Duke's predecessors, Cincinnati Gas & Electric and Cinergy, fiercely resisted EPA's efforts to force them to install modern pollution control equipment when the plant underwent major modifications until late 2000, when Cinergy settled a lawsuit with EPA by agreeing to spend millions of dollars on pollution control equipment.

In July 2011, Duke announced that it would be retiring the aging plant. The company calculated that it would take up to $850 million to comply with regulations that EPA had recently promulgated to reduce the plant's contribution to smog on the Eastern Seaboard and to reduce its emissions of mercury, a highly toxic pollutant that was accumulating in fish in the Great Lakes. It did not make economic sense to keep the plant running when it would cost much less to build a plant that produced the same amount of electricity and emit fewer pollutants by burning inexpensive natural gas. All six units were out of service by September 1, 2014. The company attempted to relocate the plant's 120 workers, but it did nothing to replace the $2 million in lost tax revenues that the local school district suffered when the plant shut down. In late 2015, Duke built a two-megawatt battery facility at the site for storing electricity to facilitate greater reliance on variable renewable energy from solar arrays and wind farms.

Big Brown, Martin Lake, and Monticello

Three of the dirtiest of the big dirties were owned by a single Texas company.[4] Luminant was a subsidiary of Energy Future Holdings, a holding company made up of Wall Street investors that made a big splash in 2007 when it purchased TXU, the largest electric utility in Texas, in an ill-conceived leveraged buyout that later resulted in bankruptcy. Its Big Brown, Martin Lake, and Monticello plants all burned lignite, an especially dirty form of coal that creates unusually high emissions of SO_2, CO_2, and mercury. The Big Brown plant was built in 1971. Its twenty-two-story-high boilers and four-hundred-foot-tall stacks stood in stark contrast to the grassy Northeast Texas prairie surrounding them. A nearby strip mine fed the plant twenty thousand tons of lignite every day to supply electricity to the Dallas area. The three units making up the Martin Lake plant were constructed between 1977 and 1979. Its 2,250-megawatt capacity was sufficient to power 1.125 million homes. The three units at the 1,880-megawatt Monticello plant came online between 1974 and 1976 at the edge of a two-thousand-acre cooling lake, which over the years gained a reputation for great bass fishing. At the outset of the Obama administration, the Martin Lake plant was the nation's largest emitter of mercury. Monticello and Big Brown followed closely behind at numbers 4 and 6. Among the fifty largest emitters of SO_2, Big Brown, Martin Lake, and Monticello ranked 13, 31, and 32 respectively.

Luminant fought several bitter battles with EPA and environmental groups over the plants. It successfully defended itself against a major EPA enforcement action accusing it of unlawfully upgrading the plants without installing pollution controls. It persuaded a Texas federal judge to order the Sierra Club to pay it $6.4 million in legal fees after convincing the court that it was not legally responsible for a single one of the tens of thousands of violations of opacity and heat input limitations that it had reported to the Texas Commission on Environmental Quality. And it fought EPA's "cross-state air pollution rule" all the way to the Supreme Court and ultimately persuaded the United States Court of Appeals for the District of Columbia Circuit (D.C. Circuit) that EPA had arbitrarily included Texas in the program. Having prevailed in its struggles with EPA and environmental groups, Luminant announced in October 2017 that it would be retiring the Big Brown and Monticello plants because "low wholesale power prices, an oversupplied renewable generation market, and low natural gas prices" made it impossible to compete in Texas's deregulated retail markets. They ceased production in early 2018. The Martin Lake plant continues to operate, but it is barely competitive in the Texas market. Two of its three units operate only during the summer months when demand for electricity is highest.

Bowen

Georgia Power's Plant Bowen rises far above the treetops in a rural valley near the small town of Cartersville, Georgia, that sometimes fills with smoke from two one-thousand-foot-tall stacks and two somewhat smaller stacks.[5] Its four units, which were built between 1971 and 1975, burn three ninety-five-car trainloads of coal daily to produce 3,376 megawatts of power. Although it imported the low-sulfur coal all the way from the Powder River Basin in Wyoming, the plant was the country's number 1 emitter of SO_2, the number 14 emitter of NO_x and mercury, and the number 3 emitter of CO_2 in 2006. Soon thereafter, the company implemented a series of changes in response to federal environmental requirements. It installed SCR technology to reduce NO_x emissions by 85 percent to meet EPA's national ambient air quality standards for ozone in the Atlanta area. It also employed scrubbers at the plant to reduce SO_2 emissions by 95 percent in response to a regulation that EPA promulgated to protect downwind states from pollution that crossed state lines. In all, the environmental improvements cost about $1 billion. As it was addressing its air pollution problems, however, it was creating another problem. In July 2002, a huge sinkhole opened up under one of the impoundments storing the coal ash near the plant, releasing around 2.25 million gallons of toxic slurry into a tributary of the Etowah River.

Cumberland

The federal government owned several of the big dirties as a result of a New Deal program to provide electricity to residences and businesses in the Southeast.[6] The Tennessee Valley Authority (TVA) was a federal agency that first built hydroelectric dams but later switched to building large coal-fired plants, which by the late 1970s were responsible for 38 percent of the SO_2 emissions in that region. After Congress overturned a Supreme Court holding that TVA was not subject to state regulation under the Clean Air Act, TVA had to deal with the statute's requirements, just like any private sector power company. And just like investor-owned utility companies, it struggled with EPA for years over its obligations to reduce emissions from its plants. Its 2,470-megawatt Cumberland plant on the Cumberland River in Tennessee was the largest producer of electricity in the TVA system. Built in 1973, its two units consumed about twenty thousand tons of coal per day.

Under heavy pressure from Kentucky officials not to switch to Wyoming coal to reduce SO_2 emissions, TVA decided in November 1991 to install scrubbers on both units, thereby reducing its SO_2 emissions by about 95 percent. Despite modifying the burners in 1998 to control NO_x emissions, it remained number 5 on the list of the nation's largest NO_x emitters and number 9 on the list of the largest CO_2 emitters. The plant's NO_x emissions decreased substantially in 2003 with the installation of SCR technology to meet its obligation to protect states along the Eastern Seaboard from unhealthy ozone levels. The Cumberland plant was part of a huge April 2011 settlement that TVA reached with EPA in which it agreed to retire eighteen of its fifty-nine coal-fired units, install additional pollution controls on the remaining plants, and invest $350 million over five years on clean energy and related air pollution reduction activities.

Navajo

Several of the biggest of the big dirties were located in the Southwest close to coal mines and not far from booming cities in Arizona, Nevada, and Southern California that were thirsty for electricity.[7] One of the largest was the massive 2,250-megawatt Navajo coal-fired plant operated by the Salt River Project. Built between 1974 and 1976 by a consortium consisting of four utility companies, the City of Los Angeles and the United States Bureau of Reclamation, the plant's three 250-megawatt units poured more than 100,000 tons per year of SO_2, 34,700 tons per year of NO_x (number 4 on the 2006 list of big dirties), and 20 million tons per year of CO_2 (number 8 on the list) into the air through three 750-foot stacks that loomed over the desert in the shadow of a multicolored butte near Page, Arizona. Both the plant and the massive Kayenta coal mine that fed it low-sulfur coal were located on the large Navajo reservation that spanned northeastern Arizona, northwestern New Mexico, and southeastern Utah. The plant and mine employed nearly eight hundred people, most of whom were members of the tribe, and it accounted for 40 percent of the tribe's income.

The location was a good one for one of the plant's primary purposes—providing power to pump 1.5 million acre-feet of water per year from the Colorado River vertically more than three thousand feet and horizontally more than three hundred miles through the Sonoran Desert to slake the thirst of residents of Phoenix and Tucson. The location was not so good for the Grand Canyon National Park, which was located twelve miles to the southwest. On some winter days, observers could see the plume from the Navajo plant roll over the lip of the canyon and fill the floor. After a bitter scientific battle with EPA and environmental groups over varying estimates of the impacts of the plant's emissions on visibility in the canyon, the plant's owners agreed in August 1991 to install $450 million worth of scrubbers capable of reducing SO_2 emissions by at least 90 percent between 1997 and 1999.

The scrubbers, however, did not solve the Grand Canyon's visibility problem, because the plant lacked controls on the NO_x emissions that also hindered visibility. Between 2009 and 2011, the plant added burner modification technology to control around 40 percent of its NO_x emissions. But environmental groups and some tribal members demanded that EPA require the consortium to install SCR technology to reduce NO_x emissions by 80 percent. The consortium strongly resisted this suggestion, arguing that the improvement would cost about $660 million and only imperceptibly improve visibility in the park. It warned that it might just shut down the plant, an option that would be devastating to the Navajo Nation's economy. The warning prompted the Republican House leadership to include a provision in a bill called the "Stop the War on Coal Act" limiting EPA's authority to regulate the Navajo Generating Station, but the bill failed in the Democrat-controlled Senate.

After lengthy negotiations, the Environmental Defense Fund, the Navajo Nation, and the consortium agreed upon a plan under which the consortium would retire one of the units and install SCR technology or its equivalent at the other two units between 2020 and 2044. EPA promulgated a plan in January 2014 that implemented the agreement, but several other environmental groups challenged it in a federal court of appeals, alleging that it was the result of a closed-door process from which they had been excluded. The court upheld the plan in March 2017. But by then the consortium had decided to retire all three units by the end of 2019 because they could not compete with electricity produced by natural gas–fired plants and renewable sources. Although the Trump administration promised to "turn over every rock" to save the plant, the Department of Interior failed to come up with a plan to make the plant economically viable. The consortium agreed to allow the tribe to use the existing transmission lines to convey electricity from a utility-scale solar facility that it plans to build at the site in the future.

STRUGGLES OVER ENVIRONMENTAL REGULATION

All of the big dirties described above are featured in struggles detailed in the following chapters over the health and environmental risks posed by power plants fired by fossil fuels, the feasibility and cost of technologies and fuel switching to reduce those risks, the threat that retiring aging power plants posed to the reliability and resiliency of the electricity grid and to the communities in which they were located, the degree to which replacing coal-fired plants with plants burning natural gas reduced their environmental impact, and the urgency with which the electric power industry should go about replacing fossil fuels with renewable technologies for meeting the nation's energy needs. These struggles took place in three broad arenas—political, institutional, and technical.

Political Struggles

The most contentious struggles played out in the political arena in battles over the content of the laws and regulations that governed how the Environmental Protection Agency and the states went about reducing the risks that power plants posed to public health and the environment. Although participants on both sides invoked scientific and technical arguments, the struggles were often ideological and economic in nature. Environmental groups claiming a right to a clean environment demanded strong laws requiring power plants to install the most effective pollution control technologies, while electric power companies, fossil fuel producers, labor unions, and supporting business groups resisted government imposition of burdensome and, they argued, "job-killing" pollution controls. At a more abstract level, academics and think tanks on the left and right debated the proper role of government in the context of activities that created "externalities"—costs that the producers impose upon others without compensation and are therefore not reflected in the price of their products.

Political struggles erupted over broad economic issues. Environmental groups urged legislatures and regulators to stop permitting new fossil fuel–fired plants and to replace existing plants with end-use efficiency and renewables, while power companies and unions pressed them to consider the impact of their decisions on employment and on the reliability and resiliency of the grid. The struggles also reflected narrow economic interests as local citizen groups concerned about property values fought with power companies concerned about profits. At times, the electric power industry divided as companies that were heavily dependent on coal fought with companies that relied more on natural gas, nuclear power, and (in later years) renewables.

The political struggles sometimes reflected geographical interests. Midwestern states joined power companies located in those states in debates with environmental advocates and downwind states along the Eastern Seaboard over interstate transport of pollutants. During one period, western states joined producers of low-sulfur coal from those states in advocating controls on SO_2 emissions from power plants that could be met by burning low-sulfur coal but not by burning high-sulfur coal from midwestern states and Appalachia. Midwestern and Appalachian states advocated either no controls or strict controls that could be met by both low- and high-sulfur coal. During another period, environmental groups joined West Coast cities in demanding that power plants in those cities stop importing electricity from coal-fired power plants in Arizona, New Mexico, and Montana, while those states joined power companies, coal interests, and Indian tribes in opposing those efforts.

Finally, political fights broke out over the best way to address the externalities that fossil fuel–burning power plants impose on society. The traditional approach of demanding the best available pollution control technology was fiercely resisted by the electric power industry and derided by economists as an especially inefficient way to go about reducing emissions. The economists' preferred approach of imposing a per-ton tax on power plant emissions was supported by environmental groups, but opposed by power plant operators and Republican politicians who viewed any tax as anathema. An alternative "cap-and-trade" approach was initially opposed by environmental groups fearing that companies would game the system. But they ultimately came to the view that the cap-and-trade approach would work for power plants, and they were joined by some electric power companies who saw the possibilities for profit in a cap-and-trade regime. The approach was strongly opposed by a coal industry that faced a decline in coal consumption and by conservative grassroots groups who remained ideologically opposed to any kind of governmental intervention into electricity markets.

The political struggles took place during a period of increased polarization. As in perhaps no other area of national policymaking, these struggles reflected the dramatic change in the two political parties that dominate our national politics. When the fountainhead statutes of the 1970s were enacted, they were supported by Republican presidents and Republican members of Congress and opposed by many Democratic members (usually from states dependent on fossil fuels). Over the past forty years, however, the Republican Party has effectively abandoned politicians who favor strong environmental regulation. As conservative Democrats migrated toward the Republican Party and the Republican Party migrated to the right, the struggles over pollution and electrical power became increasingly partisan. The Democratic Party still contains many strong proponents of the coal and electric power industries, but the Republican Party contains very few strong supporters of stringent environmental regulation.

Institutional Struggles

Struggles also broke out over which institutions should play what roles in regulating power plant pollution. Congress enacted the environmental statutes of the 1970s in response to the failure of the states to regulate power plants and other large sources of environmental pollution. They resolved the political debates over federalism, however, by assigning major roles to the states, thereby guaranteeing a continuing tension between state environmental agencies and EPA and between electric power companies and fossil fuel producers that generally preferred state regulation and environmental groups that generally favored federal regulation.

Every president during the years covered by this book has ordered EPA to send major regulations through a White House review process that provided other departments and agencies, economists in the Office of Management and Budget, and various ad hoc interagency groups an opportunity to comment on their content. The interagency review process often resulted in fierce battles, some of which had to be resolved in the Oval Office. Over the years, the process provided interest groups an opportunity to lobby White House officials, including on occasion the president himself, to kill or modify the content of the regulations.

EPA leaders have often struggled with congressional committees exercising their oversight responsibilities. EPA's administrators have spent thousands of hours before authorizing, oversight, and appropriations committees, and EPA staffers have spent much more time answering questions prepared by committee staffers, sometimes reflecting the interests of individual constituents, but more often reflecting the concerns of affected interest groups. On a few occasions, the president and his EPA administrator have struggled with powerful committee chairpersons over proposed amendments to the environmental statutes.

A final institutional struggle takes place in the courts that review EPA regulations and challenges to state and federal permits. In that setting, the struggles are over the deference that courts should afford to EPA's interpretations of its statutes and over the intensity with which they should review the agency's scientific and technical judgments. Most EPA regulations are reviewed in the D.C. Circuit Court of Appeals, but permit actions and challenges to state implementation plans are proper in any of the eleven federal courts of appeals. Although the Supreme Court rarely grants certiorari to review decisions of lower courts in challenges to administrative agency actions, we shall see that it has been quite active, if not always consistent, in reviewing cases involving power plant regulation.

Technical Struggles

The following chapters will also delve deeply into some of the hundreds of struggles over scientific, engineering, and economic issues that arose in the context of environmental and economic regulation of power plants. Scientific disputes arose in determining the levels of SO_2, ozone, and particulate matter that are "requisite" to protect public health and the environment, the impact of SO_2 emissions from Midwest power plants on acid deposition into lakes in the Adirondack Mountains, the contribution of NO_x emissions from power plants in upwind states to ozone levels in downwind states, the extent to which fine particulate matter emitted by power plants impairs visibility in national parks, and whether CO_2 emissions from power plants "endanger" public health and the environment. Engineering issues arose in determining the technological feasibility of installing scrubbers to control SO_2 emissions, baghouses to control PM and mercury emissions, SCR to control NO_x emissions, and carbon capture to control CO_2 emissions. Struggles broke out over economic issues such as the accuracy of estimates of the cost of pollution controls, the impact of EPA regulations on jobs and the economy, the desirability of employing market-based approaches to reducing pollution, and the merits of deregulating interstate wholesale and intrastate retail markets for electricity. As we shall see, many of these issues involve predictions based on complex computer models that depend heavily upon the assumptions that drive those models, not all of which command agreement in the relevant expert communities. We shall also see struggles over questions that cannot be answered with existing scientific knowledge and therefore invoke contentious policy considerations.

THE WAR ON FOSSIL FUELS

The coal industry, conservative advocacy organizations, and politicians of both political parties accused the Obama administration of waging a "war on coal" by promulgating stringent environmental regulations for power plants and supporting end-use energy efficiency and renewable power. As we shall see, that appellation does not accurately characterize the Obama administration's initiatives, which were generally sensitive to the interests of the coal industry and electric power companies that depended heavily on coal. It does, however, capture rather well the initiatives by many environmental groups, including the Sierra Club, Natural Resources Defense Council (NRDC), Earthjustice, and the Center for Biological Diversity, and a few state attorneys general to press EPA, other federal agencies, state environmental agencies and public utility commissions, Congress, and the courts to prevent companies from building new fossil fuel–fired power plants, force the retirement of existing plants, and encourage end-use efficiency and renewable resources.

In the early years, fledgling environmental groups concerned themselves mostly with ensuring that EPA adhered to statutory deadlines for promulgating regulations, insisting that new power plants install top-of-the-line pollution control technology, and demanding that existing sources retrofit available technologies instead of building tall stacks to disperse pollutants over vast areas. As scientists learned more about the role that power plant NO_x emissions played in forming smog along the Eastern Seaboard, environmental groups insisted that new and existing plants install SCR technology. Later, they supported so-called "clean coal" technologies like coal washing and coal gasification. Still later, they demanded that owners of new and existing power plants install technologies for reducing emissions of mercury and other hazardous air pollutants. As air pollution control technologies produced huge amounts of coal ash, environmental groups turned their attention to ensuring that power plants properly disposed of those dangerous solid wastes. They also pressed EPA to prevent billions of marine and aquatic organisms from being killed in cooling water intake structures. The goal in all of these instances was to make coal-fired power plants cleaner, not necessarily to prevent them from being built or to force them into retirement.

As the powerful role that CO_2 emissions from power plants played in increasing global temperatures became clear, however, the goal gradually changed. At first, the groups hoped that existing carbon capture technologies would be capable of removing a high percentage of CO_2 from the exhaust of large power plants and that suitable locations could be found to sequester the captured gas permanently in large underground caverns. But it proved very difficult to adapt those technologies to power plants, and it came at a huge cost in capital expenditures and in the percentage of the plant's energy that had to be devoted to CO_2 removal. At the same time, the George W. Bush administration was strongly encouraging the electric power industry to build more coal-fired plants, and the industry responded with a very ambitious construction agenda.[8]

Seeing this as a "once-in-a-generation opportunity" to prevent the country from being heavily coal dependent for another forty years, the Sierra Club in the summer of 2007 launched its "Beyond Coal" campaign, the goal of which was to object to every new coal-fired power plant, whether or not it complied with the relevant environmental requirements. With a $50 million boost from philanthropist Michael Bloomberg, the club hoped to force Congress to address climate change by "clogging the system" for approving new coal-fired plants. The Sierra Club was soon joined by NRDC, Earthjustice, a new group called the Environmental Integrity Project, and several regional groups that were also committed to preventing companies from building new coal-fired power plants. As we shall see, the campaign was enormously successful. Only a few of the two hundred coal-fired plants that were on the drawing boards early in the George W. Bush administration were actually built, and environmental groups often extracted promises to retire existing coal-fired units as the price for dropping their opposition to new plants. By the end of the Obama administration, nobody was contemplating building a coal-fired plant.[9]

Having achieved a degree of success in slowing down the flow of new coal-fired plants, the Sierra Club in 2009 launched phase 2 of Beyond Coal, an even more ambitious effort to force companies to retire all 525 existing coal-fired plants by 2030. In addition to ridding the air of millions of tons of SO_2, NO_x, PM, and mercury, the club believed that this initiative was required to bring about the rapid reduction in CO_2 emissions needed to prevent catastrophic climate disruption in the not-too-distant future. By 2015, the club was filing an average of one lawsuit or legal appeal every three days against new and existing coal-fired power plants and coal mines. Michael Bloomberg kicked in another $30 million in matching funds for the effort.[10]

After being burned by public disclosures that it had partially supported its Beyond Coal campaign with $26 million in contributions from a major natural gas producer, the Sierra Club declined future contributions from natural gas companies and launched its "Beyond Natural Gas" campaign in April 2012. The initial goal for this initiative was to insist that all new natural gas plants employ the most efficient possible technology. As renewable energy became competitive with natural gas, however, the club began to object to as many new gas plants as it could and to oppose the conversion of coal-fired plants to gas. This occasionally put the Sierra Club at odds with NRDC and the Environmental Defense Fund, which continued to urge companies to convert from coal to gas. In mid-2016, the Sierra Club broadened the campaign to include opposition to pipelines designed to bring gas from remote gas fields to power plants in urban areas. And by the onset of the Trump administration, the club renamed the campaign "Beyond Dirty Fuels" with the stated goal of opposing every new fossil fuel–fired power plant.[11] Chapter 10 will describe in more detail the strategies that the environmental groups employed, the industry responses, and the degree to which they succeeded.

THE TRANSFORMATION

EPA's regulatory programs, the environmental groups' war on coal, and market forces contributed to a remarkable transformation in the electric power industry from one heavily dependent on coal to one that relies on a more diverse balance of end-use efficiency, natural gas, coal, and renewable energy. By 2016, dozens of the big dirties were either retired or not utilized to their full capacity. With the added expense of pollution controls, they had become too expensive to run when the alternative was to purchase electricity from gas or renewables in deregulated markets. And the proportion of the nation's electricity supplied by coal had fallen from 51 percent in 2008 to 31 percent.[12]

The Obama administration witnessed a rapid rise in renewable power due in part to the administration's commitment of billions of dollars in economic stimulus money to renewable research and in part to the steep decline in the cost of producing solar panels and wind turbines. In March 2017, wind and solar produced more than 10 percent of U.S. power for the first time. Demand for electricity grew at historically low rates due to end-use efficiency, programs rewarding consumers for conserving energy in times of high demand, and rapid adoption of rooftop solar panels. Low demand caused electricity prices to fall, even as consumers relied more heavily on electronic devices, electric appliances, and electric vehicles. With lower electricity prices, power plants fired by fossil fuels became uneconomical.[13]

The transformation resulted in fewer jobs in the coal, electric power, railroad, and barge industries. The reduction in local tax bases due to power plant retirements had devastating effects on many communities. Local businesses suffered when laid-off workers tightened their belts, and municipalities had to scramble to find funds for essential services. At the same time, the transformation created hundreds of thousands of jobs in the pollution control and renewable power industries, and it brought about considerable environmental improvement. But that came as small consolation to unemployed miners and power plant workers who found it difficult to relocate. We will study the causes and effects of the transformation in Chapter 11.

CHAPTER 2

Electricity, Power Plants, and the Environment

It is hard to imagine life in the United States without the convenience of a dependable supply of electricity to light our businesses, homes, and recreational facilities, to power heaters, air conditioners, dishwashers, and hundreds of other consumer products, and to run an astonishing array of computers and electronic devices. In 2017, American consumers and businesses used just over 3.8 trillion kilowatt-hours of electricity, more than thirteen times the amount that they consumed in 1950. Demand has stabilized over the past decade, due primarily to energy conservation programs, but we can expect demand to increase in the next three decades as electronic devices become even more prevalent and as governments, businesses, and consumers substitute electricity for fossil fuels to provide transportation and space heating.[1]

SAMUEL INSULL'S GRAND BARGAIN

A protégé of Thomas Edison named Samuel Insull brought electricity to the households of ordinary Americans in the early twentieth century through a distribution system that charged customers for the amount of electricity they consumed. That scheme emphasized scale—a large power plant could deliver electricity to more consumers at a lower average cost than a small plant. It also relied on vertical integration under which a single company generated electricity, transmitted it at high voltages to distribution points where transformers lowered the voltages, distributed low-voltage electricity to consumers, maintained the necessary wires and facilities, and billed the consumer for all four functions.[2]

Since it was highly inefficient to string more than one set of transmission lines throughout a geographical area, electricity generation, transmission, and distribution appeared to be a "natural monopoly." Lobbyists for Insull and other large electric companies persuaded state legislatures to protect the single investor–owned company that generated, transmitted, and distributed electricity within a defined geographical area from competition in return for agreeing to supply electricity in a reliable and nondiscriminatory fashion to all consumers within a region under the supervision of a state public utility commission (PUC) charged with setting "just and reasonable" rates based on the company's reasonable operating costs and a fair return on the "used and useful" capital assets it prudently employed. By 1920, nearly every state had adopted some form of "economic regulation" of power plants for setting rates and overseeing capital investments of investor-owned utility companies (IOUs). After his overleveraged utility empire crashed during the Great Depression, Insull died a pauper, but the regulatory model that he created dominated the electric power industry for the remainder of the twentieth century.[3]

At the same time that Insull was dominating the private market for electricity, many municipalities were creating their own publicly owned entities to generate, transport, and deliver electricity to their residents. During the 1930s, when public confidence in Insull's model was fading, these municipal utilities flourished, despite warnings from IOUs that they represented "creeping socialism." Today more than two thousand municipalities, including such large cities as San Antonio, Los Angeles, and Seattle, run their own utility companies. They distribute about 10 percent of the nation's electricity, some of which they generate and some of which they purchase from IOUs and independent generators. In Nebraska, publicly owned utilities are responsible for all retail sales of electricity.[4]

Another model for generating and distributing electricity grew out of the New Deal–era Rural Electrification Administration's support for collectively owned nonprofit electric cooperatives, which sprang up throughout rural America to extend electrical services deep into the heartland. In the last several decades many of these "co-ops" have merged into larger entities with the economic resources to build large power plants and the political clout to affect the outcome of national debates on energy policy. Today, 838 electric co-ops distribute 4 percent of the nation's electricity to around forty-two million people in forty-seven states in territories covering 75 percent of the nation's land mass.[5] The New Deal also gave rise to nine federal electric utilities that are owned and operated by five federal agencies, the largest of which is the Tennessee Valley Authority (TVA). The federal utilities sell electricity to municipal and cooperative utilities at prices high enough to recoup the federal government's capital investments. As we shall see, the TVA has played a lead role in the battles over power plant pollution control.[6]

TODAY'S POWER PLANTS

The vast bulk of this country's electrical power comes from burning three fossil fuels—coal, oil, and natural gas. The United States is fortunate in having abundant supplies of all three fossil fuels, uranium to fuel nuclear power plants, and renewable resources such as hydropower, wind, solar, and biomass.[7]

Power Plant Fuels

Coal. Coal is the most plentiful fossil fuel in the United States, and it has until recently been the most stable and least expensive. Coal comes in five varieties: anthracite, bituminous, subbituminous, lignite, and peat. Peat contains the highest percentage of moisture, is quite soft, and produces the least heat per pound. At the other end of the spectrum, anthracite contains the lowest percentage of moisture, is very hard, and yields the most energy per pound. Nearly all of the coal burned in power plants is either bituminous or subbituminous, but a few plants burn lignite.[8]

Some of the largest bituminous coal beds are located in the Appalachian Basin, which extends from northern Pennsylvania to southern Alabama and includes portions of Ohio, West Virginia, Maryland, Virginia, Kentucky, and Tennessee. Another large deposit of bituminous coal is the eastern region of the interior province underlying much of Illinois and parts of Indiana and Kentucky. Easily accessible subbituminous coal can be found in the Powder River region underlying western Montana and northeastern Wyoming.

Finally, the Texas region encompasses a vast lignite bed than runs from far south Texas to northeastern Louisiana. Prior to the enactment of the Clean Air Act of 1970, the Appalachian region provided about 60 percent of the nation's coal, and the eastern interior region supplied 20–25 percent. As we shall see, coal from the Powder River region has gradually overtaken the other regions during the past two decades, primarily because of environmental requirements.[9]

Coal varies in the amount of sulfur that turns into the pollutant sulfur dioxide (SO_2) upon combustion. With some important exceptions, coal from the Appalachian region has a high heat content and high concentrations of sulfur, coal from the eastern interior region has a somewhat lower heat content and higher sulfur levels, and coal from the Powder River region has lower heat content and much lower sulfur levels. The prevalence of trace amounts of toxic metals like mercury that are released during the combustion process likewise varies from region to region. Coal from the Appalachian and Gulf Coast regions has the highest mercury content; coal from the eastern interior region contains somewhat less mercury; and coal from the Powder River region contains the least amount of mercury. Unlike electricity, coal can easily be stored. But it takes a great deal of energy to move coal long distances over rail or by barge.[10]

Oil. Since the beginning of the twentieth century, oil has been the dominant energy provider for the transportation sector, but its role in producing electricity has always played second fiddle to coal. After the oil import crises of the 1970s, oil became a "fossil fuel non grata" in the electric power industry, accounting for only 1 percent of nation's electricity production.[11]

Natural Gas. Natural gas was at one time flared as a worthless by-product of extracting oil from the ground. Following World War II, however, the electric power industry discovered that natural gas was a much less expensive alternative to fuel oil, and by 1970 natural gas was powering almost a quarter of the country's power plants. The United States has large reserves of natural gas, and estimates have grown dramatically since around the year 2000 as horizontal drilling and hydraulic fracturing ("fracking") technologies have made it easier to extract gas from previously inaccessible places. Fracking technology injects various fluids into tight geologic formations like shale at high pressure to force open and hold open fractures in the rock that allow trapped gas to escape. An extensive system of more than three hundred thousand miles of high-pressure transmission pipelines and more than one million miles of low-pressure distribution pipelines ship gas from areas containing accessible deposits to the residences, commercial buildings, power plants, and industrial facilities that burn it.[12]

Natural gas is composed of a number of lighter than air compounds dominated by methane, a simple hydrocarbon that yields far more heat per volume of resulting emissions than coal when it oxidizes into water and carbon dioxide. Although natural gas contains varying amounts of sulfur, available desulfurization technologies can reduce sulfur content to tiny levels at modest cost. Emissions from natural gas plants contain virtually no mercury. Gas-fired plants are cheaper to build than coal-fired plants, but they can be more expensive to run when natural gas prices are high. With the advent of hydraulic fracturing technologies in the twenty-first century, natural gas prices have remained relatively low. One great advantage of gas-fired plants is that they can be fired up and shut down more quickly than coal-fired plants. For that reason, utility companies have tended to use older coal-fired plants to provide continuous "baseload" power and to employ gas-fired facilities as "peaker" plants for use during periods of peak demand for electricity.[13]

Nuclear Power. Attracted by the prospect of producing electricity that was "too cheap to meter," large IOUs and two federal utilities built many nuclear power plants in the 1950s through the mid-1980s. To meet the safety requirements of the Nuclear Regulatory Commission (NRC), owners of nuclear power plants had to make huge capital investments, which, in the early years, were easily borne by ratepayers. The near meltdown at Three Mile Island, Pennsylvania, in March 1979 and the disastrous explosion at Chernobyl in 1986, however, dampened public enthusiasm for nuclear power, and the cost of building new plants to meet new restrictions made them an unattractive option. As companies are deciding whether to renew licenses for another twenty years (for a total of sixty years) or retire their nuclear units, ninety-nine nuclear reactors in thirty states still supply almost 20 percent of the nation's electricity, and they remain a highly reliable source of baseload power for the national grid.[14]

Renewable Resources. Energy from renewable resources can also be converted to electricity. With almost 2,200 power-producing dams yielding almost eighty gigawatts of power, hydroelectric power remains the most prevalent source of renewable electricity, but that is rapidly changing. Large "utility-scale" arrays of photovoltaic cells can convert energy from sunlight directly into electricity. Huge wind turbines as high as the Eiffel Tower can convert energy in the wind into electricity. Steam electric power plants can burn "biomass" consisting of wood, sawdust, prairie switchgrass, and various solid wastes. Power plants can also burn "biogas" generated from livestock wastes and landfills. Since 1992, the Internal Revenue Code has provided tax credits for investments in renewable energy, but they are scheduled to expire in the early 2020s. As of 2016, renewable resources accounted for around 15 percent of total electricity generation.[15]

How Power Plants Work

The electric generating units (EGUs) in power plants come in three basic varieties—those driven by a steam turbine, those driven by a gas turbine, and those driven by a combination of the two, called a "combined cycle." Steam turbine units have three primary components—boilers, turbines, and generators. Heat from burning fuel in a huge furnace (or nuclear reactor) converts water in the boiler into steam. The steam then passes through "superheaters" that raise the temperature of the steam to even higher levels. "Economizers" increase efficiency by using heat from exhaust gases to preheat the boiler feedwater. The expanding steam that is released from the boiler turns the blades of a turbine that is attached to a generator, which converts the energy in the rotating turbine into electrical energy. The steam is then cooled in a condenser where it is converted into water for reuse in generating steam. The heat absorbed from the steam by the condenser is then dissipated into the environment by cooling towers or by water drawn from a nearby river, lake, or ocean. In the 1960s, utility companies began to employ pressurized boilers to keep the water in the pipes at "supercritical" pressures to make more efficient use of the heat from the burning fuel. During the 1990s companies began to build "ultra-supercritical" boilers that employed even higher pressures and temperatures to attain efficiencies ranging from 44 to 46 percent. A modern coal-fired power plant can cost billions of dollars to build, assuming that the developer can overcome the opposition of consumer and environmental groups.[16]

In gas turbine EGUs, the exhaust gases resulting from the combustion of natural gas directly drive turbines that spin electricity generators. The heated gases then exit the plant through a stack. These "simple-cycle" plants can convert heat to electricity with efficiencies ranging from 35 to 40 percent. Many recently built plants have employed combined-cycle systems that use the hot gasses exiting the turbines to boil water to drive steam turbines, thereby making more efficient use of the heat. Combined-cycle plants cost more to build and run than simple-cycle plants, but they can achieve efficiencies of greater than 50 percent, thereby reducing both fuel costs and emissions. Both are far less expensive to build than coal-fired plants. Even more recently, companies have begun to build integrated gasification combined-cycle (IGCC) plants in which coal is first turned into a synthetic gas before combustion in a combined-cycle system.[17]

GENERATION, TRANSMISSION, AND DISTRIBUTION UNDER THE TRADITIONAL MODEL

A critical aspect of the alternating current electricity that power plants generate is that there are very few effective ways to store it. It can be converted into direct current and stored in batteries, or it can be converted to energy in other forms, like spinning flywheels or pumped water behind hydroelectric dams, but both of these options have historically been quite expensive and of limited practical usefulness for electric power grids. Consequently, grid operators who manage electricity grids must ensure that the amount of electricity supplied instantaneously equals the amount demanded at all times, including "peak" hours when demand is very high. To maintain that critical balance, grid operators on a daily basis forecast hourly demand for the next day, develop resource schedules for that day, and arrange for generators to be prepared to meet those schedules. Since day-ahead forecasts are not always accurate, system operators work with generators and consumers to adjust output and demand rapidly to ensure that they remain in balance. This can be tricky in the case of renewable resources, such as wind and solar power, that are by their nature variable and less predictable than power plants fired by fossil fuels. To ensure against outages, grids have historically been designed to have reserve capacity that is unused most of the time but rapidly available when needed.[18]

State Regulation of Electric Utilities

Samuel Insull's "cost-of-service" ratemaking model dominated the twentieth century, and it remains firmly embedded in the laws and institutions of most states to this day, though it applies to municipal utilities and co-ops in only a few states. Under that model, consumers purchase electricity from a single utility company at rates set by the PUC. Different rate schedules apply to different categories of consumers (such as residential, commercial, industrial). An individual consumer typically pays a single rate for a block of electricity, and the rate declines as the consumer purchases additional blocks. State PUCs approve the rate schedules with an eye toward ensuring that utilities charge reasonable prices, charge comparable prices to similarly situated consumers, and provide access to services under similar conditions.[19]

The rate formula allows the utility to recover from consumers its reasonable operating expenses (such as the amounts paid for fuel, labor, repairs, maintenance, and administration) plus a reasonable "rate of return" on a "rate base" consisting primarily of capital expenditures on "used and useful" plants, equipment, and transmission and distribution infrastructure. Because the traditional formula allows a utility to earn a larger overall profit if its rate base increases through new capital expenditures, it provides a strong incentive to overinvest in capital improvements. Consequently, state laws typically demand that capital expenditures be "prudent," and they require the utility to secure PUC approval for major capital improvements before it is allowed to include them in its rate base. Since 1992, Congress has required state PUCs to put into place "integrated resource planning" procedures under which utility companies submit periodic integrated resource plans (IRPs) containing estimates of demand growth and capacity needs, projected retirements of EGUs, and options for meeting anticipated demand through capital expenditures, purchases of electricity from other power plants, or measures to reduce demand. One of the goals of these plans is to maintain 15–18 percent reserve capacity to ensure a highly reliable supply of electricity. In many states, approval of major capital projects and facility retirements are accomplished through the IRP process.[20]

Federal Regulation of Electricity

The Federal Power Act of 1935 created a regulatory regime similar to state public utility laws for regulating interstate sales of electricity in the wholesale markets occupied by generators and distributors of electrical power. The statute empowered the Federal Power Commission (now the Federal Energy

Regulatory Commission [FERC]) to set "just and reasonable" rates and to prohibit "unduly discriminatory or preferential" practices in interstate transmission and sale of electricity. It also determines the rate that operators of transmission systems can charge generators for their services and the conditions they can impose on generators to connect to the grid. On the other hand, the act specifically divests FERC of any jurisdiction over facilities used for generating or transmitting electricity within the boundaries of a single state. The courts have, however, interpreted FERC's authority generously to reach "practices affecting" the wholesale price of electricity, which as a practical matter means any purchase of electricity by a power company from another entity on the wholesale market. And the judicially created "filed rate doctrine" holds that state PUCs must "pass through" FERC-approved wholesale rates to consumers.[21]

DEREGULATION

The electric power industry flourished during the post–World War II years as technologies improved, consumption and production increased at roughly the same rate, fuel costs remained low, and FERC and state public utility commissions kept electricity prices relatively stable. This comfortable situation changed rather dramatically during the 1970s as technological improvements slowed, oil embargoes caused oil and natural gas prices to spike, inflation increased dramatically, a recession squelched demand, newly promulgated environmental regulations increased construction and operating costs, and the retail price of electricity began to rise. Many economists concluded that the generation function of an electric utility company was not a natural monopoly after all and that competition in the market for generating electricity would lead to more efficient generating techniques and lower retail prices. Cost-of-service ratemaking, they argued, put all of the risk of problematic capital investments on ratepayers, while a market-based approach would leave that risk with the generator. A growing number of critics supported amending the regulatory laws to separate generation from transmission and distribution (which remained natural monopolies) and to allow generators of all sizes and shapes to sell electricity to grid operators, distributors, or retail consumers at prices set by the resulting market without receiving the blessing of FERC or a PUC.[22]

Deregulation at the Federal Level

Congress responded to the calls for deregulation with the Public Utility Regulatory Policies Act of 1978 (PURPA). Among other things, PURPA required incumbent electric utilities to purchase electricity from "qualifying" independent power producers at prices reflecting the "avoided cost" to the utility of generating the same amount of power. If these independent producers could generate electricity cheaper than the local utility, they would make a profit. The statute also empowered FERC to require transmission companies to allow a generator at one point on the grid to sell electricity directly to a distributor at another point on the grid by simply feeding power into the grid and allowing the customer to extract it from the grid. The Energy Policy Act of 1992 empowered FERC to exempt generators from state regulation if they sold electricity to utility companies in wholesale markets and not directly to consumers in retail markets. This resulted in a new class of "merchant generators" that built and operated conventional power plants exclusively for wholesale markets. The dam broke in 1996 when FERC promulgated Order No. 888, the so-called "open access rule," that required all public utilities to open their transmission facilities to all generators on terms and conditions identical to their own use of those facilities. It further provided guidelines for the creation and operation of wholesale electricity grids by "independent system operators" (ISOs), whether or not sanctioned by state law. Three years later, FERC promulgated Order No. 2000, which encouraged transmitters to create regional transmission organizations (RTOs) to control the transmission of wholesale electricity throughout entire geographic areas. These and related orders opened the door for competitive markets in wholesale power generation throughout the nation. Local utility companies remained responsible for distributing the purchased electricity to the ultimate consumers in retail markets subject to state PUC regulation.[23]

ISOs and RTOs assumed responsibility for the reliability of the high-voltage transmissions systems that they administered pursuant to contracts specifying the rights and obligations of the generators that connected to them. ISOs and RTOs matched demand with supply in real time, and they managed emergencies that arose because of weather, breakdowns, or other unanticipated events. Contingency reserves to meet unanticipated power losses were typically provided by small gas-fired power plants that could be brought online instantly because their turbines were constantly spinning without generating electricity ("spinning reserves"). FERC-approved RTOs and ISOs also had the power to allocate the costs of new transmission projects among the utility companies that they served, subject to the "beneficiary pays" principle. Because RTOs and ISOs were responsible for system reliability, they typically demanded "reliability must run" contracts or "system support resources" agreements to allow them to postpone retirements of older generation units. As of 2016, twenty-seven states were fully or partially covered by one or more of seven FERC-approved RTOs and ISOs. Texas long ago created its own independent transmission network, called the Electric Reliability Council of Texas (ERCOT), which is responsible for 85 percent of the electricity consumed in that state.[24]

Deregulation at the State Level

Several states experimented during the 1990s with allowing generators to sell electricity directly to consumers in retail markets. By 1999, twenty-one states had required vertically integrated utility companies to divest their generation assets, and 17 percent of the nation's electrical power was produced by nonutilities. At the same time, public utility commissions continued to regulate transmission and distribution services, because they remained natural monopolies. The deregulation bandwagon stalled during the 2001 California electricity crisis (described in Chapter 8), when critics charged that market manipulation by Enron and other electricity brokers had caused rolling brownouts and exorbitantly high electricity prices. Stunned by the California experience and unconvinced that deregulation generated significant benefits for consumers, several states reregulated power generation. By 2016 only sixteen states still allowed consumers to choose their generators, and twenty states continued to regulate retail service under the traditional cost-of-service ratemaking paradigm.[25]

THE POWER GRID AND ALTERNATIVES TO BUILDING POWER PLANTS

The current national power system consists of three large interconnected systems (the Eastern Interconnection, the Western Interconnection, and ERCOT), called "power grids," within which RTOs and ISOs manage the purchase, sale, and transmission of wholesale electricity. The nationwide grid consists of more than 9,200 electric generating units with more than a million megawatts of generating capacity connected to three hundred thousand miles of high-voltage power lines that connect to local distribution grids (managed by local utility companies) that sell the electricity in retail markets to more than 334 million consumers. Pursuant to the Energy Policy Act of 2005, FERC has designated the North American Electric Reliability Corporation (NERC) to be responsible for ensuring the reliability of the nationwide grid by promulgating and enforcing reliability standards for generators, transmitters, and distributors of wholesale electricity.[26]

Demand Response

One alternative to building more power plants is to reduce demand when it is about to exceed supply through programs designed to encourage consumers to reduce electricity consumption. Referred to as "demand response" (DR) or "demand-side management," this approach comes in two varieties: dispatchable and nondispatchable. In a dispatchable demand response regime, a consumer (usually a heavy energy user like a university or large retailer) promises to reduce its load when agreed-upon triggers are exceeded. After receiving notice of the reduction in supply from a distribution utility or system operator, the consumer can respond by reducing energy-demanding activities (as, for example, by making automatic adjustments to heating and air conditioning system settings). The generator pays the consumer agreed-upon sums for the electricity not used during the time that the reduced service remains in effect through reductions in the consumer's bills. Nondispatchable demand response encourages reduced demand in times of heavy use by increasing prices and allowing consumers to observe the price increases with "smart meters" that provide real-time price quotes. Consumers can then reduce their electricity bills by scaling back consumption. In some states, companies called "third-party aggregators" identify curtailment strategies and contract with large consumers to implement them remotely via the internet.[27]

Distributed Generation

"Distributed generation" (DG) refers to mobilizing unconventional small-scale resources at the user end to reduce demand for power from the grid or even sell power to the grid. Such resources include nonpolluting technologies like rooftop solar panels on commercial buildings and residences, backyard wind generators, and fuel cells, but they can also include inefficient polluting technologies like diesel-fired emergency backup generators and microturbines. Distributed generators that rely exclusively on wind and solar power usually need extensive battery storage, support from polluting backup generators, or access to the grid for times when the wind is not blowing and the sun is not shining.[28]

Combined Heat and Power

Combined heat and power (CHP) facilities burn fuel to produce both electricity and thermal energy for other functions such as heating buildings or supplying heat for various industrial processes. Because they put waste heat to good use, these plants are more efficient than plants that disperse heat into the environment through cooling towers. Most CHP facilities use the electricity they produce, but they can sell excess amounts to the grid.[29]

Renewable Portfolio and Energy Efficiency Standards

More than thirty-five states and several municipalities have promulgated renewable portfolio standards (RPS) pursuant to which a regulated electric utility must procure a specified percentage of its electricity from renewable resources. State RPS programs vary somewhat in what they define as a renewable resource. All of the programs include biomass, landfill methane, solar panels, and wind generators as sources of renewable energy. Most states, however, do not include hydropower, despite its renewable status. States have also promulgated energy efficiency standards requiring electric utilities to generate, transmit, and distribute electricity more efficiently, typically by a specified percentage each year.[30]

THE ENVIRONMENT

April 22, 1970, the first Earth Day, marked the beginning of an environmental movement that has had a profound impact on the nation and especially on the electric power industry. Several nationwide environmental organizations, including the Sierra Club, the Audubon Society, and the National Wildlife

Federation, were in existence long before Earth Day, but the environmental movement that grew out of Earth Day spawned many more groups, like the Natural Resources Defense Council, the Environmental Defense Fund, Earthjustice, the Environmental Integrity Project, and Greenpeace, that were quite adept at using the mainstream media to highlight environmental issues, lobbying governmental policymakers, and challenging unlawful government and private sector conduct in court.[31]

The Environmental Protection Agency

Soon after Earth Day, President Nixon created the Environmental Protection Agency (EPA) and assigned to it the pollution control and environmental research functions of eight existing departments and agencies. Since its inception, EPA has regulated power plants extensively to reduce emissions of conventional and toxic air pollutants. In more recent years, it has promulgated regulations to protect marine and aquatic species from impingement and entrainment in once-through cooling systems and to ensure that coal ash wastes do not escape from massive impoundments. At the same time, EPA has been careful not to interfere with the reliability of the nation's electricity grids.[32] We shall see that EPA has played a major role in the transformation of the electric power industry from a belligerent defender of dirty coal-fired power plants into a proponent of sustainable power through renewable resources. The coal industry has suffered a major decline as a result of the transformation, and employment in that industry has decreased rather dramatically. But those adverse effects are also attributable to market forces reflecting dramatic reductions in the cost of natural gas and renewable power.

Power Plant Pollution

When a fuel burns in a power plant, it creates gaseous by-products called "flue gas." As the gas emerges from the boiler, it is routed through the unit's pollution control facilities and ultimately through a stack into the air. The primary combustion by-products vary, depending on the fuel, but all fossil fuels yield carbon dioxide (CO_2) and various oxides of nitrogen (NO_x). Coal combustion also yields sulfur dioxide (SO_2), particulate matter (PM), and a number of hazardous air pollutants, including mercury. In 2014, the year of EPA's most recent emissions inventory, power plants emitted 37 percent of the nation's CO_2 emissions (more than any other sector), 14 percent of the NO_x emissions, 62 percent of the SO_2 emissions, and 58 percent of the mercury emissions. By 2018, however, transportation sector emissions of CO_2 exceeded power plant emissions, a trend that could reverse itself as vehicles powered wholly or partially by electricity play a larger role in transportation.[33]

These emissions can have adverse effects on human health and the environment. In promulgating health-based air quality standards EPA relies extensively on "clinical" studies that expose human subjects to carefully controlled concentrations of the relevant pollutant in an isolation chamber while engaged in varying levels of physical exertion. The studies compare the incidences of various potentially adverse effects in the exposed subjects to the incidence in subjects engaged in the same activities but who breathe filtered air that is effectively free of the pollutant. Scientists can conduct similar controlled studies on vegetation, animal species, and materials.[34]

The next-best evidence comes from epidemiological studies, which consist of statistical comparisons of cohorts of human beings who have experienced higher-than-normal exposures to a pollutant with lesser-exposed cohorts. Epidemiologists use complex statistical techniques to compare the frequency of a disease in the exposed population with the frequency in the unexposed population to determine whether relative frequency (or "relative risk") is increased to such a degree that a cause-effect conclusion is warranted. The consistency of a result across different locations, circumstances, and times likewise enhances confidence in a positive cause-effect conclusion. Controlled human studies and animal experiments can lend "biological plausibility" to the conclusions of epidemiological studies.[35]

Particulate Matter. "Particulate matter" (PM) is a generic term for a broad class of chemical and physical substances that exist as discrete particles (liquid droplets or solids). For regulatory purposes, particulate matter comes in three varieties. "Coarse" particulate matter (PM_{10}) consists of particles or droplets of greater than 2.5 microns and less than 10 microns in diameter. "Fine" particulate matter ($PM_{2.5}$) consists of particles or droplets of less than 2.5 microns and greater than 0.1 micron. "Ultrafine" particulate matter consists of particles or droplets of less than 0.1 micron in diameter. Coarse PM comes from abrasion and crushing processes, soil disturbances, plant and insect fragments, pollen, and sources other than power plants. Most fine and ultrafine particulate matter comes from combustion in power plants and various mobile sources and reactions of gaseous pollutants in the atmosphere. In general, smaller particles have the greatest impacts on human health. Short-term exposure to $PM_{2.5}$ can cause a number of adverse cardiovascular and respiratory effects, such as chronic obstructive pulmonary disease. Long-term exposure to $PM_{2.5}$ can cause adverse cardiovascular effects and premature mortality. Particulate matter can cause soiling and corrosion of painted surfaces and other building materials, and it can adversely affect vegetation and ecosystems in a variety of ways. But the most pronounced environmental effect of PM is impairment of visibility, discussed below.[36]

Sulfur Dioxide. Sulfur dioxide (SO_2) is a highly reactive gas that is directly emitted by fossil fuel–fired power plants but can also be converted to sulfates when it comes into contact with water in clouds. Short-term exposure to sulfur dioxide (ranging from five minutes to twenty-four hours) can cause bronchoconstriction and lung-function decrements in adults and children. Short-term exposures are especially hazardous for persons suffering from asthma. Long-term exposures to SO_2 in its gaseous phase can damage crops and other plant life and damage buildings and other structures.[37]

Oxides of Nitrogen. The term "oxides of nitrogen" (NO_x) refers to all forms of oxidized nitrogen. NO_x are created in power plants when molecular nitrogen in fuel or in the air interacts with oxygen in the air during the extremely hot combustion process. The higher the temperature of the combustion process, the more oxides of nitrogen are produced. Because coal contains larger amounts of fuel-bound nitrogen than natural gas, coal-fired power plants emit somewhat more NO_x per unit of energy produced than gas-fired plants. Short-term exposures to NO_x can cause adverse respiratory effects in children and increase the severity of asthma attacks. Long-term exposure to NO_x can cause respiratory illness in children and may interfere with lung development in children. NO_x can injure plants and disrupt visibility in the same way that SO_2 does. The primary environmental problem with NO_x, however, is their contribution to photochemical oxidants.[38]

Photochemical Oxidants. Photochemical oxidants are produced when NO_x from power plants and certain volatile organic compounds (VOCs) from a number of natural and anthropogenic sources react in the presence of sunlight. When these highly complex reactions occur, ground-level ozone is the predominant product. EPA therefore uses ozone as a surrogate for all photo-chemical oxidants. Ozone concentrations tend to be higher during the "ozone season," which generally runs from May through September. Short-term expo-sures to ozone can cause decreased lung function, especially in children, asth-matics, and adults who work or exercise outdoors. Short-term ozone exposure can also produce a variety of adverse respiratory symptoms, such as coughing, wheezing, shortness of breath, and airway inflammation. Asthma sufferers are at greater risk for more prolonged bouts of breathing difficulties when they suffer an attack due to other causes. Ozone exposure may also contribute to cardiovascular diseases and to cardiopulmonary-related mortality risk. Plants exposed to ozone suffer decreased growth and biomass accumulation. Food crops suffer decreased yields and sometimes decreased nutritive quality.[39]

Acid Deposition. Both NO_x and SO_2 can cause adverse effects on the ecology of lakes and streams through a process called "acid deposition." SO_2 and NO_x emitted from power plants are transported vertically by convection into the troposphere, where they travel long distances on the prevailing winds and turn into sulfates and nitrates as they combine with water in clouds. If the cloud evaporates, the sulfates and nitrates fall to the ground in a process called "dry deposition." If the cloud does not evaporate, the sulfates and nitrates form sulfuric and nitric acid and fall to earth in rain or fog in a process called "wet deposition." Acid deposition on land can kill sensitive biota and increase the susceptibility of plants to fires, drought, and wind damage. Acid deposi-tion into lakes can kill acid-sensitive species and otherwise damage aquatic and marine ecosystems.[40]

Visibility. Visibility is an important environmental asset in national parks and other remote locations featuring attractive landscapes. Fine particulate matter (including sulfate and nitrate droplets resulting from SO_2 and NO_x) impairs visibility by absorbing, diffracting, or refracting light, thereby re-ducing the distance at which objects may be seen clearly. Power plant PM emissions can impair visibility through "plume blight" (direct disruption of visibility by a power plant plume) or "regional haze" (indirect disruption of visibility caused by particles dispersed throughout the atmosphere).[41]

Mercury. Most coal contains small amounts of heavy metals, the most worrisome of which is the element mercury. Because it does not burn, the mercury in the coal either leaves the facility through the stack or remains in the fly ash collected by the facility's pollution control devices. Elemental mercury is not especially harmful, but it can go through a series of chemical reactions in wetlands, floodplains, and lakes that transform it into methylmercury, a potent neurotoxin in adults, children, and fetuses that bioaccumulates in fish. Exposure to very low levels of methylmercury in utero can cause subtle effects on memory, attention span, and language. Methylmercury can also damage the kidneys, liver, and nervous systems of children and adults. In addition, it causes adverse reproductive, developmental, and behavioral effects in fish and fish-eating birds.[42]

Greenhouse Gases. The cause-effect relationship between anthropogenic emissions of greenhouse gases (GHGs) and increasing global temperatures is now indisputable, despite the vigorous efforts of some conservative think tanks and a few scientists to raise doubts about that conclusion. The consequences of global warming for both human welfare and the environment are profound. Scientists predict that the earth will experience more drought, rising sea levels, increasingly severe hurricanes, increases in tropical disease, and massive population dislocation. Many of the world's poorest people live in areas that will be subject to inundation by rising seas and more intense tropical storms and hurricanes. A typical 1,000-megawatt coal-fired generating unit emits between 36,000 and 84,000 tons of CO_2, a potent GHG, every day. Power plants fired by natural gas emit slightly more than half that amount. Natural gas, however, consists mostly of methane, a greenhouse gas that is about seventy-two times more effective than CO_2 in trapping heat. The release of methane during the extraction, transportation, and use of natural gas for power plants may offset the advantage that natural gas has over coal in CO_2 emissions.[43]

Coal Ash. Power plants burning coal or oil produce a residue of noncombustible constituents know as coal combustion residuals (CCRs), or "coal ash." From 70 to 80 percent of the ash produced in a boiler consists of very fine particles, called "fly ash," that are entrained in the flue gas and carried out of the furnace. The remaining 20 to 30 percent of the ash is heavy enough to settle at the bottom of the furnace and is therefore called "bottom ash." Another combustion residual, called "flue gas desulfurization material," consists of the residue produced by the "scrubbers" that remove sulfur dioxide from power plant exhaust gases (see below). Power plants typically add water to the fly ash and place it and flue gas desulfurization sludge in large surface impoundments. The greatest threat posed by these retention ponds is their potential for catastrophic breach, which can inundate adjacent rivers and lakes. Toxic metals can slowly leach out of unlined or poorly lined retention ponds into groundwater and migrate to surface waters where they pose a threat to people who drink the water, to aquatic species, and to people who consume those species. CCRs can be recycled for beneficial uses. Fly ash can serve as a partial substitute for Portland cement in making concrete. Its porous nature also makes it ideal for high-volume applications like structural fills. Bottom ash and boiler slag make excellent road base materials and are also useful in manufacturing aggregates and concrete blocks. Flue gas desulfurization material can be recycled into gypsum for use in various construction products.[44]

Cooling Water Intake Structures. The "once-through" cooling systems in use at about 43 percent of the nation's power plants take water from a nearby lake, river, or ocean, use it to dissipate heat, and discharge the warmer water back into the original source. "Closed loop" systems that are used at most of the remaining power plants employ large cooling towers and recycle the cooling water to the condensers. Once-through systems can damage the environment at the "cooling water intake structures" where huge pumps pull millions of gallons of water into the cooling system. The intake pipes are protected by screens to prevent fish, crustacea, and other organisms from entering the system. Organisms that do not pass through the screens can be killed by the pressure of the water against the screens (impingement), and organisms that pass through the screens are subjected to mechanical and temperature stresses (entrainment), which can be equally deadly. Impingement and entrainment can result in substantial reductions in fish populations in affected waters, a matter of great concern to both commercial and recreational fishermen.[45]

Pollution Reduction Technologies

Power companies can address air pollutants in six basic ways: (1) dissipating air pollutants over larger air volumes by building tall stacks; (2) reducing operations or ceasing altogether when weather conditions result in high local concentrations of pollutants (intermittent controls); (3) switching to fuels that result in fewer emissions; (4) removing contaminants from the fuel before combustion; (5) capturing contaminants during combustion; and (6) removing contaminants from flue gases after combustion.[46]

Tall Stacks and Intermittent Controls. In the years before and just after Congress enacted the Clean Air Act of 1970, power companies constructed huge stacks towering as high as one thousand feet and used heat from the furnaces to eject flue gases at high velocities into the upper layers of the atmosphere. Euphemistically termed "dispersion enhancement," these tall stacks solved the problem of complaints by neighbors, but they created another problem— interstate transport of pollutants that came back to the surface as acid rain.[47]

Intermittent Controls. Intermittent controls could in theory ensure that pollutant concentrations in nearby air stayed below prescribed levels by powering down when weather conditions were favorable for high pollutant concentrations. But that technique merely dispersed pollutants over time rather than space. More important, intermittent controls were very difficult to enforce because they depended on the source's ability (and willingness) to anticipate adverse weather conditions and to power down the plant in time to avoid a violation of air quality standards.[48]

Fuel Switching. Owners of existing plants have the option of switching to less polluting fuels, but the switch may require retooling the furnace or other major capital investments. Plants can usually switch from high-sulfur coal to low-sulfur coal with only slight changes in coal handling systems if the company is willing to pay to transport the cleaner coal greater distances and if local politicians do not intervene to preserve markets for local coal producers. Power plant operators can optimize price and pollution considerations by blending clean and dirty coals before they enter the furnace.[49]

Cleaning Fuels. Another way to reduce emissions is to remove pollutant precursors like sulfur and heavy metals from the fuel prior to burning it. Easily available technologies can remove sulfur from "sour" natural gas. "Coal washing" is a process through which coal is crushed and soaked with water to allow the heavier sulfur-laden particles called "pyrites" to settle out. The industry has been less successful in removing the "organic sulfur" that is attached to carbon molecules in the coal.[50]

Coal Gasification. Technologies also exist for turning coal into a gas that behaves like natural gas and lacks most of the coal's contaminants. A large tank called a "gasifier" subjects coal to high temperature and pressure in the presence of steam, converting coal into gaseous components that are then burned under carefully controlled conditions so that only a small amount of the fuel burns completely. This produces "syngas," which is composed of hydrogen and carbon monoxide. The syngas is then burned to power a turbine.[51]

Particulate Removal Technologies. Highly effective technologies for removing particulate matter from flue gas have been available for decades. Electrostatic precipitators use electrical forces to remove more than 99.9 percent of PM from flue gas as it passes horizontally between high-voltage discharge electrodes and grounded collection plates. As the particles acquire an electric charge, they migrate toward the collection plates where they agglomerate and fall into collection hoppers. Baghouses use fabric bags to filter out particulate matter as the exhaust stream passes through the fabric. Baghouses entail high maintenance costs because the fabric can clog and suffer chemical damage. But they are even more effective than electrostatic precipitators, and they can remove some mercury.[52]

Scrubbers. A scrubber removes SO_2 from an exhaust stream by spraying an absorbent material in its path. A chemical reaction occurs in which the SO_2 reacts with a chemical in the material and is thereby removed from the exhaust stream. Wet scrubbers force the exhaust stream upward through a tower filled with a liquid spray containing water and pulverized limestone (or some other reactive material). Dry scrubbers inject pulverized limestone (or some other reactive material) directly into the flue gas where it reacts with the SO_2 as the sulfur in the coal oxidizes and is collected in the particulate removal device with the fly ash. Scrubbers typically remove around 95 percent of the SO_2 from the exhaust stream. They also remove particulate matter, which can contain mercury. Scrubbers require large capital investments, take up a lot of space, consume up to 2 percent of the plant's energy, and have historically entailed high operating costs.[53]

Fluidized Bed Combustion. Another technique for removing SO_2 from coal-fired power plants, called "fluidized bed combustion," uses a furnace that is partially filled with granules of limestone that resemble a fluid when air is forced through the bottom of the furnace causing the granules to float on the air. Coal is injected into the fluidized bed of limestone. As the coal burns, the limestone removes the SO_2 from the combustion by-products. The technology is capable of removing more than 95 percent of the SO_2 from the exhaust stream.[54]

NO$_x$ Reduction Technologies. Burning coal at low temperatures in environments that contain little oxygen can reduce NO_x emissions. Engineers have developed "low NO_x burners" that can meet these two conditions at relatively little expense, but their efficacy varies with boiler type and they tend to lower the overall energy efficiency of the boilers. A technology called "overfire air," in which the air required for complete combustion is added through special air ports located above the burner zone, can enhance the effectiveness of low NO_x burners. More recently, engineers have developed a technology called "selective catalytic reduction" (SCR) through which a reagent (usually ammonia) is added to the exhaust stream as it enters a large honeycomb-shaped ceramic filter where the reagent reduces NO_x to molecular nitrogen and water as the ceramic acts as a catalyst. SCR is capable of removing from 70 to 90 percent of NO_x in the stream. A less effective technology called "selective noncatalytic reduction" (SNCR) works in a similar fashion, but at much higher temperatures and without the stimulus of a catalyst.[55]

Mercury Removal. Scrubbers, particulate matter removal technologies, and selective catalytic reduction can remove from 60 to 98 percent of the mercury in an exhaust stream. The mercury can then be removed from the fly ash or scrubber slurry by adding chemicals to precipitate it out of the liquid, or it can be disposed of with the slurry or fly ash. A more effective technology called "activated sorbent injection" removes mercury from exhaust streams by injecting small particles of activated carbon (or other sorbent) impregnated with sulfur or iodine into the flue gas. Mercury in the flue gas attaches to the activated carbon, which is then captured and (in some cases) regenerated for reuse. Activated sorbent injection can remove from 50–99 percent of the mercury from the exhaust stream.[56]

Disposal of Coal Ash. It is not always possible to recycle CCRs either because contamination with mercury or other toxic residuals renders the useful products unsalable or because the market for those products becomes saturated. Around 40 percent of coal CCRs are beneficially recycled as described above and 60 percent are disposed of in surface impoundments. Of that amount, 80 percent are disposed of on-site. As of 2012, there were more than 310 active on-site landfills for solid CCRs averaging more than 120 acres in size and 40 feet in depth and more than 735 active on-site surface impoundments for liquid CCRs averaging more than 50 acres in size and 20 feet in depth.[57]

GHG Reduction Technologies. Unlike NO_x, SO_2, PM, and mercury, which can result from impurities in fuel or are products of incomplete combustion, CO_2 is produced in huge quantities by burning even the cleanest of carbon-based fuels. Consequently, CO_2 emissions in the United States exceed emissions of all other air pollutants combined. The most effective way to reduce fossil fuel–powered power plant emissions of CO_2 is to produce electrical energy as efficiently as possible, thereby reducing the amount of CO_2 emitted into the atmosphere per unit of energy produced. Since natural gas burns far more efficiently than coal, converting from coal to natural gas can reduce GHG emissions. Combined-cycle systems also improve efficiency.[58]

The one currently available technology for reducing CO_2 emissions from power plants is carbon capture, utilization, and sequestration (CCUS). Plants employing CCUS remove carbon dioxide from the exhaust gases and transmit it through pipelines to locations where it can be used for secondary oil recovery in oil wells or sequestered in underground storage areas. Solutions containing amines or hindered amines remove carbon dioxide from the exhaust stream. Increasing the temperature of the amine solution releases the absorbed CO_2 and renders the solution available for reuse in the absorption tower. The CO_2 is then cooled and compressed into a liquid phase that is easier to transport. CCUS is a fairly simple technology, but it is very energy intensive, consuming up to one-quarter of the plant's electricity-producing energy.[59]

CONCLUSIONS

Now that we have some sense for how electricity is generated and distributed, how electricity production and distribution is regulated to avoid monopoly power, how power plants pollute the environment, and how technology can reduce that pollution, it is time to examine the role that the federal government, state governments, and environmental and industry groups have played in the struggles to reduce the harm to human beings and the environment that power plant pollution has caused. The next eight chapters will examine many of those struggles in historical progression as they outline the evolution of environmental policy over the past half century.

CHAPTER 3

First Steps

In years right after World War II, Donora, Pennsylvania, was an exemplar of industrial America. The home of the nation's largest nail manufacturing factory and neighbor to steel mills, coke ovens, sulfuric acid plants, and a huge zinc smelter, its air was so polluted that people driving automobiles frequently had to use their headlights during the day to see. Coal supplied the energy for all of these factories, and coal was so much a part of the town that it was used to pave the roads. The smokestacks on most of the plants were not tall enough to propel their emissions beyond the six-hundred-foot-high hills that surrounded the town. But the smell of burning coal in Donora was the smell of prosperity.

On October 26, 1948, a blanket of cold air settled in over the Monongahela Valley, trapping the emissions from the town's power plants and factories. Concentrations of particulate matter (PM), sulfur dioxide (SO_2), and toxic metals grew to deadly levels. As the temperature inversion persisted for the next three days, the hospitals filled with sick and dying people. The town's eight doctors advised residents to leave town, but the factories kept up their daily routines. By the time that the inversion lifted, the funeral homes had sold out of caskets. Subsequent studies concluded that the four-day event had caused twenty deaths and almost six thousand illnesses. Despite these frightening statistics, the state of Pennsylvania took no action to prevent future disasters.[1]

December 1970 President Richard Nixon appoints Bill Ruckelshaus to head newly created EPA

December 31, 1970 President Nixon signs Clean Air Act Amendments of 1970

April 1971 EPA promulgates the first NAAQS

December 23, 1971 EPA promulgates first NSPS for power plants

April 1973 Russell Train replaces Bill Ruckelshaus as EPA administrator

June 1973 Supreme Court orders EPA to create a PSD program

June 1974 Congress enacts the Energy Supply and Environmental Coordination Act

August 1974 Gerald Ford becomes president

August 1976 EPA rejects SIPs for forty-five states

REGULATING POWER PLANTS IN THE STATES

Prior to 1967, the states had the primary responsibility for regulating power plant emissions, but, like Pennsylvania, they did very little regulating. On November 21, 1967, President Johnson signed a law authorizing the Department of Health, Education, and Welfare (HEW) to designate air quality control regions and requiring states to establish ambient air quality standards for those regions based on HEW-issued air quality criteria setting out cause-effect relationships between exposure to air pollutants and adverse effects on human health and the environment. When HEW issued its criteria for SO_2, the coal and electric power industries complained that the resulting standards would have devastating ramifications for the nation's economy because it would be impossible to meet them in many regions without severely restricting coal burning.

Two years after the statute's enactment, no state had established ambient air quality standards. Instead of requiring power plant owners to install pollution controls or switch to cleaner fuels, state environmental agencies allowed them to build tall stacks to disperse emissions over wide geographical areas. The Monsanto Corporation and Allied Chemical Corporation had developed "flue gas desulfurization" technologies to reduce SO_2 emissions, but they quickly discovered that power companies were not eager to invest in unproven technologies if they were not forced to do so. Not surprisingly, SO_2 emissions from electric power plants nearly doubled between 1960 and 1970.

More than 85 percent of the nation's electricity came from plants that burned coal or oil, and they accounted for more than 50 percent of the nation's SO_2 emissions.[2]

PRESIDENT NIXON PICKS BILL RUCKELSHAUS

President Nixon appointed William (Bill) Ruckelshaus to head the newly created Environmental Protection Agency (EPA) in December 1970. The son of a prominent Indiana Republican, Ruckelshaus had earned a reputation as a vigorous pursuer of corporate and municipal polluters as a young Indiana prosecutor. The appointment drew praise from prominent congressional Democrats, and he was easily confirmed. Ruckelshaus came to the job committed to three priorities—strong enforcement of existing laws, scrupulous implementation of the soon-to-be-enacted 1970 Clean Air Act, and reducing the costs to the agency of promulgating and enforcing regulations. Ruckelshaus later compared his efforts to implement those priorities to "running a 100-yard dash, while undergoing an appendectomy." But his efforts had strong public backing. A Harris poll taken in 1971 reported that 78 percent of the respondents were willing to pay more for electricity in order to have cleaner air and water and that 48 percent were willing to tolerate a 10 percent reduction in jobs to meet that goal.[3]

THE CLEAN AIR ACT AMENDMENTS OF 1970

Although President Nixon was deeply ambivalent about federal regulation of business, he was fully aware of the demands from the public for strong environmental protections. It was highly likely that the Democratic candidate for president in 1972 would be Senator Edmund Muskie (D-ME), a strong environmentalist, and Nixon was determined not to allow Muskie or any other Democratic politician to project an image of him as a despoiler of the environment. He therefore supported legislation empowering EPA to promulgate nationwide ambient air quality standards that would be implemented by the states. He also endorsed nationwide pollution reduction standards, requiring the electric power industry and other heavily polluting industries to install the best available demonstrated pollution control technology.[4]

In June 1970, the House passed a bill containing more stringent measures than the Nixon administration wanted by a vote of 374–1, a lopsided margin that demonstrated the political salience of clean air at the time. Senator Muskie shepherded through the Senate a bill that was far more stringent than either the administration or the House proposals. It required EPA to promulgate national ambient air quality standards (NAAQS) that the states would achieve by strict deadlines through EPA-approved state implementation plans. This put the Nixon administration in the awkward position of agreeing with the affected industries that the Muskie bill was unrealistically stringent. Their efforts to weaken the Senate bill, however, had little impact on the conference committee's deliberations. On the evening of December 16, 1970, Senator Muskie reported that the committee had agreed on a bill that differed from his bill only in minor regards. President Nixon reluctantly signed it on December 31, 1970.[5]

The 1970 Clean Air Act established a comprehensive program for promulgating, attaining, and maintaining NAAQS for a list of pollutants that "endangered" public health or welfare and that derived from "numerous or diverse mobile or stationary sources." The agency first had to prepare a "Criteria Document" (now called an "Integrated Science Assessment") that "accurately reflect[ed] the latest scientific knowledge" on the health and welfare effects of the pollutant. It then had to establish primary NAAQS for these "criteria" pollutants at a level that was "requisite to protect the public health," while "allowing an adequate margin of safety." The legislative history of the statute made it clear that the goal of the primary standards was to ensure "an absence of adverse effects on the health of a statistically related sample of persons in sensitive groups," such as "bronchial asthmatics and emphysematics who in the normal course of daily activity are exposed to the ambient environment." EPA had to set the secondary standards at a level that was "requisite" to protecting "public welfare." The statute gave the states an initial opportunity to promulgate state implementation plans (SIPs) imposing source-specific requirements and other measures capable of ensuring that the primary standards were attained within three years after the plan's promulgation. If a state failed to promulgate a SIP or promulgated an inadequate SIP, EPA was required to promulgate and enforce a federal implementation plan (FIP) meeting the statutory requirements.

The statute contained two important additions to this institutional arrangement. First, it required EPA to promulgate "new source performance standards" (NSPS) for listed categories of stationary sources reflecting the "best adequately demonstrated control technology" (BADT). These technology-based standards were automatically incorporated into SIPs and were binding on any company that proposed to construct a new source (sometimes referred to as a "greenfield" source) within the relevant category or attempted to modify an existing source. Second, Congress required EPA to write national emissions standards for hazardous air pollutants (NESHAPs), defined as pollutants that were likely to result in "an increase in mortality or an increase in serious irreversible, or incapacitating reversible, illness" and that did not come from numerous and diverse sources. Finally, the statute empowered ordinary citizens to sue the agency when it failed to perform a duty by a statutory deadline for an order requiring the agency to perform the duty in a timely fashion. It also empowered citizens to enforce the law, with any fines going to the federal treasury.[6]

IMPLEMENTING THE CLEAN AIR ACT

The Clean Air Act saddled the just-created EPA with a large number of extremely difficult tasks to accomplish within a ridiculously short period of time. It also gave the agency precious little discretion, most of which came in the form of policy judgments that the administrator would have to make in interpreting and applying the scientific and technical information relevant to real-world rulemaking exercises. Administrator Ruckelshaus chafed under the impossible requirements, but he understood that he had to make a good faith effort to implement them before Congress would be willing to consider changes to the statute.[7]

New Kids on the Block

Ruckelshaus expected pressure from the regulated industries to reduce the stringency of EPA's regulations, but the agency received pressure in the opposite direction from an unexpected source—the activist environmental groups that had popped up all over Washington, D.C. The most influential environmental group on clean air issues was the Natural Resources Defense Council (NRDC), the brainchild of five Yale Law School students who hoped to transfer the successful model of civil rights litigation to the environmental arena. Unlike existing membership groups like the Sierra Club, they wanted to specialize in using the regulatory process and the courts to advance protective environmental goals. With a generous grant from the Ford Foundation, NRDC paid its attorneys at rates that were comparable with those paid to government attorneys. They became especially adept at filing (and then settling) lawsuits aimed at holding EPA to statutory deadlines and at keeping the agency true to its statutory mission. After winning several impressive legal victories and attracting the attention of additional funders, NRDC expanded its wheelhouse to include participation in agency rulemaking, framing issues for the mass media, and lobbying Congress for and against changes to the environmental laws.[8]

The First NAAQS

EPA's first task was to promulgate national ambient air quality standards (NAAQS) for the "criteria pollutants" for which its predecessor in HEW had already prepared air quality criteria. The staff that EPA had inherited from HEW quickly proposed standards for SO_2, oxides of nitrogen (NO_x), ozone, carbon monoxide, hydrocarbons, and particulate matter (PM) with three days to spare. John Middleton, the head of EPA's Office of Air and Radiation, assured the administrator that the scientific community fully supported the proposals. Ruckelshaus signed the proposal on January 30, 1971. The final standards that EPA published in April 1971 prescribed concentrations of the pollutants in the ambient air over various averaging periods. Administrator Ruckelshaus unapologetically declared the standards to be "tough," because the Clean Air Act required the agency to "err on the side of safety." To the electric power industry's complaints that the SO_2 and PM standards could not be met on a consistent basis, Ruckelshaus responded that the industry could meet both standards by switching to low-sulfur coal and installing electrostatic precipitators or baghouses. He also warned the states that if they did not prepare state implementation plans in accordance with the statute's tight schedule, the agency would seek judicial injunctions against sources within those states.[9]

Pressed by the statutory deadlines, Ruckelshaus spent little time vetting the standards with other agencies or the White House. The press attention surrounding the final standards, however, attracted their attention, and they were not happy. At a meeting with EPA and White House staff, representatives of the Federal Power Commission (FPC) and the Department of Commerce (DOC) bitterly complained that the standards for SO_2 and PM were far too stringent. FPC ventured that some coal-fired power plants might be forced to shut down altogether, with resulting threats to the reliability of local electricity grids. Ruckelshaus responded that the statute forbade the agency from considering the cost of implementation in setting the standards. Although it was too late to change the standards, President Nixon attempted to ensure that EPA would never again sidestep interagency review by instructing EPA and several other agencies to send proposals for regulations imposing "significant costs" on regulated industries, along with estimates of their costs and benefits, to the White House Office of Management and Budget (OMB) for review prior to publication. Every president since Nixon has required executive branch agencies to follow a similar interagency review process.[10]

Despite all of the industry complaints, the sole judicial challenge to the standards targeted only the annual secondary standard for SO_2. Kennecott Copper Company argued that EPA's conclusion that a level of sixty micrograms per cubic meter ($\mu g/m^3$) was "requisite" to protecting public welfare was not adequately supported by the very brief statement that EPA issued to accompany the standards. Declining to resolve the conflict over the proper interpretation of the scientific evidence, the District of Columbia Circuit Court remanded the standard to the agency for a statement providing more detail on the basis for its action. The agency responded to the remand by revoking the standard.[11]

IMPLEMENTING THE CLEAN AIR ACT IN THE STATES

The states now had until January 29, 1972, to write SIPs containing emissions limitations and a demonstration that the requirements would achieve the primary NAAQS within three years. This required states to come up with inventories of current emissions from stationary sources like power plants and use dispersion models to determine the contributions of each source to any "nonattainment" areas in the state. They would then have to come up with emissions limitations for stationary sources and transportation controls for mobile sources and demonstrate (again with models) that those limitations would reduce emissions sufficiently to meet the standards by the three-year deadline. Given the rudimentary state of air quality modeling in the early 1970s, this analytical exercise entailed considerable uncertainty. EPA insisted that plans also ensure that emissions from sources in the state would not cause or contribute to violations in downwind states. And the Sierra Club argued that the statute required SIPs to include provisions to "prevent significant deterioration" of air quality in areas where the air did not exceed the standards.

EPA's State Implementation Plan Guidelines

Hoping to avoid an inbox full of hopelessly inconsistent and confused SIPs, EPA proposed guidelines setting out what it expected to see in those SIPs. To the suggestion in the guidelines that states require power plants to switch to low-sulfur coal or natural gas to meet the SO_2 NAAQS, the electric power industry complained that sufficient supplies of those fuels would not be available in some areas in time to meet the standards by the deadlines. FPC pointed out that if multiple state plans depended on the same supplies of low-sulfur coal and natural gas, a reliability crisis might erupt when they proved insufficient to meet the combined demand.[12]

The final guidelines that EPA prepared for publication in the *Federal Register* went through the OMB review process that the president had ordered. Documents leaked to NRDC showed that OMB, FPC, and DOC had "radically altered" EPA's final draft. Among other things, the final guidelines deleted a program for preventing significant deterioration of clean air areas and allowed plants to employ "intermittent controls" that functioned only when weather forecasts indicated that a NAAQS was in danger of being violated. The message that environmental groups heard was that "the feds aren't really serious about this."[13]

Problematic SIPs

All of the states submitted plans within two months of the deadline. To EPA's great relief, a court of appeals rejected the electric power industry's contention that affected companies were entitled to full-fledged adjudicatory hearings before EPA approved or disapproved the SIPs. After comparing "shelves and shelves" of SIPs to the statutory requirements, NRDC's Richard Ayres concluded that the SIPs were "deficient in virtually every important respect." Many states had failed to conduct the data gathering and modeling needed to demonstrate that the plans would attain the standards by the deadline. Instead, they simply assumed that power plants would install desulfurization technology or switch to low-sulfur fuels in time to achieve the SO_2 NAAQS by the deadline. Assuming that no technologies were available for reducing NO_x emissions from power plants, the SIPs focused on hydrocarbon reductions from motor vehicles and other stationary sources. A Colorado official freely admitted that the state had two plans—one "to pacify" EPA and a less "idealistic" plan "for actual use." Although EPA's staff found many defects in the hastily drafted plans, the agency fully approved plans for nine states and approved most aspects of plans for forty-one areas.[14]

Major Issues in SIPs

EPA's approvals and disapprovals of SIPs raised a number of generic issues, many of which were resolved by federal courts after years of litigation. First, the electric power industry argued that the statute required EPA to reject SIPs that were not economically or technologically feasible. EPA believed that states were free to consider feasibility in crafting SIPs, so long as they would achieve the NAAQS by the deadlines, but EPA was not obliged to take feasibility into account in deciding whether to approve or disapprove a plan. The Supreme Court, in a lawsuit brought by the Union Electric Company, upheld EPA's interpretation.[15]

Second, environmental groups argued that EPA could not approve SIPs that allowed power plants to rely on tall stacks or intermittent controls because they did not constitute "emissions limitations." They argued that tall stacks just dispersed the same load over a broader geographic area where it could fall back to earth as acid rain. They noted that Ohio Electric's James M. Gavin plant was free to burn high-sulfur coal because it had tall stacks, even though it topped EPA's list of the heaviest SO_2 emitters. Intermittent controls were difficult to enforce because of the limitations in ground-level monitoring devices, and the plants would make up for lost time when the weather was accommodating. Both strategies violated the unwritten maxim of environmental law that "dilution is not the solution to pollution." The Fifth Circuit Court of Appeals held that the statute required the states to establish emissions standards, and not tall stacks, unless emissions standards were "unachievable or infeasible" and the state adopted regulations capable of attaining the "maximum degree of emission limitation achievable." A year later, the Sixth Circuit held that the same two conditions applied to intermittent controls. The Fifth Circuit opinion gave EPA enough wiggle room that it approved SIPs allowing companies to build almost two hundred stacks greater than five hundred feet high and twenty stacks greater than one thousand feet high between 1970 and late 1977. Most of these plants became "big dirties" that EPA had to deal with in the twenty-first century.[16]

A third issue concerned the extent to which power plants should switch from high-sulfur to low-sulfur coal. Only about 10 percent of the coal east of the Mississippi River contained sulfur at low enough levels to meet EPA's suggested emission limitation of 1.2 pounds per million British thermal units (lb/MBtu), and most of that was expensive "metallurgical-grade" coal that was used to produce coke for steel plants. There was plenty of low-sulfur coal in the Powder River Basin of Montana and Wyoming, but it had a lower heat content, and it was expensive to transport it by rail to the eastern half of the country. Companies could also burn oil or natural gas, the sulfur content of which could easily be reduced, but both fuels were more expensive than coal and required pipelines. Both options were vigorously opposed by eastern coal companies and their political allies. Nevertheless, several SIPs relied on fuel switching, even though it was not at all clear that adequate supplies of alternative fuels would be available by the deadline of July 31, 1975.[17]

Perhaps the most contentious issue was whether SIPs should require coal-fired power plants to install scrubbers. Electric power companies complained bitterly about the cost of installing scrubbers in existing coal-fired power plants. Once installed, scrubbers were messy and difficult to operate. The early scrubbers frequently broke down or needed to be shut down for maintenance, thereby reducing the plant's reliability. Arguing that technology vendors were rapidly solving these problems, environmental groups urged state agencies to take a leap of faith and require existing plants to retrofit scrubbers into their operations. In EPA's view, the spotty performance was attributable to the tendency of utility companies to rely on the lowest bidder and their unwillingness or inability to hire trained operation and maintenance personnel. EPA was even willing to allow companies to include bypass systems to allow power plants to keep producing electricity when the scrubbers malfunctioned.[18]

A fifth contested issue was whether SIPs had to contain programs for the prevention of significant deterioration (PSD) of air quality in "clean air" areas that met the NAAQS. For the Sierra Club, this was the defining clean air issue, because power plants posed a major threat to visibility in pristine areas of the West. For the coal and electric power industries, it made sense to build power plants in the West near mines that provided easy access to abundant supplies of low-sulfur coal. At the Sierra Club's behest, the U.S. District Court for the District of Columbia, issued a preliminary injunction on May 30, 1972, requiring EPA to disapprove all plans that lacked PSD requirements and to write regulations to guide states in amending their SIPs to fix the problem. The court found a PSD requirement in language stating that one of the Clean Air Act's purposes was "to protect and enhance the quality of the Nation's air resources." The D.C. Circuit affirmed the holding without writing an opinion, and an evenly divided Supreme Court allowed the court of appeals' decision to stand in June 1973. EPA now had to create a PSD program from scratch before it could fully approve any SIP.[19]

THE FIRST NEW SOURCE PERFORMANCE STANDARD

The 1970 Clean Air Act required EPA to establish new source performance standards (NSPS) for categories of stationary sources that "cause[d] or contribute[d] to the endangerment of public health or welfare" based on the "best system of emission reduction" that had been "adequately demonstrated" (BADT). The states' SIPs had to apply EPA's NSPS to greenfield sources and to modifications of existing sources. The term "modification" was defined to mean "any physical change in, or change in the method of operation of, a stationary source which increases the amount of any air pollutant emitted by such source or which results in the emission of any air pollutant not previously emitted."[20]

Defining "Modification"

In 1971, EPA defined "modification" to exclude routine maintenance, repair, and replacement (RMRR) projects, even if they increased emissions. For example, in the corrosive environment of a power plant boiler, tubes can become clogged and corroded over time. Until they are replaced, the plant's efficiency decreases. Replacing the tubes will increase the plant's efficiency, thereby increasing its emissions, even though the plant looked the same before and after the replacement. The agency reasoned that these operations did not constitute a "physical change" within the meaning of the statute. This modest exemption would become the grist for many lengthy lawsuits thirty years later when EPA discovered that many of the worst performing power plants had employed it very aggressively to justify their failure to comply with the new source review requirements.[21]

Implementing a Performance-Based Approach

Since power plants were responsible for a high percentage of the nation's SO_2 emissions, it was one of the first categories for which EPA promulgated an NSPS. As EPA's attorneys read the statute, Congress meant for the agency to establish a "performance-based" standard based upon a model technology that had been "adequately demonstrated," but a company could come up with a different pollution reduction technology so long as it met the standard. If a less expensive technology performed equally well, so much the better. The first task for the agency's engineers was to decide on the model technology. The agency's economists also had to be involved, because the standard had to take costs into account.

The performance-based approach precluded the use of concentrations of pollutants in air as the unit of measurement, because such a standard could be met by forcing more air through the stack. It also precluded tall stacks and intermittent controls, because neither of those approaches actually reduced emissions during the time that the plant was operating. The standard had to be expressed in units of pollutant removed per unit of product or some easily measurable surrogate for product. For power plants, EPA's engineers selected pounds of pollutant per million British thermal units (Btu), a measure of the heat produced in a boiler. The standard was therefore expressed as "lb/MBtu."[22]

Companies could, of course, meet a performance-based standard by burning low-sulfur coal without installing any control technology, and this raised serious legal and political problems. As a legal matter, environmental groups argued that the phrase "best system of emission reduction" was limited to a technology that would reduce emissions from any source of fuel. The political problem stemmed from the geographical locations of high- and low-sulfur coals. If utility companies in the East and Midwest could meet the emissions limitation by burning low-sulfur western coal, the reduced demand for eastern coal would have serious economic consequences for high-sulfur coal-producing states like Illinois, Ohio, Pennsylvania, West Virginia, and Virginia. Administrator Ruckelshaus recognized that the NSPS for coal-fired plants would be one of the most difficult and important regulations that EPA would promulgate during his tenure.[23]

The Power Plant NSPS

On August 17, 1971, EPA proposed the first new source performance standard for coal-fired power plants.[24] The proposed emissions limitation for SO_2 was 1.2 lb/MBtu averaged over two hours, which could be met by burning low-sulfur coal. A plant burning high-sulfur coal, however, would have to remove about 75 percent of the sulfur either from the fuel or from the combustion gases. EPA lawyers later explained that the agency hoped to encourage companies to identify and use low-sulfur fuels, because there was insufficient scrubber-building capacity to meet demand for all new and existing sources. The agency proposed a NO_x standard of 0.7 lb/MBtu, which could also be met without any additional controls. The proposed standard for particulate matter was 0.2 lb/MBtu, which could easily be met by electrostatic precipitators.[25]

After receiving comments from more than two hundred parties, the agency published its final rule on December 23, 1971. The numerical standards for SO_2 and NO_x were identical to the proposed standards. This meant that power plants in the East and Midwest could burn natural gas or low-sulfur coal without installing any controls. If plants burned high-sulfur coal, however, scrubbers would be needed. The agency cut the standard for particulate matter in half to 0.1 lb/MBtu, which could still be met with electrostatic precipitators, but it also exempted steam generators burning lignite from that standard. The agency further clarified the definition of "modification" to specify that "increases in production rates up to design capacity" were not modifications, nor were fuel switches so long as the equipment was originally designed to accommodate the new fuel.[26]

The only groups that were happy with the standard were western coal producers and the railroads that shipped their coal to plants in the Midwest and the East Coast. Noting that only three plants were employing scrubbers at the time the final rule was published, electric power industry representatives argued that a workable flue gas desulfurization technology did not yet exist. New sources would therefore be forced to burn expensive low-sulfur coal or natural gas. Environmental groups also criticized the standard on the ground that the huge power plants planned for the Rocky Mountain region would impair visibility in national parks without having to install any controls because they would burn local low-sulfur coal. They also argued that the standard should have been more stringent for plants burning high-sulfur coal because scrubbers were available and could reliably reduce SO_2 emissions by at least 90 percent.[27]

Judicial Review

The D.C. Circuit in October 1973 upheld EPA's power plant NSPS in all but one regard. The evidence in the record convinced the court that EPA's emissions limitations were adequately demonstrated, that the agency had taken cost into consideration, and that the limitations were achievable. Although only three plants had successfully installed scrubbers at the time, the court was especially impressed by the "predictions and guarantees of domestic equipment manufacturers for plants under construction." The only problem with the agency's analysis stemmed from its failure to take into account the adverse environmental effects of "significant quantities of sludge byproduct" produced by scrubbers, "which present[ed] substantial disposal problems." The case was therefore remanded to the agency for further explanation of the 1.2 lb/MBtu emissions limitation in light of the adverse environmental effects of disposing of the scrubber wastes. As we shall see, the court's opinion forced the agency to focus on the looming problem of coal combustion residuals, or "coal ash."[28]

INDUSTRY PUSHBACK

The coal and electric power industries complained bitterly about the cumulative cost of EPA's regulations and the threat that they would pose to the reliability of local power systems. They also complained that EPA was issuing regulations before gathering sufficient scientific data to support its conclusions. Both refrains would be heard again and again in future struggles over EPA regulations. The industry efforts to push back against EPA's aggressive rulemaking initiatives were greatly aided by the fact that inflation and unemployment had taken hold of the American economy with such vigor that President Nixon put wage and price controls into effect. The price controls discouraged investment in energy infrastructure at the same time that they encouraged energy consumption. By the winter of 1972–1973, shortages of fossil fuels of all kinds were being experienced nationwide. As this self-imposed energy crisis began to disrupt the lives of all Americans, the coal, electric power, and petroleum industries launched a massive public relations campaign to convince the public that costly environmental controls, none of which had actually gone into effect, were the primary culprit.[29]

THE NIXON ADMINISTRATION'S SULFUR TAX

As the 1972 election year opened, President Nixon attempted to get the jump on the Democratic front-runner, Edmund Muskie, by sending legislation to Congress that included a charge of fifteen cents per pound on SO_2 emissions from power plants and other stationary sources in nonattainment areas and ten cents per pound in clean air areas beginning in 1976. It was the first serious attempt to implement a solution advocated by most resource economists. Some environmental groups supported the SO_2 tax idea, but others derided it as a "license to pollute." The coal and electric power industries and the country's major labor unions were staunchly opposed to the idea, arguing that the money would be better spent on pollution controls or wages. This, of course, missed the point that a company would in fact spend its capital on pollution controls if they were cheaper per pound than the charge. The administration could not find a single Republican in either house of Congress to sponsor the bill.[30]

THE RUCKELSHAUS / TRAIN TRANSITION

Bill Ruckelshaus left EPA in April 1973 to head the FBI and later become deputy attorney general in the midst of a blossoming Watergate crisis. He resigned that position rather than fire special prosecutor Archibald Cox in the famous Saturday Night Massacre. His replacement was Russell Train, an establishment Republican who grew up in Washington and attended Princeton and Columbia Law School. Time spent on safaris in the African wilderness had turned him into a strong conservationist, and he became the head of the Conservation Foundation in 1965. He served as the first chairman of the White House Council on Environmental Quality. Train and his wealthy wife were active on the Washington social circuit, frequently entertaining senators, representatives, and high-level executive branch officials at their well-appointed home.[31] His strong connections on Capitol Hill gave him a degree of independence not common among career civil servants. Before taking the

EPA job, he insisted on written confirmation that he would have the final word on all EPA regulations.[32]

REACTING TO THE ARAB OIL EMBARGO

The energy shortage reached crisis proportions in October 1973 when several Arab nations made good on their threat to cut off oil exports to any country that supplied military weapons to Israel in the Yom Kippur War. Although the embargo did not affect coal or natural gas supplies, President Nixon told his cabinet that any conflicts between increasing energy production and environmental considerations should be resolved in favor of energy. Administrator Train, however, rejected the assumption that environmental regulations were contributing to the energy crisis. He assured the electric power industry that no coal-burning plant would be forced to shut down if it was making a good faith effort to install proper emissions reduction technology.[33]

In November 1973, President Nixon instructed Treasury Secretary William E. Simon, EPA Administrator Train, and the staffs of OMB and the Domestic Council to come up with amendments to the Clean Air Act that would ensure the development of adequate sources of energy. Soon thereafter, the White House circulated proposed energy legislation that would have, among other things, extended the deadlines for attaining the NAAQS and relaxed controls on SO_2 emissions from power plants throughout the country. When Simon insisted that Train support the bill, he refused and threatened to resign if the proposals went forward. He then told the *New York Times* that he was "strongly opposed to most of [the] proposals" and would "fight them to the last wire" because he was convinced that they would "do substantial harm." By now Nixon and several of his top aides were mired in the Senate Watergate investigations and in no mood to stir up additional controversy by firing a prominent Republican. The White House dropped the most offending proposals.[34]

The bill that the Nixon administration sent to Congress on March 22, 1974, allowed EPA to extend the deadline for attaining the primary standards beyond the July 1975 deadline, permitted states to rely on intermittent controls as a temporary measure for rural electrical power plants, overturned the judicial requirement that SIPs contain provisions to prevent significant deterioration of air quality, required power plants to switch from oil and natural gas to coal by 1980, and required EPA to suspend emissions limitations otherwise applicable to those plants. With that, American Electric Power (AEP), the parent company of six public utility companies operating in seven eastern and midwestern states, launched a $3.7 million nationwide advertising campaign to persuade the public that environmental regulations would lead, in the words of one ad, to "widespread power shortage with diminished industrial production, fewer jobs, and all the misery that goes with them."[35]

The Democrat-controlled Congress, however, ignored the administration's proposals as it crafted emergency legislation in late 1973 to address the growing energy crisis. The conventional wisdom was that domestic oil and gas reserves were rapidly dwindling and that coal was the solution to the nation's energy problems. The solution was to force power plants to switch from oil and gas to coal to preserve the former two fuels for higher uses in powering vehicles and heating homes. But for that to happen, high-sulfur coal had to remain in the mix, and this created a classic tension between the nation's energy policies and its environmental policies. The Energy Supply and Environmental Coordination Act (ESECAct) that President Nixon signed on June 26, 1974, came down on the side of the environment. It empowered the newly created Federal Energy Administration (FEA) to order power plants to switch from natural gas and oil to high-sulfur coal for one year, but it further provided that a fuel-switching order could not go into effect until EPA certified that the plant would comply with all Clean Air Act requirements.[36]

THE SCRUBBER DEBATE

Gerald Ford became the thirty-eighth president of the United States on August 9, 1974, when President Nixon resigned in the midst of revelations of wrongdoing related to Watergate. Bringing down double-digit inflation rates was Ford's highest priority, and the environment was low on his list. The scrubber issue set off a furious debate within the new administration. The FEA, FPC, DOC, and OMB all lined up against scrubbers. Administrator Train held firm to EPA's conclusion that scrubbers were effective and reliable, and the agency issued a report showing that scrubbers owned by several utility companies had been functioning properly for over a year. In late November 1974, Train announced that the White House Energy Resources Council had agreed with EPA that scrubbers should be the technology of choice for coal-fired power plants and that EPA had agreed to seek legislation extending the deadlines to meet the SO_2 NAAQS from 1977 to 1985 to allow companies an opportunity to install them.[37]

The Democratic landslide in the 1974 post-Watergate elections clarified that Congress would not be gutting the Clean Air Act any time soon. Consequently, several companies threw in the towel and began to install scrubbers. By mid-1975, owners of nearly one hundred power plants had scrubbers under construction or were planning to install them. Other companies arranged to purchase low-sulfur coal from the Powder River Basin. Still other companies put their plans to comply with SIP requirements on hold while they litigated their legality. One critical consideration for investor-owned utilities was whether state public utility commissions (PUCs) would allow them to pass through to consumers the cost of installing scrubbers and importing low-sulfur coal as prudent expenditures. In July 1974, FEA Administrator John Sawhill wrote to all fifty PUCs to recommend that they allow such pass-throughs. EPA strongly supported the recommendation, but some environmental groups opposed it on the ground that it might precipitate a consumer backlash against scrubbers. Both FEA and EPA urged the commissions to experiment with demand-side energy conservation approaches, like marginal cost pricing, peak and off-peak pricing, and eliminating volume discounts to large users, to reduce the need for building more power plants.[38]

NONATTAINMENT

The deadline of May 31, 1975, for approval of SIPs passed without a single state having a fully approved plan in place, but EPA granted several states extensions until 1977. Of the 247 air quality control regions, 42 had not met the SO_2 standard, 74 were "nonattainment" for photochemical oxidants, 60 for PM, and 13 for NO_x. Monitoring was too sparse in much of the nation to tell whether areas did or did not meet the standards. Although 480 coal-fired power plants were subject to emissions limitations in SIPs, only twenty-one scrubbers had actually been installed. Between 150 and 200 plants were not even on an enforceable path toward compliance, and many of the others were deviating from their compliance schedules because of variances, changes in SIPs, or simple noncompliance. In August 1976, EPA formally declared that SIPs for forty-five states were inadequate in one or more regards, and it demanded revised plans by July 1, 1978. In November 1976, it proposed a "nonattainment" policy that prohibited the construction of new major sources in nonattainment areas, unless the states brought about an offsetting reduction in emissions of the same pollutant from existing plants that were already in compliance with the relevant SIP requirements. The job of enforcing the new policy, however, fell to the incoming Carter administration.[39]

INTERSTATE POLLUTION

In August 1975, seventy-five million Americans received a rude introduction to the problem of interstate transport of pollution when a stagnant mass of air laden with photochemical oxidants settled over an area ranging from Michigan to New York City and as far south as North Carolina for a full week.[40] It was becoming clear to scientists that photochemical oxidants and their precursors could travel long distances to contribute to unsafe air in cities in downwind states.[41] Similarly, the sulfates that resulted from SO_2 emissions and nitrates that resulted from NO_x emissions could travel hundreds of miles before coming back to earth as acid deposition. Both of these phenomena posed a serious problem for EPA, because the states that were responsible for controlling pollutants in their SIPs had only paid passing attention to the section of the Clean Air Act that required SIPs to protect air quality in other states, and EPA had not insisted on more during the approval process. The job of dealing with interstate transport of pollutants also fell to the Carter administration.[42]

RECALCITRANT STATES

As the recession deepened in early 1976, some states relaxed requirements in SIPs and failed to enforce other standards. The Ohio pollution control agency did such a poor job of controlling emissions from the high-sulfur coal burning power plants and other factories in the Ohio River Valley that EPA had to take over its SIP process. But EPA did not always play hardball with recalcitrant states. West Virginia initially promulgated stringent emissions limitations for its power plants, but then suspended enforcement to ensure continued consumption of high-sulfur coal from West Virginia coal mines. Instead of taking over the state's program, EPA established a task force to help the state draft lawful revisions to its plan, and it announced that it would not make enforcement of the existing requirements a high priority if the state participated in good faith in the SIP revision process.[43]

CONCLUSIONS

Under the strong leadership of administrators Ruckelshaus and Train, EPA accomplished a great deal in the six years following the enactment of the Clean Air Act. John Quarles, EPA's general counsel and deputy administrator for most of this time, acknowledged that much of this progress would not have happened but for the efforts of the environmental groups that "functioned with remarkable effectiveness as a counterbalance to the representatives of industry." And what gave those groups the political power necessary to defend EPA from fierce attacks and to advance the ball was "the force of public opinion," which "gained for them the attention of the press and the allegiance of the politicians."[44]

The Nixon and Ford administrations, however, lacked a coherent policy for reconciling environmental protection with the role that coal played in meeting the nation's energy needs. During a single week in the mid-1970s, FEA ordered a power plant operated by Florida Power & Light to convert from oil to coal, and EPA ordered Tampa Electric Company to convert a plant just across the bay from coal to oil.[45] The schizophrenic policy of the ESECAct ensured that the massive conversion from oil and gas to coal never happened. As we shall see in Chapter 4, the marketplace soon accomplished what the statute could not as the price of coal relative to oil and gas dropped and companies moved to coal.

The most remarkable thing about the first round of EPA standard setting was how quickly it was accomplished. In a sign of things to come, however, the agency suffered remands of its annual secondary standard for SO_2 and of its NSPS for power plants in two opinions that were influential in crafting the so-called "hard look" doctrine of judicial review of agency action. The agency's future attempts to prepare administrative records and write explanations capable of withstanding judicial scrutiny under that doctrine resulted in long delays and the expenditure of huge sums of money on engineering and economic analyses.

CHAPTER 4

Minor Adjustments, Major Controversy

Running as a centrist populist in the wake of the Watergate crisis, Jimmy Carter defeated President Ford by a margin of fifty-seven electoral votes, but he barely obtained a majority of the popular vote. The public interest groups that supported Carter expected him to appoint an environmentalist to run EPA and to stand behind the agency in its struggles with regulated industries. Carter fulfilled the first expectation. He appointed a thirty-seven-year-old attorney named Douglas Costle to head EPA. An experienced bureaucrat, Costle had worked at the Commerce and Justice Departments and the White House Council on Environmental Quality. He had most recently been the head of the Connecticut Environmental Protection Agency. In a bold move, the president appointed David Hawkins, a lead attorney for the Natural Resources Defense Council (NRDC), to head the agency's Office of Air and Radiation. But the aggressive EPA leadership was often stymied by equally aggressive overseers in a newly created entity called the Regulatory Analysis Review Group (RARG) during the perplexing combination of inflation and economic stagnation that plagued the economy during the final years of the 1970s. Since the Democratic Party controlled both houses of Congress, environmental groups also anticipated that much needed changes to the Clean Air Act would be more protective of the environment than the 1970 statute. This expectation was only partially fulfilled when the Ninety-Fifth Congress amended the Clean Air Act in 1977.[1]

August 8, 1977 President Carter signs 1977 Clean Air Act Amendments

June 1978 EPA promulgates regulations governing prevention of significant deterioration

November 9, 1978 President Carter signs the Natural Gas Policy Act, the Powerplant and Industrial Fuel Use Act, and the Public Utility Regulatory Policies Act

February 1979 EPA relaxes the national ambient air quality standards for ozone

May 1979 EPA promulgates new source performance standard for coal-fired power plants

December 1980 EPA promulgates Phase I visibility protection regulations

January 1981 Ohio Edison agrees to spend $500 million on pollution controls at its Sammis plant and other plants to settle first EPA enforcement action under 1977 amendments

President Carter's message to Congress on the environment, delivered only a few days after his inauguration, set the tone for his administration by emphasizing both environmental protection and economic well-being. The country should meet its energy needs from existing sources, including coal, but "in a safe and environmentally acceptable way." He therefore advocated a strong prevention of significant deterioration (PSD) program, a new program to protect visibility in pristine areas, and "strict controls on coal-burning plants to insure that they meet air quality standards." At the same time, he promised to "make increased use of economic incentives to achieve our environmental goals."[2]

Although opinion polls continued to indicate widespread public support for the environment, intense pressure from industry lobbyists, vigorous critiques from conservative think tanks, and endless haggling with White House economists took their toll on EPA's leadership. The agency slowly grew more receptive to industry demands for "flexible" regulation and academic prescriptions for "efficient" regulatory approaches to pollution control. Costle recognized, however, that he was bound by statutory dictates, which were not especially hospitable to such approaches.[3]

At the outset of the Carter administration, coal-fired power plants and industrial boilers contributed 70 percent of total SO_2 emissions, 24 percent of NO_x emissions and 33 percent of PM emissions. Although new and modified plants were subject to EPA's new source performance standards (NSPS)

of 1.2 pounds per million British thermal units (lb/MBtu), existing plants were subject to an array of standards ranging from Ohio's 9.5 lb/MBtu (which could be met by burning high-sulfur coal without any emissions controls) to New York's 3.8 lb/MBtu (which could be met by burning low-sulfur coal). Only forty plants had installed scrubbers.[4]

THE ENERGY CRISIS

One of the Carter administration's first orders of business was to address the nation's continued dependence on foreign oil and natural gas. As of March 1977, not a single plant that had received an order to shift from oil and natural gas to coal under the Energy Supply and Environmental Coordination Act (discussed in Chapter 3) had done so. The Federal Energy Administration (FEA) attributed this remarkable fact to industry recalcitrance. In a stirring speech delivered on April 18, 1977, President Carter declared that solving the nation's continuing energy crisis was "the moral equivalent of war." He offered the nation a "National Energy Plan," which prioritized energy conservation, but also envisioned a much greater role for coal in the nation's energy future. To address the resulting increases in emissions, the plan called for all new coal-fired plants to install the "best available control technology" (BACT), presumably scrubbers. It also called on Congress to enact legislation preventing significant deterioration of air quality in clean air areas.[5]

THE CLEAN AIR ACT AMENDMENTS OF 1977

The president's proposed energy legislation had to compete with clean air legislation, which also had a high priority in Congress. States with nonattainment areas wanted relief from the "no net growth in emissions" policy of EPA's nonattainment guidelines, and no one was happy with the guidelines for PSD programs that EPA had hastily promulgated in response to the *Sierra Club* litigation (described in Chapter 3). Public support for controlling air pollution remained strong. According to one poll, 81 percent of respondents said they would rather pay higher prices than save money by permitting more air and water pollution, and 59 percent felt that the government should protect the environment without regard to costs. The congressional debates over Clean Air Act amendments focused on a large number of complex and controversial issues, many of which were of direct concern to electricity generators and coal producers. The 112-page bill that President Carter signed on August 8, 1977, was a compromise document that was not entirely satisfying to anyone.[6]

Setting the NAAQS

The electric power industry, the coal industry, and the United Mine Workers took the position that the science underlying the NAAQS was so uncertain, the existing pollution reduction technologies were so unpredictable, and the economic and social impacts of achieving the standards were so poorly understood that Congress should simply call a time-out on writing SIPs until EPA had a chance to revisit the hastily promulgated NAAQS. Citing a 1974 National Academy of Sciences study concluding that the scientific evidence that had accumulated since the promulgation of the NAAQS supported those standards, environmental groups argued that the NAAQS were, if anything, too lax. The United Steelworkers union generally supported the standards, but it urged Congress to create a "dislocation assistance program" to assist workers in those "relatively few" power plants that would have to shut down in order to meet the NAAQS by the deadlines.[7]

Congress declined to call a time-out, but it did create an independent Clean Air Scientific Advisory Committee (CASAC) composed of seven members, including a physician, a member of the National Academy of Sciences, and a state air pollution control official, to assist EPA in promulgating and revising NAAQS. The committee was specifically charged with recommending to the administrator "any new national ambient air quality standards and revisions of existing standards as may be appropriate" under the statutory criteria. If a standard "differ[ed] in any important respect from any of [CASAC's] recommendations," EPA had to provide its reasons for the variance in the preamble to the rule. As we shall see, this turned out to be a mixed blessing for industry groups because CASAC frequently criticized EPA for not making the standards sufficiently stringent.[8]

Nonattainment. Coal and electric power industry representatives urged Congress to allow companies to locate major facilities in nonattainment areas without securing the offsetting emissions required by EPA's nonattainment policy (see Chapter 3). They worried that companies would not be able to purchase sufficient offsets to build much-needed new power plants near large cities and throughout much of the Eastern Seaboard. This in turn would have a detrimental impact on local economies and could result in blackouts and brownouts. The Carter administration joined environmental groups in urging Congress to give the offset program a chance to work before amending the statute to extend the deadlines. The administration, however, supported an amendment to push back the deadlines for power plants that converted from oil and gas to coal under its proposed energy plan.[9]

The amendments extended the deadline for attaining the NAAQS until December 31, 1983, with a possible extension for ozone nonattainment areas until December 31, 1987. In the meantime, states containing nonattainment areas were obliged to amend their SIPs in several important regards. First, the plans had to ensure "reasonable further progress" toward attainment by the new deadline. Second, existing sources in the nonattainment area had to install "reasonably available control technology," a term that was understood to include in-plant adjustments to reduce NO_x emissions but not scrubbers to reduce SO_2 emissions. Third, states had to create a permit program for allowing major new sources with emissions greater than one hundred tons per year or modifications of major existing sources in nonattainment areas. To get a permit, the major source would have to install technology capable of meeting the "lowest achievable emissions rate," a term that was defined to be the most stringent limitation "which is achieved in practice." Fourth, the permit applicant had to secure sufficient offsetting emissions from its own facilities or some other source in the area to ensure reasonable further progress toward attaining the NAAQS by the deadline. States failing to implement the new requirements faced a moratorium on the construction of major new stationary sources and a cutoff of federal highway funds.[10]

Intermittent Controls and Tall Stacks. Having invested so heavily in the imposing monuments to a bygone view of pollution abatement, the companies that owned tall stacks wanted Congress to overturn the judicial decisions preventing them from receiving the pollution reduction credit they thought they deserved. The entire industry urged Congress to recognize the legitimacy of intermittent controls. The United Mine Workers union argued in favor of both tall stacks and intermittent controls. Environmental groups and northeastern states remained unalterably opposed to tall stacks and intermittent controls.[11] To the delight of environmental groups, the amendments specified that the "degree of emission limitation required for control of any air pollutant" under any SIP could not be affected by (1) so much of the stack height that exceeded good engineering practice or (2) "any intermittent or supplemental control of air pollutants varying with atmospheric conditions."[12]

New Source Performance Standards for Power Plants

By 1977, it had become clear that most companies building new plants had decided to comply with the 1971 NSPS by burning low-sulfur coal. This came as a great disappointment to environmental groups, who wanted to see scrubbers installed in power plants in the West, and to coal producers in Appalachia and the Midwest, who wanted new plants constructed in those regions to install scrubbers and burn their coal. Both groups favored a "percentage reduction" requirement that would ensure that SO_2 emissions would be reduced by a set percentage, regardless of the sulfur content of the coal. Convinced that scrubbers were a proven technology, the Carter administration also supported a percentage reduction requirement, as did pollution control vendors and a few electric power companies that had already invested heavily in scrubbers.[13]

Most electric power companies joined the railroads in opposing the percentage reduction requirement on the grounds that scrubbers were not sufficiently reliable, were too costly, extracted an unacceptable energy penalty from the power plant's output, and created a huge solid waste disposal problem. Although the requirement would help coal producers in the East and Midwest, the National Coal Association sided with western coal producers in opposing the legislation.[14]

Politicians of both parties from downwind northeastern states favored the percentage reduction idea. Politicians from coal-producing states in Appalachia and the Midwest wanted power plants to burn coal from their states, but they were divided as to how to accomplish this. Conservative Republicans for the most part wanted to force EPA to ease the emissions standards. Liberal Democrats from the same region favored a percentage reduction requirement that would force utility companies to install scrubbers but allow them to burn local coal. Conservative politicians from southern states that had fairly large reserves of high-sulfur coal favored easing the standards. Representatives of both parties from western states producing low-sulfur coal strongly opposed a percentage reduction requirement, because it would reduce demand for that coal. Politicians of both parties from the West Coast generally favored the percentage reduction requirement because they wanted to protect pristine areas from the emissions of coal-fired plants in Utah, Nevada, Arizona, and New Mexico that supplied much of their power.[15]

The amendments that resulted from the interplay of these political forces defined the term "standard of performance" in the NSPS section of the bill to mean a standard "(i) establishing allowable emission limitations . . . , and (ii) requiring the achievement of a percentage reduction in the emissions which would have resulted from the use of fuels which are not subject to treatment prior to combustion." The conference committee report advised that EPA should "set a range of pollutant reduction that reflects varying fuel characteristics" if the agency found that the departure from the uniform national percentage reduction requirement would not "undermine the basic purposes" of the statute.[16] As we shall see, this language allowed EPA to adopt a "sliding-scale" approach to the percentage reduction requirement that permitted new power plants to turn off their scrubbers and burn low-sulfur coal some of the time.

Prevention of Significant Deterioration

The electric power and coal industries urged Congress to clarify that it had never intended to establish a program for the prevention of significant deterioration (PSD).[17] The current secondary standards, they argued, provided adequate protection for national parks and wilderness areas. Administrator Costle strongly preferred a PSD program that required large new sources in clean air areas to install the "best available control technology." Environmental groups supported the BACT requirement, but urged Congress to make it applicable to existing sources that were currently befouling the air in national parks and wilderness areas.[18]

The PSD program that the amendments created applied to all areas that were in attainment with the relevant NAAQS or that were unclassifiable due to the absence of sufficient monitoring information. The new program assigned every PSD area to one of three classes for SO_2 and PM. The statute specified allowable increases in concentrations (called "increments") of the two pollutants above the original "baseline" concentration for areas in each class. Every SIP had to ensure that sources within the state did not cause concentrations of the pollutants in any PSD area to exceed the increments. States also had to establish permit programs to require every new or modified "major emitting facility" (a term that included virtually all power plants) to secure a permit before beginning construction in any PSD area. To secure a permit, the owner had to demonstrate that the new or modified facility would not cause air quality in any PSD area to exceed any of the increments or any NAAQS and that the facility would install the BACT to control its emissions. The statute defined the term "best available control technology" to be the maximum degree of reduction of each pollutant that the permitting authority determined to be achievable, taking into account energy, environmental, and economic impacts. But it could not be less stringent than any applicable new source performance standard.[19]

Visibility

The amendments created a separate program to protect visibility in national parks and wilderness areas where visibility was an "important value." The goal of the program was "the prevention of any future, and the remedying of any existing, impairment of visibility."[20] States containing sources that contributed to visibility impairment in protected areas had to prepare SIPs containing emission limitations and other measures necessary to make "reasonable progress" toward meeting the national goal. In addition, each major existing stationary source that had not been in operation for more than fifteen years and that might be contributing to visibility impairment had to install "as expeditiously as practicable" the "best available retrofit technology" (BART) for controlling emissions. New sources would still be subject to the BACT requirement in the statute's PSD provisions.[21]

Emergencies

The electric power industry secured a provision in the new law establishing a process under which the owner of a power plant could ask the governor of its state to petition the president for a determination that "a national or regional energy emergency" existed that was of such severity that a "temporary suspension of any part of the applicable implementation plan may be necessary" and that "other means of responding to the energy emergency may be inadequate." Upon receiving such a determination, the governor could issue a temporary suspension of SIP requirements if the governor further found that the emergency involved "high levels of unemployment or loss of necessary energy supplies for residential dwellings." Only one suspension could be issued for a single emergency, and it could last for only four months.[22]

What the Amendments Did Not Accomplish

The amendments left a number of controversial issues unresolved. With the limited exception of visibility protection, they did not directly address emissions from existing power plants. Those plants were still being regulated by the states in their SIPs. The amendments also failed to address power plant contributions to the growing problem of photochemical oxidants in nonattainment areas. The conventional wisdom at the time was that power plant operators could do little to reduce NO_x emissions beyond modest modifications to reduce combustion temperatures.[23] Other unresolved issues included emissions of mercury and other hazardous air pollutants from power plants, the proper disposal of millions of tons of coal combustion residuals, and emissions of greenhouse gases that were contributing to climate disruption.[24]

STAGFLATION

As EPA was struggling to implement the 1977 amendments, conservative think tanks were making a persuasive case for the proposition that federal regulations were contributing to the "stagflation" that was roiling the economy by causing price increases for electricity and forcing companies to lay off workers. EPA and environmental groups countered with numerous studies demonstrating that environmental regulations were deflationary in the sense that they produced more benefits than costs. Nevertheless, President Carter in March 1978 created a new entity called the Regulatory Analysis Review Group (RARG) and charged it with holding down the inflationary impacts of major regulations.[25] The specter of inflation and job loss haunted every EPA attempt to regulate power plant emissions for the remainder of the Carter administration.

ENERGY LEGISLATION

Congress was also actively debating a number of bills designed to address the ongoing energy crisis by decreasing the nation's reliance on imported oil, increasing its use of domestic coal supplies, discovering alternative sources of energy, and increasing end-use energy efficiency. The first legislation to emerge was the Department of Energy Organization Act of 1977, which created a new Department of Energy (DOE) and the Federal Energy Regulatory Commission (FERC) as an independent five-member agency within the department that received all the powers of the old Federal Power Commission.[26]

On November 9, 1978, President Carter signed a trio of bills with profound implications for electricity generation in the United States. The Natural Gas Policy Act of 1978 deregulated the price of natural gas shipped in interstate commerce. The Powerplant and Industrial Fuel Use Act of 1978 (PIFUA) prohibited generators of electricity from burning oil and natural gas in new power plants, phased out the use of natural gas in existing power plants by 1990, and prohibited existing plants from switching from coal to natural gas after 1990. Although Congress repealed the restrictions on the use of natural gas in power plants in 1987, the statute effectively put an end to oil-fired power plants. As described in Chapter 1, the Public Utility Regulatory Policies Act of 1978 (PURPA) initiated the long and sometimes painful process of deregulating (or "restructuring") interstate sales of electricity.[27]

The result was a "record build out" of new coal-fired power plants from the late 1970s through the mid-1980s. As the coal industry faced what appeared to be inexhaustible demand, it developed new technologies for getting coal out of the ground more rapidly and efficiently. High-wall and long-wall mining techniques increased the productivity of underground mines, and huge draglines and enormous dump trucks facilitated the development of massive surface mines in the West and mountaintop removal mining in Appalachia. These improvements and more efficient techniques for transporting coal long distances by rail and inland waterways ensured that the price of coal remained low compared to deregulated natural gas.[28]

PREVENTION OF SIGNIFICANT DETERIORATION REGULATIONS

EPA published regulations implementing the PSD provisions of the 1977 amendments in June 1978. The most important aspect of the new regulations for power plants was the "new source review" (NSR) process for permitting new sources and modifications of existing major emitting facilities. The regulations exempted from NSR any change that increased emissions of an air pollutant by less than a "de minimis" amount, which it defined to be one hundred tons per year (tpy) in the case of most power plants. The agency also exempted from the BACT requirement any pollutant emitted below the same de minimis level of 100 tpy. At the urging of the agency's economists, Administrator Costle agreed to allow an existing source to avoid NSR if net emissions from an entire facility did not increase when an existing unit within the facility underwent some change that increased its emissions. This interpretation of the term "modification" effectively placed the entire plant within a bubble and subjected it to NSR only when emissions from the bubble increased more than the de minimis (100 tpy) amount.[29]

The D.C. Circuit upheld the regulations in several regards, but set aside important provisions. Although EPA had inherent discretion to promulgate regulations defining de minimis emissions thresholds for modification, it could not merely incorporate the same threshold (100 tpy) that the statute specified for determining whether a new source was a major emitting facility. Likewise, the agency could not simply incorporate the same threshold in determining de minimis levels of individual pollutants for purposes of applying the BACT requirement. On the other hand, the court upheld EPA's bubble innovation to determine when a facility had engaged in a modification. The court noted that EPA had properly qualified the bubble policy to require contemporaneous emissions reductions within the same source.[30]

EPA responded to the court's remand in August 1980. The revised regulation fleshed out the agency's approach to modifications in several important regards. First, it clarified that the relevant increase in emissions for purposes of determining whether an existing source was engaged in a modification was an increase in "actual" emissions, not allowable emissions. Second, the agency provided a table of de minimis emissions rates for determining whether changes would subject a facility to NSR. For SO_2 and NO_x, the de minimis rate was 40 tpy; for PM, it was 25 tpy; and for mercury, it was 0.1 tpy. Finally, the regulations limited BACT analysis for modified sources to those pollutants that were emitted in more than de minimis amounts as a result of the modification.[31]

VISIBILITY PROTECTION REGULATIONS

The 1977 amendments delegated a great deal of discretion to EPA to implement the new visibility protections. In November 1979, EPA determined that visibility was an "important" value in 156 areas encompassing more than twenty-nine million acres in thirty-six states. EPA decided to promulgate the visibility regulations in phases. The "Phase I" regulations covered plume blight (visibility impairment directly attributable to individual sources) and the changes that the thirty-six affected states would have to incorporate into their SIPs to comply with the statute's visibility protection requirements. The agency saved the more controversial issues concerning regional haze for the "Phase II" regulations that it would promulgate sometime in the future when better models existed.

The Phase I proposal required affected states to identify existing major stationary sources that caused or contributed to visibility impairment in the protected areas. A power plant was an "existing major stationary source" if it had the potential to emit 250 tpy of any regulated pollutant, was not in operation prior to August 7, 1962, and was in existence on August 7, 1977. Once an offending source was identified, the state had to prescribe emissions limitations based on its determination of BART, after considering the cost of the technology, the remaining useful life of the source, and the extent to which visibility would be enhanced. The source could then apply to the EPA administrator for an exemption on the ground that it did not cause or contribute to "significant impairment of visibility." Despite bitter complaints from the electric power industry that the program would entail excessive expenditures and stymie economic growth in the West, the final regulations adopted the proposed requirements with only minor changes.[32]

THE NEW SOURCE PERFORMANCE STANDARD

The amendments required EPA to promulgate a new NSPS for power plants reflecting the percentage reduction approach. The initiative was of great significance to both the environment and the economy because electric power companies were planning to build three hundred new coal-fired power plants in the next decade, and SO_2 emissions would probably increase 15–16 percent nationwide in the next fifteen years under the 1971 NSPS. At the same time, a second energy crisis (spurred by the revolution in Iran) was under way, and conservative activists were blaming environmental regulation for long waiting lines at gasoline stations. A high percentage reduction requirement would ensure that all new coal-fired power plants installed scrubbers, no matter where they were located or what kind of coal they burned. The statute, however, empowered the EPA administrator to "distinguish among classes, types, and sizes within categories of new sources" in establishing the standards. This arguably gave the agency the authority to require one percentage reduction for the class of power plants burning low-sulfur coal and another percentage reduction for the class of power plants burning high-sulfur coal, or even to recognize a continuum of classes of power plants depending on the sulfur content of the coal they burned. This interpretation was supported by the conference report language (quoted above) suggesting that EPA could employ a range of pollutant reduction levels reflecting varying fuel characteristics.[33]

As in the past, many of the struggles over the standards focused on the availability, reliability, efficacy, and cost of scrubbers. Although companies were still experiencing operational problems with existing scrubbers, a second generation of scrubbers was performing much more effectively and capital expenditures were declining. The operational cost of scrubbing could be reduced and its reliability increased if the operator were permitted to "partially scrub" exhaust gases by allowing some of them to bypass the scrubbing system. Newly emerging "dry" scrubbing technologies could be operated less expensively with fewer waste by-products, but they required even more effective and reliable particulate collection systems. And it was not clear that they could achieve the same SO_2 removal efficiencies as wet scrubbers.[34]

Pre-Proposal Deliberations

Unless cost, energy, or non-air-quality environmental considerations compelled a different result, the engineers in EPA's Office of Air Quality Planning and Standards (OAQPS) believed that the agency should require the same percentage reduction from all new power plants whether they burned low- or high-sulfur coal, an approach referred to in the interagency debates as "uniform percentage reduction." OAQPS soon produced a draft proposed rule that required all plants to remove 85 percent of the sulfur from the coal they burned, a standard that would as a practical matter require full scrubbing of emissions streams from all sources of coal. The draft also set an emission ceiling of 1.2 lb/MBtu to ensure that if plants used dirty coal, they would have to employ precombustion techniques such as coal washing before burning it. Sources burning very clean coal could avoid the percentage reduction requirement, so long as they kept emissions below a 0.2 lb/MBtu floor. Since no coal in the United States was low enough in sulfur to meet that standard, however, the floor served mainly as an incentive to come up with better precombustion technologies. The draft employed a twenty-four-hour averaging time, rather than the yearly averaging time of the 1971 standard. Although this had the effect of making the standard somewhat more stringent, the change came primarily at the behest of the agency's enforcement division, which could impose fines on a daily basis.[35]

OAQPS next convened a working group of representatives from the Office of General Counsel (OGC), the Office of Research and Development (ORD), the enforcement office, and the Office of Planning and Management (OPM). The economists in OPM wanted a higher than 0.2 lb/MBtu floor to permit plants burning low-sulfur coal to avoid the percentage reduction requirement. They further argued against a high percentage reduction requirement that would as a practical matter require full scrubbing by all new power plants. Full scrubbing would cost upwards of $700 million, but it would, in their view, produce very few benefits. OPM also objected to the twenty-four-hour averaging period on the ground that it would require extra scrubbers for use when equipment malfunctioned or the sulfur content of the coal unexpectedly increased.[36]

The Utility Air Regulatory Group (UARG) was an ad hoc consortium of trade associations and individual power companies that George C. Freeman, a lawyer from Richmond, Virginia, assembled to lobby against stringent environmental regulations. In meetings with Administrator Costle, President Carter's chief domestic policy adviser Stuart Eizenstat, numerous congresspersons, and high-level DOE officials, Freeman argued that scrubbing technology was not capable of 85 percent removal on a continuous basis and warned that a standard that power plants could not meet would result in serious threats to grid reliability. UARG suggested a "sliding-scale" approach to the percentage reduction requirement under which the required percentage reduction decreased with decreasing sulfur content in the coal. The sliding scale would provide an "escape hatch" for scrubbers that could not meet the stringent 90 percent reduction requirement by allowing the plant to blend in low-sulfur coal. UARG further suggested an emission ceiling of 1.5 lb/MBtu and a floor of 0.4 lb/MBtu. It commissioned studies from National Economic Research Associates, Inc. (NERA) concluding that the OAQPS proposal would increase average household electricity bills by $42 as compared to $20 for the UARG alternative.[37]

Working closely with UARG, DOE pressed the agency to adopt the sliding-scale approach with a floor on the percentage reduction requirement of 0.8 lb/MBtu that would allow even more partial scrubbing than the industry suggestion. Arguing that the twenty-four-hour averaging period risked wide-spread noncompliance or loss of generating capacity when boilers malfunctioned, it urged EPA to employ a monthly averaging period. DOE also warned that a scrubber requirement for new sources would prolong the retirement of dirty existing plants for as long as possible. The economists in RARG came down firmly on the side of OPM and DOE. They estimated that full scrubbing could cost the electric power industry up to $35 billion per year. The Department of the Interior, however, weighed in on the side of OAQPS, stressing that the full-scrubbing option would protect the pristine air over national forests and parks.[38]

Sensing that the passage of time was working to their disadvantage, environmental groups sued EPA when it failed to publish a proposed rule by the July 1978 deadline. This resulted in a judicially enforceable consent decree under which EPA agreed to publish a proposed rule by September 12, 1978, and to promulgate a final rule within six months after that. At the same time they were suing EPA, the environmental groups were lobbying Administrator Costle to adopt the full scrubbing option. They warned Costle that if he caved in to the industry and White House pressure, the agency would face "endless litigation."[39]

The Notice of Proposed Rulemaking

Administrator Costle decided to go forward with the OAQPS proposal, with the understanding that the public comments might convince him to move to a sliding-scale approach with partial scrubbing. In September 1978, EPA proposed an emissions limitation of 1.2 lb/MBtu, a 0.2 lb/MBtu floor with a uniform national percentage reduction of 85 percent averaged every twenty-four hours. The agency concluded that the proposed standard could be met by at least five adequately demonstrated methods of wet scrubbing. The agency recognized that the low (0.2 lb/MBtu) floor ensured that emissions from burning even very low sulfur coal would have to be scrubbed and that a floor in the range of 0.5 to 0.6 lb/MBtu would allow partial scrubbing in plants burning low-sulfur coal. Partial scrubbing would reduce scrubber costs without significantly increasing emissions nationwide. But plants burning low-sulfur coal under the partial scrubbing option would emit up to four times more SO_2 than under the full scrubbing option. Full scrubbing would also encourage use of locally available coal.[40]

Reactions to the Proposal

The electric power industry reacted to the proposal as expected. It preferred partial scrubbing to full scrubbing, but it vastly preferred the existing standard with the option of using low-sulfur coal instead of scrubbing. A sliding scale, they argued, would be far cheaper than a uniform percentage reduction approach and would encourage new technologies like dry scrubbing. Any money wasted on scrubbers would be added directly to consumer's utility bills and would ultimately contribute to inflation.[41]

The environmental groups opposed the sliding-scale approach or any approach that would tolerate partial scrubbing, which, it argued, Congress had specifically rejected in the 1977 amendments. It would undermine the purpose of the statute in at least three ways. First, it failed to maximize the use of locally available coal in the East and Midwest. Second, it encouraged power companies to relocate to the West to burn low-sulfur coal, thereby destroying the pristine air quality in parts of the West. Third, it demonstrably failed to force companies to develop new technologies. Noting that several scrubbers had reliably achieved SO_2 removals of well over 90 percent, environmental groups urged EPA to raise the percentage reduction requirement from 85 percent to at least 95 percent averaged every twenty-four hours.[42]

Post-Proposal Internal Debates

As the agency staff analyzed the comments, it became clear that one of the proposal's most controversial aspects was the averaging period. The comments revealed that there was considerable variability in the sulfur content of fuels and in the effectiveness of control equipment that made it very difficult for a source to comply with a stringent emissions limitation on a daily basis. The OAQPS staff concluded that a more rational approach would be to use a "rolling" monthly averaging period to give the source a chance to make up for excess emissions during one day by using low-sulfur coal or better controls later in the month.[43]

With the assistance of a sophisticated econometric model, the OAQPS staff concluded that lowering the emissions ceiling was an effective way to ensure that companies used both coal washing and scrubbing when they burned high-sulfur coal and that emissions would be significantly below the existing 1.2 lb/MBtu standard when they burned low-sulfur coal. A ceiling of

0.55 lb/MBtu, the model predicted, would ensure that all new plants did some scrubbing, because there was no coal with a low enough sulfur content to meet that standard by itself. It was unclear, however, whether plants burning high-sulfur coal could meet such a low ceiling even with coal washing and full scrubbing.[44]

When the trade press reported that EPA was seriously considering a 0.55 lb/MBtu cap, the political dynamics changed. Eastern coal interests now joined the western coal companies and the electric power industry in a massive lobbying campaign to persuade the agency to adopt the less stringent 1.2 lb/MBtu standard. After meeting with industry lobbyists, Senator Robert Byrd (D-WV) summoned Administrator Costle and Stuart Eizenstat, the White House domestic policy adviser, to a meeting on April 23, 1979, in his office with officials from the National Coal Association. Costle assured Byrd that he had no intention of rendering large portions of Appalachian and midwestern high-sulfur coal reserves unusable by new power plants.[45]

With the 0.55 lb/MBtu ceiling off the table, the internal debate switched to the percentage reduction requirement. OPM joined the White House economists in urging Costle to adopt a sliding-scale scrubbing approach that would save the utility industry about $1 billion, even though it would allow about one-third more power plant emissions in the West. The trick was to come up with a floor for the percentage reduction. Costle was not comfortable with either the industry suggestion (20 percent) or the DOE suggestion (33 percent), but he was not sure how much higher it should be.[46]

At this point a deus ex machina arrived in the form of dry scrubbing. The rapidly developing technology had not been seriously considered to this point, because its removal efficiency was only 70 percent. But it also appeared to be much cheaper than wet scrubbing, and it produced less solid waste. EPA could therefore justify a 70 percent floor for the sliding scale on the ground that it would encourage the development of this exciting new technology. The option would, however, result in about 200,000 tons of SO_2 emissions in the West that would not be emitted under the full scrubbing option.[47]

At an April 30 White House briefing for President Carter, Costle expressed his preference for the 70 to 90 percent sliding-scale approach with a floor of 0.6 lb/MBtu (to ensure that the less effective dry scrubbers were not used in facilities burning high-sulfur coal) and a ceiling of 1.0 or 1.2 lb/MBtu. EPA officials took the president's noncommittal response as acquiescence to EPA's proposal. Participants at another meeting with Senator Byrd two days later reported that the senator strongly hinted that a less stringent standard was the price that President Carter would have to pay for Byrd's vote on the critical Strategic Arms Limitation Treaty (SALT) that the president had painstakingly negotiated with the Soviet Union.[48]

The Final Rule

On May 26, 1979, EPA issued a final rule that retained the 1.2 lb/MBtu emissions limitation, but with a thirty-day rolling average. It further adopted a sliding-scale percentage reduction requirement that ranged from 70 percent so long as emissions remained below 0.6 lb/MBtu to 90 percent. The 70 percent minimum ensured that all low-sulfur coal would be scrubbed, but a company burning low-sulfur coal could employ dry scrubbing technology or partially scrub with wet scrubbers. The final SO_2 standard was less stringent than the proposal, because it employed a thirty-day running average, and it allowed more visibility degradation in the West. The agency estimated that the standard would result in $3.3 billion in annual costs and that electric bills could go up as much as $1.20 per month. Costle later acknowledged that the agency might have promulgated a more stringent standard but for the intervention of Senator Byrd.[49]

Judicial Review

The Court of Appeals for the District of Columbia upheld the regulation in its entirety. The court rejected the industry challenge to the standard's 90 percent removal ceiling, and it likewise rejected the environmental groups' challenge to the sliding-scale approach to percentage reduction. This was not a total loss for the groups, because they could still attempt to persuade state permitters to prescribe full scrubbing as BACT under the PSD program. Finally, the court rejected the environmental groups' claim that EPA would have established a lower ceiling had it not unlawfully succumbed to political pressure. The court noted that there was no direct evidence that EPA made any commitments in response to Senator Byrd's threats. It acknowledged that it was "always possible that undisclosed Presidential prodding may direct an outcome that is factually based on the record, but different from the outcome that would have obtained in the absence of Presidential involvement." But it was convinced that Congress did not intend for "the courts [to] convert informal rulemaking into a rarified technocratic process, unaffected by political considerations or the presence of Presidential power."[50]

REVISING THE OZONE NAAQS

EPA's first attempt to revise an ambient air quality standard came in the late 1970s when it reexamined the primary standard for ozone. The electric power industry was intensely interested in this exercise, because NO_x emissions from power plants were major contributors to ozone concentrations in most ozone nonattainment areas. The 1971 primary standard for ozone was 0.08 parts per million (ppm) calculated on an hourly average. EPA estimated that between one-third and two-thirds of the areas in the country would not attain that standard even by the recently relaxed 1987 deadline if existing sources used only "reasonably available" control technologies. Most urban regions would therefore have to limit industrial growth.[51]

In 1971, EPA had relied on a single epidemiological study showing that short-term exposure to ozone at 0.10 ppm caused an increase in asthma attacks. In a critical change of position, agency scientists reinterpreted that study to find ozone exposure did not increase asthma until it reached a level of 0.25 ppm. This did not necessarily require the agency to raise the standard to 0.25 ppm, because more recent clinical studies demonstrated that short-term ozone exposure at levels ranging from 0.15 to 0.25 ppm caused impaired lung function, chest tightness, coughing, and wheezing in sensitive individuals and healthy individuals engaged in exercise. The problem was that the clinical studies all employed healthy individuals and were not easily extrapolated to asthmatics and persons suffering from lung disease. OAQPS argued that the inconclusive data did not support a departure from the status quo. The economists in OPM argued that the standard should be relaxed to levels closer to 0.25 ppm.[52]

Adopting a middle ground, Administrator Costle in June 1978 proposed a modest relaxation of the standard to 0.10 ppm averaged hourly. The agency acknowledged for the first time a problem that would plague it in the future—the thresholds for sensitive populations were "difficult or impossible to determine experimentally" while the threshold for healthy individuals was "not likely to be predictive of the response of more sensitive groups." The agency also highlighted serious uncertainties in the scientific data. Given those uncertainties, it concluded that a standard of 0.15 ppm would not provide an adequate margin of safety. Finally, the agency concluded that while the current standard of 0.08 ppm was no longer necessary, a standard above 0.10 ppm would not "adequately protect public health."[53]

The electric power industry argued that the proposal was still far too stringent, given the cost of compliance and the uncertainties in the scientific data. Conservative think tanks argued that it would be grossly inefficient to protect a few thousand sensitive individuals from uncertain risks at a huge cost to Americans in higher prices for electricity. It would be far less costly to advise sensitive individuals to remain indoors during days in which outdoor ozone levels were high. Environmental groups adopted the OAQPS position.[54]

As OAQPS was evaluating the public comments, it ran into strong opposition from the economists in the RARG. Citing language in legislative history indicating that Congress only expected EPA to protect a "representative sample" of individuals in sensitive subpopulations, they argued that there was no principled basis for attempting to protect 99 percent of the individuals in the most sensitive subpopulations. The RARG economists also disputed EPA's interpretations of the scientific studies. Noting that all of the adverse health effects attributed to ozone were short term and reversible, they argued that a standard of 0.16 ppm would provide more than adequate protection,[55] and it would save $4–7 billion in compliance costs. OAQPS accused the RARG economists of "erroneously" interpreting the scientific studies and overestimating the costs. It further argued that the RARG's strong focus on the economic impact of the standard was irrelevant under the statute.[56]

Convinced that most of RARG's comments "were cribbed right from industry briefs," Administrator Costle concluded that the standard should be set at 0.12 ppm. Although RARG urged President Carter to overrule Costle, he declined to do so. He later told reporters that he had a "statutory responsibility and a right" to overrule executive branch agencies, but allowed that "it would be a very rare occasion whenever I would want to do so." EPA published a final rule setting the primary and secondary ozone NAAQS at 0.12 ppm averaged over one hour on February 8, 1979.[57]

Industry and environmental groups challenged the standard in the D.C. Circuit Court of Appeals. The court easily rejected the industry claim that EPA had to take cost considerations into account in setting the standard, because Congress had quite clearly subordinated economic concerns to the goal of protecting public health. Rejecting the industry challenge to EPA's interpretation of the scientific studies, the court found the rulemaking record "replete with support" for the 0.12 ppm standard. The court also rejected NRDC's argument that the 0.12 standard did not represent an adequate margin of safety. It found that EPA had reasonably interpreted the scientific studies to conclude that, despite the apparent lack of a definitive no-effect level, the probable level for adverse effects in sensitive individuals was in the range of 0.15 to 0.25 ppm, and the agency was not arbitrary and capricious in establishing a margin of safety of only 0.05 ppm. As we shall see, the Supreme Court ultimately agreed with EPA that cost considerations were irrelevant in setting NAAQS, but not for another two decades.[58]

STATE IMPLEMENTATION PLANS

EPA was also responsible for overseeing the states' revisions of their state implementation plans (SIPs) to be consistent with the 1977 amendments. As with the original SIPs, some states proved quite recalcitrant. Others encouraged their power plants to comply by allowing them to recover the cost of retrofitting expensive scrubbers in existing plants. As of March 1978, thirty-four scrubbers were in operation, forty-two were under construction, and fifty-six more were planned for both new and existing plants.[59] We will focus on struggles to write and implement acceptable SIPs for the highly industrialized Ohio River Valley and for the Tennessee Valley Authority.

Struggles in the Ohio River Valley

The many large power plants that lined the Ohio River Valley made Ohio the nation's top coal consuming state. Power plants in Ohio emitted twice as much SO_2 as all of the plants in New York, New Jersey, and New England where the SO_2 from Ohio plants fell to earth as acid rain. But Ohio was one of the most recalcitrant states when it came to controlling emissions. Five years after the 1970 statute's deadline, the state had still not submitted an approvable plan to EPA. EPA therefore had to write a federal implementation plan (FIP) for the power plants in that state.[60]

The FIP that EPA promulgated on August 27, 1976, contained emissions limitations applicable to individual power plants. The limitations varied from plant to plant and from unit to unit within plants depending on the ambient level of SO_2 in the relevant county, the size and distribution of the facilities, and the cost of reducing emissions. Companies could meet limitations in the higher ranges without SO_2 controls, and they could meet limitations at the lower end by burning low-sulfur coal without additional controls or by partial scrubbing. The FIP assigned to each affected facility a compliance schedule designed to ensure attainment of the SO_2 NAAQS within three years. With one minor exception, the Sixth Circuit Court of Appeals upheld EPA's FIP.[61]

In June 1979, President Carter announced that EPA would be waiving compliance with the Ohio FIP for Cleveland Electric's Avon Lake and Eastlake power plants, two plants that later became big dirties. New modeling information indicated that the two plants could continue burning local high-sulfur coal and emitting five times more SO_2 than allowed under the FIP without causing any violations of the NAAQS for SO_2. In reality, the agency simply switched from the urban model it had used in preparing the FIP to a less demanding rural model because the plants were located on Lake Erie. EPA predicted that the move would avoid a 63 percent drop in Ohio coal production and a 70 percent drop in employment for coal miners. The president of the Youghiogheny & Ohio Coal Co., which had recently laid off almost 1,200 workers, called the move a "new lease on life."[62]

The decision did not sit well with environmental officials in Pennsylvania, who believed that emissions from the Ohio plants were causing the SO_2 NAAQS to be exceeded in several Pennsylvania counties, or with officials in the Northeast, who blamed Ohio power plants for acid rain. In January 1980, Pennsylvania and five northeastern states challenged the waiver in the Sixth Circuit Court of Appeals. After EPA's lawyers concluded that the waiver was probably not legally supportable, the agency quietly reversed itself in June 1980 and ordered the two plants to reduce emissions by about 15 percent, which could be accomplished by blending low- and high-sulfur coals. The Eastlake plant quit producing electricity in 2014, but the Avon Lake power plant was still burning coal in 2018.[63]

Struggles with TVA

The Clean Air Act had little initial impact on the Tennessee Valley Authority (TVA), the nation's largest electric power producer and its largest emitter of SO_2, because the Supreme Court in 1976 held that as a federal agency it was not subject to requirements in state implementation plans. When EPA ordered TVA to install scrubbers in its large coal-fired power plants, TVA said that it would install electrostatic precipitators, purchase low-sulfur coal, and build coal-washing facilities sometime within the next decade. Not satisfied with this tepid response, EPA joined Alabama and several environmental groups in a lawsuit against TVA demanding that it install scrubbers. TVA's attitude began to change in mid-1978 as President Carter replaced board members with his own appointees. After several months of intense negotiations, TVA in early December 1978 announced a settlement under which it would spend around $950 million over the next ten years on pollution control equipment to bring ten of its plants into compliance with SIP requirements. It also contracted to purchase about $7 billion worth of rare Appalachian low-sulfur coal.[64]

INTERSTATE TRANSPORT OF CRITERIA POLLUTANTS

The Clean Air Act required SIPs to ensure that pollutants and pollutant precursors emitted from their plants did not contribute to nonattainment in downwind states, but it was very difficult for a downwind state to demonstrate that sources in an upwind state were contributing to nonattainment within its borders at the time EPA was reviewing the upwind state's SIP. Consequently, EPA routinely approved SIPS that made little or no effort to protect air quality in neighboring states. Congress attempted to solve the problem in the 1977 amendments by adding section 126, which permitted any state or political subdivision to petition EPA at any time for a finding that emissions from a major stationary source were contributing to a violation of a NAAQS in another state. The amendment made it unlawful to construct a new source with respect to which EPA had made such a finding or to operate an existing source for more than three months after the finding.[65]

In May 1979, the Air Pollution Control District of Jefferson County, Kentucky, which contains most of the Louisville metropolitan area, filed a section 126 petition with EPA for a finding that the Gallagher Power Station, a large coal-fired plant located just across the Ohio River in Floyd County, Indiana, caused or contributed to a violation the SO_2 NAAQS in Jefferson County. All three of the coal-fired plants that Louisville Gas & Electric operated in Jefferson County were operating under consent decrees under which they would be installing scrubbers by 1985. The Gallagher plant's emissions limitation of 6 lb/MBtu meant that the Gallagher plant did not have to control its SO_2 emissions, even though it burned high-sulfur coal. EPA's dispersion model concluded that the Gallagher plant would contribute about 3 percent to those SO_2 concentrations that violated the NAAQS in Jefferson County. The county then commissioned its own modeling exercise, which concluded that the Gallagher plant by itself would cause violations of the SO_2 NAAQS. EPA admitted that in certain parts of Jefferson County, the Gallagher plant's emissions consumed 34 percent of the primary NAAQS and 47 percent of the secondary NAAQS for SO_2, but those were not nonattainment areas. It therefore denied the petition. The Sixth Circuit Court of Appeals upheld the denial in 1984.[66] As we shall see, EPA got more serious about addressing cross-state pollution in the late 1990s.

ACID RAIN

Scientists had long ago concluded that SO_2 emissions from midwestern power plants turned to sulfates in the air and came down as rain that was twenty-five to forty times more acidic than natural rainfall in the northeastern United States and eastern Canada. Consequently, more than 100 high-altitude lakes in the Adirondack Mountains and about 140 lakes in Ontario were devoid of fish. EPA, however, was not anxious to add acid rain to its already burgeoning to-do list. It initially took the position that it was powerless to prevent acid rain without additional legislation from Congress. Just before the change in administrations, however, Administrator Costle took an action that reverberated deep into the Reagan administration. In letters to Secretary of State Edmund S. Muskie and Senator George J. Mitchell (D-ME), Costle made findings that emissions from U.S. power plants were endangering public health and the environment in Canada and that Canada provided remedies for international pollution similar to those of the United States. Under section 115 of the Clean Air Act, this triggered an EPA obligation to require the states containing the offending power plants to amend their SIPs to eliminate the endangerment.[67]

ENFORCEMENT

EPA's first enforcement action under the 1977 Clean Air Amendments was against Ohio Edison's huge Sammis plant near Stratton, Ohio, one of the big dirties. When some employees went on strike in the summer of 1977 and supervisors attempted to keep the plant running, the electrostatic precipitators

malfunctioned and the smoke was so heavy near the plant that it was difficult to see across the road. EPA claimed that the plant had routinely
violated the opacity requirements in its state-issued permit. The litigation resulted in two consent decrees in which Ohio Edison agreed to spend $400
million on dust collectors and electrostatic precipitators at the Sammis plant,
devote $100 million to pollution control projects at nine of its other plants,
and pay $1.4 million in civil penalties. It was at the time the largest settle-
ment that EPA had ever extracted out of a public utility company under the
Clean Air Act.[68]

CLIMATE DISRUPTION

At the outset of the Carter administration, many atmospheric scientists had
concluded that emissions of carbon dioxide (CO_2) were contributing to a
poorly understood phenomenon called "the greenhouse effect," the result of
which would be a gradual warming of ambient temperatures throughout the
world. As early as June 1968, government scientists were warning the Edison
Electric Institute (EEI), the electric power industry's trade association, that
increases in atmospheric levels of CO_2 from fuel combustion "might . . .
produce major consequences on the climate—possibly even triggering catastrophic effects such as have occurred from time to time in the past." Congressional hearings on energy legislation during the summer and fall of
1977 featured testimony from renowned scientists that "the overriding issue
of the coal cycle" was the contribution of CO_2 emissions to global warming.
An April 1979 report by the congressional Office of Technology Assessment
cautioned that since no technologies existed for removing CO_2 from a power
plant's exhaust stream, the only viable alternatives were to reduce fossil fuel
consumption through more efficient technologies or greater reliance on renewable sources of energy. Yet while coal produced about 1.75 times as
much CO_2 per unit of energy produced as natural gas,[69] the Carter administration pressed ahead with its coal-based energy policies as if global warming
did not exist.[70]

THE 1980 PRESIDENTIAL ELECTION CAMPAIGN

Prior to entering the presidential race, former California governor Ronald
Reagan had spent a year as a radio commentator, and EPA was one of his

favorite targets. He claimed that the bureaucrats in EPA believed in "a return to a society in which there wouldn't be the need for the industrial concerns or more power plants and so forth." He promised coal and steel industry executives that if elected, he would seek a major overhaul of the Clean Air Act and turn to them for help. Environmental groups were not thrilled with President Carter's environmental record, but they believed his heart was in the right place. They were appalled by Reagan's unabashed contempt for them and their issues. In the end, Reagan won the election in a landslide, prevailing in all but five states, one of which was West Virginia. He won in all of the northeastern states that were downwind from the big Ohio power plants except for Rhode Island. As environmental groups braced for the worst, the coal and electric power industries looked forward to the Ninety-Seventh Congress, when the Clean Air Act would be up for reauthorization.[71]

CONCLUSIONS

The Carter administration struggled with conflicting goals. To relieve American dependence on foreign oil suppliers, it sought to increase the amount of coal burned in the nation's power plants. At the same time, it hoped to decrease power plant emissions to protect human health in nonattainment areas, to protect visibility in clean air areas, and to prevent acid deposition in the Northeast. That could be accomplished by burning low-sulfur coal from the West, but the administration also wanted to protect the jobs of miners in Appalachia and the Midwest. That could be accomplished consistently with reduced emissions by installing scrubbers, but the expense of scrubbers would discourage power plants from converting from oil and gas to coal.

By the end of President Carter's term, only four of about one hundred plants capable of switching from oil or natural gas to coal had done so. Yet thirty-three new power plants were in the process of installing scrubbers and electrostatic precipitators capable of meeting the amended NSPS.[72] Although the Carter administration did not fully accomplish all of the president's conflicting goals, it made significant progress toward reducing the nation's dependence on foreign oil and its use of scarce natural gas while reducing emissions of SO_2, NO_x, and PM from power plants.

CHAPTER 5

The Lost Decade

President Ronald Reagan and his closest White House aides were committed to providing "regulatory relief" to companies that were chafing under environmental regulations. Tall stacks and intermittent controls were once again on the table as the industry crafted its legislative strategies for the upcoming reauthorization of the Clean Air Act. The coal industry was hoping that new legislative and regulatory initiatives would bring about an end to a serious economic slump that it blamed on environmental regulations. At the same time, the face of the electric power industry was changing as the Public Utility Regulatory Policy Act of 1978 (PURPA) inspired companies to build unconventional facilities to sell electricity to electric utility companies. As natural gas supplies grew and prices declined, Congress repealed the 1978 restrictions on burning natural gas in power plants, and gas-fired "PURPA machines" flourished in many states as independent "merchant" suppliers to the captive markets that PURPA created. This posed a serious threat to the coal industry because gas-fired plants were cheaper to build than coal-fired power plants. In addition, more than one thousand megawatts of wind-generated electricity went online during the 1980s, compared to seventeen megawatts during the 1970s.[1]

There were, however, positive signs on the horizon for coal-associated companies. Electricity production from coal grew almost 8 percent during

October 1981 EPA promulgates "bubble" policy allowing plant expansions in nonattainment areas

August 1982 Industry-supported Clean Air Act amendments fail in Congress

May 1983 Bill Ruckelshaus replaces Anne Gorsuch as EPA administrator

January 1985 Lee Thomas succeeds Bill Ruckelshaus as EPA administrator

June 1986 Senate committee hears testimony from Dr. James Hansen that CO_2 emissions are causing global temperatures to increase; Reagan administration uninterested

September 1986 D.C. Circuit holds that EPA is not required to reduce power plant emissions to address acid rain

June 1987 EPA revises the national ambient air quality standard for particulate matter

July 1988 D.C. Circuit holds that EPA is not required to protect air quality in downwind states from power plant emissions in upwind states

1980, a remarkable rate of growth, given the poor economy. The fact that coal was half as expensive per Btu as natural gas stimulated orders for coal-capable boilers. Another positive development for coal was the unexpected partial meltdown at the Three Mile Island nuclear facility in March 1979, which resulted in a voluntary moratorium on building new nuclear power plants. Finally, many electric power companies were firmly committed to coal. Because scrubbers were becoming less expensive and more dependable, companies were less hesitant to build new coal-fired plants under EPA's recently promulgated new source performance standards (NSPS). A study by the Electric Power Research Institute predicted that coal production would undergo a threefold increase by the end of the century.[2]

For the first time, environmental groups found themselves on the defensive as they struggled to keep Congress from gutting the Clean Air Act and to force EPA to implement existing regulatory programs effectively. They insisted that low demand due to high fuel prices and effective energy conservation was the cause of the coal industry's problems, not environmental regulation. The Clean Air Act was just the coal industry's scapegoat for the difficulties brought on by overexpansion in a troubled economy. Noting that public opinion polls continued to show strong support for environmental regulation, environmental groups recognized the need to build grassroots support for strong environmental protections.[3]

A CONTENTIOUS BEGINNING

President Reagan appointed Anne Gorsuch to be EPA's administrator. A thirty-eight-year-old attorney for Mountain Bell Telephone Company, Gorsuch had been one of the leaders of a small band of extremely conservative Colorado state legislators who called themselves "the crazies." At her confirmation hearing, Gorsuch made it clear that she was no fan of scrubbers, especially on power plants in the West that burned low-sulfur coal. It was clear from the outset that while the agency's door would be open to the regulated industries, Gorsuch would not have the time of day for environmental groups, the "vast majority" of which, in her opinion, were "anti-business" and primarily interested in obtaining "political power."[4]

Gorsuch filled the upper echelons at EPA with young professionals, most of whom had worked for the industries that the agency regulated. Kathleen Bennett, the assistant administrator for air, noise, and radiation, had been a lobbyist for Crown Zellerbach and the American Paper Institute. General counsel Robert Perry had been an attorney for Exxon. To ensure that economic considerations were a high priority in agency decision-making, she elevated the head of the economics office to the status of assistant administrator and renamed it the Office of Policy, Planning, and Evaluation (OPPE). Hundreds of seasoned employees were laid off or simply resigned to do more satisfying work.[5]

Knowing that a Democrat-controlled House would never acquiesce in radical amendments to the environmental statutes, President Reagan elected to pursue regulatory relief through EPA rulemakings subject to careful oversight and control by the Office of Management and Budget (OMB). A new executive order required agencies to prepare a comprehensive analysis of the costs and benefits of multiple alternatives for every major rule. When not prohibited by statute, the agency was obliged to choose the alternative with the greatest net benefits. The Office of Information and Regulatory Affairs (OIRA) in OMB was to review and approve all regulatory analysis documents for compliance with the order. EPA staff soon began to refer to OIRA as a "black hole" into which proposed rules descended, never to emerge again.[6]

Relaxing Requirements in SIPs and FIPs

The Reagan administration's new approach to regulation became immediately apparent in EPA's willingness to approve amendments to state implementation plans (SIPs) relaxing emissions limitations for power plants based on a very forgiving dispersion model showing that they would not cause violations of the national ambient air quality standards (NAAQS). Many of these relaxations took place in the Ohio River Valley, where the effect of the relaxation was to allow more SO_2 to be transported to northeastern states and Canada. After American Electric Power's (AEP) huge Muskingum River plant had consistently violated its emissions limitation over a four-year period, EPA entered into a consent decree with AEP allowing the plant to continue burning high-sulfur coal if it installed a coal-washing unit that reduced SO_2 emissions by only a modest amount. The agency also relaxed the 1971 new source performance standard (NSPS) to allow the three hundred coal-fired power plants built since then to average emissions over thirty days instead of three hours.[7]

Expansion of Bubble Policy

We saw in Chapter 4 that EPA's prevention of significant deterioration (PSD) regulations adopted a "bubble" policy under which a modification of a single unit within a facility would not trigger the new source review (NSR) requirement if the company offset any increase in emissions from that unit with an equal or greater decrease in emissions of the same pollutant from some other source within the plant. Although the Carter administration had disallowed the bubble in nonattainment areas, EPA in March 1981 proposed to reverse that policy. The electric power industry enthusiastically supported the move. Environmental groups strongly opposed it, stressing the importance of ensuring that modified units install advanced pollution controls in areas where pollutant levels were already unsafe. Seven months later, the agency promulgated a final rule that was identical to the proposal.[8]

In a case that has become a landmark of administrative law, the Supreme Court, in *Chevron USA v. Natural Resources Defense Council,* deferred to EPA's definition of the ambiguous word "source" in the statute in upholding the expanded bubble policy. The Court noted that the statute equated the word "source" with the word "facility," and the ordinary meaning of the latter term included "a collection of integrated elements which has been designed to achieve some purpose." The fact that the agency changed its mind after the 1980 election did not concern the Court, because an agency must "consider varying interpretations and the wisdom of its policy on a continuing basis." Finally, the bubble policy was not at variance with the Clean Air Act's policy for nonattainment areas because the statute manifested two policies for those areas—a policy favoring attaining the NAAQS by the deadlines, and a policy favoring allowing growth in nonattainment areas.[9]

Acid Rain

EPA was even more reluctant during the Reagan administration to take on acid rain than it had been at the end of the Carter administration. But the problem would not go away. An interagency task force reported in January 1981 that deposition had intensified during the previous three decades and that it was wiping out aquatic life in hundreds of lakes in the northeastern United States and Canada. It was also clear, however, that controls aimed at reducing acid rain would have a profound economic impact on power plants in the Midwest. The congressional Office of Technology Assessment concluded that an effective acid rain control program would result in a 10 percent loss in coal mining employment in Illinois, Ohio, northern West Virginia, and western Kentucky. Although environmental groups pursued multiple strategies to induce EPA to act under its existing Clean Air Act authorities, EPA adamantly refused to address the problem.[10]

Acid rain was also emerging as a serious international issue because the Canadian government blamed power plants in the United States for its dying lakes. A binational work group estimated that 85 percent of the SO_2 emissions and 81 percent of the NO_x emissions contributing to acid rain in Canada were generated in the United States. We saw in Chapter 4 that Administrator Costle found that emissions from U.S. power plants were endangering public health and the environment in Canada and that Canada provided reciprocal remedies. That triggered an EPA obligation to require SIP amendments to eliminate the endangerment. Administrator Gorsuch, however, concluded that Costle's findings were insufficient to invoke section 115's remedies, and President Reagan told her successor Bill Ruckelshaus to take no action on Costle's findings.[11]

To the great consternation of the Reagan administration and the electric power industry, the Canadian government and a coalition of Canadian environmental groups lobbied Congress to enact protective acid rain legislation. The administration struck back in February 1983 when the Justice Department formally labeled two documentary films on acid rain produced by Canada's National Film Board "political propaganda." The ruling required the board to include a warning notice at the beginning of the film and to give the Justice Department a list of everyone to whom the board had distributed the film. If the administration's goal was to reduce the number of U.S. citizens viewing the films, the action clearly backfired. After accounts of the action were reported in the media, orders for the films skyrocketed.[12]

Weakened Enforcement

Three weeks after her confirmation, Administrator Gorsuch folded the Office of Enforcement and the Office of General Counsel into a single office to reduce the autonomy of the agency's historically aggressive enforcement lawyers. She also required more high-level approvals before referring cases to the Justice Department. The predictable consequence of the changes was a precipitous drop in inspections and enforcement actions. In EPA Region 5, which included Illinois, Indiana, and Ohio, agency enforcers were told to be "sensitive to the coal miners' plight" in deciding whether to prosecute cases against power plants. EPA enforcers in the field frequently granted exemptions allowing power plants to continue to emit SO_2 in excess of the emission limitations in their SIPs, ostensibly to preserve jobs. EPA enforcement actions not resulting in exemptions usually resulted in settlements that required a modest fine and an agreement to switch to low-sulfur coal.[13]

REWRITING THE CLEAN AIR ACT

The highest legislative priority for the coal and electric power industries was to secure industry-friendly amendments to the Clean Air Act, which had to be reauthorized by October 1, 1981, to keep EPA's air program alive. With a president committed to regulatory relief and a Senate controlled by the Republican Party, there was good reason to expect that the Ninety-Seventh Congress would oblige. Yet, while most congressional Republicans shared President Reagan's free market preference, several moderate Republicans believed that government had a legitimate role to play in pursuit of public health and welfare. One of those was Senator Robert T. Stafford (R-VT), the incoming chairman of the Senate Committee on Environment and Public Works. Another hurdle for advocates of radical change was Rep. Henry Waxman (D-CA), who chaired the subcommittee of the House Interstate and Foreign Commerce Committee with jurisdiction over the Clean Air Act. Waxman represented an environmentally conscious district that included Beverly Hills, West Hollywood, and Santa Monica. He promised a "knock-down, drag-out fight" over any attempt to change the act's approaches to pollution control.[14]

The electric power industry supported amendments that would simplify the PSD permit process and make BACT no more stringent than the NSPS, delay visibility rules until more studies were completed, give states more authority to write SIPs, allow suspensions of emissions requirements during disruptions in energy supplies, and instruct courts to be less deferential to EPA when reviewing its rules. The coal industry wanted to relieve the electric power industry of the obligation to install scrubbers on new coal-fired power plants and to ensure that EPA and the states did not require them on existing power plants. Both industries urged Congress to replace the margin of safety in setting NAAQS with a cost-benefit decision criterion. And both opposed amendments giving EPA authority to address acid rain.[15]

After a leaked draft of the Reagan administration's proposed bill attracted negative attention in the press, the administration suggested eleven "principles" for Congress to follow in drafting amendments to the Clean Air Act. At an August 1981 meeting with more than three hundred representatives of corporations and trade associations, Administrator Gorsuch urged industry groups to "light up the switchboards" in Congress in support of the principles. To her chagrin, however, the White House mail was running about 200 to 1 against the principles. At a November 1981 subcommittee hearing, pollster Lou Harris reported that 51 percent of those polled wanted to retain the existing statute, 29 percent wanted to strengthen it, and only 17 percent wanted to weaken it. In the end, Senator Stafford persuaded his committee to report out a moderate bill that neither the administration nor the coal and electric power industries could support. In the House, an administration-supported bill drafted by Rep. John Dingell (D-MI) died in the Energy and Commerce Committee when Dingell, who chaired the committee, could not attract enough support from Democratic members to pass it.[16]

TURMOIL AT EPA

In late September 1981, committees in both the Senate and the House held oversight investigations into EPA's failure to implement and enforce the Clean Air Act. One House committee demanded documents relevant to allegations of malfeasance on the part of EPA political appointees, but EPA, on the advice of the Justice Department, refused to comply. The matter escalated into a full-blown constitutional crisis over executive privilege in December 1982 when the House of Representatives made Gorsuch the first cabinet-level officer ever to be held in contempt of Congress. The crisis came to an end in early March 1983 when President Reagan ordered the Justice Department to turn over all of the requested documents. Gorsuch resigned a few days later.[17] By this point public esteem for the agency had sunk to an all-time low.[18]

Anxious to get the Gorsuch scandal behind him before the 1984 election season, President Reagan persuaded former administrator William Ruckelshaus to assume the helm for a second time. Ruckelshaus was warmly welcomed by the agency's battered staff, many of whom had fond memories of working under him during the early 1970s. He possessed a gravitas that he used very effectively to achieve an unusually high degree of independence from the deregulators in OIRA and the White House.[19]

ACID RAIN

At the swearing-in ceremonies for Bill Ruckelshaus, President Reagan directed him to come up with a plan to address acid rain as soon as possible. Ruckelshaus created a task force on acid rain and charged it with conducting a thorough analysis of a wide range of policy options for addressing that problem. He joked that he could predict where a person lived within one hundred miles by listening to what he or she had to say about acid rain. But he failed to anticipate strong resistance from within the administration, despite the president's apparent desire to move forward with an acid rain bill.[20]

The debate over acid rain precipitated an unprecedented split in the coal industry. In August 1983, nineteen producers of low-sulfur coal formed the Alliance for Clean Energy to lobby for strong acid rain legislation, which they hoped would force electric power companies to switch from high-sulfur coal to their product. The breakaway caused much consternation in the National Coal Association and the American Mining Association, both of which vigorously opposed action on acid rain. The United Mine Workers union, which drew most of its membership from high-sulfur coal mines, was likewise disappointed with the move. The Edison Electric Institute hewed to its "more research before taking action" stance.[21]

In early August 1983, the acid rain task force presented a 250-page report detailing the pros and cons of eleven options for controlling SO_2 and NO_x

emissions. Among other things, the report concluded that reducing annual SO_2 emissions by 50 percent (about twelve million tons) in twenty-one states east of the Mississippi River would probably protect the Adirondack Mountains and the New England area, but would not be as efficient as reducing emissions "closest to sensitive receptor areas." In November, Ruckelshaus presented a modest plan to the White House Domestic Policy Council to reduce SO_2 emissions by 3.4 million tons per year in Ohio, Pennsylvania, West Virginia, New York, and New England. OMB director David Stockman responded with an eighteen-page memo concluding that a major SO_2 reduction program to combat acid rain was wholly unjustified. He attached an industry-prepared document listing the states where electricity rates would rise the most and where job losses would be highest and the number of electoral votes they represented in the upcoming 1984 election. After the council roundly rejected the Ruckelshaus plan, he gave up on acid rain legislation.[22]

Having steadied the ship at EPA, Ruckelshaus resigned soon after President Reagan's landslide victory over Walter Mondale in November 1984. His handpicked successor, Lee Thomas, was a career public servant who had been serving as the head of the agency's solid waste office. Thomas was careful to maintain the agency on the even keel that Ruckelshaus had set for the previous year, and he cautiously proceeded ahead on the acid rain front.[23]

By 1985, the coal industry was, in the words of National Coal Association (NCA) president Carl Bagge, in a state of "institutional chaos." It was producing more coal than ever, and its share of the nation's electrical fuel consumption had risen from 43 to 54 percent over the previous decade. Yet because of productivity gains and greater reliance on surface mining, coal mining jobs had declined by 7,000 from a peak of 63,000 in 1978. Projections for new power plants were down as the industry sought to meet anemic growth in demand by revitalizing older plants. In a break from a decade-long downward trend, annual emissions of SO_2 and NO_x increased by 2 percent during 1984, a phenomenon that environmental groups blamed on the Reagan administration's failure to implement and enforce compliance schedules in SIPs.[24]

One bright spot was the Tennessee Valley Authority (TVA). Under the leadership of chairman David Freeman and another holdover appointee from the Carter administration, TVA completed a six-year $1.2 billion program that reduced its SO_2 emissions from 2.3 million to 1 million tpy. The program included adding electrostatic precipitators or baghouses to all twelve of its coal-fired power plants, adding wet scrubbers to four of its units, installing coal-washing facilities at two plants, and purchasing low-sulfur coal to mix with high-sulfur coal at its other units. Freeman hoped that the program would serve as a model for a nationwide SO_2 control program capable of reducing emissions by 10–12 million tpy. But TVA's brief run as an industry leader in pollution control came to an abrupt end in September 1984 when President Reagan appointed a second member to its three-member board, leaving Freeman in a minority.[25]

As TVA was demonstrating that large SO_2 emission reductions were possible without large price increases, a consensus was building within the scientific community that emissions from coal-fired power plants were causing the acidification of downwind lakes and streams and that acid rain was a national problem. An alarming body of research indicated that hundreds of lakes in the Rocky Mountains were endangered by power plants and copper smelters in the Southwest and Mexico. Still other research drew connections between emissions from southeastern power plants and decreased forest growth on the Georgia Coastal Plain and red spruce defoliation in the North Carolina Appalachians. Environmental groups argued that it would be foolish to wait until there was conclusive proof that power plant emissions caused dying lakes and forests. Nevertheless, Vice President George H. W. Bush told Illinois coal miners that the Reagan administration would not be placing restrictions on power plant emissions to reduce acid rain. And Administrator Thomas hewed to the Reagan administration position that it would be "premature" to take action "based on our current understanding of the problem."[26]

A bipartisan bill with 150 cosponsors looked destined for enactment until it ran into a sophisticated lobbying campaign undertaken by Citizens for Sensible Control of Acid Rain, an "astroturf" grassroots organization consisting primarily of public relations experts at Fleishman-Hillard and financed by coal and electric power companies. Fleishman-Hillard ran a sophisticated phone bank that contacted constituents of the members of the Energy and Commerce Committee and supplied them with mailgrams urging their representatives not to pass acid rain legislation. The company also sent out 600,000 letters to the same constituents warning them that the acid rain bill would result in 30 percent rate increases and enclosing ready-for-signature "constituent" letters and a stamped, pre-addressed envelope to the recipient's congressperson. The bill died in the targeted committee.[27]

When it became clear that EPA was not going to invoke section 115's remedies, six eastern states, four environmental groups, and four individuals sued the agency in a federal district court in Washington, D.C. The court in July 1985 ordered EPA to update Costle's reciprocity determination and to require the necessary SIP changes if Canada still provided reciprocity. The D.C. Circuit Court of Appeals, however, reversed the district court. In an opinion written by Judge Antonin Scalia, the court held that Costle's findings in letters to the secretary of state and Senator George Mitchell (D-ME) were not binding on his successors, because they were not promulgated as a rule through notice-and-comment rulemaking under the Administrative Procedure Act.[28] As we shall see in Chapter 6, Congress comprehensively addressed acid rain in the 1990 Clean Air Act Amendments.

INTERSTATE POLLUTION

Interstate air pollution gave rise to a number of struggles in addition to the acid rain debates. By late 1983, EPA had received six petitions under section 126 to require upwind states to control sources of pollutants and pollutant precursors to the extent necessary to prevent them from contributing to violations of the NAAQS and impairing visibility in downwind states. The problem with these petitions, according to EPA, was the limited proof the states offered that emissions from particular upwind sources were in fact contributing to nonattainment and visibility impairment in downwind states. The available air dispersion models were not sufficiently accurate to make that kind of precise demonstration. When the states challenged EPA's denial of their petitions, the D.C. Circuit held that EPA did not have an affirmative obligation to reevaluate its previous SIP approvals and that none of the petitioners had shown that emissions from an upwind source had significantly contributed to their nonattainment or visibility impairment.[29]

PREVENTION OF SIGNIFICANT DETERIORATION

During the 1980s, Indian tribes in the West began to flex their muscles under the PSD provisions of the 1977 amendments that gave them the power to redesignate areas within their reservations. When the Northern Cheyenne Tribe of Montana redesignated its reservation from the class 2 residual category to class 1, the operators of the giant Colstrip 3 and 4 power plants were forced to burn extra-low-sulfur coal and install an extensive system of scrub-bers at a cost of about $1.7 billion to insure that the class 1 increments for SO_2 and PM would not be violated. The Spokane Tribe of Washington used a similar tactic to kill a planned two-thousand-megawatt coal-fired power plant near Creston, Washington.[30]

NATIONAL AMBIENT AIR QUALITY STANDARDS FOR PARTICULATE MATTER

In June 1981, EPA's Office of Air Quality Planning and Standards (OAQPS) recommended that the administrator retain the twenty-four-hour and annual form for the primary standard, but limit its application to particles that were smaller than ten micrometers (PM_{10}). Mounting scientific evidence demon-strated that smaller particles penetrated farther into human lungs and posed a greater health risk. The PM_{10} measure would include most particulates emitted through the stacks at power plants, but not the dust from coal piles and road dust that were included in the total suspended particles (TSP) mea-sure of the existing standard. The OAQPS staff recommended a twenty-four-hour primary standard in the range of 150–350 $\mu g/m^3$ and an annual primary standard of 55–120 $\mu g/m^3$. The scientists on the agency's Clean Air Act Scientific Advisory Committee (CASAC) agreed that the standard should focus exclusively on PM_{10}, but they concluded that the recommended ranges were insufficiently stringent to provide an adequate margin of safety.[31]

Administrator Ruckelshaus accepted the staff's recommendation that the agency promulgate a revised standard that focused on PM_{10}, and he indicated that he preferred a standard at the low end of the ranges suggested by OAQPS. When the proposal went to OIRA for review, however, OIRA urged EPA to use the standard as a vehicle to relitigate whether the statute allowed the agency to consider costs. EPA replied that since two D.C. Circuit opinions had definitively rejected that argument, it would be a huge waste of time to replay the cost issue. Noting that the Regulatory Impact Analysis that EPA had prepared for the action left open the possibility that the benefits exceeded the costs at every level considered, an OIRA desk officer suggested to his su-periors that OIRA ask the agency to look at more stringent options, but they rejected that suggestion.[32]

The notice of proposed rulemaking (NPRM) that EPA issued in March 1984 adopted the PM_{10} indicator and solicited public comment on a range of levels from 150 to 250 μg/m^3 for the twenty-four-hour standard and 50 to 65 μg/m^3 for the annual standard. To put these levels in perspective, the existing standards converted to PM_{10} would be 183 μg/m^3 and 48 μg/m^3, respectively. Reactions to the proposal were predictable. The coal industry criticized the proposal as unduly stringent and urged the EPA administrator to choose a standard from the upper end of the range. Environmental groups chastised the agency for considering levels that would effectively weaken the existing standards.[33]

The final rule that EPA published in June 1987 employed PM_{10} as the indicator and set the standards at 150 μg/m^3 and 50 μg/m^3 for the twenty-four-hour and annual averaging periods. It set the secondary standard at levels identical to the primary standard. The agency expected that power plants would use the same technologies (electrostatic precipitators, baghouses, and scrubbers) that they had been using to control PM. In an unexpected gift to the regulated industries, EPA lifted the construction bans that were in place in forty areas under the old standard for the two to three years that it would take for the states to write SIPs implementing the new standards. The D.C. Circuit later upheld the standards.[34]

A MAJOR ENFORCEMENT VICTORY

In a major victory for EPA, the D.C. Circuit rejected a claim by the electric utility industry that EPA lacked the authority to assess penalties against utility company violators. The court agreed with EPA that the fact that public service commissions typically do not allow utility companies to recoup civil penalties from consumers was irrelevant to EPA's calculation of the amount of the penalty. The court noted caustically that "[i]f public utilities really benefit so little by failing to comply and suffer so much by bearing [the cost of] fines, it is [up] to them to comply with the Act and avoid the problem altogether."[35]

GLOBAL WARMING

The Reagan administration consistently opposed efforts at the federal level to address climate disruption, despite growing scientific evidence that power plant emissions of CO_2 were causing global temperatures to increase. Proponents of government action to avert climate change achieved a public relations breakthrough in June 1986 when a subcommittee of the Senate Committee on Environment and Public Works heard National Aeronautics and Space Administration scientist James Hansen testify that the predictable doubling of greenhouse gas (GHG) emissions by 2020 would result in an increase in global temperatures of around nine degrees Fahrenheit. Senator Albert Gore (D-TN) declared that there was "no longer any significant difference of opinion" within the scientific community on whether GHG emissions caused global warming. An EPA official testified in a subsequent hearing that switching from coal and oil to natural gas in power plants would greatly reduce GHG emissions, but he warned that with most of the world's known natural gas reserves located in the Soviet Union and the Middle East, such a move was not necessarily in the national interest. In the meantime, GHG emissions in the United States continued to increase.[36]

NONATTAINMENT SANCTIONS

It was virtually impossible to write a state implementation plan that contained a credible attainment demonstration for areas like Los Angeles and Houston that were severely out of attainment for ozone. Under EPA's post-deadline nonattainment policy, the statutory sanctions did not kick in when the deadline passed in December 1987 so long as the state had an approved SIP, made reasonable extra efforts, and reduced hydrocarbon emissions by 3 percent per year. In late June 1987, EPA proposed construction bans for fourteen metropolitan areas that had failed to promulgate adequate nonattainment SIPs. To deal with areas like Los Angeles, for which the state did not even attempt to make an attainment demonstration, EPA simply ignored that fact and approved the rest of the SIP while deferring action on the demonstration. That solution, however, failed when the Ninth Circuit Court of Appeals held that EPA had exceeded its authority and ordered it to disapprove the SIP for Los Angeles.[37] This resulted in increased pressure to amend the Clean Air Act to address persistent nonattainment areas, but that would have to await the outcome of the 1988 elections.

THE 1988 ELECTIONS

Unlike most presidential elections, environmental issues played a large role in the 1988 elections. Neither the Democratic candidate, Massachusetts governor

Michael Dukakis, nor the Republican candidate, Vice President George H. W. Bush, had strong environmental records. Referring to the Reagan administration's "legacy of neglect," Dukakis promised to seek legislation sharply limiting SO_2 and NO_x emissions to prevent acid rain. Failing that, he would demand that EPA use its authority under the interstate pollution provisions of the Clean Air Act to protect downwind states from power plant emissions. Bush seized on the sad state of Boston Harbor's water quality to criticize Dukakis for failing to be a good environmental steward. Calling himself an environmentalist, Bush promised to "cut millions of tons of sulfur dioxide by the year 2000," but he assured midwesterners that no region of the country would be "hit unfairly with the cost of addressing a problem that affects all of us." After winning by a substantial majority of both the popular and electoral college votes, George H. W. Bush would have an opportunity to make good on that promise.[38]

CONCLUSIONS

The first two years of the Reagan administration were tumultuous ones for EPA as the agency ceased implementing and enforcing the Clean Air Act. Sales of scrubbers peaked in 1980 at $644.4 million and dropped to $184.4 million in 1981 before collapsing to $89.6 million in 1982. As designers and manufacturers of pollution control equipment filed for bankruptcy, they reported that several of their customers had delayed investments in scrubbers and new plants and "put a little more bailing wire on the 1940s plants" to keep them burning coal. Despite intense efforts by the coal and electric power industries and the Reagan administration to amend the Clean Air Act, however, environmental groups and their allies in Congress achieved a stalemate that persisted through the Gorsuch years.[39]

After Anne Gorsuch departed EPA, acid rain became the dominant environmental issue. Industry groups went on defense and were able to forestall aggressive legislation addressing that problem. One of Ruckelshaus's biggest disappointments during his second tenure at EPA was his failure to put an acid rain regulatory regime into place. One of his biggest irritations was the prescriptive standards and deadlines that Congress had put into the Clean Air Act. He criticized environmental groups for taking advantage of these provisions in lawsuits against EPA with little regard for the costs of regulation. The environmental groups responded that they were justifiably asserting fundamental rights to health in the face of strong resistance from the electric power and coal industries. This conflict between utilitarian concerns for ensuring that environmental regulations balanced costs against benefits and assertions of rights to a clean environment continued to characterize future struggles over controlling pollution from fossil fuel–fired power plants.[40]

The PM standard-setting exercise was emblematic of EPA's progress during the Reagan years in addressing the adverse environmental effects of fossil fuel–fired power plants. It withstood strong pressures from the electric power and coal industries and White House economists to roll back existing protections, but it did not move the ball forward. When it came to progress on critical issues like interstate pollution, visibility in national parks, and global warming, the 1980s were a lost decade.

CHAPTER 6

Major Adjustments

When George H. W. Bush became president, the American public was more concerned about the environment than it had been for many years. For the first time, a majority of the respondents in a major national poll ranked the environment as their top priority for government spending. The membership of environmental groups was soaring. The president-elect reached out to environmental groups during the transition, promising to surprise them with his commitment to the environment. They were delighted with the appointment of William K. Reilly, the patrician president of the Conservation Foundation and the World Wildlife Fund, to be EPA's administrator. A personal friend of President Bush with easy access to the Oval Office, Reilly's goal was to work cooperatively with industry and environmental groups to achieve environmental progress.[1]

Reilly had strong competition for the president's ear from White House Chief of Staff John Sununu, a former governor of New Hampshire who had no stomach for expensive environmental initiatives. An ad hoc group chaired by Vice President Dan Quayle called the Council on Competitiveness played the pro-business regulatory review role that the Office of Information and Regulatory Affairs (OIRA) had played during the Reagan administration. Composed of Quayle, Sununu, the director of the Office of Management and Budget Richard Darman, the Council of Economic Advisers (CEA) chairman

November 15, 1990 President George H. W. Bush signs Clean Air Act Amendments of 1990

September 1991 EPA approves agreement requiring Navajo plant in Arizona to install scrubbers to protect visibility in the Grand Canyon

June 1992 United States signs Rio convention that commits signatories to stabilize greenhouse gas emissions by 2000

July 1, 1992 EPA promulgates new source review regulations

July 21, 1992 EPA promulgates Title V permit regulations

January 1993 EPA publishes core acid rain regulations

Michael Boskin, Commerce Secretary Robert Mosbacher, and Attorney General Dick Thornburgh, the Council on Competitiveness was both an active intervenor in agency regulatory initiatives and the court of last resort in interagency disputes over regulations.[2]

BREAKING THE LOGJAM: THE 1990 CLEAN AIR ACT AMENDMENTS

The top environmental priority for the incoming administration was to work with the Democrat-controlled 101st Congress to amend the Clean Air Act to address acid rain, hazardous air pollutants, and the many highly populated areas in the country that had still not attained the NAAQS for ozone. One encouraging sign for acid rain legislation was the fact that Senator Robert Byrd (D-WV) had stepped down from his position as the Senate majority leader and was replaced by Senator George Mitchell (D-ME), a strong advocate of acid rain legislation. But the fact that Rep. John Dingell (D-MI) remained the chairman of the House Energy and Commerce Committee ensured that any legislation that increased the stringency of automobile emissions would face an uphill battle. And any bill imposing stringent requirements would have to overcome the active opposition of the Clean Air Working Group (CAWG), an umbrella organization representing almost two thousand companies in a wide variety of affected industries, including the electric power and coal industries.[3]

At the president's request, EPA drafted a stringent bill that included an acid rain cap-and-trade program that would cut SO_2 emissions by 10 million tons per year (tpy) from 1980 emissions and reduce NO_x emissions by 2 million tpy by 2000. Administrator Reilly, however, encountered strong

resistance from the budget director Richard Darman and other cabinet secre-
taries at a series of cabinet-level meetings. Reilly had one surprising ally,
however, in Chief of Staff John Sununu, the former governor of a state that
had been on the receiving end of acid rain for decades and a strong advocate
of market-based approaches to pollution control.[4]

Following a contentious ninety-minute meeting days before he planned
to present the administration's position to Congress, President Bush retired
to Camp David, where he decided to go forward with a version of the bill that
differed from Reilly's proposal in relatively minor regards. At that point, however, Reilly made the huge mistake of gloating in public. He boasted to the
New York Times that although he felt like he was "having his brains beat in"
during the White House meetings, he succeeded in persuading the president
to support a bill that added "50 percent or maybe more to what the country
lays out on pollution control every year." This earned him the everlasting en-
mity of Darman, who deeply resented Reilly's attempt to become a "global
rock star."[5]

Flanked by sponsors from both parties in the Rose Garden, President Bush
unveiled the administration's three-hundred-page Clean Air Act Amendments bill in late July 1989.[6] Under the bill's acid rain provisions, electric
power companies would have to reduce SO_2 emissions by 9 million tpy from
1980 levels (approximately 23 million tpy) by the year 2000. Another 1 million
tpy in reductions would come from other industrial sources, but EPA's models
indicated that they had already reduced emissions by about that much. The
bill established an innovative cap-and-trade program that would allow individual power plants to buy and sell allowances to emit one tpy of SO_2 apiece,
but suffer heavy penalties ($2,000 per ton) for emitting more SO_2 than covered
by their allowances. Power plants could comply by installing SO_2 reduction
technologies, by burning low-sulfur coal, or by purchasing allowances. The
proposal established a similar cap-and-trade program for reducing NO_x emissions by 2 million tpy by 2000. To avoid cheating, all sources in the program
would have to install continuous emissions monitoring equipment.[7]

The bill divided nonattainment areas into four groups with increasingly
stringent requirements depending on the severity of the nonattainment, ensured that smaller cities would not be penalized for nonattainment due to
ozone transport from urban areas, and allowed EPA to push back the attainment deadlines for the ozone NAAQS in six of the most polluted cities for
an additional ten years. The section governing hazardous air pollutants re-

quired new and existing facilities to install the "maximum achievable control technology" (MACT) to reduce emissions of 280 toxic air pollutants by specific deadlines, and it allowed EPA to impose even more stringent requirements to reduce any unacceptable remaining risks. MACT was defined to reflect the average of the best available technologies for a category of industrial sources, not the "state-of-the-art" technology that Reilly had advocated.[8]

One little-noticed provision established a program requiring every new and existing major stationary source to obtain a single federal operating permit incorporating all of the federally enforceable emissions limitations of the relevant SIP. A state agency would issue the permit, but it would be subject to EPA disapproval on its own initiative or as a result of a petition from any citizen. EPA would have the authority to take over a state's permitting program if it lacked the authority or the willingness to implement it.[9]

The electric power industry and high-sulfur coal companies adamantly opposed the bill, arguing that it was unworkable, did not account for increased demand over time, would force huge increases in electric bills, and would threaten grid reliability. The hazardous air pollutant provisions, they maintained, were just a backdoor requirement for scrubbers, because scrubbers would probably be MACT for mercury emissions from power plants. The United Mine Workers union strongly disputed EPA's prediction that it would have a minimal net impact on coal mining jobs. But the Alliance for Clean Energy, a trade association of low-sulfur coal producers, and railroads generally praised the bill for giving power plants the option of switching fuels.[10]

Environmental groups were thrilled that the Bush administration was advocating a vigorous acid rain program, but they were divided on the virtues and vices of the cap-and-trade approach. While the Environmental Defense Fund (EDF) was all in, other groups did not trust companies not to cheat in reporting their emissions or otherwise to game the system. The environmental groups all agreed, however, that the bill gave too much discretion to EPA to define MACT for hazardous pollutants. The pollution control industry and environmental consultants expected a bonanza as companies elected to retrofit scrubbers and low-NO_x technology on their existing plants and designed environmental controls into new plants.[11]

After the House and Senate passed two very different versions of the administration's bill, the hard choices were left to the conference committee, which spent many hours between July and late October debating the finer points of the legislation. The committee's job was rendered more difficult by its size—9 members from the Senate and 135 from the House. The committee first tackled the so-called "Title V" operating permit program. The electric power industry wanted the bill to allow companies to engage in minor changes that did not significantly increase emissions without triggering permit review, and it did not want EPA to have the power to veto permit actions. It lost both battles. The committee also retained a provision allowing citizen groups to sue EPA if it arbitrarily failed to veto a state-issued permit.[12]

The committee resolved the nonattainment issues by adopting the administration's tiered approach that classified ozone nonattainment areas into four categories (moderate, serious, severe, and extreme) depending on the extent to which air quality deviated from the ozone NAAQS. More polluted areas would have more time to meet the standard, but they would have to do more to reduce VOCs and NO_x in the interim. EPA had the discretion to design similar tiered approaches for the other criteria pollutants.[13]

The committee's definition of MACT for hazardous air pollutants (HAPs) pleased environmental groups by giving EPA less discretion. New sources had to install technology that performed as well as the top five sources in the relevant category, and existing sources had three years to achieve the level of emissions reduction achieved by the cleanest 12 percent of existing sources. EPA had ten years to write MACT standards for about 250 categories of sources of 189 HAPs. In a major coup for the electric power industry, however, the conferees agreed to exempt power plants from the requirements until such time as EPA had studied mercury and other power plant HAPs and determined that regulation was "appropriate and necessary." We will see that controversy over the meaning of these words went on for decades until it was finally resolved by the Supreme Court.[14]

Shortly before dawn on a Sunday morning in late October, after what one participant characterized as "rich, profane, embittered" negotiations, the conferees agreed on an acid rain program that mandated a 10 million tpy reduction from 1980 emissions for SO_2 emissions with half of the reductions coming from 263 units at 111 power plants in twenty-two states by 1995 and the remainder by 2000. During Phase I of the program, the first group of relatively old and dirty plants would receive allocations based on an average emissions rate of 2.5 lb/MBtu. All plants in the twenty-two states would receive allocations based on a 1.2 lb/MBtu emissions rate by 2000. The bill established an emissions trading regime with a post-2000 cap of 8.9 million tpy with 3.5 million extra "early scrub" allowances for midwestern plants that installed scrubbers in the early years. EPA would retain 3 percent of each year's allowances to sell at $1,500 apiece or auction off as a hedge against hoarding by large companies. To ensure against cheating, the bill required continuous emissions monitoring devices capable of recording emissions data at fifteen-minute intervals. The 263 units at 111 Phase I power plants would have to install low-NO_x technology, and EPA would have to promulgate a standard for the remaining power plants starting in 2000. Finally, the new law established a $250 million assistance program for displaced workers.[15]

Pleased that Congress passed a bill based on his proposal, President Bush signed it on November 15, 1990. Administration economists estimated that it would cost about $25 billion per year to comply with its requirements.[16]

INITIAL IMPLEMENTATION OF THE CLEAN AIR ACT AMENDMENTS

EPA quickly added 520 full-time positions to its staff to help with the fifty-five major rulemakings that the amendments required the agency to accomplish by November 15, 1992. The existing agency staff responded to the amendments with an "evangelic fervor" that kept them at their desks until late most nights. An army of industry lobbyists and a much smaller number of environmental activists shifted their focus from Congress to EPA and the White House as they sought to influence the critical implementing regulations. By early April 1991, EPA staff had already held more than three hundred meetings with outside groups. Administrator Reilly was determined to implement the new requirements in the most cost-effective way possible, but he encountered an unanticipated hurdle in Vice President Quayle's Competitiveness Council.[17]

The Title V Permit Rules

The first rules to emerge implemented the Title V permit process for obtaining and modifying operating permits for major stationary sources. The electric power industry's primary concern was the potential for delays in making minor operational changes if each change required a public review during which local activists could object. It argued for the flexibility to engage in any changes that resulted in de minimis emissions increases without undergoing a permit review. The industry also wanted the regulations to provide broad "permit shields" to protect permitted sources from subsequent changes in SIP requirements until the permit was renewed and to shield compliant permit holders from subsequent enforcement actions, even if they were in violation of the SIP because the permitter had misinterpreted the SIP or neglected to incorporate a requirement of the SIP into the permit. Environmental groups strongly opposed both of these suggestions.[18]

Electric power industry representatives actively participated in the drafting process at EPA, but they also urged the Competitiveness Council to rewrite the regulations when it reviewed EPA's draft in early March 1991. The draft that emerged from the council contained many of the provisions that the electric power lobbyists had demanded, including the de minimis exemption from public hearings and a broad permit shield.[19]

The notice of proposed rulemaking that Administrator Reilly signed in April 1991 allowed "minor" permit amendments authorizing emissions increases of less than the amount that triggered new source review (typically 40 tpy) to be accomplished without a hearing so long as the source provided seven days' advance notice to EPA and the state permitting authority. The proposal also contained the industry's broad permit shield. Although industry groups were generally pleased with the proposal, environmental groups were outraged, arguing that the lax provisions would render "major provisions of the new act virtually unenforceable." State air pollution control officials protested that allowing unlimited minor changes that increased emissions after only seven days' notice would not give state agencies time to determine whether the changes were prohibited by their SIPs.[20]

Administrator Reilly was "deeply disturbed" by subsequent reports of the Competitiveness Council's intrusion into the decision-making process. It mattered to him that EPA's general counsel, former Yale law professor Donald Elliott, concluded that the exemption from the statute's public hearing requirement would be unlawful. Furthermore, the dispute had attained symbolic significance as a test of who was really running EPA, and he was determined to fight future intrusions with every tool at his disposal.[21]

Unpersuaded by Elliott's conclusion, the Competitiveness Council sought a second opinion from the Department of Justice. As the November 15 statutory deadline passed, Administrator Reilly met with Vice President Quayle in an attempt to come to agreement on the public participation issue. In late April 1992, the acting assistant attorney general for environment and natural resources sent an unofficial memo to the council advising that the changes it had suggested were perfectly lawful. The memo came during what the White

House had proclaimed as "deregulatory week," reflecting the president's increasing concerns over the impact that a faltering economy would have on his reelection campaign. Undeterred, Reilly demanded that the issue be resolved by the president himself. The internal dispute finally came to an end in mid-May, when President Bush ruled in favor of the council.[22]

The final rule that EPA published in July 1992 allowed sources to make permit amendments that did not result in emissions increases of more than 40 tpy after providing the permitting authority with seven days' notice. The regulation put no limitation on the number of times a source could make minor permit modifications in a given year, but the total amount of emissions per year accomplished through minor permit changes could not exceed 245 tpy. The final rule provided a somewhat narrower permit shield than the proposal. The shield did not extend to any requirements promulgated after the permit was issued, but it did encompass all regulatory requirements specifically referenced in the permit, even if the permitter misinterpreted those requirements. The remedy in that case would be for the objector to seek an amendment to the permit to contain provisions including the correct interpretation of the requirements. A coalition of companies and three environmental groups challenged the permit regulations in the D.C. Circuit.[23] As we shall see in Chapter 7, the Clinton administration settled the litigation by agreeing to revisit the regulation.

Acid Rain

EPA published a set of "core" rules for managing the acid rain cap-and-trade program in early January 1993. By then, the Chicago Board of Trade (CBOT) had created a private market for trading allowances. This ensured a level playing field by giving access to anyone with allowances to sell and anyone with money to purchase them. The first SO_2 allowance trade, however, took place in May 1992 before the federal program even got started. Because the Wisconsin SIP required Wisconsin Electric Power Company (WEPCO) to reduce SO_2 emissions more rapidly than the federal acid rain program (accomplished largely through the purchase of low-sulfur coal), it had up to 120,000 allowances to sell. It agreed to sell 10,000 allowances to TVA, which used the allowances to buy additional time to install scrubbers at its plants. Local environmental groups in Tennessee and Wisconsin pilloried the transaction in the press as an unconscionable exchange not unlike sex for money.

The negative reactions by local groups turned the trades into a public relations disaster that only grew worse when the press reported that the White House had urged TVA to do the deal as a market starter. Utility executives blamed EDF for failing to educate local environmental groups on the virtues of emissions trading.[24]

NO$_x$RACT for Nonattainment Areas

Still another important task for EPA under the 1990 amendments was to provide guidance to the states on "reasonably available control technology" for reducing NO$_x$ emissions from existing power plants ("NO$_x$RACT") in ozone nonattainment areas.[25] The critical issue was whether EPA should base its recommendation on: (1) "low-NO$_x$" burners that redistributed air and fuel, reduced the availability of oxygen in critical NO$_x$ formation zones, and lowered the amount of fuel burned at peak flame temperatures; (2) more expensive "overfire air systems" on their boilers; or (3) even more expensive (but very effective) add-on selective catalytic reduction (SCR) systems. The electric power companies strongly opposed anything more stringent than low-NO$_x$ burner technology. Environmental groups cited predictions by experts from EPA and academia that to bring ozone levels down to the NAAQS along the Eastern Seaboard, all upwind power plants would have to install SCR technology. The final guidance that EPA issued in October 1992 after a thorough vetting by the Council on Competitiveness suggested that states require only low-NO$_x$ burners. It was a clear win for the industry, but environmental groups would still have an opportunity to persuade state environmental agencies to require existing plants to require overfire air or SCR technology in their SIPs.[26]

NEW SOURCE REVIEW

By the 1980s, many companies were depending heavily on power plants that were nearing the end of their planned lifetimes, but EPA's recently tightened NSPS and the need to purchase offsets in nonattainment areas made building a new plant a very expensive proposition.[27] Companies operating in recently deregulated electricity markets could no longer count on high electricity prices allowing them to recover large capital investments from their customers. In this environment, refurbishing the old plants looked awfully attractive.[28] The problem was that any modification that resulted in increases of pollutants by more than 40 tpy were subject to the new source review (NSR) process under which the modified plant had to install the best available control technology (BACT) or (in nonattainment areas) achieve the lowest achievable emissions rate. The NSR process was also time consuming, because states were not always expeditious in granting permits.[29]

Subsequently disclosed industry documents revealed a conscious strategy of meeting increased demand by building additional capacity into grandfathered plants.[30] With the routine maintenance, repair, and replacement (RMRR) exemption (discussed in Chapter 3) in mind, the Electric Power Research Institute advised its members in 1984 to call such projects "upgraded maintenance programs" and to "downplay the life extension aspects of these projects (and extended retirement dates) by referring to them as plant restoration . . . projects."[31] Although EPA enforcers usually lowered the boom on companies that built brand new facilities without applying for a PSD permit,[32] they focused little attention on modifications to existing plants.[33] When they did become aware of modifications, EPA and state permitting authorities made "commonsense findings" on a case-by-case basis that projects at existing plants either did or did not come within the RMRR exemption.[34]

In 1988, EPA's regional office concluded that a proposed project at Wisconsin Electric Power Company's Port Washington plant near Milwaukee had to undergo new source review because the potential emissions from the facility after completion of the project would exceed the actual emissions prior to the project.[35] Because the output of the plant's five small coal-fired units had decreased substantially due to age-related deterioration, the company concluded that "extensive renovation" of the five units and the plant's common facilities would be required to keep the facility in operation. When WEPCO presented its plan for a "life extension" project to the Wisconsin Public Service Commission for approval, the commission consulted with the state's Department of Natural Resources to determine whether a PSD permit would be required. The department referred the matter to EPA, which conferred with the company on several occasions. In September 1988, EPA concluded that the project would be subject to NSR. WEPCO then challenged the decision in the Seventh Circuit Court of Appeals.[36]

On January 19, 1990, the court in *Wisconsin Electric Power Company v. Reilly* upheld EPA's decision in all but one regard. The court observed that EPA had determined that the changes were not "routine" by "weighing the

nature, extent, purpose, frequency, and cost of the work, as well as other relevant factors, to arrive at a commonsense finding." And it agreed with EPA
that "the magnitude of the project (as well as the down-time required to implement it) suggest[ed] that it [was] more than routine." It stressed WEPCO's
own characterization of the replacement as a "life extension" project of
the sort that "would normally occur only once or twice during a unit's expected life cycle." The court, however, rejected EPA's interpretation of the
word "increase" in its regulations to require a comparison of the pre-project
baseline emissions with the plant's "potential to emit" after the project was
completed. It was inappropriate in calculating post-renovation emissions to
assume that the plant would be operating with 100 percent capacity twenty-
four hours a day for 365 days per year. If EPA wanted to apply the "potential
to emit" test to post-renovation emissions in the future, it could amend its
regulations to make that test explicit, but its current regulations required a
comparison between pre- and post-modification actual emissions.[37]

Soon after the *Wisconsin Electric Power* decision came down, EPA initiated an internal rulemaking project to respond to the remand. When industry
lobbyists learned that the internal decision-making process was not going in
a favorable direction, an Edison Electric Institute official wrote to a friend in
the Department of Energy (DOE) to solicit DOE's aid in securing "a good
WEPCO fix." The letter provided several specific proposals for revising the
NSR regulations in ways that would be acceptable to the electric power industry. Within days, DOE's acting assistant secretary for fossil energy wrote
to EPA to demand "a good and comprehensive WEPCO fix." DOE then
drafted its own proposal that included the changes that EEI had proposed,
but an EPA attorney had already rejected the industry proposal as "hogwash"
and "mostly garbage." A week later, an economist on the staff of the White
House Council of Economic Advisers who had been mediating the dispute
between DOE and EPA, penned an "urgent" memorandum ordering EPA to
prepare a new draft proposal that reflected provisions that the Bush administration had recommended for the recent clean air amendments but Congress had rejected. An attachment provided detailed instructions for "where
and how the current draft must be modified in order to achieve this." Ignoring
the concerns of agency lawyers, high-level officials in EPA promptly revised
the proposal to comply with the White House instructions.[38]

In June 1991, EPA proposed a WEPCO-fix rule that changed the NSR regulations in three significant ways. First, the agency proposed to amend the

definition of "major modification" to exclude "pollution control projects" (a term that included fuel switching) that increased emissions of another pollutant, so long as the projects were not "less environmentally beneficial." Second, the proposal "clarified" the methodology for calculating the "baseline" level of actual emissions prior to the change to allow sources to use the average tons per year during any two consecutive years within the five years preceding the proposed change. Third, the agency responded to the *Wisconsin Electric* remand by replacing the "actual-to-potential" test for determining whether there was a net increase in emissions with an "actual-to-future-actual" test that allowed the source to calculate what its actual emissions would be after the change without assuming that the unit would be operating year-round at its full operational capacity. In calculating future actual emissions, the source could use any representative two-year period within the ten years following the change. It could also take into account factors, such as "system-wide demand growth," that "would have occurred and affected the unit's operations even in the absence of the physical or operational change." Changes to allow the unit to respond to anticipated increases in demand, however, did not qualify.[39]

The electric power industry was generally pleased with the proposal. Environmental groups, however, opposed the proposed changes. They worried that the exemption for pollution control projects would allow production-driven upgrades to be disguised as pollution control projects. They further argued that allowing sources to choose the years for calculating baseline emissions and to come up with their own estimates of future emissions based on system-wide demand growth was an invitation to manipulation and a prescription for massive avoidance of the NSR requirements and the perpetuation of grandfathered sources. Sources could rely on their own speculative projections without consulting the regulatory authority, thereby making enforcement very difficult.[40]

The final rule, published on July 1, 1992, did not diverge in any significant way from the proposal. In response to concerns that sources would manipulate estimates of future actual emissions, the agency added a requirement that companies submit for five years after the change "sufficient records" to determine if the change had resulted in greater-than-predicted emissions. If so, the NSR provisions would be applicable retroactively. The agency continued to insist that emissions increases attributable exclusively to demand growth that could have been accommodated by the unit prior to the change were not caused by the change. And it continued to allow the companies undertaking changes to make NSR determinations on their own without informing the permitting authorities of those determinations.[41]

VISIBILITY

EPA's first attempt to impose controls on an existing stationary source under its decade-old plume blight regulations was aimed at protecting the iconic Grand Canyon National Park in Arizona from emissions of the Navajo coal-fired power plant located near Page, Arizona (described in Chapter 1). The consortium that owned the Navajo plant had signed an agreement with the Department of Interior in 1973 to install scrubbers as soon as they were a proven technology, but it had not fulfilled its end of the bargain. To settle a lawsuit brought by environmental groups, EPA in August 1989 issued a proposed determination that the Navajo plant was a major contributor to visibility impairment in the Grand Canyon. It relied on a controversial tracer chemical study commissioned by the National Park Service (NPS) concluding that the plant's contribution to visibility impairment in the winter was 40–50 percent and up to 70 percent on the worst days.[42]

The consortium found the NPS study to be methodologically flawed and filled with errors due to poor quality assurance. Claiming that visibility impairment was merely an "aesthetic problem," the consortium predicted that installing scrubbers to improve canyon visibility would cost as much as $1 billion. An October 1989 National Academy of Sciences (NAS) report concluded that the NPS tracer study had correctly identified the Navajo plant as a "significant contributor" to visibility-limiting sulfates in the Grand Canyon, but its quantitative calculations were unreliable. The consortium urged EPA to abandon its proposal in light of the NAS panel's findings. EDF, by contrast, argued that the report provided sufficient information to mandate scrubbers for the plant.[43]

EPA in February 1990 sent to the White House a draft recommendation that Arizona amend its SIP to require all three units to install scrubbers as the "best available retrofit technology" to reduce SO_2 emissions by 90 percent. In response, the Competitiveness Council insisted that the agency reduce its proposal to a 70 percent reduction to avoid unnecessary expenditures on scrubbers. A representative of the consortium later recounted that the council provided "a very important forum for presenting a point of view that we felt needed to be factored into the equation." The proposal that EPA published in February 1991 followed the Competitiveness Council's 70 percent recommendation.[44]

After months of intense negotiations between the environmental groups and the consortium in which EPA provided technical assistance, they entered into a memorandum of understanding on August 8, 1991, under which the consortium would install scrubbers capable of 90 percent SO_2 removal at all three units at the plant between 1997 and 1999. The consortium also agreed to conduct maintenance shutdowns in the winter when emissions from the plant most contributed to haze in the canyon. Standing on the south rim of the canyon in a mid-September 1991 photo op, President Bush announced that EPA had approved the agreement. In March 1993, the Ninth Circuit Court of Appeals upheld the memorandum of understanding. The massive scrubber project came in on time and under budget at a cost of $420 million, less than half of the consortium's original estimate.[45]

INTERNATIONAL POLLUTION

At the end of the Carter administration, EPA Administrator Douglas Costle found that SO_2 emissions from the United States had caused acid rain that endangered public health and welfare in Canada, but the Reagan administration did not act on that finding. In November 1988, the government of Ontario, several northeastern states, and environmental groups filed a lawsuit against EPA seeking to force it to begin the process of reducing SO_2 emissions. EPA responded that it would be "entirely speculative" to attempt to predict the effect of emissions from individual power plants in the United States on Canadian forests and lakes. The D.C. Circuit in September 1990 agreed with EPA that the international protection provisions of section 115 of the Clean Air Act were not triggered until the agency could "identify the specific sources" in the United States that caused harm in Canada.[46] That put an end to attempts by Canada to secure relief from U.S. courts.

CLIMATE CHANGE

During his 1988 campaign to be the nation's first "environmental president," candidate George H. W. Bush had promised to take action to address global warming. Once in office, however, President Bush yielded to pressure from the coal and electric power industries to refrain from regulating greenhouse gas (GHG) sources until the rest of the world agreed to limit such emissions. The only climate disruption legislation that Congress passed during the Bush administration was the Global Climate Change Act of 1990, which created an expanded research program and required a committee to prepare a scientific assessment. In the meantime, the United States remained by far the world's largest GHG emitter.[47]

The climate change debate shifted to the international arena as environmental agencies throughout the world prepared for a June 1992 "Earth Summit" of global heads of state convened by the United Nations in Rio de Janeiro. EPA Administrator William Reilly initially favored a "top-down" approach advocated by nearly all of the other participating countries that would establish a firm target of reducing greenhouse gas emissions to 1990 levels by the year 2000. Presidential adviser John Sununu and the staff of the Competitiveness Council, however, strongly objected to any approach that involved targets and timelines. They argued that any governmental action should proceed from the "bottom up" by way of implementing regulations already on the books to address other pollutants.[48]

By the time that Reilly attended a negotiating meeting in the Netherlands in November 1989, he had changed his tune. The U.S. government's position was that efforts to reduce GHG emissions should await further research into the causes of global warming and the economic consequences of GHG emissions reductions. At a seventeen-nation conference that the White House convened in April 1990, Reilly and the other U.S. participants pointed to uncertainties in the scientific and economic studies on the impacts of GHG emissions. Likewise, at a November 1990 meeting in Geneva to prepare for the Rio summit, the United States successfully watered down the conference declaration to eliminate specific targets for GHG reduction.[49]

In 1991, the National Coal Association, the Western Fuels Association, and Edison Electric Institute created a group called the Information Council on the Environment, which launched an advertising and public relations effort to "reposition global warming as theory (not fact)." The public relations firm it hired arranged for the sympathetic scientists on its advisory board to appear in broadcast appearances, op-ed pages, and newspaper interviews. The environmental groups responded with an all-out effort to persuade influential editors and reporters that the upcoming Rio conference was the top environmental story of the 1992 election season. The strategy worked as Democratic frontrunner Bill Clinton made a special point of criticizing President Bush for failing to commit to targets and timetables for the Rio accords.[50]

In a meeting held in New York in early May 1992 to produce a final draft of the Rio convention, U.S. delegates persuaded representatives of the other 142 participating nations to abandon targets and timelines and accept a document filled with vague assurances that they all would do what they could to reduce GHG emissions to 1990 levels by 2000. The successful bargaining chip was a threat that President Bush would boycott the Rio summit if the documents to be signed at that summit contained targets and timelines. The other participants were concerned that without the presence of the leader of the only remaining global superpower, the summit would be seen as a failure. The convention that the Rio delegates signed contained a framework for further negotiations concerning greenhouse gas reductions and a vague commitment to stabilize greenhouse gas emissions by the end of the decade.[51]

ECONOMIC REGULATION

In the early 1990s, demand for electricity soared as ordinary American households began to use home computers and other high-tech electronic devices. At the same time there were signs of change in the electric power industry. For example, the board of Southern California Edison Company, the nation's largest electrical utility company, made John Bryson its CEO and chairman of the board. Bryson was one of the four Yale law students who founded the Natural Resources Defense Council (NRDC) in 1970. Bryson promised to "provide electric service in a way that is very sensitive to the environment." Unlike many utility company executives at the time, Bryson supported programs to reduce electricity demand, and he urged state regulators to give greater weight to renewable sources of electricity.[52]

ENFORCEMENT

EPA and the Department of Justice (DOJ) filed a number of enforcement actions against power plants during the George H. W. Bush administration, most of which related to violations of emissions limitations. More serious enforcement actions against Nevada Power Company and Arizona Public Service Company alleging that they bypassed malfunctioning pollution control devices resulted in fines of $400,000 and $1.3 million. DOJ also filed a few enforcement actions along the lines of the WEPCO litigation where power plants had allegedly undertaken modifications without undergoing new source review.[53]

THE 1992 ELECTIONS

The poor economy was the dominant issue in the three-man race between President Bush, independent candidate H. Ross Perot, and Arkansas governor Bill Clinton. Bush's political advisers were confident that EPA Administrator Reilly had done such a good job that the Democrats could find little to criticize on environmental issues. They did, however, find much to criticize in Clinton's running mate, Al Gore, whom they called an "environmental extremist." Clinton won the election with a plurality of the votes (43 percent to 37 percent) and no strong mandate to regulate the nation's power plants.[54]

CONCLUSIONS

The Bush administration's crowning environmental accomplishment was the 1990 Clean Air Act legislation. That landmark law reflected a bipartisan commitment to environmental improvement that is wholly foreign to the current political culture in which the environment has become a fiercely partisan issue. It is difficult to imagine a scene today in which a Republican president would be joined by congressional leaders from both parties in announcing a proposal for strengthening environmental laws. That progressive amendments to an environmental statute would be enacted during a Republican administration may well be a historical anomaly, but it was an anomaly with considerable staying power. The 1990 amendments continued to steer the electric power industry toward a more sustainable energy sector almost three decades after their enactment.

Under Bill Reilly, EPA made a credible effort to write the regulations necessary to implement the 1990 amendments and to call out owners of power plants that had undertaken major modifications without undergoing new source review. The struggle over the Navajo plant demonstrated how companies can grossly overestimate the cost of compliance to their political advantage. In the second half of President Bush's single term, White House officials successfully pushed back against EPA's efforts to implement the new Title V operating permit program, promulgate NO_x RACT guidelines for power plants, implement a WEPCO fix to the NSR regulations, protect visibility in the Grand Canyon, and set targets and timelines for reducing greenhouse gases.

If President George H. W. Bush was not the environmental president that he promised to be, he was far more committed to protecting public health and the environment than his predecessor and (as we shall see in Chapter 8) his son. The question for the electric power industry after the 1992 elections was the extent to which Bill Clinton would build upon the strong foundation established by the 1990 amendments and take on critical unaddressed environmental issues like the effects of power plants on aquatic and marine life, the proper disposal of coal ash, and the growing problem of climate change.

CHAPTER 7

Cautious Implementation and Vigorous Enforcement

Having strongly supported the Clinton-Gore team during the election season, the national environmental groups were delighted with Clinton's selection of Carol Browner, the head of the Florida Department of Environmental Regulation, to be the next EPA administrator. Only thirty-six years old at the time of her nomination, Browner was a hardworking, no-nonsense regulator with keen political instincts. The electric utility industry was concerned by Browner's position that pulverized coal technology should be allowed for new power plants only as a last resort and only if the new plants met emissions limitations more stringent than the federal requirements. Mary D. Nichols, an attorney in the Los Angeles Office of NRDC, became the assistant administrator for air and radiation. To counterbalance Browner, Clinton appointed Hazel O'Leary, the environmental affairs director at Northern States Power Company, to be secretary of energy.[1]

As a presidential candidate, Bill Clinton had harshly criticized Vice President Quayle's Competitiveness Council, and he immediately abolished it upon assuming office. He then issued an executive order that laid out the obligations of executive branch agencies to prepare regulatory impact analyses for major rules and the procedures for interagency review of major rules through the Office of Information and Regulatory Affairs (OIRA). The new head of OIRA, attorney Sally Katzen, said that she hoped the new process,

February 1993 President Clinton proposes a Btu tax to reduce power plant emissions

August 1993 EPA decides not to regulate coal ash as a hazardous waste

October 1993 Clinton administration substitutes voluntary "Climate Action Plan" for failed Btu tax proposal

March 1994 EPA promulgates low-NO_x standards for Phase I plants

April 1995 EPA promulgates revised low-NO_x standards for Phase I plants in response to D.C. Circuit remand

June 1997 EPA promulgates revised ambient air quality standards for ozone and particulate matter

April 1998 EPA's general counsel concludes that greenhouse gases are "pollutants" subject to regulation under the Clean Air Act

September 1998 EPA issues NO_x SIP call

April 1999 EPA promulgates regional haze regulations to protect visibility in national parks

1999–2000 Deregulation at the state and federal levels precipitates a boom in natural gas-fired peaker plant construction

May 2000 California energy crisis begins and continues into 2001

December 2000 EPA publishes "appropriate and necessary" finding requiring EPA to regulate hazardous mercury emissions from power plants

January 2001 EPA finds that eleven states have failed to meet their "good neighbor" obligations under the NO_x SIP call

which allowed OIRA to become involved earlier in the rulemaking process, would be "more efficient and open to the public."[2] For the most part, it was.

ACID RAIN

The 1990 amendments required companies owning the 111 Phase I units to file plans for complying with its acid rain provisions. The plans revealed a variety of reactions to the acid rain program, including full scrubbing, partial scrubbing, fuel switching, fuel mixing, and other more novel arrangements. Only 3 percent of the Phase I boilers were scheduled for retirement.[3] EPA also faced the daunting task of implementing the Phase II requirements for the rest of the affected power plants, and promulgating Phase I and Phase II requirements for low-NO_x burners.

Phase I Acid Rain Compliance Plans

Owners of Phase I plants expected to install scrubbers in only about 10 percent of the affected units, but some of those were the largest and dirtiest plants in the country. In some cases, the prospect of receiving income from the sale of allowances was critical to the decision to install scrubbers. For example, when American Electric Power installed scrubbers at its huge 2,600-megawatt Gavin plant in southern Ohio (one of the big dirties described in Chapter 1), the 95 percent reduction in SO_2 emissions provided 213,000 allowances, which represented two-thirds of the company's Phase I reduction requirements for all fifty of the units at its twenty coal-fired plants. In other cases, state legislatures played a role. Eager to protect in-state coal mining jobs, the Illinois General Assembly passed a bill providing economic incentives to Commonwealth Edison to install two scrubbers at its Kincaid plant.[4]

Most companies, however, planned to reduce emissions by switching to low-sulfur coal or blending low- and high-sulfur coals, because that option was less expensive than installing scrubbers, reduced emissions more quickly, and allowed them to monitor emerging clean-coal technologies as compliance options. Public Service of Indiana (PSI Energy) installed five continuous online coal analyzers that provided real-time feedback on the SO_2 and energy content of the coal it was burning, allowing the operators to adjust the mix of coal to all five units to ensure that SO_2 emissions did not exceed its allocated allowances. Not surprisingly, politicians in high-sulfur coal states and the United Mine Workers union urged the permitting agencies to reject fuel-switching plans and to require scrubbers instead. Legislatures in states that contained large deposits of high-sulfur coal enacted laws aimed at discouraging local power companies from switching to low-sulfur coal and to encourage them to install scrubbers, but the U.S. Supreme Court held that an Oklahoma law requiring power companies to use local coal for at least 10 percent of their fuel requirements violated the Commerce Clause of the Constitution.[5]

Companies frequently ran into opposition in state PUCs when they asked to include the cost of pollution controls in their rate bases. When AEP sought the Ohio Public Utility Commission's (OPUC) approval to incorporate the cost of new scrubbers at its Gavin plant into its rate base, eleven large industrial and commercial customers and the Sierra Club objected that the scrubbers were just a hidden subsidy to the Ohio coal companies from which it purchased high-sulfur coal because the most cost-effective way of reducing

SO_2 emissions was to purchase low-sulfur coal. AEP was supported by the OPUC staff and the Ohio Consumers' Counsel (OCC), which argued that the OPUC should consider the social cost of the 1,258 Ohio coal mining jobs that would be lost if the Gavin plant converted to low-sulfur coal. The OPUC authorized AEP to recover up to $815 million to pay for the scrubbers, and the Ohio Supreme Court upheld that decision. Ratepayers in Ohio also argued that they should be compensated by ratepayers in other states where different AEP plants used the allowances generated by the Gavin scrubbers. That struggle was resolved by a "regulators committee" composed of representatives from all of the affected states, which came up with a compromise that the Federal Energy Regulatory Commission approved.[6]

Phase I Allowance Markets

With few regulations fully in place, markets in SO_2 allowances were slow to develop. Another reason that allowance trading got off to a slow start was opposition from local environmental groups that split with the national groups to denounce allowance trading as the sale of a "right to pollute." The few companies that did engage in trading were reluctant to disclose the details of the trades for fear of stirring up local opposition. This made it difficult for owners of the 111 Phase I plants to decide whether to rely on the availability of a future allowance market or to install scrubbers in time to meet the Phase I deadline. Since ratepayers would bear the capital cost of installing scrubbers and shareholders would bear the loss of bad trades, some risk averse utility companies opted for the former option. Others elected to sit tight and purchase allowances.[7]

Although the Chicago Board of Trade hoped to become the exclusive market maker for allowance trades, an "off-exchange" market in allowances developed as private brokers arranged bilateral trades through highly customized contracts to meet the needs of particular buyers and sellers. Enron Power Services, for example, set up private exchanges in which allowances were coupled with the sales of fuels to make the fuels more attractive to owners of power plants. The first EPA-administered auction sent a clear signal to electric power companies that the value of allowances was much lower than predicted. By mid-March 1995, the first year the program was in effect, allowance prices had plunged to the $70–75 range, far lower than the $1,500 that Congress envisioned as a cap on the price of an allowance. The low prices persuaded some companies to cancel orders for scrubbers. By the end of 1995, it was apparent that massive overcompliance with Phase I SO_2 reduction requirements was generating a huge glut of banked allowances that companies could use to forestall Phase II emissions reduction requirements for several years.[8]

The original goal for Phase I was to reduce SO_2 emissions from the 111 Phase I plants to 6.7 million tpy, but the bonus allowances that EPA awarded in the early years to states with strong energy conservation programs moved the cap up to somewhere between 7 and 8.7 million tpy. In fact, actual emissions were in the neighborhood of 5.3 million tpy. Although the cost of scrubbing declined precipitously during the five years of the Phase I program from $282 per ton of SO_2 removed to less than $100 per ton, the big story was the degree to which companies burned low-sulfur coal instead of installing scrubbers because of the flexibility afforded by the cap-and-trade program to blend high- and low-sulfur coals to optimize performance and emissions reduction. New scrubbers accounted for 35 percent of SO_2 emissions reductions from Phase I plants, fuel switching accounted for 59 percent of reductions, and 6 percent came from retirements. End-use efficiency and renewables played only a very modest role.[9]

Low-NO$_x$ Burners for Phase I Plants

The 1990 amendments required the 111 Phase I power plants to meet emissions limitations reflecting "low-NO$_x$" burner technology. To meet that requirement, the statute required EPA to set the standards for tangentially fired boilers (TFB) at 0.45 pounds per million British thermal units (lb/MBtu) and the standards for dry-bottom wall-fired boilers (DBWFB) at 0.50 lb/MBtu, unless EPA found that those rates could not be achieved using that technology. In late October 1992, the Bush administration had published a proposal that presented three alternatives: (1) a standard based on overfire air technology, which injected additional air through special ports located above the burner zone, for both TFB and DBWFB units; (2) a standard based on overfire air for TFB but not for DBWFB; and (3) a standard that ignored overfire air altogether. The electric power industry took the position that Congress contemplated in-furnace techniques that redistributed air and fuel flows within existing furnace boundaries and nothing more. The coal and electric power industries charged that the proposal was inconsistent with the statute insofar as it included overfire air on any type of furnace and warned that requiring overfire air would threaten grid reliability and make fly ash more difficult to use in concrete. Environmental groups urged the recently arrived Clinton administration to require overfire burners on all Phase I plants.[10]

The final standard that EPA published in March 1994 based the emissions limitations for both types of units on the performance of boilers with over-fire air. But it allowed a company to apply for an alternative limit by showing that it had installed low-NO_x burners with overfire air and still could not meet the emissions limitation. The electric power and coal industries successfully challenged the regulation in the D.C. Circuit Court of Appeals. In a brief opinion, the court held that the statute precluded EPA's inclusion of overfire air in its definition of "low-NO_x burner." The judicial remand meant that Phase I requirements would not kick in until EPA promulgated a response, thereby allowing companies to forestall planned controls. The revised regulation that EPA promulgated in April 1995 relaxed the emissions limitations to reflect the absence of an overfire air requirement and extended the effective date for two years until January 1, 1996.[11]

In the end, the program was a modest success. The average emissions rate for all included units dropped from 0.70 lb/MBtu to 0.40 lb/MBtu for a net reduction of more than 400,000 tons per year in NO_x emissions. Owners of 114 of the 265 Phase I units installed overfire air technology, probably because they had already commenced construction before the judicial remand. Only ten units received alternate limitations by showing that low-NO_x technology would not achieve the relevant emissions limitation in their unique circumstances. Engineers came up with an inexpensive fix for most DBWFB units that cost only $161 per ton of NO_x removed. TFB units had to be retrofitted at an average cost of $631 per ton.[12]

Phase II Allowance Markets

As utility companies filed their compliance plans for meeting the Phase II requirements of the acid rain program, it became clear that the vast majority were planning to rely on banked or purchased allowances or conversion to low-sulfur coal and that very few were planning to install scrubbers. This was not good news to producers of high-sulfur coal in Appalachia and the Midwest or the vendors of pollution control equipment. By 1999, deregulation of railroad rates made western coal affordable, and it accounted for nearly half of the coal sold in the United States, nearly all of which went to power plants to be burned or blended with high-sulfur coal. The bottom line was that SO_2 emissions were down at modest additional cost.[13]

THE 104TH CONGRESS INTERVENES

The Clinton administration did not get off to an especially fast start in regulating power plants, and it received a rude awakening in November 1994 when the Republican Party won control of both houses of Congress. Adhering to their "Contract with America," the House Republican leadership and many of the new freshmen members were committed to providing "regulatory relief" for American businesses. The most radical proponents of regulatory relief in the House wanted to repeal or rewrite the Clean Air and Clean Water Acts. But they encountered strong resistance from the chairman of the Senate Environment and Public Works Committee, John Chafee (R-RI), a strong supporter of both statutes, and a group of Republican representatives from the Northeast who opposed any legislation limiting EPA's efforts to protect their states from power plant emissions in the South and Midwest.[14]

The new deregulatory initiatives were welcomed by the coal industry and by many electric power companies. Some power company executives, however, worried that eliminating federal regulatory programs would force the states to create a patchwork of regulations that would be inconsistent with a national market in electricity. And companies that had already invested heavily in pollution controls did not want Congress to enact legislation rolling back environmental requirements for their competitors. Environmental groups strongly opposed all of the deregulatory bills. In the end, the opponents prevailed, and the bedrock statutes survived unscathed. But a chastened Clinton administration attempted to demonstrate that radical changes were unnecessary by slowing down regulatory initiatives, doubling down on its efforts to incorporate market-based concepts into regulatory programs, and placing more emphasis on voluntary approaches. EPA launched no major power plant initiatives until President Clinton was safely in the White House for a second term after the 1996 elections.[15]

NONATTAINMENT

States preparing state implementation plans (SIPs) for nonattainment areas had to specify "reasonably available control technology" (RACT) for NO_x

emissions from existing power plants. As we saw in Chapter 6, EPA provided guidance to the states in October 1992 for NO_xRACT that was slightly less stringent than the low-NO_x burner standards that it prescribed for the acid rain program. On top of this technology requirement, states had to reduce NO_x emissions from power plants to the extent necessary to meet the national ambient air quality standards (NAAQS) for ozone. This project, however, was greatly complicated in states along the Eastern Seaboard by the fact that ozone and ozone precursors migrated with the winds to those states from states in the Midwest and Southeast to such an extent that attainment was impossible without controls on power plants in those upwind states.[16]

NO_x TRADING

As it became clear that NO_xRACT alone would not bring Eastern Seaboard states into attainment for ozone, states began to experiment with trading programs. The state of Massachusetts established a trading program for NO_x in October 1993 that included all sources of NO_x emissions and allowed power plants to create allowances by installing control technology, shifting fuels, or demand-side management programs.[17] Illinois adopted a slightly different program for the Chicago area that placed a declining cap on each NO_x emitter of greater than twenty-five tons per year and allowed emitters to purchase or sell allowances. But smoothly functioning markets for offsets did not always develop because purchases were sporadic and because some companies were unwilling to sell offset credits to potential competitors.

In April 1992, California's South Coast Air Quality Management District launched an ambitious NO_x emissions trading program for the Los Angeles nonattainment area called the Regional Clean Air Incentive Market (RECLAIM). It was by far the largest cap-and-trade program in the country. The agency assigned to every source in the program allowances equivalent to its current emissions, with the number of allowances to decrease by 8.4 percent per year for ten years. To the extent that a company's emissions reductions exceeded its shrinking allotment of allowances, it could sell them to others. For the first two years, the trading system functioned smoothly, and actual emissions were well below the allocations. As we shall see in Chapter 8, the program did not survive the California energy crisis of 2001.[18]

INTERSTATE AIR POLLUTION

At the outset of the Clinton administration, interstate transport of ozone and ozone precursors had become an intractable problem to which NO_x emissions from power plants were a major contributor. It was not as easy to adapt the cap-and-trade approach to NO_x emissions, because motor vehicles also emitted NO_x and because the absence of "low-nitrogen" coal made fuel switching a less plausible option. Downwind states worried that the problem would worsen as electricity deregulation subjected power companies to competitive pressures to shift output to the cheapest and dirtiest units in their fleets. EPA proceeded on two parallel tracks—its response to a petition by downwind states under section 126 of the statute, which prohibited a source in one state from significantly contributing to nonattainment in another state, and a "SIP call" under section 110 of the statute, which authorized EPA to demand modifications in SIPs that did not protect air quality in downwind states.[19]

In August 1997, eight northeastern states petitioned EPA to take immediate action under section 126 to prescribe stringent controls for specifically named power plants in upwind states. The Edison Electric Institute (EEI), individual power companies, and a coalition of midwestern states called the Midwest Ozone Group (MOG) objected to the petitions. Maintaining that their emissions contributed negligibly to nonattainment in the petitioning states, they accused the northeastern states of using the petition to divert attention from their own failures to reduce motor vehicle and power plant emissions. EPA welcomed the petitions as "political cover" for issuing a "SIP call" to upwind states demanding that they revise their inadequate SIPs to ensure that their sources did not contribute significantly to ozone nonattainment in downwind states.[20]

The proposed "NO_x SIP call" that EPA issued in October 1997 assigned NO_x emissions budgets to twenty-two states and the District of Columbia and demanded that they amend their SIPs by the end of 1998 to ensure that total emissions did not exceed their budgets as of 2005. The agency based the allocations on the assumption that power plants would install "cost-effective" selective catalytic reduction (SCR) control technology capable of reducing NO_x emissions by about 85 percent. The agency agreed to administer a multistate NO_x allowance-trading program to reduce the cost of compliance.[21]

The National Mining Association called the initiative part of a Clinton administration "war on coal." Upwind states claimed that the SIP call was an unwarranted attempt to substitute a national regime for the state-run programs that Congress envisioned. EEI complained that the plan was "unfair, expensive and misdirected," because it relied too heavily on emissions reductions from power plants and because an 85 percent emissions reduction was unachievable. A flurry of industry-sponsored studies raised the specter of massive blackouts as utility companies shuttered old power plants to comply. An association of several environmental groups and ten Northeast electric companies that wanted plants from other states to share the burden of reducing ozone concentrations called the Ozone Attainment Coalition (OAC) supported the SIP call. Pollution control vendors strongly supported the proposal, pointing out that newer NO_x controls were routinely reducing emissions by more than 85 percent.[22]

EPA published the final SIP call in September 1998 establishing a cap of 543,800 tpy for NO_x emissions starting with the 2003 ozone season and allocating that amount among the states subject to the rule. States would have to revise their SIPS to reduce NO_x emissions by 1.1 million tpy between 2003 and 2007. States could meet the assigned caps however they pleased, but one option was to join the interstate cap-and-trade regime that would be run by EPA. A pool of 200,000 total allowances was available for EPA to allocate to the states to provide a safety valve to any power company that demonstrated that service would be disrupted by complying with the standard. The agency predicted that compliance would cost less than $1,500 per ton of NO_x removed on average. Calculating that the regulations would reduce NO_x emissions in the region by 28 percent, the agency predicted that the reductions would attain the 1979 one-hour ozone NAAQS and go a long way toward meeting its recently promulgated eight-hour standard for ozone (discussed below).[23]

In March 2000, the D.C. Circuit upheld the NO_x SIP call in all but two relatively minor regards involving the states of Wisconsin, Missouri, and Georgia. The affected states scrambled to draft acceptable plans by the October 2001 deadline, but only a few were successful. Some states, like Pennsylvania, made good faith efforts to submit adequate plans. Some defiant states, like Michigan, North Carolina, and Ohio, submitted plans that they knew would not comply with the SIP call. Still others submitted no plan at all. In the meantime, several companies arranged for installation of NO_x reduction technologies by the May 2003 deadline at a cost of hundreds of millions of dollars. In June 1998, TVA pleasantly surprised EPA with an announcement that it would spend around \$500–600 million to install SCR technology in ten generating units by 2003.[24]

As the Clinton administration was departing in January 2001, EPA formally found that eleven states and the District of Columbia had failed to submit approvable SIPs by the deadline. The finding triggered the "sanctions clock," an eighteen-month period for the states to submit adequate plans after which federal highway funds would be cut off and two-for-one offsets would be required to build new sources in nonattainment areas. The finding also triggered EPA's obligation to write FIPs for the states within two years.[25]

REVISING THE NAAQS

The Clean Air Act imposed a duty on EPA to review and, if necessary, revise the NAAQS every five years. Increasing the stringency of NAAQS for any of the three pollutants of most relevance to power plants, SO_2, NO_x and PM, had enormous political and economic implications, because it initiated a lengthy process in which the states had to revise their implementation plans to attain the new standards within five years, and this in turn required the states to ratchet down allowable emissions from mobile and stationary sources. In addition, the standard-setting process had become so laden with analytical and interagency review requirements that such reviews had become massive undertakings.[26]

Ozone/Particulates NAAQS

Because the ozone and PM NAAQS raised many similar issues and could require overlapping emissions reduction technologies, the agency decided to combine the two projects into one massive rulemaking initiative. In March 1995, the staff circulated a draft of a staff paper recommending that the EPA administrator change the averaging time for the ozone standard from one hour to eight hours and change the level of the standard from 0.12 ppm to somewhere between 0.07 ppm and 0.09 ppm. A draft PM staff paper that EPA published the following November recommended that EPA focus on small particles of no larger than 2.5 microns in diameter ($PM_{2.5}$), establish an annual standard at 12.5–20.0 micrograms per cubic meter ($\mu g/m^3$), and establish a twenty-four-hour standard at 18–65 $\mu g/m^3$. The change to $PM_{2.5}$ would allow the agency to focus on the tiniest particles (including tiny droplets containing sulfates from power plants) that could penetrate the smallest airways in the lungs and cause serious adverse effects.[27]

The Ozone and PM$_{2.5}$ Proposals. In late November 1996, EPA proposed to replace the existing one-hour ozone standard of 0.12 ppm with an eight-hour standard of 0.08 ppm. Given the uncertainties in the scientific evidence, the agency also solicited comments on standards of 0.09 ppm, which was roughly equivalent to the existent standard, and 0.07 ppm, which was so stringent that it approached what EPA believed at the time to be the "natural" background levels of ozone in the ambient air in some regions. The agency predicted that the standard would cost from $600 million to $2.5 billion to implement and that it would reduce lung function deficits in children by 1.5 million cases annually. It further proposed to establish an annual PM$_{2.5}$ standard of 15 μg/m^3 and a twenty-four-hour standard of 50 μg/m^3. It predicted that the standards would prevent 40,000 premature deaths and eliminate 250,000 cases of respiratory attacks in children annually.[28]

Reactions to the NPRMs. The National Association of Manufacturers (NAM) assembled a multi-industry coalition called the Air Quality Standards Coalition (AQSC) to lobby against tightening the standards. Recognizing that "a very strong core" of the scientific community agreed on the underlying science, the group emphasized the cost of complying with stringent standards and the threat to grid reliability. It adopted a twofold strategy of attacking EPA's credibility and organizing grassroots opposition. More than a dozen congressional hearings in the Republican-controlled Congress gave critics an ample public forum to challenge EPA's conclusions. The coalition hired a stable of seventy-five lobbyists and secured more than 250 member signatures on letters to EPA and President Clinton opposing the standards. It also sponsored radio, television, and newspaper advertisements warning that the new standards would prevent farmers from plowing their fields on windy days and jeopardize suburbanites' daily commutes, neither of which was a realistic possibility. Drawing on a $5 million war chest, Citizens for a Sound Economy, an industry-funded activist group, launched a "grass-tops" outreach program focused on local politicians, businesspersons, and community leaders in nearly every congressional district.[29]

Environmental activists and public health groups complimented the agency for taking an important first step, but they pushed for even tighter standards. The groups coordinated a letter to President Clinton from more than 1,300 health professionals urging him to support EPA's standards. The American Lung Association and the Sierra Club sponsored radio advertisements touting the health benefits of the new standards in fourteen cities. And the Sierra Club staged an event in Washington, D.C., in which asthmatic children and their mothers delivered to the White House thousands of postcards from families supporting protective standards.[30]

White House Review. In addition to the industry-coordinated opposition, Administrator Browner was also dealing with growing resistance from within the administration. National Economic Council (NEC) head Gene Sperling publicly called for Browner to adopt a less stringent annual standard for $PM_{2.5}$ of 20 $\mu g/m^3$ and a less stringent twenty-four-hour standard of 65 $\mu g/m^3$. In early June 1997, representatives of the electric power industry met for more than ninety minutes with high-level officials from OMB, EPA, and several departments. Soon thereafter, OIRA took the position that EPA should allow five more exceedances per year to the ozone standard and set the $PM_{2.5}$ standard at a 20 percent higher level. Although White House Chief of Staff Erksine Bowles sided with OMB, Administrator Browner stubbornly stood her ground. Thinking ahead to the upcoming presidential election, Vice President Al Gore sided with Browner in the interagency disputes.[31]

The Final Standards. The primary ozone standard that EPA finalized in June 1997 adopted the eight-hour averaging period at a level of 0.08 ppm, but it allowed one additional exceedance per year. The PM standards were based on $PM_{2.5}$ and were set at an annual level of 15 $\mu g/m^3$ and a twenty-four-hour level of 65 $\mu g/m^3$. Thus, EPA prevailed on the annual standard and Sperling prevailed on the twenty-four-hour standard. The regulation also allowed states an additional five years to implement the $PM_{2.5}$ standard while they put new monitors into place. President Clinton expressed confidence that the affected industries could meet the standards and "grow the economy" as well. EPA suggested that the states focus on major emitters like power plants in implementing the new standards.[32]

Congressional Reactions. Within hours of EPA's announcement, a bipartisan group of members of Congress vowed to overturn the standards legislatively. Congressional committees conducted another round of hearings on the final rules in which members from both parties grilled EPA witnesses over the same issues that the prior hearings had addressed. Efforts to enact legislation overturning the standards, however, failed to attract enough Democratic members to ensure a successful override of a promised presidential veto.[33]

Judicial Review. In May 1999, a three-judge panel of the D.C. Circuit shocked the legal world by holding that the Clean Air Act was unconstitutional as interpreted by EPA in setting the levels for the ozone and particulates NAAQS. The two-judge majority held that EPA had "failed to state intelligibly how much [health protection] is too much." The court therefore remanded the standards to the agency to provide an "intelligible principle" for drawing the line or to persuade Congress to enact legislation specifying the levels for the standards as a matter of law.[34]

The Supreme Court, however, reversed the novel D.C. Circuit holding in an opinion written by Justice Antonin Scalia. The court first rejected the industry's argument that the Clean Air Act authorized EPA to consider costs in promulgating NAAQS. It read the statute to require the administrator "to identify the maximum airborne concentration of a pollutant that the public health can tolerate, decrease the concentration to provide an 'adequate' margin of safety, and set the standard at that level." The court then dismissed out of hand the lower court's suggestion that the Clean Air Act was unconstitutional. It found that the terms "requisite to protect public health" contained an "intelligible principle" to guide EPA's discretion. It agreed with the government that "requisite" meant "sufficient, but not more than necessary," which was "strikingly similar" to the statutory language employed in a drug enforcement statute that the Supreme Court had previously upheld against a similar challenge.[35] On remand, the D.C. Circuit rejected arguments by both industry and environmental groups that the agency had not provided an adequate scientific record and reasoned explanations for its choices of the forms and levels for the ozone and PM standards.[36]

SO₂ NAAQS

The primary standards for SO_2 of 0.14 ppm averaged over twenty-four hours and 0.03 ppm averaged annually had remained unchanged since EPA first promulgated them in 1971. During the Reagan administration, EPA had considered the risk to persons suffering from asthma of exposure to short-term (five-minute) bursts of SO_2 coming from nearby power plants due to malfunctions and dramatic variations in the sulfur content of high-sulfur coal. But the EPA staff concluded that a one-hour standard would be very difficult to implement and enforce and a twenty-four-hour standard capable of preventing peak one-hour exposures would have to be so stringent that it would make compliance very difficult. The agency therefore proposed to leave the existing standard in place. When EPA failed to finalize the proposal during the George H. W. Bush administration, environmental groups sued the agency to force a decision. Early in the Clinton administration, EPA settled the litigation by agreeing to publish a final determination by April 1, 1995.[37]

After receiving several extensions, EPA in May 1996 issued a final determination that a five-minute standard was unnecessary to protect asthmatics. Addressing the clinical studies involving mild to moderate asthmatics, the agency conceded that as many as 25 percent suffered effects "distinctly exceeding" the variation in lung function routinely experienced by asthmatics and that the severity of those effects was "likely to be of sufficient concern to cause disruption of ongoing activities, use of bronchodilator medication, and / or possible seeking of medical attention." Since they had no lasting impact and were reversible, however, there was disagreement in the scientific community as to whether such effects were in fact "adverse." Nevertheless, EPA's scientists agreed that *repeated* SO_2-induced asthma attacks presented a significant health effect. But Administrator Browner concluded that the likelihood that any particular asthmatic individual would be exposed to SO_2 bursts was "very low when viewed from a national perspective." Consequently, SO_2 bursts did not present the "type of ubiquitous public health problem for which establishing a NAAQS would be appropriate."[38]

The American Lung Association and the Environmental Defense Fund successfully challenged EPA's refusal to revise the standard in the D.C. Circuit. The court read the statute to require the primary NAAQS to protect both healthy populations and "sensitive individuals." Although the court declined to second-guess the agency on the scientific issues, it noted that the administrator herself had determined that "thousands of asthmatics can be expected to react atypically to SO_2 bursts each year." By the court's calculation, this meant that "as many as 41,500 asthmatics experience[d] atypical effects from repeated SO_2 bursts each year." Given these numbers, the court could not understand why SO_2 bursts did not amount to a public health problem of sufficient seriousness to require a new primary NAAQS. The court remanded the case for a better explanation.[39] As we shall see, the agency did not respond to the remand until early in the Obama administration.

VISIBILITY

Although EPA had promulgated regulations to ensure against plume blight during the Carter administration, it had put off for more than fifteen years addressing the role that existing sources played in creating "regional haze" in national parks and wilderness areas. A March 1990 report by the General Accounting Office concluded that fully 99 percent of the stationary sources that contributed to regional haze in national parks were not covered by the prevention of significant deterioration (PSD) permit system because they were built before 1977 or emitted fewer than 250 tpy.[40] Nevertheless, the project remained a low priority for most of the Clinton administration.

Vice President Al Gore finally unveiled regional haze regulations at the Shenandoah National Park on Earth Day in 1999. The new rules expanded visibility protection to all fifty states and gave states until 2004 to submit revisions to their SIPs implementing the regional haze requirements. The ultimate goal for state plans was to restore visibility in Class 1 areas to "natural" levels (the levels that preceded human activities) by 2064. A state could take longer if it demonstrated that the sixty-year deadline was "unreasonable." The regulations defined the critical term "reasonable progress" for any given year to be the difference between "baseline" visibility (the average level of visibility on the 20 percent most impaired days) and natural visibility divided by sixty. Visibility on the 20 percent least impaired days during the year preceding the implementation plan had to be maintained throughout the sixty years. In addition, states had to impose the "best available retrofit technology" (BART) on large power plants built between August 7, 1962, and August 7, 1977. A source was eligible for BART if it emitted pollutants within a geographic area from which pollutants could be transported to a protected area, even if the state could not show that the source itself contributed to visibility impairment.[41]

The Center for Energy and Economic Development, an umbrella group of electric power and coal companies, and the Sierra Club challenged the regulations in the D.C. Circuit Court of Appeals. In February 2002, the court upheld the agency's sixty-year approach to achieving "natural visibility" as the ultimate goal for visibility plans, but it cautioned EPA that natural visibility was only a goal, and not a mandate that EPA could enforce. It rejected EPA's area-wide approach to determining which sources were eligible for BART and in determining the content of BART for particular sources, holding that the statute clearly required states to make both determinations on a source-by-source basis.[42] This left EPA with the job of ensuring that the states complied with the statutory requirements in determining BART for individual sources that contributed to visibility impairment in national parks and wilderness areas. As we shall see, the agency took up this task in a serious way during the Obama administration.

Hazardous Air Pollutants

The 1990 amendments addressed emissions of hazardous air pollutants (HAPs) from power plants in a special provision that required the agency to prepare two reports on the health effects of human exposure to power plant HAPs and alternative control strategies by the end of 1993 and 1994. If EPA found that regulating power plants was "appropriate and necessary," it had to promulgate regulations for power plants establishing emissions limitations based on the "maximum achievable control technology" (MACT).[43]

The studies, which were published in December 1997 and February 1998, concluded that the risks posed by mercury emissions from coal-fired power plants were higher than previous estimates, but they were noncommittal about how EPA should address those risks. Power plant emissions were linked to methylmercury found in soil, water, and fish, and eating mercury-contaminated fish was the most common exposure route to that potent neurotoxin. Existing technologies, however, did a poor job of reducing mercury from power plant emissions. Coal washing removed about 21 percent of the mercury in coal, and scrubbers removed around 23 percent from emissions. Electrostatic precipitators removed around 15 percent, and baghouses removed another 8 percent. Both reports declined to say whether regulation of HAPs emissions from power plants was "appropriate and necessary."[44]

EPA's failure to make the required finding precipitated a lawsuit from NRDC that led to a settlement under which EPA agreed to decide whether regulation was "appropriate and necessary" by November 15, 1998. Fearing that regulating mercury emissions would be "the ultimate coal-killer," the coal and electric power industries geared up their lobbying apparatus to persuade EPA to answer that question in the negative. On this issue, the coal industry was united because low-sulfur coal could contain just as much mercury as high-sulfur coal. The natural gas industry, by contrast, welcomed stringent mercury regulation because its product was essentially mercury-free.[45]

At the coal industry's urging, Congress passed a rider forbidding EPA from making the determination until after a panel assembled by the National Academy of Sciences (NAS) completed still another study on the health effects of mercury. The NAS report that was published in July 2000 concluded that regulation of mercury emissions from power plants was "scientifically justifiable," because methylmercury in the environment posed an unacceptable risk to children born to women who consumed large amounts of fish during pregnancy. Environmental groups maintained that it was now time for EPA to make the "appropriate and necessary" finding and get on with the business of setting MACT standards for power plants.[46]

EPA published a finding that regulation of mercury emissions from power plants was "appropriate and necessary" to protect human health as a "midnight regulation" at the end of the Clinton administration. The finding triggered a provision in the settlement with NRDC requiring EPA to propose MACT-based standards for power plants by December 2003. But that would depend upon the incoming George W. Bush administration, which, as we shall see, had very different ideas about how to address mercury emissions.[47]

Climate Change

Vice President Al Gore believed that global warming was "the world's most important environmental threat." As a senator, he had drafted climate disruption legislation and chaired hearings on global warming, and his best-selling 1992 book *Earth in the Balance* featured the growing peril. In 1990, 35 percent of anthropogenic greenhouse gas (GHG) emissions were attributable to coal combustion, and they were growing at an average annual rate of 1.4 percent. The electric power industry was the largest consumer of fossil fuels, and it relied on coal for 55 percent of its generating capacity. Yet a survey of electric utility company executives found that nearly 60 percent of the respondents admitted that climate disruption did not affect their decisions. Indeed, a determined handful of think tanks and academic scientists, most of whom were funded in one way or another by the electric power and fossil fuel industries, spent millions of dollars in sophisticated public relations campaigns aimed at sowing doubt in the public mind by discrediting scientific reports suggesting that GHG emissions were contributing to global warming.[48]

The Btu/Carbon Tax Initiative

On February 17, 1993, President Clinton announced a four-year "blueprint" for stimulating the American economy that included a "Btu tax" on the energy content of nearly all fuels. The administration estimated that it would cost the average consumer about $100–150 per year once it was fully in place at the end of a three-year phase-in period. The bill also included funding for a federal energy assistance program to offset some of the adverse effects of the tax on low-income Americans. Administration officials stressed that this aspect of the bill was meant to reduce the federal deficit and move energy consumers away from fossil fuels and toward renewable sources of energy.[49]

Fossil fuel producers, the electric power industry, and large energy consumers were united in their opposition to any new energy taxes. They argued that any tax large enough to reduce GHG emissions would have undesirable effects on the economy, impede the ability of American manufacturers to compete in global markets, and have a disproportionate impact on low-income families. Environmental groups supported the proposal, applauding President Clinton for rejecting the "false choice" between the environment and jobs. Consumer groups also supported the bill, but warned public utility companies that if they wanted to pass the tax through to consumers, they would have to address their concerns that the companies owed consumers hundreds of millions of dollars in rebates stemming from unexpected declines in interest costs on capital projects.[50]

The industries that opposed the Btu tax launched a multimillion dollar lobbying campaign to persuade the relevant congressional committees to remove the tax from the stimulus bill. The campaign included several "media blitzes" in which they took their objections directly to the public. NAM assembled a 1,300-member umbrella group called the "American Energy Alliance" (AEA) with the single goal of killing the Btu tax. The Sierra Club responded with a far less resource-intensive appeal to its members to urge their representatives to support the tax.[51]

Faced with the real possibility that the Btu tax would die in the House Ways and Means Committee, President Clinton negotiated a compromise with Rep. Bill Brewster (D-OK) under which the tax would be imposed directly upon consumers and collected by utility companies without the need for approval by state public utility commissions (PUCs). This represented a defeat for consumer and environmental groups because it would ensure that the tax showed up on consumers' utility bills where it would be obvious why their bills were increasing. Bypassing state regulators would also deprive consumer groups of opportunities to force utility companies to rebate interest costs. The committee in mid-May voted along party lines in favor of the compromise in a bill that now included exemptions for farming interests and a few energy intensive industries. The House in late May 1993 narrowly approved (219–213) the stimulus bill with the Btu provision intact.[52]

The battle then shifted to the Senate Finance Committee, which was considerably less hospitable to the tax because a large proportion of its members came from energy-producing states. Senator David Boren (D-OK), one of the majority Democrats on the committee, became the target of an intensive campaign by AEA and Citizens for a Sound Economy to assemble taxpayer rallies, commission polls demonstrating strong opposition to the tax in Oklahoma, and generate letters and phone calls from constituents urging him to oppose the tax. Newspaper ads announcing that "BTU" stood for "big time unemployment" filled local newspapers. A direct mail "blitz" to more than nine thousand Oklahoma community leaders and a corresponding telemarketing campaign generated a huge number of pre-written letters and calls to Boren's offices. The effort paid off when Boren joined fellow Democratic senator Bennett Johnston of Louisiana and Republican Phil Gramm Texas at a rally on the National Mall to urge the attendees to help them kill the tax.[53]

After hearing from George Mitchell, Senate majority leader, and Daniel Patrick Moynihan, chairman of the Senate Finance Committee, that the tax could threaten the entire stimulus package, the president threw in the towel and urged Senate Democrats to come up with a revenue-enhancing mechanism to replace the Btu tax. In a thoughtful postmortem gesture, AEA spent some of its remaining cash on newspaper ads thanking the Democratic senators who had come to its aid.[54]

The failure to pass a Btu tax did not bode well for legislative efforts to address climate change during the next seven years. Having prevailed in a face-to-face confrontation with the new president on his administration's signature climate change initiative, energy industry lobbyists correctly predicted that climate change legislation would be a nonstarter for the foreseeable future.[55]

The Voluntary Climate Action Plan

Unwilling to abandon all efforts to address climate disruption, the Clinton administration introduced a "Climate Action Plan" in October 1993 that advocated planting lots of trees, building dams for hydroelectric power, and creating voluntary government / industry "partnerships" to reduce GHGs by conserving energy. The electric power industry supported many of the programs so long as they remained voluntary. Environmental activists scoffed at the suggestion that companies would voluntarily do much to reduce emissions. They correctly predicted that the plan would fall far short of President Clinton's announced goal of achieving 1990 GHG emissions levels by 2000.[56]

Gingrich Congress Holds Hearings

The climate skeptics gained a welcome forum when the Republican Party led by Speaker Newt Gingrich (R-GA) took control of the House of Representatives in January 1995. Rep. Dana Rohrabacher (R-CA), the chairperson of the House Science Subcommittee on Energy and the Environment, declared global warming theories to be "unproven at best and liberal claptrap at worst." Rohrabacher chaired a series of hearings in 1995 on global warming featuring several prominent global warming skeptics and many witnesses from conservative think tanks who excoriated EPA for wasting resources on climate issues. For its part, the Clinton administration remained in defensive mode and had no strategy for pressing Congress to enact climate change legislation.[57]

The General Counsel's GHG Memo

After the 1996 elections, the Clinton administration decided to take limited steps toward addressing climate change administratively. EPA's general counsel prepared a memorandum in April 1998 stating that the agency had authority under the Clean Air Act to regulate GHG emissions, because GHGs were clearly pollutants. If the administrator could reasonably anticipate that GHG emissions endangered public health or welfare, EPA could regulate them as criteria pollutants or as emissions from motor vehicles. On October 6, 1999, a different general counsel reiterated that position in joint hearings before two House subcommittees. He assured the committee members, however, that EPA had no plans to exercise that authority in the near future, and he acknowledged that setting an ambient air quality standard for CO_2 was not the optimal way of going about addressing global warming.[58]

Petition to Regulate GHG Emissions from Vehicles

On October 20, 1999, nineteen environmental groups formally petitioned EPA to promulgate regulations to reduce GHG emissions from automobiles and light duty trucks under section 202(a) of the Clean Air Act. The petition contended that EPA could "reasonably anticipate" that CO_2 emissions from cars and trucks would "endanger" public health and welfare because of the adverse effects that global warming would have on both human beings and the environment. The groups considered the petition the "beginning of a process" that would ultimately result in EPA and the states placing emissions limitation for GHGs on power plants. Twenty-six trade associations, including the Edison Electric Institute filed comments on the petition insisting that EPA lacked authority to regulate GHGs, because CO_2 was not a "pollutant" subject to regulation under the Clean Air Act.[59] As we shall see in Chapter 8, the struggle over the petition wound up in the Supreme Court of the United States.

COAL ASH

As power plants began to install electrostatic precipitators and scrubbers they had to deal with an enormous volume of coal combustion residuals (CCRs, or "coal ash"). The 1980 "Bevill Amendment" to the Resource Conservation and Recovery Act temporarily exempted CCRs from that statute's hazardous waste program until such time as EPA conducted a study of those wastes, presented a report to Congress, and conducted a rulemaking to consider whether to regulate CCRs as hazardous wastes. The purpose of the Bevill Amendment was to ensure that the nation's limited capacity to treat and dispose of hazardous waste was not inundated by high-volume wastes that were not viewed as especially hazardous by the agency. The report that EPA submitted to Congress in February 1988 tentatively concluded that coal ash contained trace amounts of several toxic heavy metals, but the levels were not high enough to warrant treatment as hazardous wastes.[60]

In August 1993, EPA issued a final regulatory determination that it would not regulate coal ash as a hazardous waste. The agency found that coal ash disposal had caused adverse health impacts at only a "very limited number of sites." The agency further noted that existing state programs regulating coal ash disposal were adequate and improving. Beyond a brief reference to a single spill from a waste disposal facility, the agency's explanation did not mention the health and environmental risks posed by breaches of coal ash retention ponds.[61]

This benign state of affairs lasted until March 1998, when a coalition of environmental groups petitioned EPA to reverse its previous decision and characterize coal as a hazardous waste. When it looked like EPA might respond favorably to the petition, Senator James Inhofe (R-OK) used a hearing on the budget for EPA's hazardous waste office to warn the agency that if it did list coal ash as a hazardous waste, "there will be an effect on [the office's] budget." EPA got the not-so-subtle message. In April 2000, it reaffirmed its 1993 determination that the CCRs should not be regulated as hazardous wastes. Instead, they would be subject to state regulation under Subtitle D of the statute, and EPA's role would be limited to providing unenforceable guidelines. The agency worried that regulating coal ash as a hazardous waste might "stigmatize" a material that could be put to beneficial use in a number of ways. Environmental groups called the decision a "horrendous mistake."[62]

ENFORCEMENT

Early in her tenure, EPA Administrator Browner decided to consolidate and strengthen the agency's Office of Enforcement and have it focus on whole industries, rather than on individual companies. Many of the enforcement actions that the reinvigorated office undertook were aimed at power plants. They continued almost without interruption throughout both terms of the Clinton administration, even when the agency was having to defend itself on an almost daily basis in the Republican 104th Congress. Even after unhappy Republicans cut EPA's enforcement budget by more than 20 percent, forcing it to cancel hundreds of inspections, the agency continued to file enforcement actions against power plants. Many of these actions resulted in settlements in which the plant owners agreed to take corrective action, pay large fines, spend money on various environmental mitigation projects, and, in later years, surrender unused allowances.[63]

By the mid-1990s, EPA enforcers were puzzling over economic data suggesting that there was very little new coal-fired plant construction, but coal consumption and electricity generation were increasing. In the new competitive environment, companies wanted to keep the old plants running because they produced electricity more cheaply than new plants with up-to-date environmental controls. The companies had no obligation to tell EPA about physical or operational changes so long as they did not result in a net emissions increase of more than 40 tpy, and they had a strong incentive to conclude that the changes came within the "routine maintenance, repair, or replacement" (RMRR) exemption to the statutory new source review requirement. EPA enforcers strongly suspected that some power plants were cheating.[64]

Preparing an enforcement case against a major utility company, however, was no easy matter. First, EPA had to obtain information about the projects that the target company had completed during the relevant time period. It took the agency staff a long time to wade through the tens of thousands of documents that resulted from its information demands. After that, the agency might send a team of investigators to the plant to conduct a physical inspection of the operations and examine on-site records. If those two steps yielded sufficient information to support a conclusion that the source had undertaken a major modification without undergoing new source review (NSR), the agency would typically present that information to the company and initiate settlement negotiations. If the negotiations stalled, EPA enforcement officials would prepare a file to refer to the Department of Justice (DOJ), which determined whether it merited further prosecution. If DOJ decided to proceed with the case, it typically initiated another round of settlement negotiations before filing the case in a federal district court. This cumbersome process could take years to complete.[65]

On November 3, 1999, DOJ filed enforcement actions against seven of the nation's largest power companies, alleging that they had engaged in major modifications without undergoing NSR at thirty-two power plants. It was the largest enforcement initiative EPA had ever launched. The complaints sought both monetary penalties of up to $27,000 per day and injunctive relief requiring companies to install controls equivalent to the best available control technology or the lowest available emissions rate, depending on the plant's location. At the same time, EPA filed an administrative enforcement action against the Tennessee Valley Authority, making similar allegations and seeking similar relief. In December 2000, DOJ sued Duke Energy Co., accusing it of violating the NSR requirements at all of its eight power plants.[66]

The electric power industry responded to the lawsuits in two forums. In the courts, the individual companies played a defensive game, responding to EPA's accusations with highly technical legal arguments designed to persuade the judges to dismiss the cases and highly complex engineering arguments that forced the government to spend millions of dollars on expert testimony. In the halls of the White House and Congress, industry lobbyists employed a number of strategies designed to bring a halt to the government's enforcement initiative before the courts could rule on the merits of the claims. The amount of money at stake was so high (billions of dollars in potential fines and pollution control costs) that the companies were not reluctant to invest substantial sums in both efforts.[67]

The industry raised a host of objections to EPA's enforcement actions in both forums. The most frequently raised argument was that EPA was changing the rules in the middle of the game. Industry representatives argued that companies had been undertaking major efficiency-enhancing and life-extending projects for twenty years without EPA voicing serious objections. Industry representatives maintained that companies had been asking EPA for guidance on how to meet the NSR requirements, but the "guidance never came." EPA responded that it had issued numerous letters, guidance documents, and applicability determinations that should have made the industry well aware of EPA's positions on the issues. If nothing else, the *Wisconsin Electric Power Company* case (discussed in Chapter 6) put the industry on notice that EPA was strictly interpreting the RMRR exemption.[68]

The industry further argued (somewhat inconsistently) that EPA had issued far too many complicated and sometimes inconsistent regulations, guidance documents, memos, and letters in response to questions that came up in state permitting proceedings.[69] It was therefore difficult for companies to know in advance whether any of the many repair and renovation projects they undertook at their facilities would subject the affected units to NSR. EPA responded that the complexity of the regulations and the proliferation of guidance documents were to a large degree attributable to the industry itself, which over time had lobbied EPA for "exemptions, special rulings and interpretations to address perceived or real inequities or policy goals." In any event, the companies that were targets of the enforcement actions were very large companies with a sophisticated trade association and access to high quality legal advice.[70]

The companies argued that the lawsuits were all a big mistake, because the changes at issue were needed to produce electricity in a safe, reliable, and efficient manner through proper maintenance activities. That they might also have extended the lives of the units was merely coincidental. Power plants operate under conditions of extreme temperature and pressure where parts wear out at different rates and must be replaced periodically to ensure against unanticipated operational failures. EPA responded that the RMRR exemption was applicable only to minor maintenance activities that a source repeatedly engaged in on a regular basis. The projects that its lawsuits targeted, by contrast, entailed very large capital expenditures and were undertaken infrequently, if ever at the targeted plants. They sometimes involved years of planning in company departments that were not responsible for maintenance. And they ultimately allowed plants to operate for longer hours at higher production rates. A broader interpretation of the RMRR regulations to include major life-extension projects would effectively nullify NSR and allowed grandfathered plants to go on emitting huge amounts of pollutants for many more years.[71]

Because many companies had demonstrably engaged in far more extensive modifications than those contemplated by the exemptions, companies began to settle cases on terms very favorable to the government. In the largest settlement ever reached under the Clean Air Act at the time, Cinergy Corporation agreed to spend around $1.4 billion on reducing emissions at ten of its coal-fired power plants, including the huge Beckjord and Gibson plants in Ohio and Indiana, by repowering generating units, installing control technology, or shutting units down. It also agreed to pay a fine of $8.5 million, undertake $21.5 million in "environmental projects" unrelated to reducing emissions, and retire 50,000 SO_2 allowances between 2001 and 2005. The government's case against Virginia Electric Power resulted in a similar settlement. Neither agreement, however, was finalized before the George W. Bush administration radically changed the direction of EPA's policies regarding new source review.[72]

ECONOMIC REGULATION

When President Clinton entered office, the electric power industry was in the process of completing a rocky transition from a highly regulated industry dominated by large public utility companies to a highly competitive industry in wholesale markets and some retail markets in which unregulated "merchant" generators played an increasingly important role. The remainder of the decade continued to be a period of great ferment in economic regulation of power plants by state PUCs and the Federal Energy Regulatory Commission (FERC) as agencies experimented with deregulation and with reducing the need for new power plants by managing demand and encouraging greater use of renewable resources.

Struggles in the States

The dominant issues at the state level were: (1) "restructuring," or the deregulation of retail markets for electricity by allowing consumers to choose their energy suppliers and permitting generators to compete outside the areas they had served under the traditional cost-of-service ratemaking regimes; (2) demand-side management (demand response) to reduce the need to produce more electricity; (3) "renewable portfolio standards," or mandates that regulated utility companies derive a certain percentage of their power from renewable sources, such as wind, solar, hydropower, and biomass; and (4) the relationship between restructuring and pollution controls.

Restructuring. A great deal of the pressure to deregulate retail markets came from large industrial and commercial consumers who chafed under the regulatory system that Samuel Insull had created at the turn of the century. Because their rates were much higher than the prices they would have paid to a single independent generator that had just completed a new gas-fired plant, they wanted to be able to contract directly with generators of their choosing. This required "retail access" legislation requiring local utility companies to "wheel" the power from a generator to the purchaser over its transmission lines without discrimination against that generator. Some states went a step further to require local utility companies to divest their generation assets entirely to ensure against discrimination against outside suppliers. As utility companies sold off their power plants to merchant generating companies, the term "restructured" captured the reality of a changed electrical power industry.[73]

Demand Response. By 1994, more than one thousand utility companies had active "demand response" (DR) programs in place. These programs typically had three elements. End-use efficiency programs encouraged customers to use energy more efficiently through pricing schemes, incentives to purchase energy-efficient appliances, and educational efforts. Load management programs focused on reducing demand during periods of peak demand by making electricity more expensive during those periods or providing interruptible service to industrial and commercial consumers. Power substitution programs encouraged consumers to switch from electrical appliances to gas appliances or to install solar heating systems. Sometimes the companies adopted the programs voluntarily, but more frequently they came in response to regulatory requirements or economic incentives adopted by state PUCs. Environmental groups routinely took the position in state PUC hearings that more effective DR programs would eliminate the need for new generating units at the same time that they reduced emissions of both CO_2 and conventional pollutants.[74]

Renewable Portfolio Standards. Many states included renewable portfolio standards (described in Chapter 2) in their restructuring legislation. These laws typically required utility companies to prepare periodic "integrated resource plans" that demonstrated that they were encouraging efficiency and demand-side management and gave priority to obtaining power from renewable sources. Different states adopted different goals for renewable sources, ranging from zero up to 25 percent of the company's output.[75]

Environmental Controls. The most important question related to environmental regulation before state public utility commissions in unrestructured states was the extent to which generators could recover the cost of pollution controls (plus a reasonable return on the investment) from their customers. Several state legislatures enacted legislation allowing environmental cost recovery, and several state PUCs interpreted their existing statutes to allow such recovery. Other states were less eager to allow cost recovery. The Kentucky Public Service Commission disallowed Kentucky Power's request to recover the cost of installing low-NO_x burners at its Big Sandy power plant in 1997, because the Clean Air Act did not strictly require them until 2000. After the 2000 deadline had passed, however, it allowed East Kentucky Power to recover the cost of installing more expensive SCR technology at its Spurlock and Cooper plants.[76]

Struggles before FERC

As local utility companies in states that ordered divestiture sold off generation assets and became purchasers of power on wholesale markets, FERC
regulation of interstate wholesale markets became much more important.
Even in states that did not deregulate, local utility companies grew increasingly dependent on purchases of power from wholesale markets. As we saw
in Chapter 2, FERC was also engaged in a major effort to deregulate those
markets. Its open access rule (Order No. 888) required all public utilities to
open their transmission facilities to all generators on terms and conditions
identical to their own use of those facilities, thereby allowing purchasers of
wholesale electricity to purchase power from sellers other than the entity that
owned the transmission lines. The fly in the ointment was the effect that open
access might have on the environment as local distributors purchased the
cheapest power they could find in a multistate market without regard to
whether the source was clean or dirty.[77]

FERC reluctantly agreed to prepare an environmental impact statement
(EIS) for the open access rulemaking, but it did not agree to allow the statement to affect its approach to restructuring. FERC chairperson Elizabeth
Moler stated in no uncertain terms that FERC was not in the environmental
protection business and that energy conservation, DR, and renewable energy
sources were not a high priority in this rulemaking. She insisted from the
outset that the open access rule would not require companies to mitigate the
adverse environmental effects of increased competition, however slight or severe they might be.[78]

The draft EIS that FERC circulated in February 1996 concluded that the
proposed open access rule would have a slightly beneficial effect on air pollution because it would encourage independent companies to build more low-
NO_x-emitting gas-fired power plants that would ultimately replace high-
NO_x-emitting coal-fired power plants. EPA agreed that the open access rule
would yield "positive or indifferent environmental results" in many parts of
the country. But it would have a seriously negative impact on air quality in
the Northeast because it would give owners of retired coal-fired plants an incentive to bring them back into service and owners of in-service coal-fired
plants to boost output to take advantage of the competitive advantage they
had by virtue of their failure to invest in modern pollution controls. EPA suggested measures that FERC could include in the regulation to require mid-

western utility companies to mitigate the order's adverse environmental effects.[79]

The final EIS that FERC published in mid-April 1996 dismissed the suggestion that it take any action to mitigate adverse environmental impacts because it lacked the legal authority to do so and because EPA and the states were in the best position to address that problem via state implementation plans. EPA then appealed FERC's rejection of its suggested mitigation measures to the White House Council on Environmental Quality (CEQ). The appeal precipitated an intense lobbying effort by midwestern utilities, coal producers, natural gas producers, and members of Congress from coal- and gas-producing states to persuade President Clinton to reject the appeal and by northeastern utilities, environmental groups, and members of Congress to persuade him to side with EPA.[80]

In June 1996, CEQ issued a policy statement in which it agreed with EPA that the open access rule could lead to increased emissions and suggested a three-step process for mitigating any increases. The electric power industry did not oppose the suggestion because it was confident that it would not affect operations at most plants. EPA and environmental groups were pleased that CEQ had recommended action to reduce the adverse environmental effects of deregulation. FERC's chairperson Moler was simply "glad [the dispute was] behind us."[81]

Effects of Restructuring

The combined effect of electric utility deregulation and rapid technological change on the electric power industry was profound. By the end of the 1990s, vertically integrated utility companies throughout the country were converting their assets into generating companies ("gencos"), transmission companies ("transcos") and distributing companies ("discos"). Some of the emerging merchant gencos were large independent companies, and some were unregulated affiliates of regulated utilities. Some were tiny enterprises that hoped to sell power directly to discos that needed power at a moment's notice and were willing to pay a high price for it on spot markets, and a few were giant energy conglomerates.[82]

Deregulation inspired a buying frenzy as merchant companies bought up the generating assets of utilities in deregulated states. General Public Utilities sold twenty-three power plants to Sithe Energies in a 1998 deal that made

Sithe the largest merchant power company in the country. The industry was also consolidating as companies raced to take advantage of economies of scale in a "frenzy" that harked back to the Insull years of the early twentieth century. The mergers of the period produced new holding companies with names like Cinergy, Exelon, and FirstEnergy. Merchant gencos, however, were far riskier from a financial perspective than regulated public utilities because they faced constant competition from other generators for the same customer base and could ill afford even a single breakdown that took a significant amount of downtime to repair. Large merchant companies therefore purchased traditional utility companies to bundle stable utility company income with riskier merchant facility income.[83]

The Merchant Power / Natural Gas Boom. Deregulation brought on a boom in power plant construction by unregulated merchant gencos as engineers developed "a new generation of highly efficient, low-emission gas-fired technologies." Combined-cycle turbines were less expensive to build and far more efficient than coal-fired boilers for handling baseloads, and smaller simple cycle turbines that gencos could bring online in a matter of minutes were perfect for peak load needs and reserve requirements in anticipation of possible breakdowns. Gas-fired plants accounted for about half of all new generating capacity brought online during the Clinton administration.[84]

The merchant generating sector, however, attracted serious opposition as companies sought state permits for new plants that needed to be near the large metropolitan areas and close to natural gas pipelines. "Peaker" plants were typically no larger than a residential garage, but they required seventy-foot-high stacks that were unsightly in the rural and suburban settings where they were typically located. The Chicago metropolitan area was an especially attractive target for the new merchant companies after the Illinois legislature's enactment of restructuring legislation and the decision by Commonwealth Edison (CommEd) to sell or shut down its coal-fired and nuclear power plants. By the end of 1999, merchant gencos had proposed or constructed sixty-four new gas-fired plants, all of which hoped to sell power to CommEd. Most proposals attracted opposition from local citizen groups who worried about noise, air pollution, and reduced property values. Reassurances from the Illinois Pollution Control Board (IPCB) that the plants would emit very little pollution did little to assuage their fears.[85]

The attempt by Indeck Energy Services, a relatively small merchant genco, to build a gas-fired peaker plant in McHenry County northwest of Chicago is representative of the struggles between merchant companies and local citizen groups during this period. Indeck had received all of the necessary permits from IPCB for a new three-hundred-megawatt peaker plant on twenty-two acres just south of the town of Woodstock when a group that called itself Create Awareness for a Responsible Environment (CARE) objected to its application for a conditional use permit from the McHenry County Board of Commissioners. CARE argued that Indeck should build the plant in a suitable site that was already zoned for industrial use, not on land that was identified as agricultural in the county's land use plan. Indeck responded that the plant would not exceed the Illinois Environmental Protection Agency's (IEPA) noise and emissions standards and that it would be shielded from sight with berms and trees. At a hearing attended by about one hundred local residents wearing buttons declaring "Stop the Stacks," one of the objectors stated that the group had "no problem with generating electricity," but it was concerned about "a heavy industrial application that has no benefit but a lot of detriments," a classic articulation of the "not in my backyard" argument. The county denied the permit.[86]

Not all attempts to build peaker plants in the Chicago area over the opposition of local groups failed. For example, Reliant Energy overcame serious opposition from local residents when the City of Aurora approved a permit for a massive ten-turbine peaker plant at a local industrial park that was already zoned for use as a "public utility" and contained the famous Fermi National Laboratory. And some fortunate peaker plants encountered minimal opposition. Between 1998 and 2003, thirty-five gas-fired power plants had been built or were under construction in Illinois. During the same time period, companies responsible for thirty-five projects that had received IEPA permits had either withdrawn them or allowed them to expire.[87]

The Rise of the Power Marketers. Deregulation at the state and federal levels created a niche for a powerful new economic actor—the "power marketer." A power marketer owned no generation, transmission, or distribution assets; it merely purchased electricity in bulk from generators of all shapes and sizes and resold it to discos that needed it to service retail consumers. Power marketers were subject to FERC's jurisdiction because they bought and sold power in interstate commerce, but they were exempt from price regulation. Power marketers like Enron Corporation, LG&E Energy, and Koch Industries rapidly transformed the nation's wholesale electricity markets. Within a year after FERC's Order No. 888, an electricity futures market dominated by a few power marketers was thriving on the New York Mercantile Exchange. FERC encouraged the market with a ruling that such derivative markets were not subject to its jurisdiction. With that, Wall Street banks and energy trading companies created markets in hedges, options, "spark spreads" (the difference between the price of natural gas and electricity), and other sophisticated derivatives.[88]

Effects on the Environment. The new gas-fired power plants were undeniably cleaner than the coal-fired plants that they frequently replaced. As the price differential between coal and gas diminished, merchant companies converted a few coal-fired plants to natural gas with corresponding environmental benefits. For example, Virginia Electric Power Company settled a new source review lawsuit by agreeing to convert its Possum Point Power Station from coal to natural gas. State restructuring legislation also allowed suppliers of "green" power from wind, solar, and other renewable technologies to compete for the business of consumers who were willing to pay more for electricity than the going rate for power fired by fossil fuels.[89]

It soon became clear, however, that deregulation's impact on the environment was not entirely salutary. An extensive analysis of the environmental impacts of restructuring found that existing coal-fired plants were finding a niche as intermediate peak-load plants while generators continued to use many existing coal-fired plants as baseload generators. A January 1998 study concluded that in the two years following deregulation of wholesale markets, several power companies in the Midwest operated their highest-emitting coal-fired power plants considerably longer than in the past with resulting increases in SO_2 and NO_x emissions. Some utility companies even fired up retired and mothballed coal-fired power plants, few of which had any significant environmental controls. When Cleveland Electric Illuminating announced in June 1996 that it would restart an unscrubbed coal-fired unit at its thirty-eight-year-old Lake Shore plant to sell electricity to a northeastern utility company, local environmental groups complained that it made little sense to restart an old unit where winds from Lake Erie would blow its emissions into some of Cleveland's poorest neighborhoods solely to provide low-cost air conditioning for New England homes.[90]

The increased competition also caused regulated public utility companies to cut back on DR programs that did not yield short-term profits. As they fell into disuse at the turn of the century, demand grew sharply. The new merchant companies were happy to meet that demand with gas-fired power plants.[91]

The California Energy Crisis

California was among the first states to deregulate retail markets. Its restructuring legislation provided for open access to consumers, but a utility company could not charge retail customers more than 90 percent of the pre-enactment price for four years.[92] The legislation also required the state's three big public utility companies—Pacific Gas and Electric, Southern California Edison, and San Diego Gas and Electric—to sell off at least half of their power plants, thereby turning them into distribution companies that remained responsible for delivering electricity to retail consumers at regulated rates and dependent on generating companies and wholesale marketers for their power supplies. The legislation established an entity called the Power Exchange to establish the price of electricity for the investor-owned utilities, and it required the utilities to sell all of their power from undivested plants to the exchange and to purchase all of their power from the exchange. The California Independent System Operator (CISO) remained responsible for ensuring reliability throughout the statewide grid. When electricity was scarce, CISO could purchase wholesale power from spot markets and charge the retail distributors for it. Interestingly, the municipal utilities, like the Los Angeles Department of Water and Power, that produced about 20 percent of California's electricity were not covered by the restructuring statute.

For the first two years of deregulation, things seemed to be going fairly smoothly as public utility companies rapidly divested their power plants, wholesale spot prices stayed low, and the new ISO coordinated purchases and sales to ensure system reliability. Adequate capacity seemed assured as the new owners of the divested power plants planned repowering projects to make them more efficient and less polluting and independent generating companies filed applications for new gas-fired power plants in strategic places.

In addition, heavy rains in the Northwest allowed surplus hydropower to flow into California.

Beneath the surface, however, things were not so rosy for the California market. Only 3 percent of retail customers representing 12 percent of electricity purchases selected an alternative provider. The existing public utility companies that retained 88 percent of the retail market were required by the statute to meet demand by purchasing power. Since they could not assume that their customers would not switch in the future, they had little incentive to enter into long-term contracts to purchase power at stable prices. Instead, they purchased power on the spot market from day to day. This left them in an unenviable position when spot markets became volatile. Finally, although electricity consumption in California had grown at only modest rates during the mid-1990s because of a recession and the state's strong demand response program, rapid population growth and the booming economy of the late 1990s caused electricity consumption to increase dramatically. Yet California saw no increase in generating capacity. This meant that power plants in California had to operate at close to full capacity, thereby creating reliability risks and providing opportunities for market manipulation by generators and marketers.

California's troubles began in early 2000 when low precipitation in the Pacific Northwest wiped out surplus hydropower. Then the price of natural gas increased dramatically from $2.50 per thousand Btu in early 2000 to $60 in December. Beginning in May 2000, wholesale electricity prices on the California spot electricity market also shot up. Because the California restructuring statute capped retail prices during that period, Pacific Gas and Electric (PG&E) and Southern California Edison (SCE) paid far more at spot market prices than they were able to collect from their customers at the frozen rates. During the summer of 2000, they filed emergency petitions with the California Public Utilities Commission asking it to approve plans to enter into long-term contracts with marketers and generating companies outside of the Power Exchange. But staff resistance to the requests prevented the commission from granting them until December.

In a controversial December 2000 decision that infuriated electricity sellers, FERC put into place "price mitigation" measures designed to eliminate price spikes in California during times of heavy use. The agency established monthly "proxy" market-clearing prices and required electricity suppliers to refund any payments above those prices. Sellers could, however, justify prices over the proxy price by showing they were cost justified. This allowed electricity marketers, which were often natural gas marketers as well, to manipulate gas prices to justify higher electricity prices. Reacting to the crisis, Governor Gray Davis established an emergency program under which companies could obtain state permits for small gas-fired peaker plants without comprehensive environmental reviews and could emit pollutants for several months in amounts that violated state and federal requirements. In addition, President Clinton invoked an emergency provision of the Defense Production Act of 1950 to order out-of-state natural gas suppliers to continue to provide gas to PG&E and SCE, even though it was not at all clear that they could pay for it. Critics questioned the legitimacy of invoking a war powers statute in peacetime to address a problem that had a very tenuous connection to the national defense.

The owners of the divested power plants blamed the crisis on pollution controls and urged the California Air Resources Board to relax NO_x emissions standards until the state had weathered the reliability storm. The Ventura County Air Pollution Control District cut a deal with Reliant Energy, a Houston-based generating company, allowing it to triple the operating hours and NO_x emissions of its Mandalay peaker plant without any offsetting emissions reductions. For its part, EPA struggled to resolve the tension between the president's order and its obligation to ensure that agreements like the Ventura County deal did not violate the Clean Air Act. As we shall see in Chapter 8, grid reliability won out.

THE 2000 ELECTIONS

The fact that Vice President Al Gore was the Democratic presidential nominee guaranteed that environmental issues would play a larger role in the 2000 campaign than they had in the previous two campaigns. Republican strategists did not hesitate to paint him as an environmental extremist who favored unnecessarily stringent controls on power plants that would result in higher utility bills. Gore's opponent, former Texas governor George W. Bush, projected the image of a pro-business conservative with a benign view on the environment. But his running mate, Dick Cheney, had no use for environmentalists, and he was disinclined to compromise with people who did not share his views about the need to dramatically expand the nation's energy resources. As governor, Bush had presided over a period during which

Houston surpassed Los Angeles as the city with the worst air quality in the nation. Yet Bush had also signed electricity deregulation legislation that included provisions for limiting emissions from previously grandfathered power plants. He also professed to believe that anthropogenic emissions were causing global warming. Environmental groups pointed out, however, that Governor Bush's appointees to the Texas Commission on Environmental Quality were invariably industry-friendly officials with little commitment to protecting the environment.[93]

CONCLUSIONS

President Clinton's signature legislative initiative, the Btu tax, failed to attract sufficient support from his own party to pass in a Democrat-controlled Congress. The death of the Btu tax revealed several aspects of the politics of climate change that continue to plague governmental efforts to reduce GHG emissions to this day. First, the battle revealed the political power of energy industry lobbyists and business-sponsored grassroots organizations to peel critical Democratic senators from fossil fuel–producing states away from a policy strongly favored by a popular Democratic president. Second, the public debates over the Btu tax demonstrated that polluting industries could successfully counter appeals to the shared public interest in a clean environment with appeals to the shared public interest in jobs and economic growth. Third, the experience demonstrated the political impotence of environmental groups who could not afford to hire public relations firms to conduct sophisticated advertising and grassroots organizing campaigns. Their disappointment with the Democratic Party, however, may have been reflected in a lackluster participation in the 1994 off-year elections, which were disastrous for Democrats.[94]

   With Congress under the control of a Republican leadership determined to relieve industry from the constraints of environmental controls, the Clinton administration joined environmental groups in successfully defending the foundational statutes. Any potential to enhance environmental protections, however, had to be accomplished through administrative action. In this arena, the administration achieved some notable successes, the most prominent of which was a tightening of the ozone and PM ambient air quality standards. EPA accomplished a major shift in focus by changing the averaging period for ozone to eight hours and the indicator pollutant for PM to $PM_{2.5}$. Both changes were driven by better scientific understandings of the relationships between those pollutants and human health. The affected industries spent millions of dollars to hold the line against any changes. But Administrator Browner prevailed in the struggles within the administration, and the final standards established levels in the middle of the range of those that were consistent with the scientific studies. At the same time, EPA declined to tighten the SO_2 NAAQS to protect asthmatics from short bursts from power plants, a dereliction that was overturned on judicial review and was not corrected until late in the Obama administration.

The Clinton administration's implementation of the acid rain program was surprisingly successful. Massive overcompliance by power plants with the Phase I SO_2 emissions requirements ensured that the cost of allowances did not remotely approach the industry-predicted $1,500 per allowance or even EPA's estimate of $600 per allowance. By the end of the decade, prices for SO_2 allowances were hovering at well below $150 apiece. By the turn of the century, power plants were emitting 25 percent fewer tons of SO_2 than in 1980 while generating 41 percent more electricity. And not a single unit had failed to comply with the requirement that each ton of actual SO_2 and NO_x emissions be matched with an allowance.[95]

Several innovations were largely responsible for this unexpected result. Power plant operators discovered that they could blend high- and low-sulfur coal to achieve an optimum mix of SO_2 reduction and electrical output. Operating in a competitive environment resulting from transportation deregulation, railroads dramatically lowered coal transportation costs. Improved efficiencies in scrubbing technology reduced both emissions and the cost of removing SO_2 from exhaust streams. The cap-and-trade regime allowed companies to modify and mix strategies to meet changing cost considerations with a minimum of delay and transaction costs. It also allowed companies to cover scrubber breakdowns with allowances, rather than building redundant scrubbers to deal with those relatively rare events. Nevertheless, the allowance market was not as robust as it might have been had some state public utility commissions and state legislatures not enacted policies aimed at protecting local coal companies.[96]

To the bottom-line question whether the acid rain program was having a beneficial impact on lakes in the Northeast, the answer was not at all clear. Despite a 30 percent reduction in acid deposition over the eastern United States, an "integrated assessment" of the quality of fifty-two northeastern lakes by the National Acid Precipitation Assessment Program found little or no improvement in lakes in the Adirondacks. Virtually every study concluded that even greater reductions in SO_2 emissions would be necessary to render Adirondack lakes capable of supporting robust aquatic ecosystems.[97]

EPA's efforts to use a cap-and-trade approach to address the persistent problem of interstate transport of ozone and ozone precursors were less successful. Its attempt to use section 126 stumbled in the D.C. Circuit. The NO_x SIP call survived judicial review, but encountered resistance from several critical states that failed to come up with approvable SIPs, thereby putting the onus on EPA to implement the trading program. As we shall see, the program proved incapable of achieving the new ozone NAAQS, and EPA was forced to issue another SIP call during the George W. Bush administration.

Perhaps the most consequential of the Clinton administration's environmental initiatives was the enforcement actions that it filed against eight major power companies and TVA to remedy their failure over many years to comply with the Clean Air Act's requirement that companies undertaking major modifications to existing power plants undergo new source review. By the end of Clinton's term, this massive enforcement effort had resulted in two preliminary settlements in which companies agreed to spend hundreds of millions of dollars on new pollution controls. The effort flagged during the next administration, but it yielded spectacular results in settlements accomplished during the Obama administration.

By the end of the 1990s, deregulation of wholesale electricity markets had taken place at the federal level, and deregulation of retail markets at the state level had taken on the air of inevitability as twenty-four states representing 55 percent of the nation's consumption had deregulated or were in the process of deregulating retail markets. Deregulation, the advent of more efficient natural gas combined-cycle technology, and lower natural gas prices spurred a natural gas boom. Between 1998 and 2004, U.S. generators greatly expanded their generating capacity by building natural gas units. This led to overcapacity, lower electricity prices, and a merchant generating fleet that was at great risk when natural gas prices rebounded at the end of the 1990s.[98]

The move toward gas brought about some environmental improvements. Natural gas plants produced minimal emissions of SO_2, PM, and mercury, and they yielded far fewer CO_2 emissions per unit of energy produced than coal. If uncontrolled, gas-fired plants did produce significant NO_x emissions. And their effects on global warming were far from benign because a great deal of methane (a far more potent greenhouse gas than CO_2) was released in drilling, processing, and transporting natural gas. In retrospect, FERC's rosy prediction that deregulation would have a minimal effect on air pollution turned out to be wrong.[99]

By the end of the Clinton administration, the electric power industry was complaining bitterly about the cumulative effect of all of EPA's current and forthcoming air pollution regulations. In May 1998, EEI published a study concluding that compliance with EPA's new NAAQS for ozone and PM, the NO_x SIP call, the acid rain program, and the regional haze regulations would cost the industry $21.8 billion and cause an 11 percent increase in wholesale electricity prices by 2010. In addition, some of the predictions that the coal industry would be hard hit by EPA regulations seemed to be coming true in 1999 as mining companies closed many of their high-sulfur coal mines in the Midwest and Appalachia. As we shall see, these complaints reached sympathetic ears in the George W. Bush administration.[100]

CHAPTER 8

Retrenchment

When the Supreme Court declared the Bush-Cheney ticket to be the winner of the extremely close 2000 election, the electric power and coal industries were pleased. They had invested heavily in Republican candidacies in the 2000 campaign, and the Republican Party now controlled the White House and both houses of Congress. The coal industry had been on the skids during the Clinton administration as power companies fueled nearly all new plants with natural gas and repowered existing plants from coal to gas to meet EPA requirements. Still, about 51 percent of electric power came from coal compared to 20 percent from nuclear, 15 percent from natural gas, and 11 percent from hydropower and other renewables. Coal-fired plants remained a prominent source of SO_2, NO_x, and PM emissions and the dominant source of CO_2 emissions. And the big dirties continued to dominate. Power plants built prior to 1972 were responsible for 59 percent of the SO_2 emissions, 47 percent of the NO_x emissions, and 42 percent of the CO_2 emissions from power plants fired by fossil fuels, while producing only 42 percent of the electricity.[1]

With President Bush strongly supporting coal as part of a "balanced" energy supply, natural gas prices rising, and California suffering from a full-blown energy crisis, coal was back in the driver's seat. Determined to take full advantage of the political opportunity that the 2000 elections presented, coal, electric power, and railroad companies created Americans for Balanced

May 2001 Vice President Cheney's Energy Task Force publishes its report

June 2001 FERC imposes price caps on wholesale power sales in eleven western states, and the California electricity crisis subsides

December 2001 EPA promulgates regulations governing cooling water intake structures at new power plants

February 14, 2002 President Bush unveils Clear Skies three-pollutant bill, which Congress fails to pass

February 2003 President Bush unveils FutureGen project for capturing and sequestering CO_2 emissions from power plants

August 23, 2003 EPA promulgates "safe harbor" rule to allow power plant modifications without new source review

January 2004 EPA promulgates regulations governing cooling water intake structures at existing power plants

March 2005 EPA promulgates Clean Air Interstate Rule to require upwind states to protect downwind air quality

March–May 2005 EPA withdraws "appropriate and necessary" finding and promulgates cap-and-trade regime for mercury emissions

March 2006 D.C. Circuit sets aside safe harbor rule

October 2007 EPA and American Electric Power settle new source review litigation affecting forty-six coal-fired units at twenty-five power plants

February 2008 DOE withdraws support for failed FutureGen project

February 2008 D.C. Circuit vacates "appropriate and necessary" withdrawal and sets aside cap-and-trade program for mercury emissions

Energy Choices with an advertising budget of several million dollars to stimulate grassroots support for industry-sponsored energy legislation. By the end of 2001, around two dozen new coal-fired plants were in the planning stages.[2]

President Bush appointed New Jersey governor Christine Todd Whitman, a pro-business Republican moderate, to be administrator of EPA. After serving for less than two years, Whitman announced her resignation on May 21, 2003. In August 2003, the president appointed Governor Michael O. Leavitt of Utah to replace Whitman. Leavitt occupied the position for only a brief time, however, before he became secretary of health and human services in December 2004. To replace Leavitt, the president appointed Stephen L. Johnson, a career EPA scientist. As the administrators came and went, however, the locus of power on important environmental issues was generally in the office of Vice President Dick Cheney.[3]

THE CALIFORNIA ELECTRICITY CRISIS

High on the list of the incoming administration's priorities was the ongoing electricity crisis in California. On January 17, 2001, the California Independent System Operator (CISO) implemented the first rolling blackouts that the state had seen since World War II. When Governor Gray Davis reached out to the incoming Bush administration for help, however, the president stated unequivocally that he would not support any overture to FERC to cap wholesale prices of electricity sold to California distributors. Instead, EPA agreed in early February to waive all penalties for power plants that violated emissions limitations during the crisis. The agency also allowed companies to run heavily emitting diesel backup generators full-time when needed to prevent blackouts, despite the fact that their emissions were undoubtedly exacerbating the ozone nonattainment problem in several California cities. And President Bush ordered all federal agencies to speed up reviews of new power plant applications in California.[4]

By mid-February, Pacific Gas and Electric (PG&E) and Southern California Edison (SCE) had defaulted on $12 billion in obligations and CISO had declared bankruptcy. In mid-February, Governor Davis issued a series of executive orders aimed at increasing generating capacity within California by 5,000 megawatts by July 2001 and by 20,000 megawatts by July 2004. Among other things, the governor ordered the California Air Resources Board to create an emissions offset bank with $100 million of state money to ensure that new plants could comply with EPA's nonattainment offset requirements. Very few companies, however, took advantage of the governor's offer.[5]

When sellers refused to provide more power to the insolvent companies, the legislature passed a bill empowering the California Department of Water Resources (CDWR) to enter into long-term contracts with generators and marketers and authorizing it to issue $10 billion in bonds to finance the purchases. Soon thereafter, CDWR was spending $45 million a day to purchase power for the utilities at prices 100–200 times the average price of the previous year. In March, the California Public Utilities Commission allowed the utilities to raise rates by 40–50 percent. For the first time since the legislature enacted deregulation legislation, consumers were encouraged to conserve electricity. They were not, however, pleased by the dramatically higher rates.[6]

To the Bush administration's chagrin, the crisis spread to surrounding states as California's purchases of wholesale power drove up electricity prices in those states. But when a group of western governors demanded regional price caps for wholesale electricity, the Bush administration instinctively dismissed the idea. Blaming environmental regulations for the crisis, Energy Secretary Spencer Abraham recommended building sixty-five new fossil fuel–fired plants every year for the next twenty years. Heeding that advice, several western states rushed to remove all regulatory restrictions on the construction of new power plants.[7]

A late March 2001 warm spell and the decision by a few generating companies to cease operations until PG&E and SCE paid their bills precipitated another wave of rolling blackouts. While blackouts were often minor inconveniences for residential consumers, sudden losses of electricity could do millions of dollars' worth of damage to a production line or an office computer system. FERC began to take small steps to address rampant price gouging by sellers of wholesale electricity during emergencies. But those measures did not prevent California from experiencing more price spikes and rolling blackouts. And they did not stop PG&E from filing for bankruptcy with debts of more than $9 billion in mid-April.[8]

As wholesale power prices escalated during the summer and fall of 2000, generators in the Los Angeles area ran their grandfathered plants at full capacity and purchased allowances to meet their emissions quotas. At the same time, the overall cap for NO_x emissions was shrinking by 8.4 percent per year under the state's NO_x trading program. As allowances grew scarce, prices skyrocketed, because most other sources had also planned to purchase allowances instead of installing pollution controls. The average price of a single allowance climbed to $45,000, more than ten times the average price in 1999, and at one point they were selling for $105,000. With the markets for both wholesale electricity and allowances in chaos, the district revised its regulations in May 2001 to allow the generating units that lacked NO_x emissions controls to continue to produce electricity at full capacity for three years without having to purchase allowances if they promised to install the best available retrofit technology during that time. To compensate in some way for the additional emissions, the plants would have to pay a fee of $7.50 per pound for their NO_x emissions, which the state used to reduce mobile source NO_x emissions.[9]

In June 2001, FERC finally imposed the long-requested price caps on wholesale purchases of electricity in eleven western states. With that action, wholesale and retail prices dropped sharply, and the crisis came to an end. In the meantime, ozone levels in California were creeping upward after more than a decade of continuous declines.[10]

The California crisis spawned dozens of lawsuits and a vigorous policy debate over who or what was to blame. Defenders of electricity deregulation blamed the California restructuring legislation, which they believed to be severely flawed in several regards. As we saw in Chapter 7, California capped the rates that consumers paid for electricity at the same time that it forced the state's public utility companies to divest their generation assets and to purchase wholesale electricity in a volatile spot market from generators and marketers through a power exchange that did not permit stable long-term supply contracts. There was no incentive for retail consumers to reduce consumption, and generators and marketers had every incentive to keep supplies low (and prices high) in times of high demand. Critics also cited California's stringent environmental regulations and siting requirements as a primary reason for the lack of new power plant construction to meet the needs of a rapidly growing economy that depended heavily on electricity.[11]

Other critics blamed merchant generators and marketers who saw dramatic increases in profits during the crisis. To avoid public criticism, Enron Corporation, a major marketer, kept up to $1.5 billion in trading profits off its books during the crisis. Enron CEO Kenneth Lay told a California regulator that "it doesn't matter what you crazy people in California do, because I got smart guys out there who can always figure out how to make money." Enron's smart guys knew the natural gas and electricity markets inside out, and they specialized in making money from the fact that purchases and deliveries of electricity had to occur simultaneously. Internal company documents later revealed that Enron's traders had come up with ten ways to game the California system, many of which were designed to allow Enron to collect high "congestion fees" during times of high electricity use or low supply. The "Ricochet" strategy, for example, was a laundering scheme under which Enron purchased large quantities of electricity from the California Power

Exchange at the capped price of $250 per megawatt-hour, thereby reducing supplies in California, transferred the electricity to a nearby state, and then sold it back to desperate California distribution utilities for as much as $1,200 per megawatt-hour. FERC later found that almost all of Enron's schemes violated the FERC-approved California rules and therefore violated FERC's tariff provisions. Ten other energy marketers were engaged in similar schemes.[12]

The companies that purchased the generating assets of the California public utility companies also profited handsomely by withholding generating power in three ways. First, they physically kept power off the system during critical periods by failing to run power plants when they were needed to meet heavy demand. Second, they offered power to the market at prices far above the cost of production in violation of the FERC-approved CISO protocol. Third, they refused to bid into the Power Exchange's day-ahead system, thereby forcing the CISO to purchase the power the same day when it was desperate for power to avoid blackouts. The fact that CISO could allow itself to be sucker-punched time and again could be explained by the large number of generating company officials on its board. The California Public Utilities Commission later concluded that if five merchant generators had made all of their available capacity available for purchase on the California market, the majority of blackouts would have been avoided.[13]

By any measure, FERC fell down on the job. Having rapidly turned wholesale electricity markets into competitive markets, it failed to put in place measures to implement its statutory duty to ensure that wholesale electricity was sold at just and reasonable rates. First, it approved California's seriously flawed deregulatory program, effectively passing the buck to California regulators. Second, when the crisis hit, FERC was at best ambivalent, blaming it on California's failure to permit more power plants. Third, when it did decide to take action, it lacked the expertise and the resources to understand and monitor the swift-moving, real-time markets in electricity, and it did not catch up to the fast-acting generators and traders until it finally imposed a cap on wholesale prices. Finally, FERC's failures were in part attributable to a deregulatory ideology that pervaded Washington, D.C., during the Clinton and George W. Bush administrations that placed a great deal of trust in markets that turned out, in retrospect, to be undeserved.[14]

THE CHENEY ENERGY TASK FORCE

The Bush administration seized on the California troubles to declare that the entire nation was undergoing an energy crisis that required decisive action from the federal government. Nine days after his inauguration, President Bush created the National Energy Policy Development Group, or the Cheney Energy Task Force, as it came be known, composed of high-level governmental officials and chaired by Vice President Dick Cheney, to recommend a national energy strategy to meet the nation's growing demand for low-cost energy. The Cheney Energy Task Force met with the Edison Electric Institute on fourteen occasions and with the National Mining Association nine times to solicit their views. In a private meeting with Vice President Cheney, electric power industry lobbyists Haley Barbour (former chairman of the Republican National Committee) and Marc Racicot (soon to become the chairman) pressed the industry's case for dropping the Clinton administration's new source review (NSR) enforcement lawsuits and changing the NSR rules to allow old power plants more flexibility in undertaking life extension projects. The vice president also met with environmental groups, but only after the task force's work was essentially done.[15]

Within the task force, a power struggle erupted over how it would approach environmental regulation of power plants. On one side, Energy Secretary Spencer Abraham and National Economic Council director Lawrence Lindsey argued for relaxing existing controls, shelving the NSR lawsuits, and foregoing any new controls on greenhouse gas (GHG) emissions. EPA Administrator Christine Todd Whitman resisted relaxing controls and strongly objected to pulling back the NSR enforcement actions. She wrote a memorandum to Vice President Cheney noting that "the real issue for industry is the enforcement cases" and warning that the administration would "pay a terrible political price if we undercut or walk away from" those cases. Whitman's overtures, however, were no match for a "parade of industry groups— including the CEOs of major electric utilities" that met with Cheney and other members of the task force to complain about the lawsuits.[16]

The task force's May 2001 report projected the energy industry's message that environmental regulations had unduly constrained the nation's ability to modernize its energy infrastructure. The nation was in an "energy crisis" brought on by an imbalance between supply and demand that, if not corrected, would "inevitably undermine our economy, our standard of living, and our national security." Environmental regulation had become "overly burdensome," and "regulatory hurdles, delays in issuing permits, and economic uncertainty" were "limiting investment in new facilities, making our energy markets more vulnerable to transmission bottlenecks, price spikes and supply disruptions." The report provided a ringing endorsement for coal, and it recommended a $2 billion research program devoted to "clean coal" technologies. A section on hydraulic fracturing (fracking), which involved forcing water mixed with special chemicals into gas- and oil-bearing rock with such force that the liquid literally fractured the rock and permitted the oil and gas to escape, called it "one of the fastest-growing sources of gas production," but the final report deleted information added by EPA concerning the potential of fracking to contaminate water wells and the possibility that EPA might decide to regulate the practice under the Safe Drinking Water Act.[17]

To keep up with growing energy demand, the report announced, the country needed hundreds of additional power plants, many more miles of natural gas pipelines and greatly expanded coal production. The task force recommended that the president direct EPA to review the existing NSR regulations and direct DOJ to review the existing lawsuits to "ensure that the enforcement actions [were] consistent with the Clean Air Act and its regulations." On the legislative front, the task force urged the president to direct EPA to propose "multi-pollutant legislation" to establish a "flexible, market-based program" to significantly reduce emissions of SO_2, NO_x, and mercury from electric power generators. Echoing a memorandum that Vice President Cheney received from Haley Barbour, the report concluded that regulatory intervention to reduce GHG emissions would be unwarranted.[18]

Within days after the report was issued, President Bush signed an executive order requiring agencies to "expedite their review of permits or take other actions as necessary to accelerate the completion of" energy-related projects. He also ordered the Justice Department to review the NSR cases with an eye toward dropping cases that no longer merited prosecution and EPA to review its NSR regulations with an eye toward relaxing them. And several agencies began the task of drafting a "three pollutant" bill to send to Congress.[19]

THE RUSH TO COAL

The publication of the task force report signaled the revival of the coal industry and coal-burning electric power companies. For the first time in many years, coal industry lobbyists found attentive listeners in the Environmental Protection Agency. Recognizing that the time was ripe for major legislative initiatives, coal-related companies dedicated $10 million to a public relations campaign to burnish the image of "buried sunshine." Coal-bearing states actively encouraged the return to coal. For example, the Illinois legislature in 2001 enacted the Coal Revival Program, which authorized $500 million in general obligation bonds to provide direct financial assistance to construct or expand power plants burning the state's high-sulfur coal. By the end of 2006, 159 new plants representing ninety-six gigawatts of capacity were in the works.[20]

Environmental groups actively opposed many of the new coal-fired power plants, but they lacked the resources to go to battle against all of the plants that came online during the Bush administration. Because all of the new units were equipped with expensive controls on SO_2, NO_x, and PM to meet the Clean Air Act's best available control technology (BACT) requirement for new sources in clean air areas, environmental groups could object to them only on the ground that they did nothing to control CO_2 emissions that contributed to global warming. But at that juncture, EPA had not yet concluded that CO_2 endangered public health and welfare. In addition, some new units replaced old highly emitting units and therefore represented an improvement over the status quo.[21]

Natural gas, by contrast, was in decline throughout most of the Bush administration. The boom in construction of gas-fired peaker plants during the Clinton administration resulted in a large capacity glut in many urban areas. Comparatively high natural gas prices combined with low electricity prices due to increased competition to discourage additional gas-fired capacity. Many of the merchant power companies that sprang up in the wake of price deregulation were highly leveraged and were finding it difficult to locate desperately needed credit. By 2005, Mirant, a merchant generator spin-off from the Southern Company, was in bankruptcy, and NRG, the merchant generator that purchased the generating assets of Houston Lighting and Power Company after Texas deregulated retail markets, had just emerged from bankruptcy.[22]

THE CLEAR SKIES BILL

Environmental legislation addressing power plant emissions was high on the agenda of the president and both houses of Congress. But there was disagreement within the electric power industry over whether the bill should address SO_2, NO_x, mercury, and CO_2 (a four-pollutant bill) or just SO_2, mercury, and NO_x (a three-pollutant bill). Several large companies that had already invested heavily in nuclear power and natural gas–fired power plants and had begun to implement energy conservation measures strongly supported a four-pollutant bill. Other companies were prepared to accept modest climate change legislation as the price to be paid for protection from the attempts by a growing number of states to regulate GHG emissions on their own. The Electric Power Research Institute, however, warned that the four-pollutant bill could raise electricity rates by 43 percent by 2020, destroy mining jobs, and reduce the nation's gross domestic product by almost 2 percent by 2010. EEI head Thomas Kuhn, who had roomed with President Bush at Yale, called White House aides to press the case against a four-pollutant approach.[23]

Within the Bush administration, it looked for a time like EPA Administrator Whitman and Treasury Secretary Paul O'Neill would persuade the president to back a bill that included fairly tight caps on all four pollutants. DOE, however, argued that EPA's proposed caps for the proposed cap-and-trade program were far too stringent because they reflected flawed assumptions in EPA's economic models. After the September 11 attacks, DOE emphasized the importance of maintaining the capacity of existing coal-fired plants to ensure grid reliability in the event of another attack. EPA wanted to leave the existing NSR and regional haze programs intact, but DOE insisted on eliminating those programs as a quid pro quo for moving to a cap-and-trade regime. The White House backed DOE in nearly all of the disputes.[24]

On February 14, 2002, President Bush presented the administration's "Clear Skies" initiative to address emissions of SO_2, NO_x, and mercury from new and existing power plants. At the heart of the initiative was a "cap-and-trade" regime for power plants that reduced SO_2 emissions from the current 15.5 million tpy to 4.5 million tpy in 2010 and 3 million tpy in 2018. Emissions for NO_x were reduced from 5 million tpy to 2.1 million tpy in 2010 and 1.7 million tpy in 2018. The mercury cap would decrease from 48 tpy to 26 tpy in

2010 and 15 tpy in 2018. The bill gave EPA the authority to "readjust" the 2018 targets in light of "new scientific technology and cost information." The cap-and-trade program replaced the four programs that were currently addressing SO_2, NO_x, and mercury emissions insofar as they applied to power plants. At the same time, the administration proposed a new plan for slowing the increase in GHG emissions that relied on incentives for voluntary CO_2 reductions, tax breaks, and further research.[25]

Prospects for an administration bill suffered a serious setback in May 2001 when Senator James Jeffords, a moderate Republican from Vermont, left the party to become an independent and caucus with the Democrats. A primary reason for the move was Jeffords's growing discomfort with the position of the party on environmental issues. Suddenly the Democratic leadership controlled the Senate's agenda, and Jeffords assumed the chairmanship of the Senate Environment and Public Works Committee. Jeffords then spearheaded an effort to pass a stringent four-pollutant bill that established an ambitious cap-and-trade program for CO_2 emissions from power plants.[26]

After the 2002 elections put the Republican Party back in control of both houses of Congress, prospects for the Bush administration's Clear Skies initiative brightened. The new chairman of the Senate Committee on Environment and Public Works was James Inhofe (R-OK), the Senate's most vocal climate change denier. The electric power and coal industries offered lukewarm support for the administration bill, but stopped short of endorsing it. Environmental groups, who adamantly opposed the administration's bill, quietly supported efforts by the moderate Republican senators Tom Carper of Delaware and John McCain of Arizona to enact competing bills that included caps on GHGs. The McCain bill was soundly defeated in the Senate, and the Bush administration remained implacably opposed to adding CO_2 caps to its bill to attract the support of moderate Democrats. In the end, Congress adjourned without enacting any Clean Air Act amendments.[27]

MODIFYING NEW SOURCE REVIEW

As EPA prepared the ninety-day report required by the president's executive order, it solicited comments in four public hearings and more than one hundred meetings with various groups. The report concluded that the NSR program had "impeded or resulted in the cancellation of projects which would maintain and improve reliability, efficiency and safety of existing energy capacity." This was exactly what the industry had been arguing ever since EPA launched the 1999 NSR enforcement initiative. Revising the regulations would eliminate the threat of future enforcement actions, but it would also undermine the legal and policy foundation for the ongoing enforcement initiative. Nevertheless, the agency leadership decided to press ahead with changes to the NSR regulations aimed at providing companies greater flexibility to modify their plants without having to undergo NSR.[28]

An intense internal struggle broke out between Assistant Administrator Jeffrey Holmstead and his upper level staff in the Office of Air and Radiation (OAR) and the lawyers in the Office of General Counsel (OGC) and the Office of Enforcement (OE). Having served as a lawyer / lobbyist for the electric power industry, Holmstead was quite familiar with the legal issues. The most contentious issue was an OAR-proposed amendment to the routine maintenance, repair, and replacement (RMRR) regulation to provide a "safe harbor" for any physical or operational change that fell below a specified cost threshold, even if it greatly increased emissions. Holmstead insisted on a threshold of 20 percent of the replacement cost for the entire electric generating unit for the safe harbor. OE and OGC predicted that such a safe harbor would render NSR inapplicable to the vast majority of future plant modifications. It would also limit DOJ's selection of remedies in the pending lawsuits, because judges were unlikely to require a source to install pollution controls to remedy violations of regulations that the safe harbor superseded. As it became clear that OGC and OE were not going to prevail, three mid-level career OE officials who had been directing the NSR litigation resigned in protest. In his resignation letter, Eric Schaeffer wrote that the agency was "about to snatch defeat from the jaws of victory" in the litigation. Schaeffer went on to create the Environmental Integrity Project, which became a vigorous participant in the Sierra Club's "Beyond Coal" campaign against coal-fired power plants (discussed in Chapter 10).[29]

In late December 2002, EPA proposed amendments to the RMRR regulations that provided a safe harbor for any modification to an existing source that cost less than a prescribed percentage of the replacement cost of the affected unit. The proposal asked for public comment on what the percentage should be, but it alluded to data from the Internal Revenue Service that could support a threshold as high as 20 percent of replacement cost.[30]

The proposal ran into a storm cloud of opposition from environmental groups, state air control agencies, and attorneys general from downwind states who argued that thousands of modifications that were clearly covered by the statute would be exempted and grandfathered facilities would therefore escape new source review in perpetuity. NRDC suggested that "a source owner, staffed with reasonably sentient accountants, could reconstruct an entire process unit through a series of five projects carried out sequentially over a relatively short period of time without triggering NSR." One especially troubling aspect of the proposal for state air control agencies was its willingness to allow sources to make their own determinations of whether a project came within the exemption with no oversight from the permitting authority.[31]

The electric power industry supported the proposal as a "critical first step" toward NSR reform. They criticized the agency, however, for not making the rule immediately effective so as to cut off future lawsuits from DOJ, state attorneys general, and environmental groups. Breaking with the rest of the industry, Calpine Corporation, a California merchant power company that relied heavily on natural gas and renewable resources, opposed the proposal as an "economic subsidy" to grandfathered coal-fired power plants "in the form of indefinite regulatory relief from environmental compliance."[32]

The draft of the final rule that Assistant Administrator Holmstead circulated within the agency in May 2003 exempted from NSR any project that cost less than 20 percent of the unit's replacement value. When EPA enforcement officials saw the 20 percent safe harbor, they concluded that its real purpose was simply to eviscerate NSR. They pointed out that even a 5 percent safe harbor would have exempted all of the projects that the agency was challenging in the ongoing NSR enforcement litigation. Assistant Administrator Holmstead, however, had a powerful ally in Vice President Dick Cheney whose office put enormous pressure on Administrator Whitman to sign off on the 20 percent safe harbor. In the end, Whitman did not sign the rule, noting that the president had "a right to have an administrator who could defend it, and I just couldn't." Administrator Whitman resigned on May 21, 2003.[33]

As President Bush was boasting about his administration's strong environmental policies on a trip to the Pacific Northwest, the acting EPA administrator Marianne Horinko on August 23, 2003, issued the final safe harbor rule. The final regulation allowed life extension and equipment replacement projects that cost up to 20 percent of a unit's replacement cost per year even if the modifications resulted in increased emissions. The preamble to the rule echoed the industry position that "the expense and delay associated with NSR scrutiny" had caused many facilities "to forego needed and beneficial maintenance, repair, and replacement activities," including many projects that might have reduced emissions. The agency attributed little environmental harm to the changes, and in fact concluded that the expanded exemption "may well produce environmental improvements." At its Halloween party later that year, the OGC staff held a wake for the Clean Air Act featuring "wanted" posters seeking the apprehension of Jeff Holmstead for bringing about the untimely death of that statute.[34]

Three years later, the D.C. Circuit vacated the safe harbor rule, holding that it clearly departed from the plain meaning of the statutory definition of "modification," which included "any physical change" that resulted in an increase of emissions. The court noted that the term "physical change" included life extension and equipment replacement projects, and it held that Congress's use of the modifier "any" demonstrated that the term "physical change" included "any activity at a source that could be considered a physical change that increases emissions." The agency was empowered to exempt activities that resulted in de minimis amounts of emissions, but the emissions permitted by the 20 percent safe harbor were by no means trivial.[35] During the three years that the rule remained in effect, however, it had a powerful impact on new source review enforcement.

ENFORCING NEW SOURCE REVIEW

At the outset of the George W. Bush administration, grandfathered power plants were responsible for about 84 percent of the NO_x emissions from the electric power sector and more than 88 percent of the SO_2 emissions. On average, they emitted twice as much SO_2 and about 25 percent more NO_x than plants employing the "best available control technology" that would have to be installed when modifications triggered NSR. Only 200 of the country's 1,100 coal-fired units had undergone NSR. DOJ had sued thirty-four power plants, entered into one major settlement, and was on the verge of settling two others. Many of the original defendants were in serious settlement negotiations, and EPA was also negotiating with sixty facilities that had received notices of violation but had not yet been sued. EPA enforcement officials were busily investigating more than one hundred additional facilities for potential NSR violations.[36]

After President Bush issued his executive orders in response to the Cheney Task Force, however, both EPA and DOJ froze new NSR enforcement actions pending completion of DOJ's review of the ongoing litigation. Energy Secretary Spencer Abraham urged DOJ to drop at least some of the lawsuits, but EPA Administrator Whitman strongly supported moving ahead with the suits that had already been filed so as not to send a message to all EPA regulatees that they could violate the environmental laws with impunity. Not surprisingly, the companies in settlement negotiations were not nearly as eager to agree to expensive upgrades as long as DOJ was seriously reevaluating the bona fides of the original lawsuits.[37]

The ninety-day report that the Justice Department unveiled with little fanfare on January 16, 2002, concluded that EPA's interpretation of its NSR regulations and its position in the ongoing litigation were both "reasonable." Attorney General John Ashcroft told the press that DOJ would continue to pursue the litigation that had already been filed unless EPA changed either its enforcement policy or the regulations. For the next several months, DOJ pursued the lawsuits quite vigorously, and the EPA enforcement staff began to assemble new cases against violators of the existing rules. Resources at both EPA and the Justice Department, however, were spread very thin as the Bush administration pursued other priorities. Citing resource limitations, the Justice Department agreed to stays in cases against Georgia Power and Alabama Power that were extended for several years.[38]

In August 2003, the Justice Department secured its first NSR victory when a District of Columbia federal court held that Ohio Edison, a subsidiary of FirstEnergy, had violated the NSR regulations by spending $136.4 million on eleven upgrade projects at its huge W. H. Sammis plant in Jefferson County, Ohio, without undergoing NSR. The judge rejected the defendant's claim that the money was spent on routine maintenance, but he also criticized EPA for allowing the violations to continue for years without taking any enforcement action against the company. Rather than appeal, Ohio Edison entered into a consent decree in which it agreed to spend more than $1 billion on scrubbers and SCR technology at the Sammis plant and three other plants in Ohio and Pennsylvania, to pay an $8.5 million fine, and to spend $25 million on supplemental projects to improve the environment.[39]

In the ensuing years, DOJ brought several of the Clinton EPA-initiated cases to trial. Those trials often revealed internal documents demonstrating that the companies were consciously disregarding EPA guidance documents and making it clear that the projects at issue would trigger NSR.[40] DOJ won several of these cases and in the process set some important precedents.[41] In some cases, however, DOJ lawyers were unable to convince courts and juries that the companies had engaged in unlawful life extension activities. And in one case, the Seventh Circuit reversed a jury verdict for EPA.[42]

In September 2003, however, DOJ announced that it would no longer prosecute violations if they came within EPA's proposed safe harbor rule. Then, in November 2003, EPA's assistant administrator for enforcement instructed the EPA staff to refrain from bringing new enforcement actions against companies for modifications that violated existing regulations but came within the safe harbor exemption. The agency staff ceased work on forty-seven cases for which it had already issued notices of violation, and it dropped ongoing investigations of potential violations at seventy additional power plants. The agency persisted in its new enforcement policy even after the D.C. Circuit in late December 2003 stayed the safe harbor rule. This was, of course, precisely the outcome that the coal-burning power companies had desired when they vigorously lobbied the agency to amend the NSR rules. Outraged environmental groups complained that the agency was effectively giving scofflaws a get-out-of-jail-free pass for ongoing violations of laws that remained in effect. The environmental groups and state attorneys general who were also parties to the lawsuits vowed to pursue the existing cases to completion, with or without EPA.[43]

The criticism from environmental groups had an impact on the newly arrived EPA administrator Michael Leavitt. In late January 2004, he announced that the federal government would aggressively bring new actions to enforce the existing NSR regulations until the D.C. Circuit resolved the legal challenges to the safe harbor rule. Shortly thereafter, the Justice Department filed the Bush administration's first new NSR enforcement case against the East Kentucky Power Cooperative, accusing it of modifying three coal-burning units at its Spurlock and Dale plants in the late 1990s without undergoing NSR. The company later entered into a settlement with DOJ in which it agreed to spend $650 million on pollution control equipment and pay a $750,000 fine. In July 2004, EPA referred fourteen new cases to the Justice Department for prosecution. The number of notices of violation that the agency issued for avoiding NSR also picked up dramatically, and some of the notices involved conduct that would have come within the safe harbor rule.[44]

After Leavitt left the agency in December 2004 to become the secretary of health and human services, however, NSR prosecutions came to a screeching halt. In November 2005, Deputy Administrator Marcus Peacock ordered the enforcement office to stop working on potential new cases that involved projects that would have come within the safe harbor rule. The pipeline of cases slowed to a trickle. Even after the D.C. Circuit set aside the safe harbor rule in March 2006, the Peacock order remained in effect until well into the election year of 2008.[45]

EPA officials who were directly involved in the NSR enforcement initiative from its inception were convinced that the rule changes had a devastating impact on the settlement negotiations. Two large electric power companies, Dominion and Cinergy, walked away from settlement negations in which they had already agreed in principle to spend a total of about $1.9 billion on additional pollution controls. Dominion later agreed to a settlement with EPA and five states that was virtually identical to the earlier agreement in principle. But Cinergy never returned to the negotiating table, preferring instead to litigate every conceivable issue in a war of attrition that lasted for more than a decade. With the demise of the safe harbor rule, settlements picked up. Over the course of the Bush administration, EPA entered into settlements with fifteen electric power companies in which they agreed to pay $57,450,000 in fines, spend $224,450,000 on supplemental environmental projects, and spend just over $11 billion to install pollution control equipment or repower coal-fired plants with natural gas. The largest settlement was an agreement on October 9, 2007, with American Electric Power in which the company agreed to pay a fine of $15 million, support supplemental environmental projects worth $60 million, and spend $4.6 billion to clean up forty-six units at twenty-five power plants. The predicted SO_2 emissions reductions alone exceeded the total SO_2 emissions of forty-five states.[46]

HAZARDOUS AIR POLLUTANTS

As its sweeping Clear Skies legislative initiative encountered difficulties in Congress, the Bush administration moved along a parallel track to accomplish the same results through the regulatory process.[47] The strategy was readily apparent in the administration's evolving approach toward regulating mercury emissions from power plants. Power plants accounted for 50 percent of domestic mercury emissions, and 28 percent of U.S. watersheds were contaminated with mercury as a result of deposition from power plant emissions. And the Centers for Disease Control and Prevention reported that mercury was present in 10 percent of women of childbearing age at levels associated with reduced cognitive performance in offspring.

As we saw in Chapter 7, EPA published Administrator Browner's finding that regulating mercury emissions from power plants was "appropriate and necessary" as a "midnight regulation" in December 2000. The statute required EPA to promulgate national emissions standards for hazardous air pollutants (NESHAPs) for mercury emissions from power plants reflecting the maximum achievable control technology (MACT) within two years. But the incoming Bush administration faced several difficult implementation problems, the most intractable of which was the absence of commercially demonstrated technologies for removing mercury from power plant exhaust or for continuously monitoring the mercury concentrations in that exhaust. Another difficulty stemmed from the fact that the mercury content of coal varied considerably, making it difficult to choose fuel with mercury emissions in mind.[48]

In December 2003, EPA published two proposed rules, one to withdraw the Clinton administration's "appropriate and necessary" finding and the other to establish a cap-and-trade program for mercury under a novel interpretation of its power to promulgate new source performance standards. The latter proposal required coal-fired power plants to reduce mercury emissions from 48 tpy to 34 tpy by 2010 and down to 15 tpy in 2018, a total reduction of 70 percent. The 34 tpy interim goal was based on the mercury reduction that would result as a co-benefit from complying with the agency's pending "Clean Air Interstate Rule" (CAIR), discussed below. The proposal also allowed sources to "bank" credits by reducing emissions early in the process and drawing on them to extend the 2018 deadline.[49]

The radical shift in direction was largely the handiwork of Assistant Administrator Jeffrey Holmstead. He forced the proposal through the EPA review process over the strong objections of attorneys in the Office of General Counsel, who warned Holmstead that the statute did not authorize a cap-and-trade approach to controlling emissions of hazardous air pollutants. Holmstead apparently relied on his former colleagues at Latham & Watkins for legal support. The preamble to the proposed rule contained several paragraphs that tracked almost word-for-word language in memos that the law firm had submitted to the agency on behalf of the electric power industry.[50]

EPA received 680,000 comments on the proposal. EEI and the Electric Reliability Coordinating Council defended the cap-and-trade concept, noting that it would be far less expensive to implement than requiring each individual unit to install MACT. The coal industry's primary problem with the proposal was the 34 tpy interim cap, which it predicted would rule out most bituminous and lignite coal. A few eastern coal companies favored the stringent cap, because it was easier to remove mercury from some eastern coals than competing coals from the Powder River Basin. But the cap was strongly opposed by western coal producers and the railroads that hauled their coal to power plants in the Midwest and the East.[51]

Environmental activists called the proposal a cruel joke because it was so obvious that the statute did not authorize a cap-and-trade approach to controlling hazardous air pollutants. A coalition of electric power companies that were not heavily dependent on mercury-laden coal agreed that EPA lacked authority to establish a cap-and-trade program for mercury emissions. Environmental justice advocates warned that adverse effects of the cap-and-trade proposal would fall disproportionately on indigenous peoples and other minority groups for whom fish constituted a significant source of annual protein intake. Worse, giving large mercury emitters the choice of purchasing allowances instead of controlling emissions would create "hot spots" of high mercury concentrations, many of which would be located in minority neighborhoods in urban areas. Critics further warned that banking reductions could draw out the ultimate attainment date to well after 2025.[52]

On March 15, 2005, EPA promulgated a final regulation finding that regulating power plant mercury emissions was not "appropriate and necessary" after all and removing power plants from the list of sources for which it planned to issue NESHAPs. Two months later it promulgated a final rule establishing a cap-and-trade program for mercury emissions from power plants that established total emissions caps for individual states and suggested a voluntary cap-and-trade regime. The regulation established an interim cap of 38 tpy for 2010 (less stringent than the 34 tpy proposal) and a final cap of 15 tpy for 2018 (the same as the proposal). Assistant Administrator Holmstead conceded, however, that the 15 tpy cap might not actually be achieved until after 2020 because sources would be able to bank early emissions reductions.[53]

In an opinion that found EPA's logic comparable to that of the Queen of Hearts in Lewis Carroll's *Alice's Adventures in Wonderland,* the D.C. Circuit in February 2008 vacated the mercury rule. The court held that the agency violated the Clean Air Act when it delisted power plants without making the required finding that emissions from no source in the category exceeded a level that protected public health. It also rejected EPA's argument that it had the inherent authority to reverse its "appropriate and necessary" finding. The court therefore vacated the cap-and-trade program that EPA had established under section 111 of the statute because that section was inapplicable to sources that were listed under section 112.[54]

INTERSTATE AIR POLLUTION

When EPA began enforcing the NO_x SIP call program (discussed in Chapter 7) in May 2004, dozens of plants had already installed SCR technology, and companies were in the process of installing SCR technologies in around 190 additional units to meet the requirements of the SIP call and in some cases to meet their obligations under consent decrees in the ongoing NSR litigation. Some of these were huge structures like the twenty-story-tall unit at Ameren's Coffeen, Illinois, station that required 1,800 tons of structural steel and 1,100 tons of ductwork. Two major brokerage firms in June 2001 announced that they had brokered the first purchases and sales of NO_x allowances under the NO_x SIP call. EPA reported that NO_x emissions in the SIP-call states were 50 percent lower in 2004 than in 2000. And in 2006, only 4 out of 2,579 affected units were out of compliance.[55]

Despite this success, pressure mounted on EPA to withdraw the SIP call and come up with a less ambitious approach to achieving the national ambient air quality standards (NAAQS) for ozone along the Eastern Seaboard. To the great consternation of the electric power industry, Administrator Christine Todd Whitman announced that the rule would remain in effect. It was, however, becoming increasingly clear that more would be necessary. EPA had tightened ozone NAAQS in June 1997, and the NO_x SIP call would not be sufficient to bring the downwind states into attainment with the new standard. In addition, interstate transport of particulate matter and particulate precursors was causing many of the same areas to violate the 1977 NAAQS for fine particulate matter ($PM_{2.5}$). Finally, although sources in the acid rain program had decreased SO_2 emissions by more than 40 percent and NO_x emissions by almost 50 percent from levels at the outset of the program, acid deposition was still causing a great deal of harm to New York and New England lakes.[56]

In mid-December 2003, EPA proposed a new "Clean Air Interstate Rule" (CAIR) that combined the NO_x SIP call with the acid rain program. The proposed CAIR expanded the existing cap-and-trade program for power plants to twenty-nine states and the District of Columbia, reduced the cap for NO_x emissions by 1.4 million tpy by 2010 and by another 1.7 million tpy by 2015 (a 50 percent total reduction), and reduced SO_2 emissions by 3.5 million tpy from the acid rain cap by 2010 and by another 2.3 million tpy by 2015 (a 70 percent total reduction). To cushion the blow, EPA later proposed a "safe harbor" that would allow compliance with the CAIR to constitute compliance with the "best available retrofit technology" (BART) requirement of EPA's regional haze program. Together with the proposal issued the same week to control mercury emissions from power plants (discussed above), the proposal aimed to accomplish administratively the goals of the administration's Clear Skies bill, only three years sooner. EPA estimated that compliance with the rule would cost the electric power industry $3 billion in 2010 and $5.7 billion by 2015.[57]

The electric power industry generally supported the CAIR, but it warned that companies would struggle to meet the deadlines because of the amount of time it took to finance, plan, permit, and build emissions controls on hundreds of coal-fired generating units. It therefore demanded a "safety valve" that would ensure that the cost per ton of pollutant removed for any given unit would not exceed a predetermined amount. Environmental groups were harshly critical of the proposal, arguing that it would actually slow down emissions reductions already required under EPA's existing rules. They urged the agency to require greater emissions reductions and move up the deadlines to 2009 and 2012. They also opposed the proposed safe harbor for the BART requirement on the ground that it would result in less than one-third the visibility improvement of the BART rule. And they opposed the safety valve, arguing that it would only delay attainment of the ozone standard and the recovery of lakes in New York and New England.[58]

EPA promulgated the final CAIR in March 2005. The CAIR required twenty-eight upwind states and the District of Columbia to revise their SIPs to include control measures aimed at reducing emissions of SO_2 and NO_x in two phases. The NO_x cap would decline by 1.7 million tpy between 2005 and 2009 (more stringent than the proposal) and by another 0.2 million tpy by 2015 (much less stringent than the proposal). Emissions of SO_2 would decline by 5.8 million tpy from the acid rain cap between 2005 and 2010 (far more stringent than the proposal) and by another 3.3 million tpy by 2015 (less stringent than the proposal). The plan allocated a budget of emissions allowances to each of the upwind states pursuant to a complicated formula preferred by the owners of older coal-burning plants, relying on heat input. The rule also created a "cap-and-trade" program for both pollutants in which every power plant in the twenty-eight states would have to participate unless the state opted out of the program and promulgated an implementation plan that otherwise adequately controlled power plant emissions. Power plants in states opting into the cap-and-trade program could purchase or sell allowances in an open market, and they were exempt from the "reasonably available control technology" (RACT) requirement that was otherwise applicable to existing sources in nonattainment areas and from the BART requirement of the regional haze program. EPA estimated that the rule would cost $17 billion.[59]

While the CAIR was undergoing judicial review, many companies began to undertake the fuel conversions and install technologies to achieve the necessary emissions reductions. Some had already planned to invest in pollution controls to meet the terms of NSR consent decrees. And for others, the prospect of having to comply with the CAIR made it easier to reach settlements with the plaintiffs in NSR cases. Companies hoped that meeting CAIR requirements would have the "co-benefit" of reducing mercury sufficiently to comply with the upcoming mercury rule. Other companies undertook modest attempts to retrofit pollution reduction technology but planned to rely heavily on purchases of allowances.[60]

North Carolina challenged the CAIR on the ground that its cap-and-trade programs did not adequately assure that an upwind state would in fact prohibit all emissions within that state that significantly contributed to downwind nonattainment. In July 2008, the D.C. Circuit agreed with North Carolina. The court pointed out that although the CAIR allocated initial emissions budgets to upwind states, sources had the option of purchasing allowances from sources in other states, thereby allowing states to "emit more or less pollution than their caps permit[ted]." Furthermore, since EPA based each state's budget on whether modeled emissions reductions were "highly cost effective" at the region-wide level, "it never measured the 'significant contribution' from sources within an individual state to downwind nonattainment areas." This was clearly inconsistent with the statutory prohibition on emissions from individual sources that contributed significantly to nonattainment in other states.[61]

The court's decision had an immediate and devastating impact on the markets for SO_2 and NO_x allowances. NO_x allowances, which were trading for almost $5,000 apiece prior to the decision, fell to $1,000. The spot SO_2 allowance market was still alive because the acid rain program remained in effect, but prices for allowances dropped precipitously to record lows of $90 per allowance before rebounding to $150 per allowance by the middle of August. The markets remained in a state of turmoil through September 2008 as Wall Street investors who had provided liquidity through their purchases and sales sold off their allowances. On December 23, 2008, the D.C. Circuit stayed its mandate to allow the CAIR to remain in effect until EPA promulgated a new regulation to replace it. The court's action had an immediate effect on the allowance markets where prices for both NO_x and SO_2 allowances increased dramatically. The states covered by the CAIR quickly swung into implementation mode as the start date for Phase I of the rule, January 1, 2009, was only a week away. Representatives of environmental groups hoped that the stay would provide an opportunity for the incoming Obama administration to craft a stronger regulation governing interstate air pollution.[62]

REVISING THE NAAQS

EPA undertook major revisions to the ozone and PM national ambient air quality standards (NAAQS) during the George W. Bush administration. Since both exercises involved similar struggles, we will focus exclusively on the ozone standard.[63] Although the Clinton administration's eight-hour standard should have gone into effect in 1998, the old 1979 one-hour standard for ozone remained in effect while the Supreme Court decided the fate of the new standard. In the meantime, the agency settled a lawsuit brought by environmental groups by agreeing to publish a final rule revising the standard or leaving it in place by March 12, 2008.[64]

The Criteria Document

The criteria document that EPA published in February 2006 summarized and assessed the relevant scientific information that had become available since

1996. It concluded that controlled human (clinical) studies provided "clear evidence of causality" for associations between acute ozone exposure and lung function decrement observed in numerous recent epidemiological studies. Two very recent clinical studies, conducted by Dr. William Adams of the University of California at Davis and sponsored by the American Petroleum Institute, found significant lung function decrement in human subjects exposed to 0.08 ppm (the level of the current NAAQS) and some "adverse lung function effects" in individuals exposed to 0.06 ppm, but no statistically significant differences between exposed and unexposed subjects at 0.04 and 0.06 ppm.

A number of epidemiological studies published since 1996, like Michelle L. Bell's 2004 study of ninety-five communities, revealed "robust" associations between daily ozone exposure and increased mortality risk. In reaching that conclusion, EPA scientists defied a demand by the White House Office of Information and Regulatory Affairs (OIRA) that the agency first put the question to a National Academy of Sciences panel. The agency commissioned an NAS report, but it did not delay the standard-setting process. The report, which recommended that EPA include ozone-related mortality in its analysis of the health effects of ozone, did not arrive until after EPA had completed the revision. The criteria document concluded that "no clear conclusion can now be reached regarding possible threshold levels" for ozone-induced health effects. If a threshold existed, it was "likely near the lower limit of ambient [ozone] concentrations in the United States."

Addressing the secondary standard, the criteria document concluded that many studies published since 1996 provided "strong evidence" that ozone concentrations under the existing standard adversely affected annual, perennial, and woody plants, decreased crop yields, and impaired the aesthetic quality of many native plants and trees.[65]

The Staff Paper

A staff paper prepared by the Office of Air Quality Planning and Standards (OAQPS) in January 2007 noted that the agency's Clean Air Scientific Advisory Committee (CASAC) had unanimously concluded that the current 0.080 standard was definitely not protective of human health and recommended a standard in the 0.060 to 0.070 ppm range. It also noted that the World Health Organization had recommended a standard in the range of 0.051 to 0.061 ppm. The staff paper found that Professor Adams had employed a statistical approach designed to avoid a false conclusion that there was an association between ozone exposure and lung function decrement. The staff's reanalysis of the Adams data using a different statistical technique showed a small but statistically significant decrease in lung function decrement at 0.06 ppm. The staff paper also noted that the published version of the 2002 Adams study reported that "some sensitive subjects experience notable effects at 0.06 ppm." The staff further concluded that the available evidence supported both increasing the stringency of the secondary standard and changing its form to reflect cumulative, seasonal exposures. The paper noted that a cumulative, seasonal secondary standard in the range of 7 ppm-hours (a measure of total exposure to ozone over time) to 21 ppm-hours would conform to the recommendations of CASAC and a recent report published by the National Academies of Sciences.[66]

The Proposal

In June 2007, EPA proposed to lower the primary standard from 0.084 ppm to between 0.070 and 0.075 ppm, using the same eight-hour averaging period. In discussing lung function decrement, the preamble to the proposal highlighted the Adams studies and the staff's reanalysis of those studies. Relying on the recent Bell study and many earlier epidemiological studies, the preamble found the association between ozone exposure and mortality to be strong, robust, and consistent. The preamble recognized that the existing studies neither supported nor refuted the existence of a threshold. It was therefore important to "balance concerns about the potential for health effects and their severity with the increasing uncertainty associated with our understanding of the likelihood of such effects at lower ozone levels." Finally, the preamble suggested two options for revising the secondary standard. One option, recommended by CASAC and the staff, was to adopt a cumulative, seasonal standard set at an annual level "in the range of 7 to 21 ppm-hours." The other option was to make the secondary standard identical to the proposed eight-hour primary standard.[67]

Initial Responses to the Proposal

Public health and environmental groups chastised EPA for failing to follow CASAC's advice to set the primary standard at the low end of the 0.06 to

0.07 ppm range. The environmental groups argued against retaining the current eight-hour form for the secondary standard. They noted that damage to vegetation had persisted in areas that attained the current eight-hour standard, and it would undoubtedly persist if that standard were lowered by a mere 0.10 ppm. They urged the agency to adopt a cumulative, seasonal form and to set the standard at the low end of the 7 to 21 ppm-hour range.[68] Affected industry groups and a handful of states argued that the existing scientific studies did not support a more stringent primary or secondary standard. Highlighting the uncertainty and variability in the existing scientific studies, they argued that attaining a more stringent standard would cost tens of billions of dollars while yielding very little demonstrable public benefit.[69]

The White House Intervenes

As the deadline of March 12, 2008, approached, a broad coalition of trade associations launched a massive lobbying campaign, spearheaded by the former Republican Party chairman Haley Barbour, to persuade the Bush administration to leave the current standard in place. After several meetings with the industry groups, OIRA Administrator Susan Dudley wrote a letter to EPA expressing OIRA's "concerns" with the idea of establishing a secondary standard for ozone that was different from the primary standard. The letter criticized EPA for focusing exclusively on detrimental impacts on plant species to the exclusion of other "economic values." EPA Deputy Administrator Marcus Peacock responded that considering "economic values" would be unlawful under the Clean Air Act as interpreted by the Supreme Court in the recent *American Trucking* case (discussed in Chapter 7). Furthermore, EPA's criteria document and staff paper had fully considered the scientific concerns raised in Dudley's letter. The next day, Dudley wrote to EPA that "the President has concluded" that the agency should set the secondary standard at the same level as the primary standard.[70]

The agency had to scramble literally overnight to come up with a rationale to support dropping the cumulative, seasonal secondary standard. Outraged EPA staffers called the change a matter of "pure politics." Solicitor General Paul D. Clement warned administration officials late that night that the final version of the regulations contradicted the government's previous submissions to the Supreme Court. CASAC chairperson Rogene Henderson later testified that "willful ignorance" had "triumphed over sound science" in the decision-making process.[71]

The Final Rule

The final rule that the agency published in late March 2008 changed the level of the eight-hour primary standard to 0.075 ppm. Administrator Johnson concluded that because of the uncertainties in the available data at low exposure levels, "the likelihood of obtaining benefits to public health with a standard set below 0.075 ppm . . . decreases, while the likelihood of requiring reductions in ambient concentrations that go beyond those that are needed to protect public health increases." As directed by President Bush, Johnson set the secondary standard equivalent to the primary standard. He acknowledged that newly available evidence compelled a revision of the secondary standard to make it more protective, but, after "a robust discussion within the Administration" on the "strengths and weaknesses" of the new studies, he decided not to change the form of the standard. Gamely attempting to explain why he was rejecting the advice of his staff and CASAC, he explained that adopting a cumulative secondary standard at the high end of the range suggested in the staff paper would result in very little improvement over making it identical to the new primary standard.[72]

Delayed Judicial Review

A large number of environmental and industry groups challenged the final standards. As we shall see in Chapter 9, however, the court stayed its consideration to give the Obama administration an opportunity to reconsider the standard. When EPA abandoned that exercise without coming to a conclusion, the court allowed the parties to renew their challenges. In July 2013, the D.C. Circuit upheld the 2008 primary standard against attacks from both sides, but it remanded the secondary standard. The court rejected the industry's argument that EPA had to set the primary standard at exactly the level that was neither higher nor lower than requisite to protect health. EPA was required to explain why the level it selected was requisite; it did not have to explain why the previous standard was "no longer up to the task." To the industry's contention that EPA's staff had misinterpreted the Adams studies, the court held that EPA's independent interpretation of scientific data was well within the realm of its expertise. The industry's challenge to EPA's interpretation and use of the epidemiological studies was equally unavailing.

At the same time, the court rejected the environmental groups' contention that the primary standard was insufficiently stringent. The court held that EPA had "rationally treated the 0.06 ppm results as inconclusive." The court had previously recognized "the impossibility of eliminating all risk of health effects from 'non-threshold' pollutants like ozone." It likewise disposed of the environmental groups' contention that EPA had not given sufficient weight to epidemiological studies concluding that ozone had adverse effects on people at levels as low as 0.06 ppm. With respect to their contention that EPA had not allowed a sufficient margin of safety, the court noted that the margin of safety was a policy choice to which it had in the past given "wide berth" to the agency, and EPA's determination did not represent an abuse of that broad discretion.

The court did, however, agree with the environmental groups' contention that EPA's decision to reject a cumulative, seasonal secondary standard was arbitrary and capricious. It noted that in setting the secondary standard equivalent to the primary standard, the administrator had merely compared the primary standard to a cumulative, seasonal standard at the upper end of the range that the staff and CASAC had recommended. He never found that a cumulative, seasonal standard at the upper end of the range was "requisite" to protect public welfare. A cumulative, seasonal standard at the lower end of the range would clearly offer more protection than the primary standard, but the administrator failed to explain why a more protective standard was not requisite to protect the environment.[73]

NONATTAINMENT

One Clinton administration initiative that the Bush administration was pleased to continue was its program for encouraging states to employ emissions trading techniques in their state implementation plans (SIPs). The "open market" programs that EPA approved in Illinois and New Jersey allowed power plants and other sources that exceeded targets for reducing NO_x emissions in the SIPs to sell allowances to other facilities or bank for their own use at a later time. Environmental groups were not opposed to market-based programs in principle, but they opposed open-market programs. Among other things, the programs employed a system for self-reporting emissions that, in their view, was virtually unenforceable and therefore primed for abuse by companies selling fictitious credits for which there were no actual reductions. A year after EPA approved the programs, the agency's inspector general sharply criticized them for allowing companies to claim allowances on the basis of inaccurate monitoring data and questionable calculations. In August 2002, the head of the New Jersey Department of Environmental Protection wrote to EPA requesting that it approve the discontinuation of the program because it had encountered so many flaws that it was not resulting in reduced emissions. Among other things, the department discovered that companies were claiming emission reductions achieved years prior to the initiation of the program.[74]

CLIMATE CHANGE

After a White House–commissioned NAS report concluded that GHG emissions were "accumulating in earth's atmosphere as a result of human activities, causing surface air temperatures and subsurface ocean temperatures to rise," the administration in February 2002 announced a plan for achieving "voluntary" reductions in GHG emissions with the goal of achieving an 18 percent reduction in "greenhouse gas intensity" by 2012. The plan, called the "Global Climate Change Initiative," was based almost to the word on a proposal prepared by a group of fossil fuel–dependent companies. The primary incentives in the plan came from tax credits for various energy conservation and alternative energy measures that companies could undertake. A year later, President Bush announced a new permutation of the initiative called the "Voluntary Innovative Sector Initiatives: Opportunities Now" (VISION) program.[75]

It soon became apparent that industry commitments were far less ambitious than the administration had envisioned. By January 2004, only fourteen of the fifty companies participating in the program had actually set voluntary GHG reduction goals for 2010 as required by the program. EEI pledged only a 3–5 percent reduction in GHG emissions per kilowatt-hour, much less than the 7 percent reduction that DOE's Energy Information Administration predicted would result from end-use conservation efforts undertaken without the program. None of the trade associations that signed up for the VISION program required individual companies to establish CO_2 reduction goals, and many of the heaviest polluting electric power companies avoided both programs. A report issued by the EPA's inspector general at the end of the Bush administration concluded that the voluntary programs were ineffective and produced few reductions in emissions because of difficulties EPA encountered in "convincing companies to spend money on activities that are entirely optional."[76]

Action-Forcing Lawsuit

Unmoved by the administration's voluntary initiatives, a coalition of environmental groups on December 5, 2002, sued EPA for failing to respond to their October 1999 petition to regulate GHG emissions from motor vehicles (discussed in Chapter 7). Reversing the interpretations of two previous general counsels, EPA responded that it did not have authority under the Clean Air Act to limit greenhouse gas emissions because they were not "pollutants." Even if the agency were authorized to regulate GHGs, EPA explained, it would be unwise as a policy matter because the agency was already undertaking voluntary initiatives.[77]

Failed Legislation

The 2006 elections brought a dramatic shift in the legislative climate. With both the House and the Senate controlled by the Democrats, the prospects for climate change legislation seemed bright. In addition, several large companies, including Cinergy, Duke Energy, Entergy, and Exelon, joined in the calls for government action on climate disruption. Some of the new proponents were concerned about a rapidly developing patchwork of state GHG regulations. Others hoped their support for controls would guarantee them a place at the table when decision makers hammered out the details of a federal climate disruption regulation.[78]

The picture was not entirely rosy for proponents of regulation, however, because many of the newly elected Democrats were moderates for whom environmental issues were not a high priority, and a large number of Democratic members still came from energy-producing states. Moreover, the election took a huge toll on moderate Republicans from the Northeast, thereby dimming the prospects for truly bipartisan legislation. President Bush retained his power to veto any legislation that he judged to be too aggressive, and the administration made it clear that it remained deeply opposed to any legislation providing for mandatory GHG emissions reductions. The legislative efforts failed when opponents successfully filibustered a modest cap-and-trade bill sponsored by Senators Joe Lieberman (I-CT) and John Warner (R-VA).[79]

Massachusetts v. EPA

Although Congress appeared incapable of addressing climate disruption, the Supreme Court in April 2007 delivered a landmark opinion that opened the door to action by EPA to reduce greenhouse gases from power plants. In *Massachusetts v. EPA,* the Court held that the Clean Air Act gave EPA authority to regulate GHG emissions from automobiles because GHGs easily came within the statute's "capacious" definition of "air pollutant." If EPA found that GHG emissions could reasonably be anticipated to endanger public health or welfare, it could begin regulating GHGs not just in automobiles but also in the permits required for new and modified power plants. Likewise, EPA could promulgate technology-based new source performance standards for GHG emissions from new and modified power plants.[80]

President Bush reacted to the opinion by issuing an executive order instructing EPA to work together with the National Highway Traffic Safety Administration (NHTSA) to increase automobile fuel efficiency by the end of 2008. But he made it clear that the United States would not act to reduce GHGs from power plants until China and India agreed to undertake similar steps. EPA and NHTSA coordinated two independent rulemakings aimed at reducing GHG emissions from motor vehicles and improving automobile fuel economy. Before the rulemakings could go forward, however, the agency had to decide whether to make an endangerment finding, and Administrator Johnson was in no hurry to cross that Rubicon.[81]

EPA's Air Office dedicated fifty-three staff members and $5.3 million in contractor services to supporting an endangerment finding and coming up with a global warming program. The effort produced a three-hundred-page technical support document concluding that greenhouse gas emissions from mobile sources endangered public health and welfare. After Administrator Johnson signed off on the document, Deputy Associate Administrator Jason Burnett emailed it to his counterpart at the White House Office of Information and Regulatory Affairs (OIRA). Well aware of the content of the missive and of the likelihood that the document would become available under the Freedom of Information Act, the OIRA official declined to open the attachment to the email. Instead, he called EPA Administrator Johnson to demand that the agency withdraw the submission. Although Johnson initially resisted, he ultimately yielded to the White House demand. Appalled by Johnson's lack of fortitude, Burnett resigned in protest.[82]

After reworking the document to remove references to the devastating impacts of global warming and to soften its conclusions, Johnson resubmitted the document to OIRA. White House officials were still not satisfied, and they demanded that the agency delete sections of the document that supported the conclusion that greenhouse gas emissions endangered the public health and welfare. At this point, EPA shelved the endangerment finding and published a notice soliciting public comment on the endangerment issue. In a press conference accompanying the release, Johnson opined that "if our nation is serious about regulating greenhouse gases, the Clean Air Act is the wrong tool for the job."[83]

State Climate Change Initiatives

While EPA dithered, much regulatory activity was taking place at the state level to reduce GHG emissions. Massachusetts enacted a four-pollutant bill in May 2001 that required the state's four coal-fired power plants to reduce CO_2 emissions by 10 percent by 2008. The California legislature passed a bill that established a goal of reducing GHG emissions statewide to 1990 levels by 2020 (a 25 percent reduction) beginning in 2012 and required the California Air Resources Board to promulgate mandatory requirements capable of achieving that goal. After the state of Washington's legislature enacted a bill requiring operators of all new coal-fired plants to begin capturing CO_2 emissions within five years, it became virtually impossible to permit new coal-fired plants in that state. By 2008, twenty states had enacted legislation establishing GHG emissions reduction goals for power plants. And some rural states enacted CO_2 trading platforms to facilitate sales of allowances by agricultural and silvicultural interests.[84]

In addition to individual state attempts to regulate GHG emissions, two regional efforts were initiated during the last half of the Bush administration. By far the most significant was the Regional Greenhouse Gas Initiative

(RGGI), a cap-and-trade regime for fossil fuel–fired power plants established in 2003 by ten New England and mid-Atlantic states over the opposition of the Bush administration and the electric power industry. The program limited overall emissions to 2005 levels through 2014. After that, the cap shrank by 2.5 percent per year through 2018, at which point emissions would be at approximately 1990 levels.[85]

Carbon Capture, Utilization, and Sequestration

When engineers discussed technologies for reducing GHG emissions from power plants, they generally focused on "carbon capture, utilization, and sequestration" (CCUS), a process in which CO_2 was removed from power plant exhaust, liquefied, and piped to a location where it could be used in recovering oil from deep underground or otherwise sequestered permanently. Formidable obstacles lay along the road to widespread employment of CCUS, including the cost of the technology in terms of both capital investment and energy devoted to the process and the absence of a pipeline infrastructure to transport liquid CO_2 from power plants to appropriate sequestration sites. The electric power industry urged policymakers not to force companies to adopt the technology until it was commercially available. Environmental groups were split on the issue, with Greenpeace opposing CCUS on the ground that it was an unproven energy hog and NRDC willing to accept it if adequately regulated.[86]

In February 2003, President Bush proposed a $1 billion, decade-long project to build the world's first coal-fired power plant with essentially zero emissions that he labeled "FutureGen." A major component of the project was a state-of-the-art CCUS facility to sequester CO_2 from a large integrated gasification combined-cycle power plant fueled by hydrogen generated by a large coal gasification unit. To assist in the conceptual work and the financing, DOE helped create a nonprofit corporation called the FutureGen Industrial Alliance, composed of coal and electric power companies from around the world. In December 2007, the alliance announced that it had selected Mattoon, Illinois, as the site for the FutureGen project. But by then, alarming cost overruns, smaller congressional appropriations than expected, and insufficient commitments from private entities persuaded DOE that the project was no longer viable. In February 2008, DOE announced that it would no longer be supporting the FutureGen project.[87]

COOLING WATER INTAKE STRUCTURES

Most fossil fuel–fired power plants had a direct adverse effect on marine and aquatic life by virtue of their need to pull large amounts of water from nearby rivers or lakes into their cooling systems through cooling water intake structures (CWISs). Every year, billions of larvae and millions of fish, shellfish, and other aquatic and marine organisms were entrained with water inflow into the pumping systems or became impinged on the screens that intake structures employed to keep debris out of the cooling waters. Section 316(b) of the Clean Water Act required EPA to promulgate regulations for CWISs reflecting the "best technology available" (BTA) for "minimizing" their "adverse environmental impact."[88]

New Power Plants

The "Phase I" regulations for new power plants that EPA published in December 2001 established a flexible two-track approach to establishing requirements on a case-by-case basis by permitters. The company could elect the fast track by installing a closed-cycle system that used little external water and therefore posed no risk to aquatic organisms, or it could choose the slow track by convincing the permitting agency that the technology it selected performed 90 percent as well as a closed-cycle system. In addition, a company could apply for a variance from a requirement by showing that compliance would result in costs "wholly out of proportion" to the costs the EPA considered in establishing the standards. At OIRA's insistence, EPA added a provision that allowed a company to obtain a permit if it employed "restoration measures" like restocking killed fish or creating alternative habitats to compensate for losses of other organisms.[89]

Both the electric power industry and environmental groups challenged the regulations in the Second Circuit Court of Appeals. The court rejected the industry's argument that EPA's focus on impingement and entrainment as the measure of environmental harm was unlawful. The industry's contention that killing some aquatic organisms could be good for the environment was almost laughable, and it had not suggested a better indicator of adverse environmental impact. The court also rejected the environmental groups' argument that it was arbitrary and capricious for the agency to allow companies to demonstrate that the technology they selected was only 90 percent as effective as closed-cycle technology. It agreed with EPA that the available tools for measuring impingement and entrainment were so inexact that a 10 percent margin of error was warranted. Although the statute was silent on the issue of variances, EPA was not unreasonable in interpreting it to tolerate an administrative variance. The only aspect of the rule that the court found unlawful was the OIRA-inspired provision allowing new sources to meet the standard through restoration measures. That provision's focus on water quality was "plainly inconsistent" with the technology-based BTA standard.[90]

The regulations stimulated creative research on intake structures. Some solutions, like the "modified traveling screen" that employed various cues to warn fish to keep away from intake structures, were "brilliant in their simplicity" and approached 100 percent effectiveness in preventing impingement. By 2007 several new power plants without closed-cycle systems were complying with the standards.[91]

Existing Power Plants

While it was litigating the Phase I regulations, the agency proceeded ahead with the Phase II regulations for the 550 existing power plants, each of which withdrew at least 50 million gallons of water per day from the nation's surface waters. These facilities accounted for more than 50 percent of the electrical power generated in the United States. They used more than 214 billion gallons of water daily, and they caused the impingement or entrainment of more than 3.4 billion aquatic organisms per year. Environmental groups supported a uniform nationwide standard based on closed-cycle technology, and industry groups advocated a "site specific" approach to determining BTA in which permitters would consider various factors, including costs and benefits.[92]

The regulations that EPA published in January 2004 allowed a source to comply with BTA by installing a closed-cycle cooling tower, but the agency rejected the environmental groups' demand that it require closed-cycle systems nationwide. Requiring all affected plants to meet a standard based on closed-cycle systems would cost the industry around $3.5 billion per year and require so much energy that output would go down by 2.4 to 4.0 percent. Instead, the regulation adopted the site-specific, cost benefit–based option that the industry had requested.[93]

In April 2009, the Supreme Court upheld EPA's site-specific approach. In an opinion written by Justice Antonin Scalia, the Court held that EPA was "reasonable" in interpreting the words "best technology available for minimizing adverse environmental impact" to encompass a cost-benefit decision criterion. Relying on *Webster's New International Dictionary,* Justice Scalia interpreted the word "best" to mean "most advantageous." The "best" technology might well be one that "produces the most of some good," but it could also "describe the technology that most efficiently produces that good." EPA could reasonably interpret the words "best technology" to "refer to that which produces a good at the lowest per-unit cost, even if it produces a lesser quantity of that good than other available technologies." Furthermore, the phrase "for minimizing" "admits of degree and is not necessarily used to refer exclusively to the 'greatest possible reduction.'" Interestingly, Justice Scalia did not refer to the dictionary this time. The court below had relied on *Webster's,* which defined "minimize" to mean "to reduce to the smallest possible extent." That definition seemed to dictate the technology-based approach that the environmental groups had advocated. Although the agency prevailed on the critical cost-benefit issue, it still had to respond to several issues that were the subject of remands from the Second Circuit Court of Appeals but were not taken up by the Supreme Court.[94] We will follow the progress of the regulations in Chapter 9.

ECONOMIC REGULATION

The California electricity crisis brought the rush to deregulate in the states to a halt. By the end of 2001, seven states that had enacted restructuring legislation had delayed its implementation. No state enacted new retail restructuring legislation during the next decade, and consumer advocates in deregulated states called for reregulation. This was a welcome development for environmental groups who worried that deregulation would put irresistible pressures on companies to extend the lifetimes of the dirtiest old power plants, burn more coal, and cut corners when it came to environmental compliance. Illinois remained committed to its restructuring legislation, but it also put into place a mercury reduction program that was more stringent than EPA's regulations. Texas, by contrast, cut back environmental regulation of power plants as it vigorously deregulated retail markets.[95]

The California crisis also inspired a renewed focus on demand response as an inexpensive way to ensure grid reliability. As of 2002, twenty-three states had enacted some form of demand-side management program. A few distribution utilities implemented sophisticated regimes employing "smart meters" that provided real-time usage and price information so that customers could cut back usage during peak load times when prices were higher. The results, however, were decidedly mixed. For example, Puget Sound Energy's program got off to a successful launch with more than 300,000 customers signing up for the voluntary program, but they quickly abandoned it after the media reported that they were paying on average one dollar more per month than flat-rate users.[96]

By 2007, more than twenty states had enacted renewable portfolio standards (RPSs) requiring utility companies to derive a specified percentage (varying from 3 to 25 percent) of their capacity from renewable resources like wind, solar energy, hydropower, and biomass. Legislatures often attached RPSs to restructuring legislation to ensure that market pressures did not force distributors to rely exclusively on cheap coal-fired sources of electricity. The standards caused a large increase in demand for wind power, which also benefited from a federal production tax credit of 1.5 cents per kilowatt hour (and adjusted upward over time), enacted as part of the Energy Policy Act of 1992. By 2005, six hundred utility companies in thirty-four states were offering green power options. But many states lacked significant renewable energy resources, and most states declined to enact mandatory requirements. Although RPSs in the states appeared to be working well, the Bush administration opposed a national RPS, and Congress failed to include a national RPS requirement in the Energy Policy Act of 2005.[97]

INDUSTRY CONSOLIDATION

The Energy Policy Act of 2005 repealed parts of the Public Utilities Holding Company Act and assigned to FERC the lead role in approving public utility company mergers and acquisitions. Concluding that larger companies could more easily absorb the fixed cost of installing EPA-mandated pollution controls, industry experts accurately predicted that the new law would bring on a "feeding frenzy" as large power companies and private equity firms gobbled up smaller power companies in pursuit of economies of scale. In the largest merger of the decade, Duke Energy acquired Cinergy for $9.1 billion.

Duke had a major presence in North and South Carolina, while Cinergy was dominant in Ohio, Indiana, and Kentucky. The merger allowed Duke to diversify its heavy reliance on natural gas with Cinergy's many coal-fired plants. Although FERC formally approved the merger in December 2005, the companies found it more difficult to persuade the public utility commissions in the many states in which they both operated to approve the merger. For example, the Indiana commission approved the deal only after Duke agreed to give Cinergy's Indiana customers $40 million in rate credits for a year and provide $5 million in low-income energy assistance. After obtaining all of the approvals and closing the merger in April 2006, Duke Energy became the second largest electric utility company in the country.[98]

THE 2008 ELECTIONS

During the 2008 election season, most of the major candidates supported government action to address global warming. After the campaign boiled down to John McCain and Barack Obama, it was difficult to distinguish the two on environmental issues. McCain had been the more prominent environmental proponent in the Senate, having cosponsored with Senator Joe Lieberman (I-CT) a national cap-and-trade bill that EEI opposed. Both candidates promised to depart dramatically from the Bush administration's environmental policies. But McCain expressed greater reluctance than Obama to use EPA's power under the Clean Air Act to regulate GHG emissions in the absence of congressional action. Both McCain and Obama wanted to see nuclear power play a greater role in domestic energy production. But Obama wanted the federal government to play a larger role in developing renewable energy technologies. Environmental groups grew more hostile to McCain after he chose Sarah Palin as his running mate, because she had a reputation as a pro-development Alaska governor who had little use for environmentalists. The groups were therefore pleased to see Senator Obama prevail, and they looked forward to working with him and a Democrat-controlled Congress.[99]

CONCLUSIONS

The George W. Bush administration ushered in a period of retrenchment. The California crisis left two of the nation's largest electric utility companies near bankruptcy, and large utility companies in other deregulated states faced a credit crisis as the major rating agencies lowered the status of their bonds. Many of the newly arrived merchant power companies were likewise fending off bankruptcy as poor credit, low electricity prices, and high natural gas prices took their toll. On the other hand, companies like Consolidated Edison and the Southern Company, which had not ventured into the deregulated world, were much better off. Most power cooperatives and municipal utilities also emerged from the crisis largely unscathed. With the merchant power industry in retreat, some regulated utility companies initiated programs to build new coal-fired power plants under the traditional ratemaking model.[100]

Despite the Bush administration's vigorous efforts to relieve coal-fired power plants of environmental restrictions, the coal revival envisioned by Vice President Cheney's task force never fully materialized. While the Bush administration's enthusiasm for coal never waned, opposition to coal-fired power plants at the state and local levels became increasingly difficult to overcome. Furthermore, the uncertainty hovering over efforts to control greenhouse gas emissions also cast a pall over new projects. More than 120 coal-fired power plant projects were canceled between 2001 and 2009. At the same time, companies were retiring older power plants. Some retirements were driven by the simple economics of keeping aging plants functioning efficiently. But many resulted from new source review lawsuits brought by EPA, state attorneys general, and environmental groups. Despite perennially high prices for natural gas, some companies replaced coal-fired boilers with natural gas–fired turbines.[101]

The Bush administration successfully resisted any action to reduce greenhouse gas emissions for eight years while taking the politically palatable position that the uncertainties in the underlying science precluded governmental action. The industry therefore had little reason to develop new technologies other than the modest incentives that DOE and some states provided. A few advanced integrated gasification combined-cycle plants were in operation, and several more were on the drawing boards, but no carbon capture and sequestration projects had reached even the pilot plant stage.[102]

The Bush administration's modest attempt to improve air quality in nonattainment areas came to an ignominious end in July 2008 when the D.C. Circuit set aside the administration's Clean Air Interstate Rule establishing a multistate cap-and-trade program for NO_x emissions. Even the utility companies that brought the challenge were shocked that the court threw out the entire program, some aspects of which they had relied upon in purchasing and selling emissions credits. That decision and the same court's rejection of the administration's odd approach to regulating mercury emissions from power plants ensured that EPA did not impose a single additional control on power plants during George W. Bush's eight years in office.[103] These failures left a great deal on the table for the incoming Obama administration.

CHAPTER 9

The Obama Administration Takes a New Approach

The 2008 elections appeared to cap a major shift in the politics of energy and the environment. Both houses of Congress remained under the control of the Democratic Party, and the Democratic caucus in the Senate reached the magic number of sixty, enough to cut off Republican filibusters. Newly elected president Barack Obama announced that one of his highest domestic priorities was to enact cap-and-trade legislation with the goal of reducing greenhouse gas (GHG) emissions by 80 percent from 1990 levels by 2050. His inaugural address specifically mentioned modernizing the electric power grid and encouraging the use of renewable resources.[1]

To head EPA, the president appointed Lisa Jackson, the soft-spoken but strong-willed former head of the New Jersey Department of Environmental Protection. The top choice of the business community, she pleased environmental groups in her first major address when she alluded to her eagerness to begin shaping the regulatory environment for the electric power industry in a more environmentally friendly fashion. The president's choice to head EPA's Office of Air and Radiation was Gina McCarthy, the straight-talking Republican head of Connecticut's Department of Environmental Protection. The president hoped to demonstrate his commitment to climate change legislation by making the former EPA administrator Carol Browner a White House special assistant and charging

February 2009 EPA and DOJ begin a new National Power Plant Enforcement Initiative

July 2009 House passes climate change bill

December 2009 EPA finds that greenhouse gas emissions endanger public health and welfare

May 2010 EPA promulgates tailoring rule to limit the number of power plants subject to greenhouse gas regulation

July 2010 Senate abandons efforts to enact climate change legislation

August 2010 DOE relaunches FutureGen carbon capture and sequestration project

March 2011 FERC promulgates Order No. 745 requiring system operators to pay market price for electricity saved by demand response programs

July 2011 EPA promulgates Cross-State Air Pollution Rule to replace the Clean Air Interstate Rule

August 2012 EPA promulgates a federal implementation plan for the Four Corners power plant in Arizona to reduce regional haze in sixteen national parks and wilderness areas

December 2012 EPA publishes Mercury and Air Toxics Standards to control emissions of hazardous air pollutants from power plants

June 2014 Supreme Court overturns tailoring rule, but leaves most power plants subject to best available technology requirement for greenhouse gas emissions

August 2014 EPA promulgates cooling water intake structure regulations that allow site-specific cost-benefit balancing

February 2015 DOE cancels FutureGen project for the second time

April 2015 EPA promulgates regulations establishing nationally applicable minimum criteria for states to use in regulating coal ash disposal

August 2015 EPA promulgates new source performance standards regulating greenhouse gas emissions from new fossil fuel–fired power plants and Clean Power Plan providing guidelines for states to use in reducing greenhouse gases from existing fossil fuel–fired power plants

her with directing the administration's efforts on matters relating to energy and the environment.[2]

At the outset of the Obama administration, the prospects for coal were fairly bright. Coal-fired plants were prospering in the Midwest. Twenty-five new coal-fired plants representing almost 15,000 megawatts of capacity were under construction. More new coal-fired plants came into service during 2009 than in any year in the previous two decades, and more than 120 proposals for building coal-fired plants were in the works. And coal mines in the Powder River Basin were struggling to meet demand for low-sulfur coal. In her confirmation hearings, Lisa Jackson called coal a "vital resource," and most observers expected that coal would remain the most heavily relied upon source of power plant fuel for the foreseeable future.[3]

During the spring and summer of 2009, Jackson and McCarthy invited electric power industry executives to a series of informal meetings at EPA's headquarters to tell them what was in the regulatory pipeline and to probe the possibility of implementing several regulatory programs in a single comprehensive deal. The straight talk did not yield the hoped-for grand compromise, but it did reveal noticeable divisions in the electric power industry. A group consisting primarily of companies with significant investments in gas-fired and nuclear plants was willing to negotiate with EPA over pollution controls. A second group consisting of companies that were heavily dependent on coal planned to fight EPA every step of the way. Still another group consisted of companies that were so confounded by the huge technical and political uncertainties surrounding the industry's future that they could not make up their minds whether to join the first or second camp.[4]

CLIMATE CHANGE LEGISLATION

As carbon dioxide levels in the ambient air reached record highs, the stars appeared to be aligned for Congress to enact robust climate change legislation capable of rearranging the nation's energy production and consumption priorities.[5] After wresting the chairmanship of the House Energy and Commerce Committee from longtime chairman John Dingell (D-MI) in an audacious coup, Rep. Henry Waxman (D-CA) announced that the 111th Congress had "an opportunity that comes only once in a generation" to enact landmark climate change legislation. Although Waxman had the energy of a person half his sixty-nine years, the committee had a full plate of important bills, including the president's signature Affordable Care Act, that taxed even his formidable capacity for hard work.[6]

The committee's starting point was a "Blueprint for Legislation" that had been drafted by the Climate Action Partnership (CAP), a coalition of environmental groups, large manufacturing companies, and non-coal-dependent electric utility companies. The blueprint's goal was to achieve a 42 percent reduction in emissions from 2005 levels by 2030 and an 80 percent reduction by 2050 through a cap-and-trade program that would allow large emitters to trade greenhouse gas (GHG) allowances nationwide. A "substantial portion" of the allowances would be allocated on the basis of historical emissions, and the rest would be auctioned off. Various "cost containment" measures would act as safety valves to avoid price spikes in allowance prices. The blueprint also allowed companies to purchase "offset credits" from companies that agreed to improve the efficiency of their operations or farmers who agreed to plant more carbon absorbing vegetation.[7]

The coal industry and coal-dependent electric utility companies opposed all climate change legislation, but argued that any bill should include an easily triggered safety valve and a transition period of fifteen to twenty years to allow companies time to develop workable carbon capture, utilization, and sequestration (CCUS) technologies. Representatives of small businesses wanted to receive a portion of the allowances to offset predictable increases in electricity prices. Farmers argued that the bill should place responsibility for managing offset credits in the United States Department of Agriculture (USDA).[8]

The 1,400-page bill that the House took up after several committees had marked it up reflected dozens of deals that House leaders made with wavering Democrats and persuadable Republicans. It was chock full of giveaways and inducements to utility companies, energy-intensive manufacturing industries, agricultural interests, and small businesses. But it did establish a fairly stringent cap-and-trade regime based on the CAP model. President Obama vigorously lobbied Democratic members to vote for it. Prominent environmental and consumer groups held their collective noses and urged members to support it. Most of the industry groups supported the bill but hoped to get a better deal in the Senate. The coal industry and coal-dependent power companies opposed the bill. The Republican leadership characterized it as a massive tax on consumers of electricity that would result in the loss of millions of jobs.[9]

The bill passed the House by a narrow 219–212 margin. It replaced EPA's existing GHG regulatory authorities with a multisector cap-and-trade program that capped GHG emissions at 17 percent below 2005 emissions by 2020, 42 percent by 2040, and 83 percent by 2050. FERC would oversee the allowance markets, and USDA would oversee the offset markets. The bill set aside allowances amounting to about 60 percent of current emissions for EPA to distribute to various public and private beneficiaries. Beginning in 2026, the free allowances would gradually be phased out until they ended in 2030, at which point all allowances would be allocated by auction. Instead of reducing emissions, companies could create or purchase offset credits by increasing energy efficiency, planting vegetation to take CO_2 out of the atmosphere, capturing methane emissions from cow manure, or other forms of permanent carbon sequestration. The bill established a $25 per megawatt-hour safety valve to ensure against price spikes. Finally, it contained a national renewable portfolio standard under which utilities were required to generate 15 percent of their electricity from renewable resources and achieve a 5 percent reduction in demand from energy efficiency by 2020.[10]

Attention then shifted to the Senate where many groups that had supported the House bill now hoped to have their favorite provisions included and their least favorite provisions deleted. During the congressional recess of August 2009, a river of money flowed into efforts to build support for and against climate change legislation at the grassroots level. The National Association of Manufacturers and the National Federation of Independent Businesses spent several million dollars on television ads in thirteen swing states. Americans for Prosperity, an advocacy group created by the Koch brothers, hosted eighty events for Tea Party groups at which speakers asserted (erroneously) that backyard barbecues would be taxed if Congress enacted the House bill. Another industry-funded grassroots advocacy group called the American Energy Alliance (AEA) arranged a bus tour through coal-producing and manufacturing states featuring "free lunches, free concerts and speeches" to generate public opposition to climate change legislation.[11]

To rally support for climate change legislation, environmental groups purchased television ads, operated phone banks, and sponsored public events. The Alliance for Climate Protection, a group assembled by former vice president Al Gore, and the Blue-Green Alliance, an umbrella organization of environmental groups and labor unions, undertook a twenty-two-state "Made in America" tour to demonstrate how climate legislation would create good jobs. A coalition of sixty-eight environmental, labor, civil rights, and consumer groups calling itself the Clean Energy Works Campaign launched a $20 million media buy and a major grassroots campaign to generate calls, letters, and emails to key members of Congress.[12]

In late October 2009, Senators John Kerry (D-MA) and Barbara Boxer (D-CA) unveiled a 923-page draft bill containing a cap-and-trade regime with a more stringent 2020 cap than the House bill and providing for auctioning 25 percent of the initial allowances. The allowance market would be overseen by the Commodity Futures Trading Commission. All of the remaining giveaway allowances would be phased out by 2030. To ensure a stable market in GHG allowances, the bill created a "soft collar" on the price of allowances that required EPA to allocate allowances from a "strategic reserve" when the price hit $28. As in the House bill, companies could purchase offset credits from farmers. In a departure from the House bill, the Senate bill did not disturb EPA's existing authority to regulate GHG emissions.[13]

A quick series of three hearings in the Senate Environment and Public Works Committee revealed serious disagreements among its Democratic members over critical aspects of the Kerry-Boxer bill. Senator Max Baucus (D-MT) argued for a less stringent 2020 cap, and he strongly supported adding a provision to prevent EPA from regulating GHGs under its existing authorities. Committee Republicans criticized the bill as an overly complex and unduly stringent mandate that would harm the economy, kill jobs, and favor some regions at the expense of others. But it was clear that without a handful of Republican votes, the bill's supporters would not be able to block a Republican filibuster.[14]

Senator Kerry then reached out to Senator Lindsey Graham (R-SC) and Joe Lieberman (I-CT) to hammer out a compromise bill. The Edison Electric Institute and the CEOs of Exelon and Duke Energy joined the Environmental Defense Fund and a number of other environmental groups in endorsing the Kerry-Graham-Lieberman bill. Even the U.S. Chamber of Commerce suggested that it might support the bill. But the project attracted opposition from the AEA, which sponsored a series of radio, television, and online advertisements in South Carolina warning that one of the "scary stories" coming out of Washington that Halloween was Senator Graham's support for "a national energy tax called cap-and-trade." Environmental groups responded with a more modest ad campaign asking why "out-of-state interests" were attacking Graham for "backing an energy plan that produces more power for America."[15]

Three days before the bill's much-anticipated rollout, Senate Majority Leader Harry Reid decided to move immigration reform ahead of climate change legislation on the Senate's calendar. Characterizing the move as "a cynical political ploy" aimed at attracting Hispanic voters in the upcoming off-year elections, Senator Graham announced that he would no longer work to pass a climate change bill. Because he was enduring vicious attacks in his home state from Tea Party activists, he may well have welcomed the opportunity to separate himself from the bill. Senators Kerry and Lieberman gamely introduced the 987-page bill on May 12, 2010, but it attracted no Republican support. In late July, Senator Reid announced that neither the Democratic leadership nor the president had been able to cobble together sixty votes for a climate disruption bill of any size or shape, and he called a halt to their efforts.[16]

CLIMATE CHANGE REGULATION

Climate disruption was now in the hands of EPA, which was exercising its limited power under the Clean Air Act, and states that were willing to take on that controversial topic. The stakes were high. A multiagency report concluded that the effects of human-induced climate changes were already under way and included "heavy downpours, rising temperatures and sea levels, rapidly retreating glaciers, lengthening growing seasons, earlier snow melt and alterations in river flow."[17]

The Endangerment Finding and the Tailoring Rule

On April 24, 2009, EPA proposed to find that six greenhouse gases, including CO_2 and methane, were "pollutants" that "may reasonably be anticipated to endanger public health and welfare." A lengthy technical support document relied heavily on summaries of the scientific information prepared by the Intergovernmental Panel on Climate Change, the U.S. Climate Change Science Program, the U.S. Global Change Research Program, and the National Academy of Sciences, which in turn relied on hundreds of peer-reviewed studies. The agency also had to contend with false accusations that the Clean Air Act would force EPA to regulate methane emissions from cattle, sheep, and pigs. No matter how often Administrator Jackson flatly denied that the agency was planning to regulate flatulent animals, Republican congresspersons from rural states confronted her with the accusation that EPA was planning a "cow tax."[18]

In mid-September 2009, EPA proposed limits on CO_2 emissions from motor vehicles based on the proposed endangerment finding. As a result of the emergency bailout of financial institutions in late 2008 and early 2009, the federal government owned much of the stock of two of the major automobile companies. This gave EPA significant leverage in negotiations with the auto industry, and the stakeholders quickly agreed to a phased program of increased fuel economy and reduced emissions for new motor vehicles. Because the transportation sector was second only to the electric power industry in GHG emissions at the time, these regulations had the potential to move the country significantly toward President Obama's GHG reduction goal as new automobiles replaced old ones.[19]

Finalizing both regulations would trigger the duty of permitters in clean air areas to consider GHG emissions in implementing the statute's new source review (NSR) provisions for power plants. Among other things, major new and modified power plants would have to install the "best available control technology" (BACT) for CO_2 emissions. This posed a major dilemma for the agency. The statutory "majorness" threshold for subjecting a new or modified stationary source to NSR was 100 tpy for most power plants and 250 tpy for the rest. Since no effective controls on CO_2 emissions other than greater combustion efficiency were available, even the tiniest fuel-burning sources emitted more than 250 tpy. If it wanted to avoid a regulatory regime that was impossible to manage, it would have to finesse the threshold issue. The agency's solution was a proposed "tailoring" rule that set the NSR threshold at 25,000 tpy for new (greenfield) plants and 10,000 tpy for modifications of existing plants. The electric power industry questioned the lawfulness of the proposal, arguing with considerable plausibility that it was hard to interpret "100" to mean "25,000." Environmental groups supported the EPA proposal, arguing that it was legally justified in the circumstances.[20]

In December 2009, EPA published a final finding that greenhouse gas emissions "endangered" human health and the environment. The final tailoring rule that EPA promulgated in May 2010 set the thresholds at 100,000 tpy for greenfield plants beginning in July 2011 and 75,000 tpy for existing sources already subject to the PSD program because emissions exceeded the threshold for another pollutant (so-called "anyway sources") beginning on January 2, 2011. The 75,000 tpy threshold would apply to modifications that did not otherwise trigger NSR in July 2011. The D.C. Circuit unanimously upheld the endangerment finding and found that industry petitioners lacked standing to challenge the tailoring rule because it was doing them a favor.[21]

On appeal, the Supreme Court held that EPA had impermissibly interpreted the statute to require a source to undergo NSR solely on the basis of its potential to emit GHGs. The Court had previously held in *Massachusetts v. EPA* that GHGs fell into the statutory category of "air pollutants," but it now found that the pollutants that triggered NSR requirements were in a smaller category of *regulated* air pollutants. EPA therefore had the discretion to limit that category to exclude GHGs if including them would lead to absurd results. EPA acknowledged that applying NSR requirements to GHGs would have "calamitous consequences" if it interpreted the statute's 100 and 250 tpy thresholds literally. The agency's tailoring rule did not solve the problem, in the Court's view, because it represented an impermissible interpretation of the clear language of the statute. The good news for EPA was that it had reasonably interpreted the statute to require permitters to include GHGs in determining BACT for the so-called anyway sources that crossed the 100 or 250 tpy thresholds for other pollutants. The bottom line was that most electric power companies would have to install BACT to control GHG emissions in greenfield sources and when they undertook modifications that triggered the de minimis thresholds (usually 40 tpy) for regulated pollutants other than GHGs in existing plants.[22]

FutureGen 2.0

In August 2010, Energy Secretary Steven Chu gave carbon capture, utilization, and sequestration (CCUS) technology a shot in the arm when he announced that DOE had awarded $1 billion in stimulus money to a resurrected "FutureGen 2.0" project at a seventy-year-old, two-hundred-megawatt unit operated by Ameren Corporation in Meredosia, Illinois. The project would repower the plant with advanced oxy-combustion technology to capture and sequester 90 percent of its CO_2 emissions and eliminate nearly all SO_2, NO_x, PM, and mercury emissions. The Clean Air Task Force supported the project because it was convinced that generators would have to retrofit existing power plants with CCUS technology if the nation were to have any hope of reducing GHG emissions. The Sierra Club, however, challenged the air permit that the Illinois Pollution Control Board (IPCB) awarded to the plant, and local groups challenged EPA's first-ever permit under its new regulations governing the sequestration aspect of the project. While all of the appeals were pending, DOE in February 2015 announced that it was abandoning the FutureGen project for the second time.[23]

The 2010 Elections

Environmental regulation was a high-profile issue in the 2010 off-year elections as Republican candidates blamed many of the nation's ills on EPA's resurgent regulation. Tea Party groups like Americans for Prosperity purchased millions of dollars' worth of advertising, launched social media campaigns, and held town hall events that warned of a "train wreck" in the electric power industry as EPA finalized the regulations on its agenda. Advertisements were "laser targeted" at constituents of Democratic representatives in coal-producing states who had voted for the Waxman-Markey bill. The Republican Party captured 242 seats to the Democrats' 193 seats in the House and picked up 5 seats in the Senate to leave the Democrats with a 53–47 majority. One-half of the eighty-seven newly arrived House Republican freshmen questioned whether human activities contributed to global warming. And most of the remaining Republican members were simply unwilling to discuss the issue out of respect for the power of the climate change–denying Tea Party faction of the party and the politically powerful coal and electric power industries. A chastened President Obama pivoted away from promulgating new regulations in anticipation of the 2012 elections.[24]

Legislative Attempts to Forestall GHG Regulation

The House Republican leadership assigned top priority to legislation designed to stop EPA in its tracks. The "Energy Tax Prevention Act" that the House passed in April 2011 would have prohibited EPA from taking any action to address climate change and retroactively repealed all of the GHG regulations that EPA had recently promulgated. The bill failed in the Senate in April 2011 by a vote of 50–50, even though four Democrats from fossil fuel–dependent states voted in favor of it. The House leadership then attached the bill as a rider to must-pass bills. By the end of the first session of the 112th Congress, the House had cast twenty-seven votes to block or inhibit EPA's efforts to reduce GHG emissions, none of which succeeded in the Democrat-controlled Senate.[25]

CO_2 Controls in Individual Permits

The focus now shifted to individual NSR permit proceedings where the issue was how permitters would determine the "best available control technology"

(BACT) for CO_2 emissions. This turned out not to be the cumbersome exercise that industry officials had feared, and predictions that the tailoring rule would freeze new permit applications were quickly disproven. The guidelines that EPA published in November 2010 suggested that maximizing combustion efficiency was the best technology for reducing CO_2 emissions, but they also suggested that permitters consider fuel switching and CCUS technology as it became available. Since companies had a natural incentive to improve efficiency to conserve fuel, many state permitters concluded that the original plant design was BACT without any additional requirements. Environmental groups focused most of their attention on EPA's forthcoming new source performance standard (NSPS) for CO_2 emissions from power plants, which would set a floor for BACT in future permits.[26]

The Proposed New Source Performance Standard

In late March 2012, EPA proposed to limit CO_2 emissions from new (greenfield) electric power plants to 1,000 pounds per megawatt-hour (MWh), about one-half the rate of a typical coal-fired unit, averaged over a thirty-year period. The model technology upon which the agency based the standard was a well-operated gas-fired combined-cycle (GFCC) power plant. Although a GFCC plant could easily meet that standard, a new coal-fired plant could only comply by capturing at least 50 percent of its CO_2 emissions. The thirty-year time horizon, however, allowed a company to build a carbon capture–ready plant on the assumption that CCUS technology would be available in time to meet the thirty-year average. The proposal set aside for another day the thorny issue of modifications of existing power plants.[27]

The proposal attracted strong criticisms from the coal industry, mine worker and electrical worker unions, and coal-dependent power companies. They argued that it was unlawful for EPA to create a "supercategory" of fossil fuel–fired power plants that would effectively force the industry to abandon coal-fired plants altogether. They further argued that since CCUS was "decades away from commercial availability," EPA's thirty-year averaging proposal did not sufficiently reduce the uncertainties regarding its future availability to warrant investing in new coal-fired plants. They predicted that the standard would lead to a systemic overreliance on gas-fired plants that would in turn threaten the reliability of the grid if natural gas prices increased rapidly as they had in the past. Environmental groups and some power companies that were not heavily invested in coal argued that EPA should tighten the standard because many existing GFCC plants were achieving much lower emissions rates. They noted that Canada had recently imposed a standard for new and existing coal-fired plants of 926 lb/MWh.[28]

As the EPA staff pored over the public comments, the nation entered a presidential election year in which coal states would be critical to the outcome. The Obama administration was in no hurry to promulgate a final rule that might adversely affect the president's chances in those states.[29]

The 2012 Elections

President Obama downplayed environmental regulation during his 2012 election campaign. The president did not mention coal at all in his February 2012 energy speech, focusing instead on the jobs that new investments in clean energy would create. On election day, voters returned Barack Obama to office by a lopsided 303–206 electoral college margin. With the Republican Party still controlling the House by a substantial margin, however, the prospects for climate disruption legislation remained quite low. Most observers saw Obama's victory as a mandate for EPA to proceed ahead with its environmental initiatives. Congressional Republicans, however, did not read the elections as a signal to back away from their adamant opposition to EPA's climate change initiatives. The House scheduled a series of "messaging" hearings on climate change and EPA's proposed NSPS to generate public opposition.[30]

When Administrator Lisa Jackson resigned in December 2012, President Obama replaced her with Gina McCarthy, a straight-talking environmental health specialist with a dry wit and an openness to dialogue with regulated industries. Senate Republicans used McCarthy's confirmation hearings as a forum for castigating EPA's "regulatory onslaught" and its "garbage can of regulations and failures."[31]

President Obama's Climate Action Plan

No longer willing to allow Republican climate deniers to frame the debate, President Obama on June 25, 2013, announced an ambitious "Climate Action Plan" providing the steps his administration would take to address climate disruption. Among other things, the president pledged to reduce GHG emissions by 17 percent from 2005 levels by 2020. He ordered EPA to publish a new proposal for establishing emissions limitations for new power plants by September 20, 2013, and to finalize the rule "in a timely fashion." He further ordered EPA to issue proposed guidelines under section 111(d) of the Clean Air Act for regulating GHG emissions from existing plants by June 1, 2014, to finalize the standard by June 1, 2015, and to require states to modify their implementation plans to incorporate the new regulations by June 30, 2016.[32]

To the coal industry and many electric power companies, the Climate Action Plan was a call to arms. The National Mining Association increased its public relations campaign through radio advertising, media outreach, and a grassroots petition drive in coal-dependent states. The American Coalition for Clean Coal Electricity purged its staff, hired a new public relations firm, and launched a multimillion-dollar advertising campaign on television and social media to bring pressure on EPA to forego regulating existing sources. Environmental groups responded with rallies, screenings of films, and an advertising campaign targeting climate disruption deniers in Congress.[33]

The Revised Proposal

The firestorm of criticism that the agency encountered from the electric power industry following its March 2012 NSPS proposal convinced EPA's leadership that the single 1,000 lb/MWh emissions limitation standard for both gas- and coal-fired plants was too aggressive. The reproposal that EPA unveiled in September 2013 set separate standards for large gas-fired plants (1,000 lb/MWh) and coal-fired plants (1,100 lb/MWh). Since the standard for coal-fired plants was about 700 lb/MWh lower than emissions from the average supercritical boiler burning pulverized coal, new plants would still have to rely partially (30–50 percent) on carbon capture. The preamble pointed to four power plants with CCUS that were undergoing construction or soon would be as proof that the industry was capable of meeting a standard that did not require full-time reliance on that technology. The new proposal eliminated the thirty-year averaging period, but it allowed a new coal-fired plant to average emissions over a seven-year period if it agreed to an emissions limitation of 1,050 lb/MWh. Administrator McCarthy then began a series of road trips to more than a dozen cities to listen to comments on the rule.[34]

Republican Party operatives saw the reproposal and the prospect of section 111(d) restrictions on existing plants as a gift from the president that Republican candidates could use in the 2014 off-year elections to seal the party's majority in the House and win control of the Senate. An hour after Administrator McCarthy finished announcing the reproposal, thousands of emails went out to voters in the home states of seven vulnerable Democrats entitled "Democrats Side with Obama's Radical EPA over Local Workers, Business and Industry." The House held another series of messaging hearings with titles like "EPA Power Plant Regulations: Is the Technology Ready?" In January 2014 a group of eighteen Democrats called the Senate Climate Action Task Force took the offensive on the floor of the Senate, where they staged an overnight talkathon in support of EPA's GHG regulations.[35]

In March 2014, Murray Energy Corporation and a number of other companies sued EPA in a West Virginia federal district court alleging that the agency had failed to prepare an analysis of the impact of the reproposal on jobs as required by section 321(a) of the Clean Air Act. That section, which was not addressed to any particular rulemaking, required EPA to "conduct continuing evaluations of potential loss or shifts of employment which may result from the administration or enforcement of the provisions of this chapter." The lawsuit asked the court to enjoin EPA's proceeding ahead with the Climate Action Plan until after it had evaluated whether its Clean Air Act regulations over the past six years had resulted in job losses in the coal industry. EPA responded that section 321(a) required EPA to undertake periodic evaluations of the impact of its regulations on employment, but it did not require a separate employment impact evaluation for every major rulemaking, nor did it require EPA to focus on any particular industry.[36]

In October 2016, the West Virginia district judge ordered EPA to undertake a fresh analysis of the impact of all of its air and climate regulations on employment in the coal industry, and it gave the agency fourteen days to come up with a plan to accomplish that task. The court did not, however, grant Murray's request for an injunction against proceeding ahead with pending rulemaking initiatives. The following June, the Fourth Circuit Court of Appeals overturned the district court, holding that the decision how and when to implement the statutory employment evaluation was a matter that was committed to the agency's discretion by the Clean Air Act.[37]

The Proposed Clean Power Plan

Under section 111(d) of the Clean Air Act, EPA must promulgate guidelines for categories subject to new source performance standards for states to use in adopting standards of performance for existing sources to address pollutants, like carbon dioxide, that are not regulated as criteria or hazardous air pollutants. The standards have to reflect the "best system of emission reduction" that has been "adequately demonstrated," but the state may take into consideration additional factors, including "the remaining useful lives" of the affected sources. The agency had previously issued such guidelines on only nine occasions, one of which was the Bush administration's failed attempt to use section 111(d) instead of section 112's hazardous emissions standard authority to regulate mercury emissions.[38]

Once again, EPA initiated "public listening sessions" at cities throughout the country. Industry efforts to influence and, if necessary, combat the guidelines kicked into high gear. Lawyers, lobbyists, and Republican strategists met regularly in Washington, D.C., at the offices of the U.S. Chamber of Commerce, to put together a legal and political strategy. The small group rapidly expanded to a large network of attorneys, lobbyists, state officials, and Republican congressional leaders. Prior to one of EPA's listening sessions, coal companies transported several thousand miners to Washington, D.C., for a "Rally for American Energy Jobs" where they carried derogatory signs with slogans like "Save America, Impeach Obama."[39]

In a rare reversal, White House officials pressed EPA staff to complete the guidelines expeditiously. Determined to make climate disruption regulations a legacy issue for the Obama administration, White House aide John Podesta was "laser focused" on promulgating strong regulations on the president's ambitious schedule. The EPA staff spent late nights at the office trying to come up with guidelines that were sensible but also scientifically defensible. It also reached out to energy experts in DOE and FERC in an attempt to ensure that the regulations were consistent with the overall reliability of the national electricity grids.[40]

On June 2, 2014, Administrator McCarthy rolled out a proposal for "state-specific rate-based goals for carbon dioxide emissions from the power sector" based on the "best system of emission reduction" (BSER) and "guidelines for states to follow in developing plans to achieve the state-specific goals." The agency called the package the "Clean Power Plan" (CPP). It interpreted BSER to include far more than just installing an end-of-pipe pollution control technology. It believed that a "system of emission reduction" could include: (1) increasing the efficiency of coal-fired units; (2) shifting load from coal-fired units to gas-fired units; (3) shifting load from fossil fuel–fired plants to nuclear and renewables; and (4) demand-response programs. Using these "building blocks," EPA calculated the number of tons per megawatt-hour that each state could achieve from applying the most efficient combination of the four approaches and set mandatory interim (2022 though 2029) and final (after 2030) goals for each state. The state could achieve those rate-based goals by imposing emissions limitations on individual sources or by converting them to mass-based goals and establishing an in-state or multi-state cap-and-trade program. State plans could also meet their goals by enacting renewable portfolio standards and end-use energy efficiency programs. The states that already had such programs in place would get credit for reductions they had already accomplished.[41]

The coal industry and states with large fossil fuel production strongly opposed the proposal. The United Mine Workers and several other unions urged EPA to withdraw it, or, failing that, to make extensive changes aimed at softening the blow and lengthening the time for compliance. Coal-dependent electric power companies opposed the plan, predicting that it would precipitate another energy crisis. Small rural cooperatives that were almost totally dependent on coal worried about their ability to provide reliable power if they had to close older plants to meet the standard. A coalition of non-coal-dependent power companies generally supported the proposal. Splitting the difference, the head of the Edison Electric Institute said that most electric power companies could support the proposal, with "some important tweaks," chief of which would be softening the targets for the 2020 deadline.[42]

The mainstream environmental groups supported the proposal and rallied to EPA's defense. But they made it clear that it was just a first step toward a more comprehensive effort to address climate disruption. They urged EPA to add a provision suggesting that states establish programs to assist workers by protecting wages and benefits, providing training and education, and recognizing basic worker rights.[43]

Republicans in Congress leapt to the attack with a number of riders and stand-alone bills to stop the CPP from going into effect, but they went nowhere in the Democrat-controlled Senate. The Republican-controlled House also conducted another series of messaging hearings on various topics related to the proposed CPP with titles like "The Administration's Climate Plan: Failure by Design." The Democratic leadership in the Senate scheduled a competing series of hearings designed to bolster EPA's proposals. At one of the latter hearings, four former Republican EPA administrators—William Ruckelshaus, Lee Thomas, William Reilly, and Christine Whitman—testified in support of rapid government action to reduce GHG emissions.[44]

The 2014 Elections

The 2014 off-year elections delivered another blow to the Obama administration's efforts to address climate change. The Republican Party gained thirteen seats to give it a 247–188 majority in the House, and it picked up eight Senate seats to gain a majority of 54–46. It was now up to the White House to push its power plant initiatives through to completion over the opposition of the leadership of both houses of Congress. In his January 2015 State of the Union address, President Obama vowed that he would not let Congress "turn back the clock on our efforts." Since the prospects for enacting legislation aimed at stymieing EPA's efforts were dim, much of the Republican energy went into messaging hearings.[45]

The Final New Source Performance Standard

EPA unveiled the final power plant NSPS and CPP in early August 2015. EPA left the standard for new baseload gas-fired turbines at 1,000 lb/MWh. The standard for new coal-fired EGUs was 1,400 lb/MWh based on an "efficient new supercritical pulverized coal (SCPC) utility boiler implementing partial carbon capture and storage." The standard was considerably less stringent than the 1,100 lb/MWh proposal, a change attributable in part to EPA's conclusion that integrated gasification combined-cycle (IGCC) technology was not BSER and in part to cost concerns. To meet the 1,400 lb/MWh standard, a new supercritical pulverized coal boiler burning subbituminous coal would have to capture 16 percent of the unit's CO_2. Alternatively, a new coal-fired unit could meet the standard by co-firing with natural gas, an option that allowed companies to build new coal-fired plants without CCUS. EPA concluded that the "best system of emission reduction" did not have to be in place at any single plant, so long as EPA could demonstrate that the technology would work on new plants. SaskPower's Boundary Dam Project in Canada had come online in late 2014, and its CCUS unit was working better than expected at about 80 percent removal capacity with less parasitic energy loss than predicted. The agency also pointed to three other projects that were still in the works in California, Texas, and Mississippi. (Only the Texas project was ultimately successful.)[46]

The Final Clean Power Plan

The Clean Power Plan that EPA unveiled on the same day attracted more than 4.3 million comments, a new EPA record. EPA learned a great deal from the comment process, and it made important changes as a result. The agency adopted the same approach as the proposal, except that it did not include the fourth benchmark (demand response) in determining state goals. EPA explained that the term "system of emission reduction" was necessarily "limited to a set of measures that work together to reduce emissions and that are implementable by the sources themselves." It also concluded that CCUS was not yet available as a retrofit technology for existing plants and therefore not included in the building blocks. At the same time, EPA rejected industry arguments that building blocks 2 (shifting load from coal-fired plants to gas-fired plants) and 3 (shifting load from fossil fuel–fired plants to nuclear and renewables) were unlawful because they reached "beyond the fenceline." It concluded that the phrase "best system of emission reduction" was "capacious enough" to include off-site actions taken by the plant's operator and third-party actions taken pursuant to a commercial relationship with the operator. Indeed, the two building blocks were already being implemented by many generators as part of their day-to-day dispatch decisions.

Although EPA did not consider demand-response programs in setting state goals, they were still appropriate tools for a state to use (along with emissions limitations and cap-and-trade programs) in meeting their goals so long as they met the guidelines' general eligibility requirements. On the other hand, "out-of-sector" emissions reductions, like planting trees to sequester CO_2 and reducing mobile source CO_2 emissions, could be used only in states that adopted mass-based goals and trading programs. A state plan also had to include a process for reporting progress to EPA and for taking corrective action if the state did not reach the required goals in a timely fashion. To aid the states, EPA proposed model rules for both rate- and mass-based programs.

The agency recognized the "paramount importance of ensuring electric system reliability," but it noted that commenters had raised reliability concerns in the context of every standard that it had promulgated for the electric power industry since 1970. And in every case, the industry found ways to comply without posing any threats to reliability. Nevertheless, the agency provided a reliability safety valve for individual sources. When a conflict arose between the state plan's requirements for a specific EGU and system reliability because of an "extraordinary and unanticipated event" that presented "substantial reliability concerns," the "reliability-critical" EGU could meet an alternative standard established by the state for a period of up to ninety days.

EPA predicted that the regulations would reduce CO_2 emissions from the electric power sector by 32 percent below 2005 levels in 2030. It estimated that the combined climate benefits and health "co-benefits" for the rate-based approach would be between $32 and $54 billion per year in 2030. The overall compliance cost for the rate-based approach would be $2.5 billion in 2020, $1.0 billion in 2025, and $8.4 billion in 2030. The final guidelines would cause a net decrease of up to 34,000 job-years in the construction, electric power, coal, and natural gas industries. But they would cause an increase of 52,000 to 83,000 jobs in end-use efficiency industries. EPA encouraged the states to mobilize existing education and training resources and employ economic and labor market analyses to identify strategies to provide training assistance to displaced workers.[47]

Congressional Efforts to Defeat the Plan

With both Houses of Congress under Republican control, EPA had to endure another series of messaging hearings. A frustrated Democratic member, Rep. Bobby Rush (D-IL), complained at one hearing that the committee was examining EPA's Clean Power Plan "for the exceedingly umpteenth time" for no obvious purpose. Both Houses passed a joint resolution of disapproval under the Congressional Review Act using a special procedure in that law that prevented Senate filibusters, but President Obama quickly vetoed it.[48]

Judicial Review

The D.C. Circuit received a "torrent of lawsuits" challenging various aspects of the CPP. On January 21, 2016, the court denied the industry petitioners' requests for a stay of the rules. Citing the likelihood of massive power plant retirements and resulting electrical blackouts, the challengers petitioned the

Supreme Court to issue the stay. On February 9, 2016, the Supreme Court issued the requested stay by a 5–4 vote in a one-page order that did not provide any reasons for that unprecedented move. Although the D.C. Circuit agreed to expedite its consideration of the case, it was clear that the legal issues would not be resolved and the stay lifted until after the end of the Obama administration. Three days after the court issued the stay, Chief Justice John Roberts announced that Justice Antonin Scalia had passed away. With the court now evenly balanced between Republican and Democratic appointees, the outcome of the case on the merits would likely depend on which president appointed Scalia's replacement. After the Republican leadership in the Senate vowed to block any Obama nomination to the court for the remainder of the Obama administration, the significance of the 2016 election for climate change grew dramatically.[49]

State Implementation

Most state agencies started the drafting plans to implement the guidelines. California, which had a cap-and-trade program in place and was heavily invested in renewables, thought it could comply with EPA's guidelines with only minor adjustments. Montana concluded that the plan's flexibility allowed it to reach its goal without curtailing or closing any coal-fired EGUs. Many states that were challenging the proposal in court hedged their bets by preparing plans on the assumption that Hillary Clinton would win the 2016 election. Some states, however, dragged their feet in the hope that the courts would overturn the plan or that a Republican would become the next president. Six states enacted legislation requiring legislative approval of any state plans. Kentucky passed legislation prohibiting state agencies from submitting a plan to EPA, and Oklahoma's governor issued an executive order to the same effect.[50]

INTERSTATE AIR POLLUTION

The Obama administration inherited the D.C. Circuit's remand of the Bush administration's Clean Air Interstate Rule (CAIR). In addition to meeting the court's objections, the agency hoped to improve upon it. The proposed Cross-State Air Pollution Rule (CSAPR) that EPA published in August 2010 included more states, required greater emissions reductions, and established a more rapid pace than the CAIR. In a controversial move, EPA decided to promulgate federal implementation plans (FIPs) at the outset and to allow states to opt out of the FIPs only if EPA approved SIPs that protected downwind states as effectively as the FIPs. To meet the court's objections, the proposal used a "multi-factor analysis" that took into account "both air quality and cost considerations" to identify the portion of the state's contribution to downwind nonattainment that was in fact significant. To ensure that significant contributions were in fact eliminated, EPA developed emissions budgets for each of the states consisting of the emissions left after eliminating the emissions that significantly contributed to downwind problems. The proposed FIPs implemented the emission reduction requirements required to meet the state budgets with cap-and-trade programs.[51]

Environmental groups praised the proposal as a step in the right direction, but they urged EPA to tighten the caps because the technology to achieve lower caps was already widely available. Some electric power companies that did not depend heavily on coal also supported the proposal. The coal industry and coal-dependent power companies, however, were sharply critical of the proposal, arguing that compliance costs would far exceed EPA's $2.8 billion estimate. As always, the industry cautioned that it might not be possible to comply with the proposed rule without sacrificing reliability.[52]

Upwind states argued for more time to come up with their own SIPs.[53] Some criticized EPA's complex modeling exercises. For example, Texas objected to including it in the program on the basis of the model's prediction that power plants in Texas would significantly contribute to a single nonattainment area in Wisconsin. The Sierra Club, however, used a different model to demonstrate that coal-fired power plants in Texas had contributed to sixty nonattainment days in Oklahoma, twenty in Louisiana, and sixteen in Arkansas during the previous year.

A coalition of coal and utility companies spent around $35 million on television advertising criticizing EPA's proposal. One ad featured a businessman with a briefcase struggling to stay aboard a bucking bull while the narrator observed that "too many Americans are just trying to hang onto their jobs" and wondering why EPA was "in a rush to push regulations that would saddle Americans with higher energy costs and throw even more of us out of work?" Environmental groups conducted their own public relations campaign featuring an inflatable eighteen-foot inhaler for use in a traveling road show.[54]

The final CSAPR that EPA released in July 2011 retained the basic structure of the proposal. The agency found that emissions from twenty-seven states, including Texas, significantly contributed to nonattainment in downwind states. The rule identified state-specific emissions caps and corresponding emissions reductions for those upwind states, and it promulgated FIPs capable of achieving the required emissions reductions through "cost-effective and flexible requirements for power plants." Each of the twenty-seven states, however, had the option of replacing the regulations in the FIPs with state rules to achieve the required amount of emission reductions from any sources that the state cared to select. The FIPs allowed interstate trading, but, unlike the CAIR, they included provisions to ensure that the necessary emissions reductions occurred within each covered state.[55]

The battle now shifted to Congress, where the Republican leadership in both houses vowed to do what they could to keep the CSAPR from going into effect. The House committees held more messaging hearings with provocative titles like "Lights Out: How EPA Regulations Threaten Affordable Power and Job Creation." Senator Rand Paul (R-KY) introduced a joint resolution under the Congressional Review Act to overturn the CSAPR rule. The effort, however, flew in the face of public opinion. According to one poll 67 percent of likely voters supported the CSAPR and only 16 percent opposed it. The attempt failed when the Senate voted the resolution down by a 56–41 vote, with six Republicans voting against it and two Democrats voting for it.[56]

To the surprise of many observers, the Supreme Court upheld the CSAPR in April 2014. The court concluded that EPA had reasonably determined that the upwind states were significantly contributing to nonattainment in the downwind states. It rejected the D.C. Circuit's conclusion that EPA had to allocate responsibility for reducing emissions to states in a manner exactly proportional to each state's contribution to the nonattainment, because that ideal could never be achieved in a general regulation. If a state believed that EPA's allocation required too much, it could file an "as applied" challenge to the CSAPR in the D.C. Circuit. Furthermore, EPA was reasonable in considering cost as well as pollutant amounts in determining whether a state's emissions contributed significantly to downwind state nonattainment. Since EPA could not avoid the problem of choosing amounts to eliminate, the agency reasonably chose the amount that was less costly to eliminate. Employing cost as a criterion was also sensible because it subjected to stricter regulation the states that had done relatively less in the past to control pollution. Finally, the Court held that EPA did not have to give the states yet another opportunity to revise their SIPs to protect downwind states before promulgating its own FIPs.[57]

The case then moved back to the D.C. Circuit to resolve several remaining issues, including "as applied" challenges by individual states. In late July 2015, that court held that EPA's cost-based formula would cause thirteen states, including Texas, to reduce emissions of SO_2, NO_x, or both by more than was necessary to eliminate a significant contribution to nonattainment in downwind states. It therefore remanded the case with respect to those states for a new determination by EPA of fair emissions budgets.[58]

HAZARDOUS AIR POLLUTANTS

Abandoning the Bush administration's attempt to establish a cap-and-trade program for power plant mercury emissions, EPA decided to promulgate standards for mercury emissions reflecting the "maximum achievable control technology" (MACT) standard pursuant to section 112 of the Clean Air Act. At the outset of the Obama administration, the nation's 491 coal-fired power plants emitted forty-eight tons of mercury per year. The good news was that power plant mercury emissions had declined by 6.5 percent between 2000 and 2008. The bad news was that about half of the nation's lakes still contained fish with levels of mercury exceeding the levels necessary to protect public health.[59]

The Proposed Mercury and Air Toxics Standards

In March 2011, EPA proposed standards for new and existing coal-fired power plants that it referred to as the Mercury and Air Toxics Standards (MATS). The agency proposed to reaffirm the Clinton administration's finding that hazardous air pollutant standards were "appropriate and necessary" to control emissions of mercury, acid gases, and various other heavy metals. The standard for mercury emitted from large existing coal-fired units was 1.0 pound of mercury per trillion Btu (lb/TBtu) of heat input, and the standard for new sources in the same subcategory was 0.00010 pounds per gigawatt hour (lb/GWh) of electrical output. Both were based on dry sorbent injection, a technology that involved injecting activated carbon (or some other sorbent) impregnated with sulfur or iodine into the flue gas. EPA predicted that compliance with these standards would reduce mercury emissions from power plants by about 91 percent. Under the statute, existing sources would have up to three years from the date of promulgation to comply with the standards with the possibility of a one-year extension if the agency determined that it was needed to install the necessary controls.[60]

A coalition of fifteen environmental and public health groups were generally supportive of the stringent technology-based standards. Companies that were heavily invested in nuclear power and natural gas were well positioned to comply with the standards and tended to support them in the hope that they would force the retirement of older plants with a corresponding drop in the supply of electricity and increase in the price. The coal industry, coal-dependent power companies, organized labor, and conservative think tanks opposed the proposal. The Edison Electric Institute, however, sought a middle ground, accepting the need for regulation but suggesting detailed changes in the proposal.[61]

Opponents argued that EPA's "appropriate and necessary" finding was unlawful because it was not based on a careful balancing of the costs of compliance against the environmental benefits.[62] Environmental groups, some states, and some natural gas producers responded that the record clearly showed that mercury emissions posed a hazard to human health and the environment, and that was sufficient to justify the appropriate and necessary finding. Opponents argued that three to four years was not nearly enough time for existing power plants to develop and install the technology necessary to meet the standards. Delaying the deadlines would enhance grid reliability by giving companies more time to construct complying units and by giving managers of regional power grids sufficient time to evaluate and adjust to the impacts of retirements on grid reliability. Supporters, however, were confident that power companies could meet the three-year deadline in most cases. In a pinch, EPA could grant a one-year extension or even issue an administrative order permitting temporary violations of emissions reduction requirements.

Several coal-dependent companies threatened to retire dozens of existing coal-burning units, rather than install expensive technologies, if the MATS rule went into effect, and this would force them to lay off workers. Environmental groups acknowledged that utility companies would have to retire some old units, but they maintained that many of the retirements were likely to happen in the absence of the EPA rules because of low electricity prices and much less expensive natural gas due to the hydrofracturing revolution that was greatly increasing domestic supplies. Some jobs would be lost as a result of the rules, but retrospective studies of previous EPA rules concluded that they had had little, if any, impact on overall employment rates. Moreover, spending on environmental controls necessary to comply with the rules would create thousands of new jobs that would more than offset the jobs lost due to retirements.[63]

As the deadline for publishing the final rule approached, the American Coalition for Clean Coal Electricity ran a full-page advertisement in dozens of newspapers warning that the MATS was "the most expensive rule ever written for power plants" and that it would destroy 183,000 jobs. The American Lung Association responded with an ad featuring a young boy catching a snowflake on his tongue but leaving a deposit of pollution in his mouth that he exhaled as a cloud of smoke.[64]

The Final Mercury and Air Toxics Standards

In December 2012, EPA promulgated the final MATS. Rejecting the industry contention that it was obliged to take costs into consideration in making the "appropriate and necessary" finding, it reaffirmed the Clinton administration's finding. It left the 1.2 lb/TBtu mercury standard for existing coal-fired power plants in place, but it doubled the standard for new sources from 0.00010 lb/GWh to 0.00020 (lb/GWh). A number of other measures were included to ease the burden of compliance. At $9.6 billion in compliance costs, however, it was still the most expensive rule that EPA had ever promulgated under the Clean Air Act. EPA estimated that it would require the industry to retire about 4.7 gigawatts of power, but those older plants represented less than one-half of 1 percent of the industry's current generating capacity. The agency predicted that the rule would prevent around 11,000 premature deaths, 4,700 heart attacks, and 130,000 asthma attacks per year, mostly as a by-product of reductions in $PM_{2.5}$ emissions that would result from retirements and the installation of technologies needed to reduce mercury emissions. Although it found it difficult to monetize the benefits of reduced mercury emissions, it calculated that they would come to at least $6 million. When it added in the "co-benefits" of reduced $PM_{2.5}$, however, its estimate came to $37–90 billion. Administrator Jackson encouraged states to make "very liberal use of the fourth year" extension, and she suggested that the agency would exercise its enforcement discretion to give some companies a fifth year to bring reliability-critical units into compliance.[65]

Congressional Reversal Efforts

As the states began to implement the MATS, politicians and coal-dependent power companies broke out the tried and true "train wreck" and "war on coal" metaphors to attract public support for legislation overturning the rule or extending its deadlines. Senator James Inhofe (R-OK) introduced a joint resolution under the Congressional Review Act to nullify the regulation. As the resolution came to the Senate floor, an advocacy group called American Commitment spent $1 million on television ads in Tennessee and West Virginia aimed at generating pressure on Senator Lamar Alexander (R-TN) and Senator Jay Rockefeller (D-WV) to support the resolution. The effort backfired. In a stirring speech against the resolution, Senator Rockefeller chastised the coal industry for "refusing to entertain any middle ground, and denying even a hint of legitimacy for the views on the other side." The Senate defeated the resolution in June 2012 by a 46–53 vote. Several freestanding bills were introduced in Congress to extend the MATS deadlines and create a safety valve for reliability-critical units, but none of them passed.[66]

Implementation of the MATS

Companies that relied mostly on nuclear power and natural gas were not greatly affected by the regulations. As they put pen to paper, large coal-dependent companies discovered that installing the necessary controls would not be nearly as expensive as they had predicted. For example, First Energy's estimate of overall compliance costs fell from early predictions of $3 billion to $975 million, and American Electric Power cut its estimate for its Ohio plants from $1.1 billion to $400 million. One reason for the surprising decrease was the fact that dry sorbent injection was as effective as scrubbers and considerably less expensive. Another reason was the willingness of states to grant fourth-year extensions. By the end of 2014, states had granted 421 requests for extension to plants representing over 142 gigawatts of capacity. FERC recommended fifth-year extensions for a handful of plants, and EPA granted all of them. As we shall see, many companies decided to retire older coal-fired units, rather than comply with the MATS.[67]

Since the standard for new sources was so stringent that only a single plant in the country achieved it, companies expressed skepticism about building new coal-fired power plants in the future. As natural gas prices continued to decline, it was easier and cheaper to build gas-fired plants that did not emit mercury. Companies undertook no new coal-fired projects during President Obama's second term.[68]

Judicial Review

The Supreme Court in a 5–4 opinion held that EPA had unreasonably interpreted the word "appropriate" to allow it to disregard the cost of compliance. The Court found that the word "appropriate" invited EPA to consider all of the relevant factors in deciding whether to regulate power plants, and one of those relevant factors was cost. Alluding to the industry argument that the MATS imposed $9.6 billion in costs for benefits of $4 to $6 million in benefits (excluding the co-benefits of reduced PM emissions), the Court could not say that it was "even rational, never mind 'appropriate,' to impose billions of dollars in economic costs in return for a few dollars in health or environmental benefits." In the Court's view, "reasonable regulation ordinarily requires paying attention to the advantages *and* the disadvantages of agency decisions."[69]

In mid-December 2015, the D.C. Circuit remanded the rule to EPA without vacating it. EPA now had to make the "appropriate and necessary" determination once again, but this time it had to consider cost. EPA published a supplemental "appropriate and necessary" finding in April 2016, concluding that "under every metric examined, the cost of MATS [was] reasonable." Based on the formal cost-benefit analysis in its previously prepared regulatory impact analysis, Administrator McCarthy concluded that "the benefits (monetized and non-monetized) of the rule are substantial and far outweigh the costs." Several states and trade associations challenged the supplemental finding in the D.C. Circuit, but, as we shall see, the Trump administration decided to do its own cost-benefit analysis.[70]

NAAQS

The Obama administration was more active than any of its predecessors in revising the national ambient air quality standards (NAAQS). Although it encountered serious resistance from the Office of Information and Regulatory Affairs (OIRA) and White House staff, it tightened the standards for ozone, PM, SO_2, and NO_x. As in Chapter 8, we will focus primarily on the ozone standard, but the agency also addressed the lingering problem of the effects of short-term bursts of SO_2 from power plants on persons suffering from asthma.

Ozone

Not long after Administrator Jackson was sworn in, EPA announced that it would reconsider the 2008 NAAQS for ozone.[71] After persuading the D.C. Circuit to postpone its consideration of the standards, the agency stayed implementation for the duration of the review. Trusting the Obama administration to do the right thing, the states and environmental groups that had challenged the Bush administration standard acquiesced in the delay.[72]

Since this action was technically a reconsideration of the previous standards, Administrator Jackson decided not to consider any scientific studies that became available in the intervening years. In January 2010, EPA published a proposal to replace the Bush administration's identical primary and secondary standards of 0.075 ppm with a primary standard at some point between 0.060 and 0.070 and a "cumulative, seasonal" secondary standard of somewhere between 7 and 15 ppm-hours. The agency predicted that the proposal would increase the number of counties that did not meet the standard from 322 to 515 counties under a 0.070 standard and to 650 counties under a 0.060 standard. This would place a heavy burden on the states to amend their state implementation plans (SIPs) to impose new restrictions on sources of ozone precursors like power plants. It would also require another round of reductions under the just-promulgated CSAPR rule. The agency estimated that the overall cost of meeting a 0.060 standard would be between $52 and $90 billion in the year 2020. On the other hand, the proposal would yield between $13 billion and $100 billion in benefits by reducing ozone-induced morbidity and mortality.[73]

The proposal attracted predictable responses from the affected groups. Public health and environmental groups praised the agency for improving on the Bush administration standard, and they urged EPA to set the standard at the 0.060 ppm level. They also supported EPA's adoption of the cumulative, seasonal form for the secondary standard and advocated a standard at the low end of that range. A united electric power industry rejected both proposed standards as "lacking in scientific justification." One industry-prepared report estimated that complying with the standard could cost more than $1 trillion and result in the loss of more than 7.3 million jobs. Pointing to areas in Wyoming where background ozone levels were reported to exceed 0.060 ppm and 0.070 ppm, Arch Coal wondered how such areas could "come into attainment when they were already at the background level for ozone."[74]

Knowing that a rule with a high price tag would not be well received at some offices in the White House, Jackson met three times with White House Chief of Staff William E. Daley, a former lobbyist for the Chamber of Commerce who President Obama had brought on board to smooth relations with the business community after the 2010 elections.[75] At that point, the agency's draft regulation had set the level for the eight-hour primary standard at 0.065 ppm. Daley queried Jackson extensively on the costs and the benefits of the proposal as well as on timing issues. Jackson and the agency staff then worked out what she believed to be a compromise that set the primary standard at 0.070 ppm and included additional measures to provide some flexibility to the states in implementing the standard. But she failed to anticipate the clout of OIRA's director Cass Sunstein, a former colleague of President Obama at the University of Chicago School of Law, who was never persuaded that it was a good idea to revisit the ozone standard in the first place.

When word got out that a draft of the final rule had arrived at OIRA, the affected industries put on a full-court press to force the agency to withdraw it. The multilevel strategy involved a massive lobbying campaign aimed at the White House and Congress, a letter-writing campaign from industry leaders, and an advertising campaign aimed at swaying Washington policymakers. The ads claimed that the new rules would impose $90 billion in unnecessary regulatory costs on American businesses, kill thousands of jobs in a time of high unemployment, and bankrupt much of the industrial heartland.

White House Chief of Staff Daley, OIRA Administrator Sunstein, and EPA Assistant Administrator Gina McCarthy met in the West Wing of the White House with representatives of industry groups on August 16, 2011, to hear their complaints about the ozone standard. During the meeting the head of the American Petroleum Institute produced maps showing that much of the burden of compliance would fall on states that President Obama had won in the 2008 election, some by narrow margins. Before an industry spokesperson had a chance to spell out the regulation's political implications, Daley cut him off with a terse "I got that." Later that day the same officials met with public health and environmental groups to hear their arguments in support of a more stringent standard. After listening to their presentation, Daley asked whether they were aware of the health aspects of unemployment. When one participant related the results of polls demonstrating strong public support for clean air, Daley cut him off with an expletive.

On September 1, 2011, President Obama summoned Administrator Jackson to the Oval Office where he told her that he had decided against going forward with a more stringent ozone standard because of the cost and uncertainty that it would impose on industry and states containing ozone nonattainment areas. Despite this direct public repudiation of thousands of hours of extra effort by the EPA staff, a three-page White House press release praised "the hardworking men and women at the EPA as they strive every day to hold polluters accountable and protect our families from harmful pollution." Administrator Jackson had little to say about the clear usurpation of her statutory authority to establish NAAQS. She considered resigning but concluded that it would be a "futile gesture." Environmental groups challenged the action, but the D.C. Circuit declined to hear the appeal on the ground that the action was not final.[76]

The agency staff now faced the task of incorporating dozens of studies that had been published since the 2008 standard into a new integrated science assessment and deciding whether to issue a new standard by October 2015. The final standard that EPA promulgated on October 1, 2015, lowered the primary standard from 0.075 ppm to 0.070 ppm, the very top of the range suggested by the agency's Clean Air Scientific Advisory Committee (CASAC) and the EPA staff. Abandoning the cumulative seasonal approach as it had during the George W. Bush administration, the agency made the secondary standard identical to the primary standard. Given the improvements that coal-fired power plants were already making to meet the MATS rule and EPA's forthcoming climate change regulations, the agency predicted that the new standard would affect only five electric generating units, and the owners of those units could comply by installing readily available selective catalytic reduction technology.[77]

Sulfur Dioxide

At the outset of the Obama administration, the annual primary NAAQS for SO_2 of 30 parts per billion (ppb) and the twenty-four-hour standard of

140 ppb had been attained in all areas of the country, but short-term bursts of SO_2 from power plants still posed a threat to public health. The risk assessment that the agency staff circulated in August 2009 concluded that a one-hour standard in the range of 50–75 ppb would offer far greater protection against five-minute SO_2 bursts than the existing twenty-four-hour standard. When EPA sent a draft notice of proposed rulemaking proposing a one-hour standard in the range of 50–100 ppb to OIRA, the economist assigned to the rule objected on the unlawful ground that compliance would be too expensive for the electric power industry. Ignoring OIRA's objections, Administrator Jackson in December 2009 proposed to revoke the existing twenty-four-hour and annual standards and to substitute a one-hour standard in the range of 50–100 ppb. EEI joined other industries in criticizing the science that EPA relied on to support the proposal. Environmental groups applauded the decision to establish a one-hour standard, and they urged EPA to promulgate a standard at the lower end of the range. The final rule that EPA published in June 2010 established a one-hour standard of 75 ppb. It was the first revision of the primary standard in almost forty years. The agency predicted that the standard would cost about $1.5 billion over the next ten years and prevent 2,300 to 5,900 premature deaths and 54,000 asthma attacks annually. As usual, the electric power industry challenged the standard, but the D.C. Circuit upheld EPA's action in every regard. Asthmatics could breathe easier as the states implemented the standards over the next few years.[78]

COAL ASH

On December 22, 2008, a dike at the Tennessee Valley Authority's Kingston Power Plant ruptured, and the contents of an eighty-four-acre impoundment containing more than a billion gallons of coal combustion residuals (CCRs, or coal ash) poured down a hillside and into the Emory River.[79] The spill was so forceful that the liquid waste flowed across the riverbed and inundated more than three hundred acres of land on the opposite shore, uprooting trees, destroying three homes, and damaging dozens of others. Although no one was killed, the catastrophic breach caused millions of dollars in property damage and resulted in incalculable environmental damage to the Emory River and the Clinch River into which it flowed.[80]

It was the worst CCR spill in U.S. history, but it was not the first, nor would it be the last. The Kingston catastrophe did, however, focus the nation's attention for a brief period on the health and environmental risks posed by coal ash, and it stimulated environmental groups to demand greater protection from the Congress and EPA, both of which had, until the spill, treated the electric power industry's CCR disposal problem with kid gloves. EPA Administrator Lisa Jackson announced in March 2009 that her agency was in the process of developing regulations to regulate coal ash.[81]

The 136 million tons of CCRs that the nation's 495 coal- and oil-fired power plants generated in 2008 represented a significant share of the nation's total solid waste stream. Environmental groups wanted EPA to characterize coal ash as a hazardous waste that had to be disposed of in facilities that contained liners, impermeable covers, leachate detection systems, and other requirements in regulations that EPA had promulgated under Subtitle C of the Resource Conservation and Recovery Act (RCRA). But that designation would require EPA to conduct a rulemaking to lift the 1980 Bevill Amendment's exemption for CCRs (discussed in Chapter 7).[82]

Industry groups warned that stringent regulations could have devastating economic consequences and could stigmatize beneficial reuse of CCRs in cement, wallboard, and other uses and cause the secondary market for recycled ash to dry up. Since recycling absorbed about 45 percent of the coal ash produced annually, loss of the recycling market would mean that the country would have to come up with space to dispose of almost twice as much waste. Adding that much waste to the two million tons per year that were currently disposed of at the nation's twenty-one licensed hazardous waste disposal facilities would overwhelm existing disposal capacity and raise the cost of disposing far more dangerous materials in those facilities. Characterizing CCRs as hazardous waste would also impose greater costs on power plants that would result in higher prices for electricity. The industry expressed a willingness to accept further regulation of CCR disposal, but only if it was accomplished by the states pursuant to guidelines issued by EPA under Subtitle D of RCRA.[83]

Environmental groups responded that there was no objective evidence that calling coal ash a hazardous waste would stigmatize beneficial reuse of that substance. In any event, some so-called "beneficial" uses, like using coal ash as fill to "reclaim" abandoned quarries, were in fact detrimental because they posed a high risk of leaching toxic metals into the environment. As demonstrated by the Kingston spill, state regulatory programs were at best uneven and at worst nonexistent. For example, Texas (the largest producer of coal ash) did not require a permit for CCR waste managed on-site, which meant that coal ash impoundments were subject to less regulation than landfills that received household garbage.[84]

The draft proposed rule that EPA forwarded to OIRA in October 2009 characterized CCRs as hazardous wastes. It based that determination on the potential for toxic constituents to migrate from impoundments into groundwater and on the need to avoid catastrophic spills like the Kingston breach. During the seven months that OIRA spent studying the proposal, its staff met on forty-seven occasions with stakeholders, approximately two-thirds of whom were with companies and industry groups.[85]

The notice of proposed rulemaking (NPRM) that the agency published in June 2010 looked nothing like the draft that EPA had submitted to OIRA. The preamble announced that the agency was "revisiting" the 1993 Bevill determination for CCRs in light of the Kingston catastrophe. EPA's assessment of the risk of another catastrophic release revealed that 109 of the 584 existing surface impoundments had either a "high" or a "significant" hazard potential rating. In a survey conducted by the Association of State and Territorial Solid Waste Management Officials, twenty-four of the thirty-six responding states did not require CCR impoundments to have liners, and only thirteen regulated the structural integrity of surface impoundments. Although groundwater monitoring requirements were "a minimum for any credible regulatory regime," twenty-two of the thirty-six states lacked such requirements.

The NPRM set out three alternatives for regulating CCR disposal. First, the agency could reverse the 1993 Bevill Amendment conclusion and regulate CCR waste as a hazardous waste under Subtitle C. Beyond the generally applicable requirements for hazardous waste disposal facilities, the NPRM proposed requirements for dam safety and stability for impoundments. Second, EPA could promulgate national standards under Subtitle D for states to apply in their programs for disposing of nonhazardous waste. At the very least those standards would require liners, groundwater monitoring, postclosure care, and stability requirements. The third proposal would be identical to the second proposal, except that existing surface impoundments would be grandfathered.[86]

Proponents of stringent regulation received a boost in February 2014 when the nation's attention turned to another massive coal ash spill, this time from an impoundment at Duke Energy's recently retired Dan River power plant into the Dan River. EPA had issued a technical report in 2009 concluding that the Dan River impoundments were "significant hazard potential structures," but lax oversight by the state agency allowed Duke to take no action to abate that potential. Duke later pleaded guilty to nine misdemeanor counts of unlawfully discharging pollutants and failing to maintain pollution control equipment at five of its North Carolina plants.[87]

The final regulations that EPA published in April 2015 did not characterize CCRs as hazardous wastes. Instead, they established nationally applicable minimum criteria under Subtitle D for the safe disposal of CCRs. The minimum criteria addressed structural failures of CCR surface impoundments, groundwater contamination from improper management of CCRs, requirements for liners in new impoundments, and fugitive dust emissions. The criteria applied to on-site and off-site facilities that were actively receiving CCRs and to inactive impoundments at active power plants, but they did not apply to inactive off-site landfills and inactive impoundments at retired power plants.[88]

EPA's reliance on Subtitle D meant that states did not have to establish permit programs and that the federal criteria could only be enforced by citizen groups under the untested theory that noncomplying disposal constituted prohibited "open dumping." In late 2016, however, Congress amended RCRA to allow states to seek EPA approval of permit programs for coal ash storage and disposal facilities and require EPA to establish permit programs for states that failed to submit their own. States with EPA-approved programs could file enforcement actions that, if diligently prosecuted, would preempt citizen lawsuits. The amendments also authorized EPA to undertake inspections and file enforcement actions. In the judicial challenges to the rule that both industry and environmental groups filed, EPA settled part of the litigation by agreeing to a voluntary remand of four industry challenges. And in November 2017, the D.C. Circuit agreed with environmental groups that several aspects of the regulations, including the exemptions for inactive off-site landfills, were arbitrary and capricious, and it remanded those aspects of the regulations to the agency. It would be up to the Trump administration to decide how to respond to the amendments and the remand.[89]

COOLING WATER INTAKE STRUCTURES

It took all of President Obama's first term for EPA to prepare a proposed response to the Second Circuit's remand of its cooling water intake structures (CWISs) rule for existing power plants. The proposal that EPA issued in late

March 2011, just after the 2010 elections, required all existing facilities withdrawing more than two million gallons per day to meet either a design or a performance standard for impingement mortality based on state-of-the-art traveling screens. Alternatively, the source could install closed-cycle cooling technology. The preamble explained that the agency did not propose to require closed-cycle cooling because it was more than ten times as expensive as traveling screens. The proposal allowed permitters (typically state agencies) to establish entrainment controls reflecting the best technology available (BTA) on a case-by-case basis, taking into account factors specified in the proposal, including cost and "monetized and non-monetized benefits of controls." EPA estimated that more than half of the affected facilities already complied with the standards.[90]

Deeply disappointed by what they called a cost-dominated approach that would do little to protect marine and aquatic organisms, environmental groups urged EPA to prescribe closed-cycle systems as BTA. They further argued that state permitting agencies lacked the resources and expertise to make case-by-case BTA judgments that would minimize harm to imperiled organisms. They also opposed requiring permitters to base entrainment requirements on cost-benefit balancing, which, they believed, systematically undervalued benefits. A unified electric power industry was pleased that EPA did not mandate closed-cycle technology, but it urged EPA to allow permitters to make impingement BTA determinations as well as entrainment determinations. At most, EPA should provide a "safe harbor" list of pre-approved technologies that companies could install in lieu of meeting the numerical requirements.[91]

The final regulations that EPA promulgated in mid-August 2014 gave owners of power plants seven technology options for reducing impingement mortality that they could employ by demonstrating to the permitter that the selected technology would perform as well as traveling screens. Four "pre-approved" technologies provided the safe harbor the industry had requested. The regulations allowed the permitter to balance costs and benefits on a site-specific basis in determining whether to require any additional efforts to protect against entrainment. EPA expected that existing sources could meet the standard without retrofitting closed-cycle cooling systems. Therefore, the rule would have "relatively minor economic impacts."[92]

Environmental groups were outraged that EPA had so thoroughly capitulated to the industry in making a weak proposal even weaker. So much was left to the discretion of state permitters that the rules would "do almost nothing to protect our waterways and fisheries from the destructive impacts of power plants." The electric power industry was generally pleased with the regulations. But it remained concerned that the regulations did not give permitters enough discretion. The Second Circuit upheld the regulations against challenges from both sides.[93]

REGIONAL HAZE

As the Obama administration assumed office, most of the struggles over EPA's regional haze guidelines were over, and attention had shifted to the states as they modified their state implementation plans (SIPs) to put the rules into effect. EPA hoped to set an example for the states as it wrote a federal implementation plan (FIP) for Arizona Public Service's (APS) forty-five-year-old, 2,000-megawatt Four Corners power plant in northwest New Mexico, one of the big dirties of the West. The largest single source of NO_x emissions in the country, its uncontrolled NO_x emissions significantly impacted more than sixteen national parks and wilderness areas. The plant was of huge interest to the Navajo Nation because the associated coal mines employed hundreds of tribe members and because the mining companies and APS paid it royalties and taxes.[94]

In October 2010, EPA proposed selective catalytic reduction for NO_x emissions as the best available retrofit technology for all five units at the Four Corners plant and new baghouses for three of them. It estimated that SCR would reduce NO_x emissions from 39,000 tpy to 6,000 tpy, or an 80 percent reduction. Environmental groups preferred that APS retire the plants, but they supported EPA's proposal. APS argued that SCR technology would provide very little more protection to the Class I areas than much less expensive combustion modification. In early November 2010, APS announced that it would install SCR in the plant's two newest units at a cost of around $426 million and close the three remaining units in 2013 or 2014. In August 2012, EPA published a federal implementation plan for the plant that required APS to do just that, and the Tenth Circuit Court of Appeals upheld the FIP in July 2014. The Four Corners litigation set a precedent for post-combustion controls like SCR as BART, rather than less expensive combustion modification.[95]

As the agency reviewed the forty regional haze SIPs that the states had submitted and began to write FIPs for states that did not submit adequate SIPs, it approved SIPs for a few states like Colorado and Oregon that took serious steps to protect national parks and wilderness areas. In March 2010, Colorado's governor Bill Ritter and a coalition of legislators, energy companies, and environmental groups agreed on pathbreaking legislation called the Clean Air–Clean Jobs Act that required Xcel Energy to secure approval by the Colorado Public Utilities Commission of a plan to reduce NO_x emissions from its coal-fired plants on the front range of the Rocky Mountains by up to 80 percent from 2008 levels. The expectation was that the NO_x reductions would enable the Denver area to attain the 2008 ozone NAAQS and to meet the state's obligations under EPA's regional haze rule. Over the strong objections of the coal industry, the commission approved a plan in December 2010 that required Xcel to close five coal-fired units at its Cherokee, Arapahoe, and Valmont plants by the end of 2013; convert one coal-fired unit at each of its Cherokee and Arapahoe plants to natural gas; and spend $340 million on SCR technology at the remaining coal-fired units at its Pawnee and Hayden plants. EPA easily approved Colorado's regional haze SIP based on that plan in September 2012.[96]

Other states like Texas and Oklahoma got into vigorous and lengthy disputes with EPA over their implementation plans. The Oklahoma Department of Environmental Quality (ODEQ) submitted a SIP revision that reflected an agreement between ODEQ and Oklahoma Gas & Electric (OG&E) under which OG&E would install low NO_x burners with overfire air on all seven of the BART-eligible units at its Muskogee, Sooner, and Seminole stations. But the agreement also allowed OG&E to continue to burn low-sulfur coal at the plants instead of installing scrubbers based on ODEQ's finding that scrubbers would not be cost-effective. In March 2011 EPA approved the plan's NO_x and PM provisions, but it disapproved the SO_2 provisions because ODEQ used unrealistically high cost estimates in concluding that scrubbers were not cost-effective. EPA then promulgated a federal implementation plan that required OG&E to achieve emissions limitations for SO_2 that could be achieved either by installing scrubbers on four units or switching them to natural gas. Oklahoma and OG&E challenged the FIP in the Tenth Circuit Court of Appeals, but the court upheld EPA's action in all regards.[97]

ENFORCEMENT

The Obama administration moved rapidly to reinvigorate the new source review enforcement initiative that had remained dormant throughout much of the George W. Bush administration. In February 2009, EPA and DOJ announced a new National Power Plant Enforcement Initiative under which EPA issued another round of information requests, notices of violation, and lawsuits. The agency went after some of the big dirties that had not already been sued, like Luminant's Big Brown and Martin Lake plants in Texas. But it also focused for the first time on smaller power plants, including some owned by municipalities. By mid-2012, EPA had investigated most coal-fired units in the country and found that more than half of them had violated the NSR regulations.[98]

The initiative also included a major effort to settle pending cases. Between January 2009 and December 2011, EPA settled nineteen cases in which electric power companies agreed to pay $38.2 million in fines, spend $102.5 million on supplemental environmental projects, spend $8.6 billion on pollution control equipment, and retire or repower twenty-nine coal-fired units. In addition, EPA entered into an April 2011 consent decree with TVA, under which TVA agreed to pay $10 million in civil penalties, invest $350 million in supplemental environmental projects, retire eighteen of its oldest units at three plants over the next six years, and spend up to $5 billion to add SCR and scrubbers to its fifty-nine remaining units or repower or retire them between 2012 and 2018. Some of the retirements and pollution control expenditures may have been undertaken anyway to comply with EPA's Cross-State and MATS rules. But the supplemental environmental projects, like Dairyland Power's commitment to spend $5 million over a five-year period on a major solar energy project, remediation of national forests and parks, and home weatherization projects, were attributable exclusively to the lawsuits.[99]

The lawsuits were not always successful. The enforcement initiative hit a major stumbling block when several courts dismissed some or all of EPA's claims for civil penalties because the violations occurred more than five years prior to DOJ's filing the lawsuits and were therefore barred by the general five-year federal statute of limitations. DOJ took the position that a modification was a permanent change that constituted a continuing violation of the NSR requirements so the statute of limitations was not triggered as long as the modification remained in place. But six courts of appeals rejected that theory, holding that the relevant modification was a single event that triggered the five-year filing period. Employing this "single event" theory, some courts also dismissed claims against successor corporations to the companies that authorized the challenged modifications.[100]

CHANGING ECONOMIC REGULATION

The Obama administration put a high priority on demand response (DR), distributed generation (DG), and smart grid technology. In March 2011, FERC promulgated a regulation (Order No. 745) governing demand response compensation in wholesale markets that required independent system operators (ISOs) and regional transmission organizations (RTOs) in most cases to pay the market price for the electricity that participants in demand response programs did not consume. In demand response programs, consumers enter into arrangements with distributors under which they limit their use of electricity during peak hours and in other situations in which overall demand on the grid is high. Consumer and environmental groups joined large electricity consumers in supporting the regulation as "critical" to reducing barriers to demand response, which would in turn reduce GHG emissions and increase system reliability. Critics of the proposal, including system operators who did not want to administer the program and merchant generators who would sell less power if more consumers participated in the program, argued that the one-size-fits-all regulation was an unwarranted subsidy for consumers who took advantage of the program. The Supreme Court decision upholding the regulation in January 2016 represented a major victory for environmental groups, the manufacturers of smart meters and energy-efficient appliances, and the nascent industry of demand response providers.[101]

While the Obama administration was occupied with reducing GHG emissions from new and existing power plants, many state legislatures and public utility commissions were consumed by "net metering wars" over DG. DG reduces demand on power plants by allowing end users to generate electricity and sell what they do not use to the grid in states with net metering laws that require distribution utility companies to pay for the electricity whether they need it or not. End users can rely on rooftop solar panels and backyard wind turbines, which do not emit pollutants, for DG. But in most states they can also employ fossil fuel–powered internal combustion engines and biomass generators that emit both conventional pollutants and CO_2.[102]

During the early years of the Obama administration, DG soared as large commercial users invested heavily in solar rooftop panels and backup generators. Large solar energy vendors offered to install rooftop solar panels in residences for free in return for a share of the amount that the utility company paid the customers for electricity not consumed or returned to the grid.

By early 2016, forty-four states had established net metering policies, and net metering wars raged in at least twenty-two of those states as electric utility companies and industry-funded advocates like the American Legislative Exchange Council (ALEC) spent millions of dollars lobbying state legislatures and public utility commissions for radical changes to those policies.[103]

Distribution companies complained that customers who employed DG paid a smaller share of the distributor's fixed costs than other customers, even though they needed the distributor's power to be available on a 24–7 basis to provide electricity when the sun was not shining or the wind was not blowing. If distributed generators did not pay their fair share, grid reliability would erode as older plants retired and could not be replaced with new plants. Worse, compensating DG customers whether or not their electricity was needed provided an incentive to non-DG customers to install generation technologies, and that could lead to a "death spiral" in which a smaller and smaller contingent of non-DG users were forced to bear a larger and larger share of the utility company's fixed costs.[104]

Proponents of DG responded that consumers providing DG should only have to pay for so much of the electrical infrastructure as they actually used. Some cross-subsidization was warranted because all customers benefited from the lower GHG emissions, reduced congestion, deferred investment in new generation capacity, and increased reliability that DG provided. Talk of a "death spiral" was overblown, they argued, because consumers would not migrate to DG nearly as rapidly as opponents suggested. To the contrary, the rapid move toward greater electrification of transportation and heating would ensure healthy markets for utility-supplied electricity for the foreseeable future.[105]

Several state regulators reacted to the pressure from utility companies by adopting steep tariffs for DG. The Texas Public Utility Commission approved a minimum monthly charge of $30 for residential and $39 per month for commercial DG providers. The Hawaii Public Utilities Commission and the Arizona Corporation Commission voted in 2015 and 2016 to end net metering altogether. Several other states, however, resisted calls for adding charges to the bills of distributed generators.[106]

Environmental groups strongly supported DG, but they vigorously opposed the use of gasoline- and diesel-fueled backup generators, because their air emissions were generally uncontrolled and they were more difficult to monitor than large power plants. Indeed, specialty companies were remotely controlling multiple internal combustion generators to create "virtual" power plants that could sell power to system operators for less than conventional power plants that had to pay for emissions controls. Supporters of greater use of backup generators argued that emissions were actually lowered by 15–30 percent when diesel generators substituted for natural gas-fired power plants because they were subject to fewer start-up emissions than large gas turbines.[107] Although the net metering wars subsided somewhat during the Trump administration, they are still continuing in some states.

THE 2016 ELECTIONS

The 2016 presidential election presented a sharper contrast on environmental issues than any since the 1980 election that sent Ronald Reagan to the White House. Four of the five top Democratic candidates pledged to continue President Obama's environmental initiatives. The party's eventual candidate, Hillary Clinton, offered an energy and climate platform in July 2016 that called for greatly expanding renewable power during her first term and expending $30 billion to aid communities that suffered employment losses from power plant shutdowns and coal mine closures. All but one of the Republican candidates promised to roll back EPA's power plant regulations, and most denied the basic science underlying global warming. When he emerged as the frontrunner, Donald Trump expressed his view that "the concept of global warming was created by and for the Chinese in order to make U.S. manufacturing noncompetitive." Former vice president Al Gore predicted that the Republican Party would be badly damaged by the refusal of most of its prominent figures to acknowledge the reality of human-caused climate disruption. He was wrong. Trump won the election by a margin of 306 to 232 electoral votes.[108]

CONCLUSIONS

Like President Bill Clinton, President Barack Obama supported environmental regulation in principle, and he appointed strong women to head EPA. Like President Clinton, he pulled back in the wake of a disastrous off-year election before pushing forward with major programs after winning a second term.

The major effort at the outset of the Obama administration to enact cap-and-trade legislation for reducing GHG emissions could not overcome the opposition of all Senate Republicans and a few rust-belt and coal-state Democrats. Any colorable claim that the opponents of the legislation could make that it would result in the destruction of jobs had to be taken seriously by members of Congress whose constituents were suffering from the worst economic downturn since the Great Depression. Opponents of the legislation gained traction with money-conscious voters by framing the program as "cap and tax." Another serious problem was that Representative Waxman's negotiating strategy produced a bill that was so larded with giveaways and exceptions that it was easily attacked in the Senate as a special interest bill. Despite spending hundreds of millions of dollars on the campaign, environmental groups lacked the political clout to persuade a Senate with a Democratic supermajority to enact their flagship legislative offering.[109]

Despite the legislative failures, the Obama years witnessed an unprecedented outpouring of highly significant regulations relevant to the electric power industry. EPA refashioned the failed Bush administration CAIR into a more stringent and successful CSAPR. A May 2014 analysis of EPA monitoring data concluded that all of the upwind states had reached their 2012 CSAPR budgets and eleven states had met their 2014 budgets. Much of the decline, however, was attributable to low natural gas prices. By mid-2018, many areas in the Northeast had still not attained the 1997 ozone standard, and a somewhat smaller number of areas had failed to attain the 2008 ozone standard. Yet air quality in the downwind states has improved, and it is unlikely that it would have happened in the absence of the still evolving CSAPR.[110]

The MATS played a significant role in a wave of retirements of coal-fired power plants in 2015 and 2016. Although the Supreme Court remanded the standard to EPA to balance compliance costs against benefits, the real world overtook the litigation. Lower-cost technologies for removing mercury and low natural gas prices made compliance much less difficult than anticipated. Of the two hundred units that received extensions through April 15, 2016, only twenty-two that lacked compliant MATS controls remained online by that date.[111]

EPA went to great lengths to address climate change during the Obama administration. Although the Supreme Court stayed the Clean Power Plan, electric power companies undertook many of the load shifting and energy conservation programs that the plan suggested anyway. TVA announced that the stay would not affect in the slightest its plans for retiring coal-fired plants and relying more heavily on natural gas, nuclear, and renewable power. Many companies were operating under PUC-approved multiyear integrated resource plans that factored CO_2 emissions–reducing changes into future capital expenditures. Companies also recognized that renewables would be playing a large role in future power generation because their customers were demanding it.[112]

The president disappointed environmental groups when he ordered Administrator Jackson to withdraw EPA's proposal to tighten the ozone standard. Rather than attempting to persuade the public that environmental regulation could create jobs at the same time that it protected the lungs of vulnerable children, the president made the ozone standard a sacrificial lamb to demonstrate his commitment to limiting costly government regulations.[113] During his second term, EPA revised the ozone NAAQS to make it a bit more stringent. It also tightened the PM, SO_2, and NO_x standards and successfully defended them in the D.C. Circuit. No previous administration had revised so many of the NAAQS to reflect new scientific information.

EPA's renewed emphasis on NSR enforcement resulted in some very impressive legal victories and settlements that brought about huge reductions in emissions and a large number of retirements of coal-fired power plants. Some companies may also have been inspired by the prospect of upcoming EPA regulations, but most of the settlements were probably driven by the prospect of very high fines and/or injunctions.

Although environmental groups were generally pleased with EPA's efforts during the Obama administration, the electric power industry prevailed in some struggles. EPA's cooling water intake structure regulations for existing power plants represented a clear victory for a unified electric power industry. Despite public outcries following two massive coal ash spills, the electric power industry persuaded EPA to forgo listing coal combustion residuals as hazardous wastes and to allow states to manage coal ash impoundments as ordinary solid wastes pursuant to broad criteria that gave state agencies a great deal of discretion. Because there are many industrial waste streams that are more hazardous than coal ash, EPA's concern for preserving scarce hazardous waste disposal capacity was sensible. Its worries about stigmatizing beneficial reuse of coal ash were less realistic. Characterizing coal ash as a hazardous waste may have had a modest impact on its use in wallboard for homes, but some reuses, like using unregulated coal ash to fill empty holes in the ground, are not especially "beneficial," considering their environmental risks.

Throughout the first half of the Obama administration, coal and coal-dependent electric companies and their allies in Congress complained loudly about a coming "train wreck" in the electric power industry as power plant owners absorbed the cumulative effects of the CSAPR, the MATS, EPA's revisions to the NAAQS for ozone, its climate disruption regulations, its cooling water intake structure regulations, and its coal ash regulations. By the end of 2012, however, most companies had abandoned that metaphor because EPA's regulations were not as burdensome as they had predicted, demand for electricity was down, companies were converting to natural gas for economic reasons unrelated to EPA's regulations, and renewable resources were making substantial contributions to electricity supplies. Reliability concerns still existed, but they were localized and limited to the impacts of EPA's regulations on a few power plants.[114]

Natural gas did well during the Obama administration. By mid-2012, the price of natural gas had declined so dramatically due to increased supplies from hydraulic fracturing that gas-fired peaker plants in some places were replacing coal-fired plants for baseload generation. Natural gas passed an important milestone in April 2012 when power plants produced as much energy nationwide from natural gas as from coal. Coal and gas traded places as the most used fuel during the ensuing years, but the trend was clearly toward gas. By the beginning of President Obama's second term, a new wave of gas-fired power plant construction and coal plant repowering was under way.[115]

The eight years of the Obama administration saw coal's fortunes rise slightly and fall again as the electric power industry was buffeted by winds of change driven by declining natural gas prices, greater efforts to reduce demand through end-use efficiency and demand-response, EPA regulations, and an extraordinary campaign by the Sierra Club and other environmental groups (described in Chapter 10) to prevent companies from building new coal-fired power plants and to force them to retire existing plants. The last two years of the Obama administration were especially unsettling for the coal industry as companies retired dozens of coal-fired plants prior to the MATS deadlines. Louisville Gas and Electric's Cane Run plant, which President Jimmy Carter had praised as "a testimony to the technological genius of Americans" when it was built in July 1979, converted to natural gas in

May 2015 because state of the art in 1979 was incapable of meeting the stringent requirements of the MATS standard. In 2015, the coal industry experienced a wave of bankruptcies that included large companies like Walter Energy, Patriot Coal, and Alpha Natural Resources, as well as many smaller concerns. Yet, despite the industry's troubles, coal-fired power plants remained a critical component of the electric power industry as a source of stable, dependable baseload power.[116]

CHAPTER 10

The War on Fossil Fuels

Spearheaded by the Sierra Club, a coalition of national environmental groups, including NRDC, EDF, and the Environmental Integrity Project, commenced a major initiative in the summer of 2007 called "Beyond Coal" to oppose each of the 150 new coal-fired power plants that were on company drawing boards in response to the George W. Bush administration's call for a thousand new plants. The initiative employed litigation, advertising, letter writing, protests, and grassroots organizing to fight new coal-fired power plants throughout the country. The goal was to challenge every plant at every level of government employing every legitimate tactic. Sometimes a campaign would focus on a single company with many projects in the works, and sometimes it would attack an individual proposal on multiple fronts. By "clogging the system," they hoped to force Congress to address climate disruption legislatively. They were, however, willing to negotiate over new plants that were already well on the way to completion, if they could extract retirements of heavier polluting existing plants in return. The initiative received a huge shot in the arm in the summer of 2011 when the former New York mayor Michael Bloomberg donated $50 million to the campaign.[1]

The electric power companies that were the targets of the initiative called it an abuse of the legal system. In their view, it was grossly unfair to oppose every coal-fired power plant, regardless of its merits under the relevant laws.

They responded with a public relations campaign of their own featuring the virtues of coal and a legal campaign aimed at stifling the lawsuits. In virtually every challenge launched by the environmental groups, the project's proponents stressed the prospect of jobs and other economic benefits that construction and operation of the power plant would bring to the local community. And in many cases, state governors and local politicians jumped on the bandwagon.[2]

So long as the environmental groups trained their sights on proposed coal-fired plants, the natural gas industry supported their efforts. Aubrey Mc-Clendon, the CEO of Chesapeake Energy, contributed $26 million to the Sierra Club in 2007 to support litigation and millions more to the American Lung Association to finance its "Fighting for Air" advertising campaign. Many environmental activists believed that forming strategic alliances with less polluting industries was a necessary strategy in the battles to reduce pollution, but they drew the line at taking money from companies that they might be opposing in future battles. After receiving complaints from two of its chapters that were opposing natural gas fracking operations and a change in leadership, the Sierra Club stopped accepting natural gas money in 2010. The head of the initiative later admitted that taking the money was a mistake.[3]

The environmental groups' war on coal expanded in 2009 to include a similar campaign aimed at retiring all existing coal-fired plants by 2030. In this larger struggle, power plant owners could point to existing jobs held by real people that would be lost if their plants closed, and they could argue that removing the most reliable source of baseline power from the grid would undermine its reliability and resiliency. In 2012, the Sierra Club went a step further to launch a "Beyond Natural Gas" campaign that focused on new natural gas plants and the pipelines that transported gas from wells to power plants. All three campaigns are still waging a broadscale war on fossil fuels.

STRUGGLES OVER NEW COAL-FIRED PLANTS

The first phase of the Sierra Club's Beyond Coal campaign focused exclusively on new coal-fired power plants. Under the Clean Air Act's prevention of significant deterioration (PSD) program, every new power plant had to have a permit. To get that permit, the applicant had to demonstrate that the plant would employ the "best available control technology" (BACT), protect downwind PSD increments, and not impede visibility in national parks and wilderness areas. Under EPA's program for regulating hazardous air pollutants, an applicant had to demonstrate that it would employ the "maximum available control technology" (MACT) for reducing emissions of hazardous air pollutants, including mercury. At the state level, a public utility company had to demonstrate to state public utility commissions (PUCs) that a new plant was necessary to meet demand and that the investment was prudent. All of these steps presented opportunities for environmental groups to object to projects on multiple grounds. They seldom persuaded the relevant decision makers to disapprove the projects, but in a surprising number of cases the companies either canceled the projects or entered into settlements with the environmental groups on terms very favorable to the environment.

Struggles over BACT

The George W. Bush administration's heavy emphasis on increasing the number of coal-fired power plants resulted in a raft of new applications for permits under the Clean Air Act's prevention of significant deterioration program, virtually all of which were opposed by environmental groups. In the permit hearings, the most controversial issue was the permitter's highly discretionary BACT determination. When environmental groups failed to convince the permitters, they usually challenged permits before EPA's Environmental Appeals Board (EAB) and in state and federal courts. They did not prevail every time, but they did win a surprising number of appeals. And on many occasions, the constant opposition caused companies to abandon projects that might otherwise have been completed.

Fuel Switching. Environmental groups in several permitting proceedings argued that the permitting agency should consider switching to low-sulfur coal or natural gas as BACT. The Seventh Circuit Court of Appeals, however, rejected this contention in the Sierra Club's challenge to Peabody Energy's permit to build a large mine-mouth power plant near Lively Grove, Illinois. The court upheld the EAB's determination that switching to low-sulfur coal would constitute a change in the plant's design and was therefore not an "available technology" to reduce emissions from the project as designed.[4]

IGCC. The most promising approach to reducing power plant emissions during the George W. Bush Administration was "integrated gasification combined cycle" (IGCC) technology. In an IGCC plant, coal was gasified, and any remaining pollution-producing impurities were removed from the gas. The gas was burned to power a combustion turbine, and the turbine's exhaust created steam in a boiler to fire a separate turbine. According to the Clean Air Task Force, SO_2 emissions were one-fifth, mercury emissions half, and CO_2 emissions one-quarter of the emissions of an equivalent plant burning pulverized coal. NO_x emissions, however, were about the same for both. IGCC plants were also more compatible with carbon capture, utilization, and storage (CCUS), a technology that the environmental groups advanced as BACT for removing CO_2 from the exhaust stream. IGCC construction costs, however, were about 50 percent higher than for conventional plants, and some companies had concerns about its reliability.[5]

Environmental activists were split on the desirability of IGCC. Greenpeace and some local groups refused to support any coal-fired technology, arguing that a new generation of IGCC plants would commit the nation to at least fifty more years of burning coal. During most of the Bush administration, however, the Sierra Club, NRDC, and the Clean Air Task Force argued that IGCC was BACT for new coal-fired plants. EPA took the position that permitters did not have to consider IGCC technology for coal-fired power plants designed to use pulverized coal, because, like fuel switching, that would constitute a redefinition of the basic design of the source. Relying on EPA, state agencies and courts consistently rejected environmental group attempts to establish IGCC as BACT. By the end of the Bush administration, the Sierra Club had changed its position and begun to object to new projects, even when the applicants agreed to employ IGCC technology. That opposition, combined with unexpectedly high capital costs and low prices for natural gas and renewables, removed IGCC from the drawing boards of companies needing new capacity.[6]

CO$_2$ Controls. A related battle broke out over whether PSD permitters had to consider technologies like CCUS for reducing CO_2 emissions as BACT for power plants after the Supreme Court held in *Massachusetts v. EPA* that CO_2 was a pollutant. In early 2005, Tenaska Energy, a large merchant power producer, proposed to build a 677-megawatt IGCC power plant near Taylorville, Illinois. In June, Governor Rod Blagojevich proudly announced that the Illinois Environmental Protection Agency (IEPA) had issued an air permit for the $2 billion project as a "critical part" of his energy plan for the state. It was the first air permit issued for a commercial IGCC plant in the country, and it contained the most stringent pollution control requirements for conventional pollutants ever placed on a coal-fired power plant. The governor's plan included a pipeline to transport CO_2 from Taylorville to places where it could be used in secondary oil and gas recovery. The permit did not, however, include a technology for capturing CO_2.[7]

In July 2007, the Sierra Club appealed the IEPA permit to EPA's Environmental Appeals Board, arguing that it should have included a carbon capture requirement. It was the first such challenge after the Supreme Court's decision. A spokesperson for the Clean Air Task Force, which normally joined with the Sierra Club in its challenges, criticized the club for impeding a project that would result in much lower CO_2 emissions than existing coal-fired plants. It also worried that if the Sierra Club planned to challenge even the cleanest plants, companies might just keep the dirty ones running. In late January 2008, the EAB dismissed the Sierra Club's appeal on the technical ground that it had failed to raise the issue during the public comment period. The board also noted, accurately, that in other proceedings the Sierra Club had taken the position that IGCC alone was BACT. In spite of the victory, Tenaska later abandoned the project after concluding that it was not economically viable when compared to natural gas and renewable alternatives.[8]

States were always free to impose controls on CO_2 emissions in PSD permits that went beyond EPA's definition of BACT, but none did until Kansas took a leap of faith. In February 2006, Sunflower Electric Corporation, a generator for rural electrical cooperatives, applied for a PSD permit from the Kansas Department of Health and Environment (KDHE) to add two new 700-megawatt pulverized coal units to its Holcomb power plant in western Kansas. In May 2007, the Sierra Club petitioned KDHE to require CO_2 reduction technology in the plant's PSD permit and sued when it refused to grant a full-fledged evidentiary hearing on the petition. Members of several state and local environmental groups then packed the informal hearing that the agency held in the college town of Lawrence.

The state's secretary of health and environment, Roderick Bremby, shocked the electric power industry in October 2007 when he denied Sunflower Electric's permit because it lacked controls on CO_2 emissions. It was the first instance of a state agency rejecting a permit application for that reason. Bremby was no environmentalist, having spent most of his career as a city administrator, but he was persuaded by the club's arguments, and he noted that much

of the plant's electricity would be sent to other states. Governor Kathleen Sebelius, a Democrat, supported the decision.

Sunflower Electric sought to overturn Bremby's decision in state and federal courts. At the same time, it launched a million-dollar lobbying campaign
to persuade the state legislature to reverse Bremby's decision. A coal company ran a series of newspaper advertisements claiming that the KDHE decision benefited Vladimir Putin, Hugo Chavez, and Mahmoud Ahmadinejad
by forcing Kansas to "import more natural gas from countries like Russia,
Venezuela and Iran" (the United States did not import natural gas from those
countries at the time). Both houses of the Republican-controlled legislature
passed three bills allowing the project to go forward, but Governor Sebelius
vetoed all of them.

The political dynamics changed when Governor Sebelius resigned to become President Obama's secretary of health and human services and was replaced by Lieutenant Governor Mark Parkinson. After meeting privately
with Sunflower and Republican leaders, Parkinson agreed to a compromise
bill under which Sunflower agreed to develop a wind farm and permanently
retire two oil-fired units that it had not operated in a decade and KDHE would
be prohibited from imposing more stringent emissions limitations on permits than those mandated by EPA without legislative approval. Within days,
both houses passed the necessary implementing legislation, and the Holcomb
project was back on track.

On November 3, 2010, Governor Parkinson unceremoniously fired Secretary Bremby. Concluding that the KDHE secretary was fired because he was
not prepared to rush out a permit for the plant, the Sierra Club called for EPA
to take over the PSD permitting process for Kansas. After KDHE issued a
final permit that contained no CO_2 limitations on December 16, 2010, the
Sierra Club challenged the action in a state court of appeals. It argued, quite
plausibly, that the permit was invalid because KDHE was constrained by the
Parkinson deal and the legislation implementing it, neither of which were approved by EPA as an amendment to the Kansas state implementation plan.

The project took a serious hit when the *Kansas City Star* published a story
based on emails that it had obtained from KDHE indicating that during the
time that its staff was rushing to put together the final permit, Sunflower had
effectively prepared the agency's responses to many of the public comments.
The beleaguered project then suffered a second serious blow in October 2013
when the Kansas Supreme Court overturned KDHE's award of the PSD

permit to Sunflower. The court agreed with the Sierra Club that the department had to consider the recently revised ambient air quality standards for SO_2 and NO_x in determining whether the new unit would cause a violation of the NAAQS. The court further held that on remand, KDHE would have to incorporate EPA's new MATS limitations for mercury and other hazardous air pollutants.

In May 2014, KDHE issued a final addendum to Sunflower's permit that purported to address the new standards for SO_2 and NO_x and to incorporate EPA's MATS but did not actually change any of the permit's emissions limitations. The Sierra Club challenged the amended permit, but the Kansas Supreme Court in March 2017 upheld KDHE's action. As of April 2019, the company was still evaluating whether a decision to go forward with the project made economic sense.[9]

Struggles over Protecting Increments

An applicant for a PSD permit had to show that its emissions would not cause a degradation of air quality in downwind areas to exceed the PSD increment for that area. A struggle broke out in 2006 when merchant generator Sithe Global Inc. sought a PSD permit for a $2.5 billion, 1,500-megawatt power plant adjacent to a coal mine on Navajo land just south of Shiprock, New Mexico. Because EPA was responsible for issuing PSD permits to sources located on tribal lands, EPA proposed to issue the permit in July 2006. According to EPA's regional office, the emissions limitations in the permit were some of the most stringent in the nation. Environmental groups pointed out, however, that the plant would add more pollution to the Four Corners PSD area that was already pressing against its statutory PSD increments. To meet that objection, Sithe agreed to offset the new plant's emissions by 110 percent by paying for SO_2 and NO_x control projects at other plants in the region.

The project divided the Navajo tribe into two factions. Having just taken a huge $40 million per year economic hit with the closure of the Mohave Generating Station on its land in part because of its adverse impact on visibility in the Grand Canyon, the tribe's president, Joe Shirley, enthusiastically supported the project. He predicted that it would bring in upwards of $50 million per year in revenues for the tribe along with a large number of $60,000 per year jobs. When the chapter of the tribe where the plant would be located voted to ban it, the central tribal government adjusted the chapter boundaries to put the project in a chapter that supported it. Shirley blamed outside environmental groups for stirring up dissent within the tribe, but the primary objector was a group of tribe members called Dine Citizens Against Ruining Our Environment (Dine CARE) led by a young Dartmouth-educated tribe member who came home to fight the plant. Dine CARE pointed to the tribe's abundant solar and wind resources as alternatives to the power plant.

The opposition to the project attracted national attention in December 2006 when several of the tribe's respected grandmothers formed a blockade to prevent Sithe Global contractors from conducting exploratory drilling. As supporters arrived from around the country to assist the grandmothers, police forced the protesters off the road. Undeterred, the protesters established a campsite and kept a vigil over the road for more than a year. The project suffered another setback in August 2007 when New Mexico's governor (and presidential candidate) Bill Richardson expressed his "grave concerns" over the fact that CO_2 emissions from the plant would increase the state's overall GHG emissions by 15 percent. The project suffered a fatal blow in 2010 when California's Public Utility Commission prohibited the state's utility companies from purchasing electricity from coal-fired sources, thereby eliminating the prospective plant's primary source of revenue.[10]

Struggles over Visibility

To obtain a PSD permit, new sources also had to demonstrate that they would not adversely affect visibility in national parks and wilderness areas. A major visibility struggle broke out in 2003 over Longview Power's proposed 600-megawatt coal-fired plant in Monongalia County, West Virginia. The National Park Service and the Forest Service joined environmental groups in opposing a draft PSD permit for the project out of concern for its impact on two nearby wilderness areas. The company then modified its permit application slightly and recalculated the impact in light of an agreement by EPA and Dominion Virginia Power to reduce SO_2 emissions by 95 percent at Dominion's nearby Mount Storm power plant. In addition, Longview Power agreed to purchase and retire sufficient SO_2 allowances from local sources to more than offset the remaining emissions. That was sufficient to satisfy the federal agency objectors, and a commitment to donate $500,000 per year for ten years to a tax-exempt entity to fund projects to diminish acid deposition and reduce CO_2 emissions from power plants won the environmental groups over. When construction commenced in June 2007, it was the first greenfield coal-fired plant to be built in the eastern United States in twenty years.[11]

Struggles over Technology

During the time between when the D.C. Circuit set aside the Bush administration's mercury rule and the time the Obama EPA promulgated its MATS rule, several struggles broke out between power companies and the Sierra Club over whether states were obliged to prescribe MACT for new power plants on a case-by-case basis.[12] After the courts answered that question in the affirmative, struggles broke out over the level of mercury reduction that the statute required. A good example is the struggle over a $1 billion 600-megawatt supercritical pulverized coal plant that the South Carolina Public Service Authority (Santee Cooper), a state-owned electric utility company, proposed to build at a site along the Great Pee Dee River near Kingsburg, South Carolina. After it filed its application for a PSD permit with the South Carolina Department of Health and Environmental Control (SCDHEC), the Sierra Club and other environmental groups objected to the permit because it lacked controls on mercury emissions. Joining them were members of the Pee Dee tribe, who consumed fish from the river. The fish contained such high levels of mercury that they were the subject of state-issued warnings.

The SCDHEC issued a draft PSD permit containing mercury limits to Santee Cooper in October 2007 and held a public hearing in early November. The high school gym in the tiny town of Pamplico was packed with Santee Cooper employees and environmental activists testifying for and against the plant. The mayor of Pamplico and more than one thousand residents in the area where the plant was to be built signed a petition asking the agency to approve the permit. The owners of several nearby factories and smelters also supported the project. The environmental groups argued that the draft permit's mercury emissions limits were far too high.

After the D.C. Circuit set aside the Bush administration's mercury rule in February 2008, Santee Cooper produced a MACT analysis showing that the baghouse that it planned to install was sufficient because it would reduce mercury emissions to fifty-seven pounds per year (a 90 percent reduction). The company also launched an "educational initiative" emphasizing the facts that mercury came from many sources (including volcanoes), that it could travel thousands of miles, and that only about 1 percent of the world's mercury emissions came from U.S. power plants. The environmental groups responded that the MACT standard required Santee Cooper to perform at the level of power plants in other states that were emitting far less mercury per unit of energy than the proposed Pee Dee plant.

In December 2008, SCDHEC issued a final PSD permit for two 600-megawatt units with a mercury cap of 46.3 pounds per unit and a requirement that the authority complete a study on technologies to reduce emissions even further. The environmental groups appealed the decision to an administrative appeals board, which upheld the permit in all regards in mid-February 2009. Then, in August 2009, the Santee Cooper board of directors announced that it was suspending its efforts to obtain permits for the Pee Dee plant. On top of the resistance it was encountering, the authority's largest customer had announced that it would be purchasing more of its power from Duke Energy in North Carolina.

The local newspaper recounted the many times Santee Cooper had warned that if it did not build the plant, "the Grand Strand, South Carolina's tourist economic engine, won't have enough electricity to keep its beachfront towers aglow." Now the company's president was calling the plant's cancellation a "triple win" for Santee Cooper, its customers, and the public. Three years after Santee Cooper canceled the plant, millions of dollars' worth of unused equipment remained at the site, and it was paying $13 million a year to keep it from deteriorating.[13]

Struggles before the Rural Utilities Service

Rural electrical co-ops frequently received loans, grants, or loan guarantees from the U.S. Department of Agriculture's Rural Utilities Service (RUS) to assist in building power plants. The Sierra Club insisted that RUS prepare environmental impact statements (EISs) detailing the adverse environmental effects of the projects before providing assistance for power plants. In the battle over Sunflower Electric's Holcomb addition, for example, a federal district court held that RUS had to prepare an environmental impact statement for the Holcomb project before it could go forward. In April 2010, RUS suspended all direct financing of coal-fired power plants, forcing the co-ops to seek financing for those projects in unsubsidized capital markets.[14]

Struggles in State Public Utility Commissions

Environmental groups joined consumer groups and low-income advocates to object to dozens of new power plant projects before state public utility commissions in states that had not deregulated retail generation. They argued that companies building expensive new plants should not be allowed to pass all of the costs through to ratepayers. The enormous cost of building plants that complied with current and upcoming pollution control requirements, they maintained, would far exceed the cost of producing the same amount of electricity from gas-fired plants, renewable resources, or demand response programs. The groups were occasionally joined by natural gas providers and rival electric power generators. Arguing that the plants would create thousands of jobs, local chambers of commerce and construction unions often intervened on behalf of the power companies.[15]

Struggles before state commissions often spilled into the public arena as participants filled local media outlets with advertisements.[16] Contests also featured public demonstrations and counterdemonstrations aimed at persuading commissioners and the public that a new facility was or was not warranted. At first, it appeared that public utility hearings provided "the illusion of involvement while rubber stamping coal projects." By the end of the George W. Bush administration, however, environmental groups were more successful in persuading public utility commissions to disapprove projects.[17]

The mother of all PUC struggles took place before the Mississippi Public Service Commission. In December 2006, Mississippi Power, a subsidiary of the Southern Company, received a certification from DOE and approval from the Internal Revenue Service for a $133 million tax credit to build a $1.8 billion IGCC power plant with carbon capture, utilization, and sequestration (CCUS) in Kemper County, Mississippi. The project would create 540 temporary construction jobs and 260 permanent jobs in a low-income county with an unemployment rate hovering at twice the national average. The company would sell the captured CO_2 to an oil company for use in enhanced oil recovery.[18]

With Governor Haley Barbour's strong encouragement, the Mississippi legislature passed a bill authorizing the state's Public Service Commission to allow utility companies to recover "prudently incurred" costs for new base-load power plants while the plants were still under construction, thus permitting the commission to shift some or all of the risk that the plant would not be completed from Southern's shareholders to its ratepayers.[19]

Although the Sierra Club had argued in previous PSD permit proceedings that IGCC met the test for BACT, the Mississippi chapter of the Sierra Club expressed unequivocal opposition to the plant. It contended that IGCC technology was inefficient, that CCUS was an unproven technology, and that the plant's high cost would be reflected in very high electric bills. Despite the planned mercury controls, it argued, the plant would still emit "enough toxic mercury to contaminate thousands of waterbodies and millions of pounds of fish." The club also urged the commission to consider the environmental risks of the company's plan to bury thousands of tons of coal ash in reclaimed strip mines.[20]

Another strong opponent of the project was the state's populist attorney general, Jim Hood, a rare Democrat among elected state officials. Among other things, he maintained that Mississippi law required the company to present far more detailed information in support of its preconstruction rate increase. A group of owners of land over which the plant's transmission lines were likely to be located also opposed the plant, as did merchant power companies in Mississippi and nearby states with underutilized capacity that they were anxious to sell to Mississippi Power. Even the American Association of Retired Persons opposed any rate increase for the project because of the impact that it would have on retired and low-income customers.[21]

In January 2009, Mississippi Power applied for a certificate of public convenience and necessity from the Mississippi Public Service Commission (MPSC). By then, it had already received a PSD permit from the Mississippi Department of Environmental Quality. In early June 2009, MPSC established a two-phase hearing arrangement under which it would receive testimony on the need for the project first and then receive testimony on its costs and environmental impacts. In early November 2009, the commission published its unanimous determination that the plant would in fact be needed by 2014, when it was expected to come online, because of growing demand in the state and a number of retirements that were likely in response to EPA's upcoming power plant regulations.[22]

At the completion of the second hearing, the commission by a 2–1 vote concluded that the "uncertainties and risks, relative to their potential benefits," were too high to warrant a finding of "public convenience and necessity." But it offered several conditions under which the project would meet that test and gave Mississippi Power twenty days to accept or reject the conditions. The dissenting commissioner, Brandon Presley (a cousin of deceased pop star Elvis Presley) would not have approved the project, even if the company agreed to the conditions. Because none of his northern Mississippi constituents were served by Mississippi Power or would benefit from the economic development in Kemper County, he had no political stake in the outcome.[23]

A very disappointed Mississippi Power announced that the conditions made it impossible to finance the plant. It then filed a motion for rehearing in May 2010 in which it proposed a $2.88 billion cost cap instead. Later that month, the commission by another 2–1 vote agreed to the company's suggested cap and allowed it to recover ongoing construction costs from ratepayers after January 1, 2012. Mississippi Power said the conditions were acceptable, and it proceeded ahead with the project with a 2014 target completion date. Claiming that the commission's "flip-flop" came only after considerable pressure from the governor's office, the Sierra Club filed an appeal in a state chancery court.[24]

Armed with the PSC's approval and the necessary environmental permits, Mississippi Power broke ground on the project, which it named Plant Ratcliffe after Southern Company's recently retired CEO, in mid-December 2010. Before a crowd of two hundred invited guests, Governor Barbour suggested that "prayers of thanksgiving are very appropriate here." Overhead, a small airplane hired by the Sierra Club pulled a large banner with the words: "Dirty Expensive Unnecessary."[25]

The project suffered a serious setback in March 2012 when the Mississippi Supreme Court unanimously held that the commission had failed to provide substantial evidence to support its decision to allow early cost recovery. The Sierra Club then petitioned the commission to convene a meeting to come up with a plan for the orderly shutdown of construction activities at the site. Escalating the rhetoric, a Mississippi Power spokeswoman called the petition "just another in a series of scare tactics by a group of extremists."[26]

In late April 2012, MPSC, by a 2–1 vote, issued a new order finding once again that the plant would serve the public convenience and necessity under the conditions established in the May 2010 order. The Sierra Club once again appealed the decision to the chancery court. In July, however, the commission indefinitely delayed the company's request for a 13 percent rate increase. The company immediately appealed directly to the Mississippi Supreme Court. At that point, the project was already $366 million over budget. After the court refused to order the commission to allow it to raise rates during the appeal, bond rating companies began to downgrade the company's credit ratings. The project was clearly in trouble.[27]

In late January 2013, a divided Public Service Commission and Mississippi Power entered into a settlement agreement under which the company agreed to drop its lawsuit and the commission agreed to allow the company to begin recovering costs immediately and escalate rates gradually over the next seven years up to a $2.4 billion cap (a bit lower than the previous $2.88 billion cap). Mississippi Power's lobbyists then persuaded the legislature to pass a bill empowering the company to issue tax-free bonds to finance project costs above $2.4 billion. The Southern Company could locate the obligation off its books in an Enron-style "special purpose entity," and the bonds would be secured not by the plant or any other Southern Company assets, but by Mississippi ratepayers.[28]

As the May 2014 deadline for completing the project approached, the site was crowded with more than six thousand workers, but rainy weather and unanticipated work kept it behind schedule. In October 2013, the company announced that it would not meet the deadline and would repay the $133 million in tax credits it had received at the project's outset. The unfinished plant was an impressive structure, rising "like a fantastic industrial castle from the pinewoods and cow pastures of eastern Mississippi." And it now had symbolic value to clean coal proponents extending well beyond the local fights over who was going to pay for it. Splitting with the Sierra Club, the Clean Air Task Force supported the project for the potential that it offered as a model for future EPA regulation of coal-fired power plants.[29]

In May 2014, Mississippi Power announced the plant's cost was now up to $4.4 billion and that Southern's shareholders had absorbed $1.5 billion in overruns. The good news was that the plant started generating electricity in the summer of 2014 by burning natural gas in the plant's combustion turbines. In early August, the Sierra Club and the company announced a settlement under which the club would cease opposing the Ratcliffe plant and the company would cease burning coal at its 870-megawatt Plant Watson in Mississippi by April 2015 and at its 500-megawatt Greene County plant in Alabama by April 2016. In addition, Southern Company would provide $15 million for increasing energy efficiency in the homes of its low-income consumers and for supporting clean energy demonstration projects at schools. The challenge to the Mississippi Public Service Commission's previous approval of rate increases in the Mississippi Supreme Court, however, remained alive, because a single ratepayer named Thomas Blanton had filed a challenge of his own along with the Sierra Club.[30]

In February 2015, the project suffered a devastating blow when the state supreme court held that in granting the rate increases, the commission had violated the 2008 statute that allowed Mississippi Power to recover its costs prior to completion of the project. The court noted in a footnote that the cost of the project now exceeded the annual budget for the state of Mississippi. As a remedy, the court ordered the commission to disallow any additional rate increases until the company had refunded the unlawful rate increases.[31]

By early 2017, the plant was almost three years behind schedule and the estimated cost had ballooned to more than $7 billion. The Southern Company was facing a Securities and Exchange Commission investigation, a shareholder action from a Michigan pension fund, and a lawsuit by ratepayers alleging that it had fraudulently obtained rate increases. Documents and secret recordings released by a whistleblower showed that Mississippi Power had grossly underestimated the timing and cost of the project, mismanaged its execution, and attempted to conceal serious quality issues when they arose. Worst of all, Brandon Presley was now the chairman of the Mississippi Public Service Commission.[32]

In June 2017, the commission ordered Southern Company to enter into negotiations with its staff to bring about the orderly termination of work on the plant's two coal gasifiers. A week later, the company suspended all further work on the CCUS unit at the plant and announced that it would be adding $3.4 billion to the $2.9 billion that it had already written off on the $7.5 billion project. In late November 2017, the Southern Company and the Mississippi Public Service Commission staff reached an agreement under which Southern agreed to reduce its annual collection from ratepayers for the plant by $14 million and to limit the charges to the portion of the plant that burned natural gas.[33] The risky campaign to revive coal had come to an ignominious end.

Struggles in Multiple Jurisdictions

In some cases, companies had to secure the approval of public utility commissions in several states, thereby increasing the opportunities for environmental and consumer groups to scuttle the projects. Otter Tail Power Company and six Minnesota and South Dakota utility companies signed an agreement in June 2005 to build a $1 billion 600-megawatt super-critical pulverized coal unit at Otter Tail's Big Stone I plant near Milbank, South Dakota. The majority of the unit's output would be sent to the Twin Cities area in Minnesota, but some of the power would go to customers in North and South Dakota. The project therefore needed approvals from PUCs in all three states to write the cost of the plant into the utilities' rate structures and to build the transmission lines necessary to transport the electricity. At the same time, the companies were exploring the possibility of building excess capacity into the transmission lines to allow them to carry wind power to their customers from wind farms that could be located nearby.[34]

Despite its potential to advance wind power, several environmental groups (but not the Sierra Club) opposed the Big Stone II project when it came before the South Dakota Public Utilities Commission (SDPUC) on the ground that its CO_2 emissions would unnecessarily contribute to global warming. The SDPUC approved the project in July 2006, and the environmental groups appealed the decision to a state district court. The South Dakota Supreme Court ultimately upheld the approval, holding that the commission did not have to consider the unit's CO_2 emissions in awarding a certificate of convenience and necessity.[35]

The project faced even stronger opposition from environmental groups in the proceedings before the Minnesota Public Utilities Commission (MPUC), which had recently promulgated stringent criteria for justifying new power plants. Minnesota law also required the commission to factor externalities into the determination of public need for power plants. In August 2007, a panel of two Minnesota administrative law judges (ALJs) recommended that the MPUC approve the transmission project, but only if the companies offset all 4.7 million tons per year of the CO_2 the plant would emit and came up with 120 megawatts of power from wind energy. This prompted two of the seven partners to abandon the project. Concerned about securing MPUC approval, the remaining partners scaled down the project to a 500-megawatt plant. This required a new hearing before a panel of Minnesota administrative law judges, and this time the ALJs recommended against approval. The ALJs concluded that the developers had justified at most 160 megawatts of new power, not 500 megawatts, and they had not adequately considered the project's impact on global warming.[36]

At the same time, the North Dakota Public Service Commission approved the project, concluding that a pulverized coal-fired plant with demand response and energy conservation was more cost effective than wind, natural gas, or IGCC technology. The commission did not address CO_2 emissions because North Dakota law prohibited it from employing "environmental externality values" in determining whether projects were "reasonable and prudent." The environmental groups appealed the decision to a North Dakota district court.[37]

At that point the only approval lacking was that of the Minnesota Public Utilities Commission, which had been sitting on the case for many months. In an effort to ease the commission's concerns about CO_2 emissions, the developers offered to use an ultra-supercritical boiler and to leave space for installing CO_2 capturing technology in the future. In mid-January 2009, the MPUC unanimously approved the project, so long as the new unit obtained offsets of its CO_2 emissions from other sources. But it also placed a $3,000 per kilowatt-hour cap on the cost of the project and a $26 per ton cap on the cost of CO_2 offsets, and it required the companies' shareholders to foot the bill for any costs incurred above the caps.[38]

In early September 2009, Otter Tail announced that it was withdrawing from the project, citing reduced demand due to the economic downturn and increased uncertainty over where EPA was going with GHG regulation during the Obama administration. The remaining four companies attempted to find a new partner, but to no avail. They canceled the project in early November 2009. Unfortunately for the partners' customers, state laws allowed them to pass through the $30 million in development costs as rate increases. The cancellation also meant that the transmission lines would not be upgraded to allow them to carry wind power to the Twin Cities area.[39]

Struggles over Additions to Existing Plants

When owners of existing plants applied for a PSD permit to build additional units, they had to demonstrate that their ongoing operations were in compliance with applicable permit requirements. That requirement sent opponents searching through agency files for evidence of ongoing violations. For example, the Sierra Club in March 2007 objected to the PSD permit that the Missouri Department of Natural Resources issued to Kansas City Power & Light (KCP&L) for a second 800-megawatt coal-fired unit at its Iatan power plant near Weston, Missouri, on the ground that numerous unaddressed permit violations by Unit 1 at the facility precluded its award of a permit for Unit 2. Soon thereafter, the company and the club announced a settlement under which the company agreed to purchase hundreds of windmills, undertake a major conservation program, and take such other actions as were necessary to offset all of the CO_2 emissions from the new unit. It was the first time an electric power company had ever agreed to fully offset all of its CO_2 emissions from a new unit, and it came long before the federal government formally recognized that CO_2 emissions endangered the planet.[40]

Struggles in the Boardrooms

Environmental groups even got involved in corporate boardroom struggles over mergers and acquisitions. By far the most significant deal during the George W. Bush administration was the takeover of TXU by Energy Future Holdings. TXU was the largest power company in Texas with 2.4 million customers, more than fifty mostly coal-fired generating units, and a large wholesale business. In April 2006, Texas governor Rick Perry joined executives from TXU at a gala announcement of the company's plans to build eleven supercritical units burning pulverized coal as quickly as possible to meet anticipated demand for electricity in Texas's booming economy. The massive project soon attracted opposition from an international coalition of environmental groups, activist shareholders, and members of the Texas legislature who had introduced a bill to declare a moratorium on new power plant construction. The controversy drove TXU's stock prices down far enough to make it an attractive target for private takeover specialists.

On Monday, February 12, 2007, NRDC's David Hawkins and EDF's Fred Krupp received calls from Bill Reilly, a former EPA administrator who was currently working for a large private equity fund called the Texas Pacific Group. Reilly wanted to know what it would take to persuade their organizations to support an effort by Texas Pacific and another private equity fund, Kohlberg Kravis Roberts (KKR), to take over TXU in a massive $45 billion buyout. The companies were willing to go to some lengths, Reilly explained, to avoid opposition to the deal by environmental groups. Texas Pacific had been foiled by opposition from environmental and consumer groups in its 2003 attempt to purchase Portland General Electric Company during the Enron bankruptcy, and KKR had likewise failed to obtain approval from the

Arizona Corporation Commission for its planned purchase of UniSource Energy Corporation because consumer and environmental groups had objected. This time, they wanted the groups on board before they launched the takeover bid.

On Tuesday, Jim Marston, the head of EDF's Texas office, was on a plane to San Francisco to meet with Reilly, partners in the two funds, and their lawyers. By one o'clock Wednesday morning, they had a deal. Texas Pacific and KKR agreed to cancel eight of the eleven coal-fired units in Texas and drop TXU's plans for new coal-fired units in Maryland, Pennsylvania, and Virginia. Any new units in the future would employ IGCC technology and be compatible with CCUS. They further agreed to accomplish a reduction in the company's CO_2 emissions to 1990 levels by 2020 through expanding its renewable energy portfolio and spending $400 million on energy efficiency measures. To mollify consumer groups, the firms promised to reduce rates by 6 percent immediately and by another 4 percent when the deal was consummated and to leave the rates in place through 2008. In return, the environmental groups promised not to oppose the deal or the three plants that remained in the project.

The deal received bipartisan praise from members of Congress. The *Wall Street Journal* suggested that the buyout might "signal a broader remaking of private equity's image in the utility industry" from that of "temporary, profit-driven caretakers not answerable to public shareholders or sensitive to consumers" to stewards of resources and protectors of the environment. EDF and NRDC hoped that the deal would serve as a precedent for future financers of power plant construction projects. What they did not know at the time was that TXU's board had already decided to scale back the project to just six plants. The deal was not, however, roundly applauded by environmental groups. Some suggested that EDF and NRDC should have held out for cancellation of all eleven plants, and they later suggested that the two national groups were taken to the cleaners by a company that had already decided to cancel half of the plants. Noting that the deal effectively killed grassroots pressure on the state legislature to enact moratorium legislation, the Sierra Club vowed to keep challenging the three plants that remained in the project.

After the buyout closed in October 2007, TXU was broken up into a generating company, a transmission company, and a distribution company under a holding company called Energy Future Holdings Inc. The generating subsidiary was named Luminant Energy. The transmission company was called

Oncor Electric Delivery, and TXU Energy was responsible for retail distribution. Despite the continued opposition of the Sierra Club and local residents, the three units were completed in 2009 and 2010 at a cost of $3.25 billion. The new companies now had to figure out how to raise revenues to the extent needed to service the huge $40 billion debt load. This necessity was vastly complicated by the sudden decline in natural gas prices, which made coal comparatively less attractive, and a drop in electricity demand that accompanied the 2009 recession. As we shall see, Luminant still had to deal with environmental group opposition to its existing lignite plants.[41]

Struggles across Changing Political Terrain

The struggles over new coal-fired power plants extended across such large timescales that the political terrain sometimes changed as the struggles continued. The struggles over three unrelated projects that developers launched in Michigan during the George W. Bush administration rush to coal is a good example of how changing politics changed the legal outcomes of particular disputes but could not change the fundamental economics of electric power generation. The three projects were MidAmerican Energy's 750-megawatt pulverized coal project near Midland, Consumers Energy's 800-megawatt addition to its Karn-Weadock plant near Bay City, and Wolverine Power Cooperative's 600-megawatt project near Rogers City.

Faced with this onslaught of coal-fired plants, the Sierra Club in December 2007 announced that it had formed a coalition with NRDC, local environmental groups, and small businesses called Clean Energy Now to object to any new coal-fired power plants in Michigan. The coalition had a powerful ally in Michigan's Democratic governor Jennifer Granholm, who believed that renewable energy and end-use efficiency would create more in-state jobs than importing coal from the Powder River Basin. In February 2009, she instructed the Michigan Department of Environmental Quality (MDEQ) and the Michigan Public Service Commission (MPSC) to ensure that new coal-fired power plants in the state were in fact needed before approving them.[42]

Governor Granholm's order had an immediate effect on MidAmerican's project. MidAmerican was a joint venture of LS Power Group, an investment firm, and Dynegy Inc., a merchant generator. Soon after the project was announced in late 2007, the Sierra Club announced its firm opposition. In January 2009, Dynegy dissolved the joint venture, and in May 2009, LS Power canceled the Midland project, citing slow growth in electricity demand and regulatory uncertainty due largely to the Granholm directive.[43]

Although the Republican attorney general Mike Cox issued an opinion stating that Governor Granholm had no authority to order MDEQ to condition permits on the need for the plant, the agency determined that the opinion did not restrict its authority to consider alternatives to the proposal, including the no-need alternative, in evaluating the proposal. Consumers Energy then filed a "needs analysis" with MDEQ showing that another baseload plant would be needed to meet projected demand and to fill the gap left open by the upcoming retirement of several of the older plants in its fleet. NRDC responded with a consultant's report concluding that the state could easily cut end-use demand by 30 percent and that the Consumers Energy plant was not needed to meet the state's future energy needs. The MPSC staff then weighed in with a report concluding that the new plant was not economically justified unless the company retired about 950 megawatts of existing coal capacity.

MDEQ issued an air permit for the plant at the end of December 2009, but it required the company to retire up to seven of its aging coal-fired units representing 958 megawatts of power by the time the new unit came online. The plant also had to be "sequestration ready" in anticipation of a CCUS requirement when that technology became commercially available. Consumers Energy was happy with neither of these conditions. And it still had to obtain a certificate of need from MPSC before it could make ratepayers shoulder the project's cost. In December 2011, the company threw in the towel.[44]

Wolverine Power also had to submit a needs analysis with MDEQ. The MPSC staff objected to the analysis, concluding that another baseload plant was unnecessary in the absence of the imminent retirement of an existing plant of equivalent output. Unlike the Consumers Energy project, MDEQ denied Wolverine's application in May 2010 on the ground that it was too expensive. In addition, Wolverine had not demonstrated that it could not meet its future needs through long-term power purchase agreements from other generators. Wolverine challenged the decision in a county circuit court, which reversed the permit denial on the ground that the Granholm order was unconstitutional.

In the meantime, the politics of power had changed dramatically in Michigan after the 2010 elections put the Republican Party in control of the legislature and sent a Republican, Dale Snyder, to the governor's office. Stung by the election results, many Democrats in the House became strong advocates of "clean coal" power plants and urged the new governor to rescind the executive order that the court had just found to be unconstitutional. Told by Governor Snyder that it had an "obligation not to impede business," MDEQ reversed itself and approved the permit. The Sierra Club challenged the permit, but the county court and a court of appeals upheld MDEQ's action.

It looked for a time as though the changed political climate had saved one of the three projects. Then, to the surprise of everyone, Wolverine announced in late May 2012 that it had suspended the bid process for major plant components, citing EPA's recently issued MATS standard. In December 2013, Wolverine announced that it was abandoning the project altogether. The environmental groups' challenges delayed the project long enough to allow a newly elected Republican governor to change the composition of the MDEQ, but also long enough for a newly elected Democratic president to change the applicable federal regulations.[45]

Death by Delay

In many instances, opponents were able to slow down the approval process to such a degree that the new plants were no longer economically viable by the time that the developers secured all of the necessary approvals.[46] For example, in the summer of 2006, a joint venture consisting of Dynegy and LS Power proposed to build a $1.3 billion 750-megawatt supercritical coal-fired power plant with state-of-the-art controls for SO_2, PM, NO_x, and mercury just east of Waterloo, Iowa. The project attracted opposition from local residents concerned about the coal dust that additional rail traffic would spread over town. Noting that the merchant plant would not be providing electricity to nearby customers, they also worried that the plant would result in lower property values. Local businesses and labor unions, however, supported the plant because it would stimulate the local economy by creating one thousand temporary construction jobs and one hundred permanent jobs.[47]

The first local hurdle for the project was a request that the city annex the site and rezone the property, an operation that would require the approval of the Waterloo Planning, Programming, and Zoning Commission and the Waterloo City Council. Since several rural landowners between the site and the city did not want to be included, the annex took on an odd shape resembling a flag on a flagpole in which the distant plant was connected to the city by a long quarter-mile-wide stretch consisting of the road between them and a small strip of land on either side. Despite strong opposition from local residents at the hearings, the commission approved the request in April 2007, and the city council approved it the next month.[48]

That precipitated a September 2007 hearing before the state's City Development Board, which had the authority to overturn the city's annexation decision if it did not comply with state law, was inconsistent with orderly urban development, or was otherwise not in the public interest. In anticipation of the hearing, the local opposition groups sponsored a well-attended rally at Merle Bell's farm near Waterloo. At the five-hour hearing, an overflow crowd of more than three hundred people packed the Waterloo Center for the Arts to argue for and against the plant. When the board rejected the city's annexation plan by a 3–2 vote, the local Sierra Club called it a "major victory" for the citizens of Waterloo.[49]

In March 2008, LS Power filed a revised request for the city to annex and rezone its property that deleted the property along the road to the plant and included a narrow band of property on either side of a Union Pacific Railroad right-of-way that ran from the city to the plant. Since the company had purchased an option to buy Merle Bell's property several years earlier, it was able to silence his opposition to the annexation by threatening to buy his property outright. Because the request included no other property owners who objected to the annexation, the state's City Development Board no longer had jurisdiction to review the city's action.[50]

In early April, the Waterloo Planning, Programming, and Zoning Commission once again approved LS Power's request by an 8–2 vote. Later that month, the Waterloo City Council voted 6–1 to approve the annexation and made it clear that it would approve the rezoning request after the obligatory three readings. Then, in January 2009, Dynegy announced that it was pulling out of its joint venture with LS Power. Within days, LS Power announced that it was halting all further activities in connection with the project because "load growth is slowing in the region due to the downturn in the economy." A spokesperson for the local citizens' groups acknowledged that the opponents did not stop the project, but they did prolong the process until the financial crisis of late 2008 killed demand for additional electricity from merchant plants. And Merle Bell got to keep the family farm.[51]

Cutting Deals

The Sierra Club was prepared to tolerate a company's building a new plant if it employed state-of-the-art pollution control technology and agreed to retire existing coal-fired units. For example, in the first ever agreement between a major national environmental group and a municipal electrical power provider, the Sierra Club agreed not to challenge a PSD permit for a new 200-megawatt pulverized coal unit at the City of Springfield, Illinois's Dallman power plant to replace the municipality's ancient 76-megawatt coal-fired unit. The new unit would produce 99 percent fewer SO_2 emissions and 90 percent fewer mercury emissions and reduce CO_2 emissions by 600,000 tons by 2012. The city also agreed to add 120 megawatts of wind capacity to the city's electrical supply and increase the city's budget for investments in energy efficiency by \$4 million over the following decade. Similarly, in return for ending its prolonged struggle against Duke Energy's new supercritical pulverized coal unit at its Cliffside facility in Rutherford County, North Carolina, the Sierra Club extracted a promise from the company to retire four of the five existing coal-fired units at the facility when the new unit came online and to retire all of the company's other units that were not equipped with scrubbers by 2015. And the club dropped its opposition to MidAmerican Energy's nearly completed Walter Scott addition near Sioux City, Iowa, in return for the company's agreement to convert six coal-fired units to natural gas.[52]

Plants That Escaped

A few new coal-fired projects escaped the environmental groups' war on coal. The major national environmental groups did not object to IGCC projects, like AEP's Mountaineer Plant in New Haven, West Virginia, and the Summit Power Group's Clean Energy Project near Penwall, Texas, that included carbon capture technology. Both projects, however, encountered strong resistance from consumer groups objecting to the cost of CCUS, and both ultimately dropped plans for carbon capture. Similarly, no one objected to Great River Energy's ninety-nine-megawatt lignite-fired combined heat and power facility near Spiritwood, North Dakota, which featured a circulating fluid-ized bed boiler, SCR, and a baghouse, because environmental groups at the time were encouraging investments in those technologies. Cleco Power Company "bet [its] balance sheet" on a $1 billion 600-megawatt unit using state-of-the-art circulating fluidized bed boiler technology at its Rodemacher Power Station near Alexandria, Louisiana. The Sierra Club criticized Louisiana governor Kathleen Blanco for supporting the project, but did not actively oppose it because it was also capable of burning biomass.[53]

STRUGGLES OVER EXISTING POWER PLANTS

As the pipeline of new coal-fired power plants emptied, the big dirties were still emitting millions of tons of pollutants into the air. Fully 75 percent of the SO_2, 64 percent of the NO_x, and 54 percent of the CO_2 emitted by fossil fuel–fired plants came from plants built before 1978. In 2009, the Sierra Club launched phase 2 of its Beyond Coal campaign with the goal of forcing companies to retire all five hundred existing coal-fired plants by 2030. Recognizing that EPA's CSAPR and MATS rules would bring about substantial reductions in emissions over time, the club feared that the regulations might not be promulgated in a timely fashion, might not be sufficiently stringent, or might not survive judicial review. It therefore decided to take action immediately to bring about substantial reductions in emissions and to force the retirement of plants for which expensive emissions controls were not economically justified. It pursued this goal at the federal, state, and local levels with special attention to persuading local municipally-owned utility companies to wean themselves from coal. As with its opposition to new coal-fired plants, however, the Sierra Club was prepared to resolve individual battles with settlements that allowed plants to continue burning coal if it could extract significant concessions from companies in return. Its efforts received a boost when Bloomberg donated another $30 million to the campaign in April 2015 to be matched by $30 million from other committed sources.[54]

Four tactics that proved successful in this campaign were: (1) objections to renewals of Title V operating permits when the plants were contributing to nonattainment or regional haze; (2) filing new source review lawsuits for past failures to install pollution controls when the operators attempted to preserve or upgrade the plants; (3) filing citizen enforcement actions to require coal-fired power plants to adhere to the limitations in their permits;

and (4) objecting to utility company requests to recover the cost of pollution controls in public utility commission proceedings. One observer predicted that environmental groups were "going to keep on nipping at your ankles until you cry uncle."[55]

Operating Permit Objections

One issue that environmental groups raised repeatedly in hearings over operating permit renewals was the absence of enforceable limitations on heat input. Because heat was a critical determinant of emissions, the absence of a heat input limit could make it difficult to ensure that emissions of pollutants like NO_x and SO_2 remained below the limits set for them in the permits, and heat input limits remained the only enforceable way to ensure that emissions of pollutants not otherwise subject to emissions limitations were controlled. Environmental groups also insisted that state agencies require fuel switching to reduce SO_2 and NO_x emissions, and they objected to provisions in state laws allowing companies to average emissions across a fleet of facilities within a single state. After the D.C. Circuit overturned the Bush administration's Clear Skies regulation, environmental groups insisted that permitters engage in case-by-case MACT determinations to reduce mercury emissions. Two other issues the groups frequently raised were the absence of adequate monitoring to enforce opacity limitations and the lack of any time limitations on startups, shutdown, and maintenance exemptions for violations of opacity limitations. Finally, environmental groups sometimes alleged that companies had undertaken major modifications without undergoing new source review (NSR) as required by statute. Many of these were successful, but the environmental groups lost about as many challenges as they won. And the courts were generally reluctant to overturn EPA failures to object to Title V permits. But the groups often persuaded companies to enter into settlement agreements that required more controls than the state permitters had proposed.[56]

New Source Review Enforcement Actions

Hoping to stimulate more activity on the part of the government, environmental groups and downwind states sent sixty-day notices of intent to sue many power plants alleging that they had violated EPA's new source review regulations. In several cases, the strategy worked as EPA intervened and took over the litigation. When they could not persuade EPA to file an NSR enforcement action, environmental groups (often joined by state attorneys general) filed their own lawsuits. Many of these lawsuits resulted in settlements in which companies agreed to retire existing units, install additional controls, purchase renewable power, and fund supplemental environmental projects.[57]

Not all of these resource-intensive lawsuits were winners. The lawsuit that the Sierra Club filed in 2001 against TVA alleging that it had avoided NSR when it engaged in a major upgrade of its Bull Run power plant in Tennessee went to trial in early 2009 after eight years of fighting. TVA maintained that the renovation was "routine maintenance, repair, and replacement" and therefore exempt from NSR. At the trial, however, a former manager at a TVA plant testified that the $6 million that the authority spent on the renovation had to get approval from the top management. Nevertheless, the judge agreed with TVA and dismissed the case.[58]

Citizen Enforcement Actions for Permit Violations

The Clean Air Act allows any citizen to bring an enforcement action against the operator of a power plant seeking civil penalties or injunctive relief against violations of the plant's permit requirements, and environmental groups aggressively used this power in their efforts to force owners to retire aging power plants.[59] Occasionally, a citizen lawsuit would inspire state regulators to tighten restrictions in power plant permits.

In the absence of an EPA standard for mercury emissions, the groups could not directly address them in citizen enforcement actions because power plant permits typically lacked emissions limitations for mercury. But they could achieve some reductions in mercury emissions by insisting that plants comply with permit requirements for the opacity of emissions, especially during startups, shutdowns, and malfunctions. Opacity served as a surrogate for PM emissions for plants that lacked continuous particulate monitors, and mercury clung to particulates. Since power plant operators had to report opacity readings on a regular basis, it was fairly easy to comb through a plant's reports and identify violations. The power companies responded that because opacity was merely a surrogate for PM emissions, it was not necessarily true that a company reporting an opacity violation had actually emitted too much PM.

Most of the lawsuits resulted in settlements that brought about genuine environmental improvements, but environmental groups lost some cases. And some resulted in bitter battles that went on for years. For example, the Sierra Club assigned a high priority to closing down three lignite-fired plants in East Texas owned by Energy Future Holdings' subsidiary Luminant—its Martin Lake plant near Longview, its Big Brown plant in Freestone County, and its Monticello plant near Mount Pleasant. As we observed in Chapter 1, the first two plants were notorious "big dirties," and they were responsible for more than 46 percent of the state's coal-fired power plant emissions. Between August 2008 and October 2011, the club filed notices of intent to sue Luminant alleging tens of thousands of violations of opacity and heat input limitations in the plants' permits. In all three instances, the company responded that the alleged violations occurred during startup, shutdown, or malfunction (SSM) events that were lawful under the state's EPA-approved state implementation plan (SIP).

In September 2014, federal judge Walter S. Smith Jr. ruled that the Sierra Club lacked standing to bring the case and that the Texas SIP provided valid SSM affirmative defenses for every one of the tens of thousands of violations. In an unprecedented blow to the club, the judge found the lawsuit to be frivolous and ordered it to pay Luminant's $6.4 million in litigation fees. Overburdened with debt from its takeover of TXU in October 2007 (discussed above), Energy Future Holdings declared bankruptcy in April 2014. In December, Luminant and the Sierra Club entered into a settlement in which Luminant agreed not to collect the $6.4 million in legal fees and the club agreed to drop its appeal of the court's decision and terminate ongoing litigation against the three plants. It was a crushing defeat for the Sierra Club and a powerful indication of the chilling effect of attorney fee awards on even well-financed public interest groups. During the Obama Administration, EPA required states to remove the SSM affirmative defense from their SIPs.[60]

Protecting Pristine Areas

Environmental groups paid special attention to power plants that threatened visibility in national parks and wilderness areas.[61] A major struggle broke out in the mid-1990s over the contribution to visibility impairment in the Grand Canyon of the giant Mohave plant in Laughlin, Nevada. Owned by Southern California Edison (SCE) and several smaller partners, the two-unit 1,500-megawatt plant provided electricity for one million homes in Nevada and Southern California, and it was a mainstay of the economy of the Navajo and Hopi tribes that supplied coal to the plant through a long slurry pipeline. But its uncontrolled SO_2 emissions of 40,000 tpy made it the largest single source of SO_2 upwind of the Grand Canyon. In February 1998, the Sierra Club and the Grand Canyon Trust sued SCE and its partners, demanding that they install scrubbers on both units to reduce SO_2 emissions. The lawsuit alleged that the plant had routinely violated several provisions of the Nevada SIP relating to SO_2 and particulate emissions. Soon thereafter, a seven-year EPA monitoring and modeling study concluded that the plant was a relatively small (16 percent) contributor to overall visibility loss in the canyon, but it was the largest single source of emissions causing the problem. After much negotiation, the parties entered into an October 1999 settlement in which SCE and its partners agreed to install scrubbers, baghouses, and low-NO_x burners at the Mohave plant by January 1, 2006, or retire the plant. When the cost of installing the controls ballooned from $300 million to around $1 billion, SCE retired the plant at the end of 2005.[62] It was a major win for the environmental groups, but a huge loss of income for the tribes.

Objections to Recovering Costs for Pollution Controls

A major theme of the Sierra Club's phase 2 was the argument that retiring old power plants and replacing their capacity with renewable energy and end-use efficiency was far less expensive for consumers and better for the environment than installing new environmental controls. The Sierra Club and other environmental groups objected to utility company requests to recover costs expended on pollution controls in state public utility commission proceedings, even though such controls would unquestionably improve air quality. The groups occasionally achieved partial victories through sheer persistence and efforts to sway public opinion. Sometimes local politicians and coal interests fought harder for PUC approval of pollution control projects for dirty power plants than the operators, who preferred to retire or repower big dirties in anticipation of EPA's MATS and climate change regulations.[63]

Environmental groups usually failed to persuade state PUCs, some of which were industry-dominated, to deny cost recovery for pollution control expenditures.[64] But they did achieve some notable victories. For example, the

Sierra Club, the American Association of Retired Persons, and a group of large energy consumers persuaded the Oklahoma Corporation Commission in December 2015 to deny Oklahoma Gas & Electric Company's request to include in its rate base a massive $1.1 billion project to install scrubbers at both coal-fired units at its Sooner plant and convert two coal-fired units at its Muskogee plant from coal to gas. The commission agreed with the Sierra Club that OG&E had not adequately considered alternatives such as building new wind farms, demand-side management, and power purchase agreements.[65]

The groups were sometimes willing to compromise environmental concerns to support advocates for low-income consumers. For example, the Sierra Club and NRDC agreed to a settlement in which Kentucky Public Service Commission allowed Louisville Gas & Electric to recover $2.25 billion expended on multiple pollution control projects, because it also provided for two annual contributions of $250,000 to home energy assistance programs. The groups said they agreed to the settlement "to support their low-income housing advocate allies."[66]

Struggles at the Local Level

The Sierra Club and local environmental groups also engaged in several struggles at the local level before city councils. For example, the Sierra Club in September 2011 launched an advertising campaign in the Chicago area against Midwest Gen's ancient Fisk and Crawford plants accusing the plants' emissions of causing illness and premature death and pointing out that more people lived near those plants than any other coal-fired plants in the country. In late February 2012, Midwest Gen announced that, pursuant to an agreement among the company, Chicago mayor Rahm Emanuel, and the environmental groups, it would retire the Fisk plant by the end of 2012 and the Crawford plant by the end of 2014. Both plants were off-line by September 2012.[67]

The Sierra Club also brought its struggle against municipally owned coal-fired power plants to the city councils that were responsible for them. For example, the club in 2006 objected to the Springfield City Water, Light, and Power's decision to add a $500 million Unit 4 to its Dallman power plant near Springfield, Illinois. To avoid delays that would have cost the city around $100 million, it entered into an agreement with the club under which it would purchase 120 megawatts of wind power and undertake an end-use efficiency program. Springfield later became the first city in the United States to bind itself to the Kyoto Protocol by cutting CO_2 emissions.[68]

Although the Sierra Club was quite successful at the local level, it did not win all of its local battles. It failed to persuade the City of Austin's municipal utility company, Austin Energy, to sell its stake in the Lower Colorado River Authority's (LCRA) Fayette plant near LaGrange, Texas, or to use its partial ownership of the plant to force LCRA to shut it down. The Austin City Council reluctantly concluded that pulling out of the Fayette plant would cost too much because Austin Energy would still have to pay off around $260 million in debt and interest that was largely attributable to a recently installed $400 million scrubber.[69]

Struggles over Coal Ash Impoundments

After the disastrous coal ash spill at TVA's plant near Kingston, Tennessee, environmental groups began to object to new coal ash impoundments and proposed expansions of existing impoundments. In addition, the groups filed citizen suits under section 7003 of the Resource Conservation and Recovery Act against power companies alleging that their coal ash impoundments posed an imminent hazard to public health and the environment and asking federal courts to order the companies to take action to reduce the risks posed by the sites and to remediate contamination that had already occurred. They also filed administrative complaints with state agencies alleging that existing coal ash impoundments were polluting groundwater. Finally, the groups monitored coal ash impoundments near rivers and streams to detect discharges through groundwater that lacked federal permits and filed lawsuits to prevent such discharges when it found them. After EPA promulgated its coal ash regulations, under which private citizens were the only entities other than the states capable of enforcing their requirements, environmental groups were on the front line of the struggles to protect the environment from leaking coal ash impoundments.[70]

The initiative achieved some notable victories. In January 2013, GenOn Energy settled a lawsuit by agreeing to pay $2.2 million in fines and spend $1.9 million remediating ground and surface water contamination caused by coal ash impoundments at three of its plants in Maryland. The Sierra Club made its case that the almost daily discharges of large amounts of

contaminants from Louisville Gas & Electric's Mill Creek coal ash impoundment did not come within the company's permit allowing "occasional" discharges by surreptitiously filming the outfall from across the river for a
year. The Southern Environmental Law Center persuaded a federal district
judge to order TVA to remove coal ash from storage ponds at the site of its
Gallatin plant and dispose of it in lined landfills far away from the adjacent
river at a cost of around $2 billion. The Sierra Club settled its lawsuit
claiming that the huge coal ash impoundment at the Public Service Company of New Mexico (PNM) San Juan plant near Farmington, New Mexico,
created an imminent and substantial endangerment when PNM agreed to
build a slurry wall to protect groundwater and to spend $2 million on proj-
ects to improve water quality on Navajo lands.[71]

Struggles on Multiple Fronts

Environmental groups frequently fought battles on multiple fronts. In its
campaign against PNM's San Juan plant, the Sierra Club filed a citizen enforcement action against PNM in federal court alleging thousands of opacity
violations, objected to the New Mexico Environment Department's regional
haze plan for the plant on the ground that its NO_x emissions limitations were
insufficiently stringent, challenged PNM's proposal to the New Mexico Public
Regulation Commission (NMPRC) to purchase only forty megawatts of renewable power to replace the loss of nine hundred megawatts occasioned by
retiring two coal-fired units, and objected to PNM's negotiating with BHP
Billiton to purchase its San Juan coal mine.[72] The Sierra Club's battle against
Midwest Gen's six power plants in the Chicago area raged on for more than
a decade as the club focused on the plants' air emissions, groundwater contamination from their coal ash impoundments, water pollution resulting
from their effluent, and damage to aquatic species due to their cooling water
intake structures.[73]

Reliability Trumps Environment

Travelers to Washington, D.C., sitting in a left window seat of an airplane as
it proceeded northward along the Potomac River just south of Reagan National Airport prior to 2012 could easily see a coal-fired power plant incon-
gruously nestled among row houses, condominiums, and hotels. This was the

Potomac River Generating Station, which over its sixty-year life span was owned by several companies. During the George W. Bush administration, it was owned by Mirant Corporation. After two local residents raised concerns about fly ash settling on downwind neighborhood buildings, the Virginia Department of Environmental Quality and Mirant commissioned a study that concluded that emissions from the plant could contribute to serious violations of the national ambient air quality standards for SO_2, NO_x, and PM in the Alexandria area through a phenomenon known as "downwash" as winds blew emissions from the plant's short stacks against recently erected tall buildings and spiraled rapidly downward. Unwilling to spend hundreds of millions of dollars for pollution controls, Mirant temporarily halted production at the plant in August 2005.

The PJM Interconnection, the manager of the electric grid for mid-Atlantic states, worried that taking Potomac River off-line could cause reliability problems because Washington, D.C.'s business district was in a "load pocket" that was connected to the grid by only two inadequate transmission lines. In late December 2005, Secretary of Energy Samuel Bodman ordered Mirant to resume limited production at the plant to ensure reliability during what he declared to be an emergency. EPA followed with a June 2006 administrative order easing the emissions limits pending the completion of additional transmission lines to import electricity from elsewhere. The plant was permanently closed in the fall of 2012.[74]

THE "BEYOND NATURAL GAS" INITIATIVE

The Sierra Club did not think that building more natural gas–fired plants was an appropriate response to the looming climate disruption crisis. Among other things, a big build-out of natural gas–fired plants would lock the nation into another fossil fuel and take away market share from wind, solar, and other renewable resources. The "Beyond Natural Gas" campaign that the club initiated in April 2012 targeted new natural gas plant projects for opposition. It did not initially take the position that the plants should not be built, insisting instead that they employ the most efficient technology possible. As the campaign progressed, however, the club objected to many new gas-fired plants and to most new natural gas pipeline projects to motivate investors to move to renewable energy. By October 2016, the club's goal was to eliminate all gas-fired plants by 2030.[75]

The Sierra Club's first outright opposition to a new gas-fired power plant took place in Southern California, and it got the campaign off to an inauspicious start. In June 2010, a group of developers called the Energy Investors Funds Group (EIF) applied to the California Energy Commission to build a 300-megawatt plant consisting of three gas-fired simple-cycle units in anticipation of selling the power to San Diego Gas & Electric Company (SDG&E) during peak hours. The Sierra Club failed to persuade EPA that the best available technology for the plant was a far more efficient combined-cycle design and that the 4,000-hour-per-year operational cap was far more than needed for a true peaker plant. It also failed to persuade the California Public Utilities Commission to reject SDG&E's application to enter into a power purchase agreement with EIF for the plant's output.[76]

The Sierra Club's campaign suffered several setbacks before EPA's Environmental Appeals Board and state public utility commissions, but it did achieve a few modest successes. In March 2014, the club reached a settlement with Footprint Power under which it agreed to drop its opposition to a $1 billion 630-megawatt gas-fired plant and the company agreed to declining caps on the plant's CO_2 emissions and to retire the plant at the end of its useful life in 2050.[77]

Opposition by local groups to gas-fired plants was generally more successful. We saw in Chapter 7 how local groups in the Chicago area caused companies to cancel proposed plants during the boom in peaker plant construction following deregulation in Illinois. Sometimes developers overcame local opposition by accepting changes like high walls and additional pollution controls to blunt local impacts.[78] When residents of the Haciendas del Norte subdivision in East El Paso, Texas, objected to El Paso Electric's proposed natural gas–fired peaker plant consisting of four eighty-eight-megawatt simple-cycle turbines next to a petroleum storage tank terminal just outside the subdivision, they obtained the services of an attorney with Texas Rio-Grande Legal Aid. One of the many low-income colonias along the border between the United States and Mexico, the subdivision lacked paved roads and natural gas service of its own. The residents feared that the plant's emissions would have an adverse effect on their health, and they were concerned that a power plant built so close to large petroleum storage tanks would increase the risk of explosions that could level the neighborhood. On December 11, 2013, the residents and El Paso Electric reached an agreement under which the company agreed not to construct more than four units at the facility, to use special fixtures to reduce light pollution, and to create a fund to provide energy efficiency measures in homes within a mile radius of the plant.[79]

In the later years, the club also argued against converting coal-fired units to natural gas, a position that was somewhat inconsistent with its previous support for converting coal-fired plants to natural gas. Its opposition to conversions sometimes put the club at odds with other environmental groups, like NRDC and EDF. In mid-2016, the Sierra Club broadened the focus of its campaign still further to include opposition to oil and gas pipelines, liquid natural gas export terminals, hydraulic fracturing wells, and other operations that collected and supplied oil and natural gas. By the end of 2016, one of the goals of its renamed "Beyond Dirty Fuels" campaign was to oppose every new fossil fuel–fired power plant.[80]

Of particular relevance to power plants was the Sierra Club's opposition to pipelines, which were essential for delivering natural gas from areas that contained natural gas to areas in need of gas-fired electrical power. In July 2016, environmental groups calculated that if the expanded gas production contemplated by pending pipeline projects in Appalachia were completed, the United States could not meet its GHG reduction commitments under the Paris Agreement. Noting that new pipelines were designed to last forty years, they argued that leaks from the pipelines and associated compressor stations would be releasing methane into the air long after the nation would have to dramatically reduce its reliance on fossil fuels. In October 2016, the Sierra Club launched a $5 million campaign to object to gas pipelines throughout the country.[81]

Two of the most heated struggles concerned two large pipeline projects totaling $4.6 billion that were designed to carry 3.5 billion cubic feet per day of natural gas from the Marcellus shale formation in western Pennsylvania and West Virginia to plants in Virginia and North Carolina—the Mountain Valley project and the Atlantic Coast project. The Sierra Club and local environmental groups objected to both pipelines in a variety of forums. Before state public utility commissions and FERC (which had to approve all interstate pipelines), they argued that the pipelines should be disapproved because they were unnecessary, would result in increased electric power rates, and would contribute to climate disruption. In eminent domain proceedings in state courts, they supported landowner claims that the pipelines were not legitimately taking property for a public use. They urged state environmental agencies to veto the U.S. Army Corps of Engineers permits to cross hundreds of rivers and streams under section 401 of the Clean Water Act and to deny air emissions permits to pipeline compressor stations. Before FERC and the Fish and Wildlife Service, they argued that the pipelines would destroy habitat of endangered species. They objected to Forest Service permits to cut down trees in national forests along the pipelines' routes. And they protested at the banks that financed the pipelines and in the trees that were scheduled for felling.[82]

Although they were generally unsuccessful in persuading state and federal agencies to disapprove the pipelines, they achieved a degree of success in court. After hearing four challenges to the two pipelines on the same day, the Fourth Circuit Court of Appeals in August 2018 upheld the state of Virginia's certification that the Mountain Valley pipelines complied with its water quality standards, but held that the Fish and Wildlife Service's "incidental take" permit authorizing the destruction of endangered species habitat and the National Park Service's grant of a right-of-way beneath the Blue Ridge Parkway for the Atlantic Coast Pipeline were arbitrary and capricious.[83] It later held that the Corps of Engineers lacked authority to ignore West Virginia's requirement that construction of river and stream crossings be accomplished within seventy-two hours. That ruling sent the West Virginia Department of Environmental Protection scurrying to amend the seventy-two-hour rule for natural gas pipelines. And in December 2018, the court held that the Fish and Wildlife Service had not properly conducted its review of the project under the Endangered Species Act and that the Forest Service had improperly issued permits to construct the Atlantic Coast pipeline across national forest lands because it had not adequately analyzed the pipeline's adverse environmental effects under the National Environmental Policy Act.[84]

CONCLUSIONS

The environmental groups' war on coal was audacious in its conception, comprehensive in its scope, and thoroughgoing in its execution. The groups were at times highly ideological in their appeals to citizens to rally against coal-fired power plants and highly innovative in the ways that they framed those rallies to appeal to public concerns about polluted air and water and the future of a warming planet. High-level officials in the electric power industry conceded that the environmental groups were effective advocates. But the groups were also pragmatic in their willingness to compromise to make progress or secure assistance for low-income consumers, even if that meant allowing a new plant to be built or an old one to keep running for a few more years. And it could not have happened without unprecedented infusions of resources from wealthy donors and progressive foundations.

The campaign was remarkably successful, often with the help of local groups that were naturally inclined to oppose large industrial facilities that made a lot of noise, attracted heavy traffic, and caused local property values to drop.[85] Its comprehensive and persistent battles against existing coal-fired plants contributed greatly to the wave of retirements that took place during the 2010s. The Sierra Club, in particular, was in an ideal position to draw on local opposition for assistance, because it was made up of hundreds of local chapters throughout the United States.[86]

The more recent initiative to halt the move to natural gas has thus far been less successful. Environmental group attempts to establish combined-cycle technology as BACT for new gas-fired plants failed in EPA's Environmental Appeals Board. Efforts by environmental groups to persuade PUCs to disallow cost recovery for new gas-fired plants and conversions of existing plants from coal to gas have failed more often than they have succeeded, and the groups have been able to extract only modest concessions from power plant developers. Local environmental groups, however, have had more success in blocking new gas-fired plants, or at least in forcing their relocation to areas where they attracted less resistance. Thus far, there have been few efforts by national environmental groups or local groups to force the retirement of existing gas-fired plants. And despite some notable successes in court, the jury is still out on the environmental groups' campaigns against natural gas pipelines.

CHAPTER 11

The Transformation of the Electric Power Industry

One proposition that commands general agreement among electric power industry observers across the political spectrum is that the industry is in the midst of a great transformation. Technological change, changing consumer expectations, changes in the relative costs of fuels and renewables, deregulated markets, environmental regulation, and the environmental groups' war on fossil fuels have all played roles in this transformation. The consensus disappears, however, when the discussion shifts to the relative prominence of these forces. Some view the transformation with alarm, predicting that it poses a fundamental threat to the reliability of the national grid. For others, the transformation offers exciting possibilities for improving both economic well-being and environmental protection.[1]

EVIDENCE OF TRANSFORMATION

Perhaps the strongest evidence of the transition is the dramatic decline in the amount of electricity derived from coal and the equally dramatic increase in electricity from natural gas and renewables. In addition to diversification of the sources of power, the nationwide grid is experiencing a remarkable transformation in the number of generators as rooftop solar, backyard wind turbines, and diesel generators provide power to individual homes and microgrids. Further evidence of transformation is the major role that demand response (DR), distributed generation (DG), and other energy conservation measures play in reducing demand for electricity. Electricity markets are more competitive than they were during the twentieth century, with large system operators bearing primary responsibility for wholesale markets and state public utility commissions (PUCs) playing a less prominent role in retail markets. Finally, there is strong evidence of significant reductions in emissions of air pollutants from the electric power sector.

Decline of Coal

The "rush to coal" at the outset of the George W. Bush administration slowed to a crawl by the end of his tenure. By mid-2008, projects representing more than 50 percent of the new capacity announced during the Bush administration had been canceled. Things only got worse for coal during the Obama administration. By 2013, electric power companies were commissioning no new coal-fired power plants, except for a few rare projects for specific highly advantageous purposes. The consensus among electric power executives at an August 2014 conference on the future of the industry was that companies would not be building any new coal-fired plants in the foreseeable future. The nation also witnessed an extraordinary acceleration in the pace of coal-fired power plant retirements during the Obama administration. The proportion of the nation's electricity supplied by coal fell from 51 percent in 2008 to 31 percent in 2016. During that time, companies retired fifty-nine gigawatts (17 percent of the country's coal-fired capacity) and switched thirteen gigawatts (4 percent) to natural gas.[2]

Coal consumption declined 29 percent between 2007 and 2015. Domestic production hit a thirty-year low in 2015 at 900 million short tons, down from nearly 1,200 million short tons in 2008. For a time, exports of U.S. coal helped offset the losses due to declining demand for coal to burn in power plants, and in the last half of 2016, coal production actually increased slightly because of an increase in demand for exports. But it was not enough to revive the ailing coal industry. Half of all of the operating coal mines in 2008 were closed by the end of 2017. Between 2012 and 2016, twenty-seven coal mining companies, mostly located in Appalachia, filed for bankruptcy protection. Five of the largest coal production companies sought bankruptcy protection in

2017. In addition to putting mine workers out of work, the bankruptcies sometimes left retirees without pensions and health benefits.[3]

Ascent of Gas

Betting that natural gas prices would remain stable, many power companies shifted from coal to natural gas to provide baseload energy, relegating coal-fired plants to peak load providers.[4] Electricity production from natural gas increased by 51 percent between 2008 and 2015.[5] By 2016 electricity from natural gas-fired plants exceeded that from coal-fired plants for the first time.[6] In Texas, where coal had been the dominant fuel for decades, inexpensive natural gas produced 1.5 times as much electricity as coal in June 2017.[7]

Greater Role for Renewables

Another prominent contributor to the decline of coal was the rapid ascent of renewable power as the cost of wind and solar energy decreased 60 and 80 percent, respectively, between 2009 and 2016. In 2015, wind accounted for 41 percent of all new capacity additions and solar accounted for 26 percent, with natural gas accounting for 30 percent and coal adding virtually nothing to the mix. The proportions were much the same in 2016. In 2016 alone, new wind power alone accounted for $14 billion in investment and created 14,500 full-time jobs. Solar was the largest source of new electricity in the country, providing 14.8 gigawatts of new renewable power. In Pacific Gas & Electric's service area alone, rooftop solar homes increased from 163 in 2000 to 9,500 in 2005 to 60,000 in 2012 to more than 240,000 in early 2016. As of 2017, more than 77,000 small "backyard" wind turbines were producing power for individual facilities in all fifty states.[8]

Growth in wind generation was especially strong in power-hungry Texas, more than doubling from 9,200 megawatts in 2010 to 18,589 megawatts in 2017 when more than 11,500 turbines were spinning. In the Midwest, wind farms provided almost 20 percent of consumed electricity. Iowa derived 40 percent of its electricity from wind and solar installations. The Kansas legislature's repeal of the state's 20 percent renewable portfolio standard in 2015 proved irrelevant as wind generated 30 percent of the state's power anyway. Competition from less expensive wind and solar forced power companies to retire old gas-fired plants, cancel new ones, and, in the case of one merchant generator, declare bankruptcy.[9]

Slower Growth in Demand

A major contributor to the transformation of the electric power industry was the unexpected drop in demand for electricity, despite the fact that electricity was powering an increasing number of appliances and devices that Americans relied upon in their everyday lives. During the twenty-first century, economic growth and population growth have far outpaced energy consumption growth. Demand for electricity grew at a very slow rate of 0.7 percent during the first decade of the twenty-first century, compared to a 9.8 percent rate in 1949–1959.[10]

Lower Electricity Prices

Dire predictions from coal industry executives that "citizens on fixed incomes will not be able to pay their electric bills, and our manufacturers of products for the global marketplace will not be able to compete" as a result of EPA regulations never came true. Competition, low demand, low natural gas prices, and the declining cost of renewable electricity have had a salutary effect on the price of electricity at the retail level, despite the retirement of hundreds of coal-fired power plants and the failure to build new coal-fired plants to take their place.[11]

Merchant Generators Suffer

Reduced demand and low energy prices have taken their toll on independent ("merchant") generators who do not have the luxury of captive markets and sympathetic public utility commissions. With wholesale prices for electricity in 2017 at half of their 2008 levels, merchant generators were struggling to pay for a large build-out of gas-fired plants during the intervening years. They also found it hard to attract sufficient capital to build new power plants and install expensive pollution control technologies in existing plants required by Obama administration EPA regulations. Merchant companies with large coal and nuclear power plants found it difficult to compete even in capacity markets where their need to run full-time was less disadvantageous. Some companies with both regulated and merchant units hastened to divest their merchant assets, while two of the largest merchant generators, Dynegy and Vistra Energy (the parent company of Luminant) merged to create a $20 billion behemoth that hoped to withstand future market pressures. Other companies, like Exelon's merchant subsidiary in Texas, sought bankruptcy protection.[12]

Fewer Jobs in the Coal and Electric Power Industries

Reduced demand for coal translated into fewer jobs for miners and other coal industry employees as well as fewer jobs on the railroads that hauled coal. The number of coal miners nationwide fell from more than 80,000 in 2008 to about 53,000 in 2018. Coal miners received higher than average wages, and many incurred high levels of debt that they will not be able to repay on the much lower wages they can earn in service industry jobs. And most lack the skills to compete for jobs in the technology and renewable energy industry without additional training. It turned out that building new coal-fired power plants did not produce nearly as many jobs as their proponents predicted when state agencies were considering them. An April 2011 study of six recently built plants concluded that only 56 percent of the promised jobs were realized in the real world. Moving from coal to gas also reduced jobs in the electric power industry because new gas-fired plants are more highly automated and do not have to transfer load, dispose of coal ash, and clean up dirty facilities. And the move from fossil fuels to renewables resulted in thousands of job losses in companies, like General Electric and Siemens, that produced large gas turbines.[13]

Devastated Communities

When employment in a locality's dominant industry shrinks, the effects are felt throughout the community as restaurants and car dealerships close, lower tax revenues prompt layoffs of government employees, and local vendors go bankrupt or leave town. Support structures, like churches and other nonprofit charitable institutions, also suffer from lower donations. Coal regions in Appalachia have suffered acutely because many of the highest quality and most easily accessible coal seams are now exhausted. The decisions by many power plant operators to burn low-sulfur coal from the Powder River Basin during the 1980s and 1990s and the move away from coal toward natural gas and renewables in the 2010s have had devastating effects on the central Appalachian region. By 2016, the region was ravaged with low employment and an epidemic of opioid abuse.[14]

The situation is not much different in coal-producing regions of the West where coal has been the dominant economic force. For several decades, the Powder River Basin benefited from easy access to coal through surface mining and EPA requirements that could be met by burning the region's low-sulfur coal. Several large power plants and associated mines are located on tribal lands and are major contributors to the tribal employment and tax bases. As tighter sulfur dioxide (SO_2) control requirements and new requirements for reducing carbon dioxide (CO_2) emissions have kicked in, however, coal from the region has become less competitive with natural gas and renewable energy. Communities that experienced economic booms at the turn of the century are now struggling with an economic bust.[15]

Job Growth in the Renewable Energy and Energy Efficiency Industries

The increase in jobs in the renewable energy sector offset and even exceeded the job losses attributable to declines in the coal and electric power sectors. In 2016, more than 260,000 employees worked in jobs in which they spent most of their time creating solar energy, and wind power accounted for another 100,000 jobs. Fully 2.2 million Americans were employed in designing, installing, and manufacturing energy efficiency products and services. Most of these renewable energy and efficiency jobs were in red states that voted for Donald Trump, and a large number of them were in the rust belt where coal miners and electric power employees were being laid off.[16]

Environmental Improvement

The forty-year effort to reduce the environmental externalities attributable to fossil fuel–fired power plants has yielded some noticeable successes. According to the Edison Electric Institute, electric power companies reduced NO_x emissions by 74 percent and SO_2 emissions by 80 percent between 1990 and 2014, while electricity use increased by 36 percent. DOE's Energy Information Administration reported that NO_x emissions from power plants decreased from 3,799,000 tons per year (tpy) in 2006 to 1,630,000 tpy in 2016, a 57 percent reduction. Mercury emissions from power plants declined 69 percent between 2000 and 2015. CO_2 emissions from power plants declined from 2,488,918,000 tpy in 2006 to 1,928,401,000 tpy in 2016, a modest 23 percent reduction, but ahead of schedule to meet the clean power plan's goal of a 26 percent reduction by 2025. While emissions of CO_2 from coal-fired plants decreased, emissions from gas-fired plants increased and for the first time exceeded emissions from coal-fired plants in 2016. Owners of coal-fired power plants are now paying attention to how they dispose of coal combustion residuals (coal ash). Although they do not have to dispose of coal ash in licensed hazardous waste disposal facilities, they can no longer dump it into unregulated impoundments. The ever-expanding unlined impoundment that leaches toxic metals into groundwater and runs the risk of catastrophic spills should be a thing of the past. In addition, companies are busily remediating closed impoundments to reduce the risks that they pose to the environment.[17]

REASONS FOR THE TRANSFORMATION

Throughout the Obama administration politicians and experts debated whether the transformation of the electric power industry resulted from EPA regulations, the comparatively low price of natural gas, the declining price of renewable energy, or reduced demand brought on by the economic recession, end-use efficiency, demand response, and distributed generation. In truth, all of these factors played a role in the ongoing evolution of the industry. Indeed, the factors are intimately related.

Reduced Demand for Electricity

The economic recession following the financial meltdown of 2008–2009 had the effect of reducing demand for electricity. Low prices brought on by reduced demand and difficulties in securing credit during the recession played a large role in the cancellation of many of the coal-fired projects that were on company drawing boards during the George W. Bush administration. As the economy slowly recovered from the recession, DR and DG programs in most states continued to stifle demand for electricity. Sophisticated smart meters allowed consumers, their utility companies, or third-party DR aggregators to adjust or turn off heating and air conditioning systems, appliances, and other equipment as part of DR programs. Responding to the Energy Policy Act of 2005, which required electric utility companies to offer smart meters to customers, nearly all of the states implemented some form of net metering under which the amount of energy from the grid that a consumer uses is offset by the amount of energy that the consumer feeds back to the grid from rooftop solar or other DG devices. As of 2016, 9.8 million customers were enrolled in DR programs nationwide, 8.7 million of which were residential consumers. This represented energy savings of 1.3 billion megawatt-hours of electricity. Some large entities like hospitals, universities, and prisons with a critical need for power at all times have created "microgrids" that employ solar power and backup diesel generators that are capable of rendering them autonomous in emergencies. By reducing peak load demand, demand response and distributed generation have reduced the need to build centralized fossil fuel–fired power plants.[18]

Reduced demand also resulted from far more efficient consumer appliances, flat-screen televisions, and smaller electronic devices. Many electric utility companies have established programs for providing energy efficiency rebates and low-interest loans to make high-efficiency lighting, air-conditioning, and electrical appliances more affordable. They also provide free energy audits to analyze homeowner energy use and make recommendations for more efficient adjustments. These programs depend upon the willingness of state public utility commissions to include their expense in the utility company's rate base. In many states, such programs are authorized by state laws establishing energy efficiency resource standards, which set long-term energy savings targets for electric utility companies. In the late 1970s, policymakers encouraged energy conservation to reduce the nation's dependency on foreign oil, but by the turn of the century it was the primary tool for combating the growing threat of global warming.[19]

An Aging Coal-Fired Fleet

Lower demand and resulting lower electricity prices forced power companies to retire some of their older coal-fired units because they were uneconomical. The average age of the more than 13.5 gigawatts of power that companies retired in 2015 was fifty-four years. The plants had simply outlived their usefulness. For example, FirstEnergy relegated its 2,233-megawatt Sammis plant in Ohio to peaking status in August 2012 because the slow economy and low electricity prices had rendered the forty-year-old plant unprofitable. By that time, the company and its predecessor had invested $1.8 billion in pollution controls for the plant, and it complied with its permit. But it made no sense to keep it running full-time.[20]

The Dash to Gas

The shale gas fracking revolution that began in 2007 and continues to the present brought about a dramatic increase in domestic natural gas production and a corresponding decrease in gas prices. Since at least 2008, natural gas prices have remained so low that burning gas is more economical than burning coal in most parts of the country. The lower volatility of natural gas prices in the 2010s encouraged companies to depend far more heavily on natural gas than they had in the past. In addition, the efficiency of natural gas plants improved by 34 percent during the twenty-first century while coal plant efficiency remained relatively static. While the cost of building a coal-fired plant has escalated dramatically during the 2010s, the cost of building a gas-fired generator has steadily declined. Many experts have concluded that low natural gas prices played a greater role in coal's decline than did EPA's regulations.[21]

Less Expensive Renewable Power

The move to renewables was made much easier by "technological advances, economies of scale, and the experience curve," which presaged dramatic reductions in the cost of renewable energy during the twenty-first century. The cost of electricity from utility-scale solar power declined 60 percent between 2010 and 2016, and the cost of onshore wind declined 51 percent. It is cheaper per megawatt of capacity to build new utility-scale wind generating facilities than to build fossil fuel and nuclear power plants. Manufacturing and installation costs for rooftop solar panels have likewise declined dramatically. By 2017, electricity from wind farms, utility-scale solar facilities, backyard wind turbines, and rooftop solar panels was less expensive than that produced by existing nuclear and coal plants, even without the tax credits they received. Consequently, power companies have been investing heavily in renewable power, even in states that lack mandatory renewable portfolio standards.[22]

Public Support for Renewables

Public opinion has consistently favored renewable power during the twenty-first century. This has translated into consumer demand for renewable power from both residential customers and from commercial establishments determined to establish a green public image. More than two-thirds of Fortune

100 companies and more than half of Fortune 500 companies have implemented renewable energy or sustainability polices with specific targets. Large retailers and energy users like Walmart, Costco, Ikea, Google, Apple, and Amazon have installed solar panels on their buildings and entered into long-term supply arrangements with suppliers of wind and solar energy, often to fulfill pledges to their customers that they would reduce their carbon footprints. Some states, like Minnesota and Texas, have created popular programs under which residential customers can enter into long-term contracts to ensure that all of their power comes from utility-scale renewable sources.[23]

Government Support for Renewables

Public support for renewable power has also translated into government policies favoring renewable energy, and electric power companies have cited competition from subsidized renewable energy as a reason for canceling fossil fuel–fired projects. The renewable portfolio standards (RPSs) that twenty-nine states have enacted requiring electric distribution utilities to obtain some percentage of their power from renewable resources have provided a strong incentive for utility companies to invest in renewable power. According to a Case Western Reserve University study, renewable portfolio standards have played a larger role in the decline of the coal industry than have EPA regulations. States have also provided direct and indirect subsidies for renewable power. The Texas Public Utility Commission in 2008 initiated a $7 billion state-financed project to facilitate transmission of power from wind farms in seven "competitive renewable energy zones" in West Texas to the state's large cities. This massive project paid for itself as the state blew past its 2025 renewable portfolio standard's goal in 2012. Texas leads the nation in wind power by a considerable distance. State governments have also provided subsidies and tax incentives to homeowners and businesses to install renewable power.[24]

Federal subsidies and tax credits have provided strong incentives to harness power from renewable resources. The production tax credit for wind energy has proved especially effective in advancing that form of renewable power, especially in deregulated markets. By one estimate, the federal tax credit for wind investments has resulted in at least $10 in private sector investment for each federal dollar invested. DOE's loan programs have also supported dozens of private sector renewable energy projects over the years.

Although there were a few spectacular failures like the notorious Solyndra loan, the program was instrumental in the rapid development of the renewable power industry. The $23 billion that the American Reinvestment and Recovery Act of 2009 appropriated to promote wind, solar, and geothermal power helped double production of renewable power in the nation during President Obama's first term. The federal government has also devoted many billions of dollars to renewable energy research and development.[25]

Changing Patterns of Capital Investment

The investment community, including big banks like Goldman Sachs, Citibank, and Bank of America, have been channeling billions of dollars into producing electrical power. Capital investment in the electric power infrastructure has increased dramatically from $69 billion in 2008 to a record high of $115 billion in 2015, partly as a consequence of reregulation of retail markets. Most of the money went to upgrading and reinforcing electric and gas transmission and distribution infrastructures, including smart grid technology, to make them more efficient and more resilient to bad weather and cyberterrorism. Some went to building new gas-fired power plants and switching coal-fired plants to gas. But some also went to pollution controls and developing renewable resources.[26]

Competition in a Deregulated Environment

In some cases, coal-fired power plant retirements and repowerings and nuclear plant retirements reflected the fact that merchant coal and nuclear plants could not compete with regulated plants in a weak economy. In competitive markets, natural gas generators bid against each other to determine the marginal price for electricity, and for many hours of the day that price covers operating costs but not fixed costs for gas generators. When renewables are sending a great deal of power to the grid, however, the marginal price can dip below the bids of natural gas generators, and gas plants shut down. Coal and nuclear plants, however, cannot easily turn on and off. They not only fail to cover fixed costs, but they may also have to pay customers to take their power. Coal and nuclear plants can make up the difference during peak demand hours when demand is high and prices go up. But those hours are declining as end-use efficiency and distributed generation reduce peak load demand and battery storage fills the gaps. Unable to sustain losses in competitive deregulated markets, owners of coal and nuclear plants are retiring them.[27]

Environmental Regulation

Environmental regulation has played an important role in the transformation of the electric power industry. Strong technology-based emissions limitations pushed the electric industry to adopt the best available technologies. As the industry gained experience with technologies, installation and operating costs declined.[28] The agency leadership was not always anxious to tangle with the industry, but citizen suits by environmental groups to enforce statutory deadlines often forced its hand. The unsung heroes were the career EPA staffers who devoted long hours to writing and defending regulations, overseeing permit applications, pursuing lawsuits, and doing what they could to protect the public health and the environment.

Largely as a result of environmental regulations, coal-fired power plants installed billions of dollars' worth of pollution reduction technologies. By 2016, 688 units with 228 gigawatts capacity were equipped with scrubbers, 942 units with 253 gigawatts capacity had electrostatic precipitators, 605 units with 110 gigawatts capacity had baghouses, 1,477 units with 361 gigawatts capacity had selective catalytic and noncatalytic reduction systems, 477 units with 152 gigawatts capacity employed activated carbon injection systems, and 120 units with 26 gigawatts capacity employed other sorbent injection systems to control emissions of SO_2, NO_x, PM, and mercury.[29] An industry that had fought EPA tooth and nail over several decades had by 2016 reluctantly accepted the need to do the best that it could to protect public health and welfare.

In many cases, however, it was not worth the investment to install expensive environmental controls in plants that were nearing the end of their useful lives, and plant owners shuttered old plants or converted them to natural gas. The MATS standard had the most powerful impact on unit retirements. A wave of retirements in 2015 reflected decisions by power companies to run aging plants right up to April 15, 2015, the deadline for the MATS rule, and then shutter them. More than 12 gigawatts of the existing 299 gigawatts of coal-fired capacity came offline in that year, a significant, but by no means intolerable reduction. EPA's regional haze regulations also had an impact on coal-fired power plant retirements in the West.[30]

Although compliance with environmental regulations was often quite costly for electric power companies, it did not result in large increases in consumers' electric bills. Power companies and politicians warned of price spikes and blackouts when EPA was preparing the CSAPR and MATS in 2011, but the price of electricity did not skyrocket as power companies replaced coal with gas and renewables. In fact, the price of wholesale electricity on the mid-Atlantic grid, the nation's largest, plummeted 40 percent between 2011 and 2016 as 346 coal-burning units were retired. Steep declines in the cost of natural gas and renewables and more rapid moves toward demand response, distributed generation, and greater end-use efficiency more than offset the increased cost of installing controls in coal-fired plants.[31]

The federal government's new source review program brought about large reductions in power plant pollutants. The settlements that electric power companies agreed to required them to reduce pollution from, repower, or retire specific units at named power plants by dates certain. Of the 881 electric generating units (EGUs) that EPA investigated prior to 2012, it took some kind of enforcement action against 467, representing about 45 percent of the coal-fired EGUs in the United States. The initiative resulted in twenty-two major settlements covering 263 units, about 30 percent of the 881 units. According to the calculations of the Government Accountability Office, the companies that were parties to the settlements agreed to spend around $12.8 billion on pollution controls. EPA estimated that the settlements would reduce SO_2 emissions by more than 1.8 million tons per year and NO_x emissions by more than 596,000 tons per year.[32]

Some states have also been active in enforcing strict state implementation plans. For example, PPL Corporation entered into a settlement with the New Jersey Department of Environmental Protection in which it retired the two coal-fired units at its Martins Creek Generating Station in September 2007. To comply with North Carolina's strict Clean Smokestacks law, Progress Energy Carolinas retired three unscrubbed coal-fired units with a combined capacity of 397 megawatts at its H. F. Lee plant near Goldsboro and replaced them with a 950-megawatt combined-cycle gas-fired unit at a cost of around $900 million.[33]

Environmental Group Opposition

By any measure, the Sierra Club's Beyond Coal initiative was a spectacular success. It began as a nationwide campaign to oppose the construction of 150 new coal-fired power plants that were on the drawing boards during the George W. Bush administration. During the Obama administration, the campaign's goals expanded to include forcing the retirement of all existing coal-fired plants by midcentury and opposing all new fossil fuel–fired power plants. By the end of 2015, the club claimed responsibility for the retirement of more than two hundred coal-fired plants. When environmental regulators were reluctant to pursue violations of emissions limitations or new source review requirements, environmental groups stepped in with their own lawsuits. When power companies pressed state PUCs to allow them to include expensive new power plants in their rate bases, environmental groups objected. They even opposed requests by owners of aging coal-fired plants to recover the cost of additional pollution controls on the ground that the controls would only extend the lives of plants that should be retired. Local grassroots movements sprang up all across the country to oppose fossil fuel–fired power plants, coal mine expansions, and natural gas pipeline projects, all of which increased the cost of fossil fuel–fired power relative to energy conservation and renewables.[34]

Changing Attitudes in Some Public Utility Commissions

At the same time that environmental regulators were stiffening pollution reduction requirements, attitudes were changing in the staffs of state PUCs. Instead of focusing exclusively on reliability and keeping rates low, commissions were factoring clean energy into the mix. Sometimes this was inspired by legislative mandates requiring PUCs to take environmental considerations into account. Electric utility companies saw another opportunity to increase profits by making capital investments in environmental control and renewable energy and receiving a guaranteed profit on those investments. At some PUCs, it was easier to justify investment in green projects than investment in large fossil fuel–fired power plants.[35]

A Combination of Factors

The coal industry and politicians from coal-dependent states attribute the rather dramatic decline in coal consumption by power plants to EPA's "war on coal." Most experts, however, agreed that cheap natural gas and renewables and low electricity prices brought on by reduced demand also played roles in the retirement and repowering of coal-fired power plants. In reality, there was a good deal of truth to all of these explanations as companies weighed the complex economics of shifting load from coal-fired plants to gas-fired plants versus installing pollution controls in existing coal-fired plants versus building new coal-fired plants equipped with the best available control technology (BACT) versus repowering existing coal-fired plants to burn gas versus building new gas-fired power plants with BACT versus investing in a wind farm or utility-scale solar plant versus purchasing electricity on the open market.[36]

CONCLUSIONS

The transformation of the electric power industry is still a work in progress. Some of it is subject to reversal, but much of the transformation has by now become permanent. Scores of heavily emitting coal-burning units no longer exist. With powerful institutional actors heavily invested in the deregulated wholesale market, it has by now achieved a degree of permanence that will be very difficult to undo. Deregulation did not take as well in retail markets, which are still heavily regulated in most states, and there are occasional efforts to reregulate retail markets in deregulated states. Demand response and distributed generation have been integrated into the everyday lives of many companies and homeowners, and they are increasingly accepted by distribution utility companies and state PUCs. Renewable energy occupies an increasing share of the national grid. It is, of course, possible that changed conditions could motivate the electric power industry to abandon natural gas and renewables and begin building new coal-fired power plants. One condition that changed dramatically in 2017 was the resident of the White House, a matter to which we now turn.

CHAPTER 12

The Trump Effect

Donald Trump's election in 2016 came as an unpleasant shock to environmental activists. During a presidential campaign in which climate change was a major point of contention, he had promised to repeal environmental regulations that, in his view, caused power plant retirements and impeded the construction of new plants. Environmental groups worried that the incoming Trump administration would abrogate the United States' commitments to reduce greenhouse gas (GHG) emissions under the Paris Agreement and roll back the Obama administration's power plant regulations, many of which remained in litigation. For the coal industry, however, Donald Trump's election offered a "spark of hope." Grateful fossil fuel companies contributed millions of dollars to the gala inauguration festivities.[1]

The activists' fears were soon realized. President Trump appointed Oklahoma attorney general Scott Pruitt to head EPA. After being elected attorney general in 2011, Pruitt had disbanded the office's environmental enforcement division and replaced it with a new unit to challenge EPA regulations. One of its first cases was a challenge to EPA's regional haze plan for southwestern Oklahoma discussed in Chapter 9. Fearing retaliation from environmentalists, Pruitt demanded around-the-clock bodyguard protection. Pruitt's concerns for security extended to first-class air travel and a secure phone booth that cost the agency $43,000. His demands on his staff to serve his private needs, such as seeking a $200,000 per year job for his wife, gave rise to a number of internal and external investigations that ultimately resulted in his resignation in July 2018.[2]

Trump's appointee to be EPA's deputy administrator, Andrew Wheeler, had spent the previous eight years as a lobbyist for the electric power industry and Murray Energy Corporation. Wheeler succeeded Pruitt in late 2018. Trump appointed William Wehrum, an attorney who represented the Utility Air Regulatory Group and other industrial clients, to be the assistant administrator for air. Wehrum had promoted several of the unsuccessful changes to EPA's new source review (NSR) regulations when he was an assistant to Jeffrey Holmstead at EPA during the George W. Bush administration.[3]

Former Texas governor Rick Perry became the new secretary of energy. Perry had advocated abolishing the department when he was running for president in 2012. In his campaign book, he called the scientific consensus that human activities caused global warming a "contrived phony mess." As governor, he had strongly opposed EPA's efforts to regulate GHG emissions, and he had issued an executive order to speed up permitting eighteen new coal-fired power plants. But he also supported a massive publicly funded build-out of transmission lines to West Texas called the "Competitive Renewable Energy Zone" to enable wind power to develop into a booming local industry.[4]

A RADICAL CHANGE IN DIRECTION

Within hours of the inauguration, the incoming Trump administration purged the White House website of all mention of climate change and government attempts to deal with it. Over the next several months, EPA modified its website to drop all mention of President Obama's Climate Action Plan and delete descriptions of climate disruption. DOE revamped its website to emphasize fossil fuels over renewables and to remove a chart showing the link between coal and GHG emissions. After announcing to the Conservative Political Action Conference that his administration would be lifting environmental restrictions on "beautiful, clean coal," President Trump signed an executive order in late March 2017 requiring all departments and agencies to review existing regulations affecting energy production and "suspend, revise, or rescind" any that "unnecessarily burden the development and use of domestically produced energy resources." In particular, the president ordered EPA to suspend, revise, or rescind the Clean Power Plan (CPP).[5]

Pulling Back the Clean Power Plan

The Justice Department immediately asked the D.C. Circuit to hold the pending litigation over the CPP in abeyance until EPA completed the regulatory review that the executive order required. Over the objections of environmental groups, the court in early August 2017 postponed the proceedings for sixty days. It continued to put off deciding the case through mid-2018, when three of the court's judges warned that they would support no further stays. As we saw in Chapter 9, the Supreme Court had already issued a stay preventing the CPP from going into effect. The plan therefore remained in limbo.[6]

Reversing the Endangerment Finding

The Competitive Enterprise Institute (CEI) filed a formal petition in February 2017 asking EPA to rescind the endangerment finding that formed the basis for all EPA regulation of GHG emissions. Although the White House considered a positive response, Pruitt knew that it would be exceedingly difficult to reverse the endangerment finding in a credible way, given the overwhelming consensus among climate scientists and the even stronger evidence of human-induced global warming that had become available since the original 2009 endangerment finding. At his confirmation hearings, Pruitt testified that the endangerment finding was the law of the land and therefore "must be respected."[7]

Rather than tackle the endangerment finding head-on, Pruitt hit upon a backdoor approach to bypassing the scientific consensus and putting the endangerment finding back on the agenda. Prominent climate change skeptic Steve Koonin had suggested that EPA might assign a "red team" of scientists who were skeptical of climate change claims to review and critique the primary documents supporting human-induced climate change and assign a "blue team" of mainstream scientists to respond to the red team's assessment. An EPA-appointed commission then would moderate public hearings to highlight points of agreement and disagreement. After meeting with Koonin, Administrator Pruitt endorsed the red team/blue team idea without consulting the agency's scientists. Energy Secretary Rick Perry quickly endorsed the idea, as did prominent climate skeptics.[8]

Public interest groups opposed the idea as "an act of false equivalence" designed to give climate skeptics more credibility in the public debates than they had in the scientific community. Ignoring the criticism, Pruitt asked the Heartland Institute, a prominent critic of EPA's previous climate initiatives, to identify prospective members for the red team. The institute submitted a list in late October 2017, but by then upper-level White House officials were worried that the media display that Pruitt had in mind might backfire or distract the agency's attention from its other deregulatory initiatives. The electric power industry was also unenthusiastic about the spectacle of a Scopes trial over global warming. In March 2018, White House Chief of Staff John F. Kelly told Pruitt to drop the idea. The still-pending CEI petition offered a more conventional route to overturning the endangerment finding, but the agency has failed to act on it as of this writing.[9]

Replacing the Clean Power Plan

On October 9, 2017, EPA Administrator Pruitt flew to Hazard, Kentucky, to announce that "the war against coal is over." The next morning, he signed a notice proposing to repeal the CPP. EPA now believed that the CPP was based on the erroneous legal premise that the "best system of emission reduction" could encompass "measures that would generally require power generators to change their energy portfolios" by creating or subsidizing power sources like solar and wind energy at locations that were beyond the power plant's fence line. According to Pruitt, the proposal demonstrated that the Trump administration was "committed to righting the wrongs of the Obama administration by cleaning the regulatory slate."[10]

Environmental groups and several states were highly critical of the move. A group of 233 mayors from forty-six states and territories warned that rolling back the CPP would have devastating impacts on their fifty-one million constituents. Religious leaders from the Catholic, Protestant, Evangelical, Jewish, and Quaker faiths condemned the move as inconsistent with the responsibilities of human beings to the earth and to each other. Several electric power companies that did not rely heavily on coal were also disappointed by the action. Coal companies and coal-dependent power companies, however, praised the reversal. Worried that a future administration might promulgate a draconian plan to make up for lost time, the U.S. Chamber of Commerce and the National Association of Manufacturers urged EPA to replace the CPP with a plan that was "narrowly tailored and consistent with" the statute.[11]

Administrator Andrew Wheeler complied with their demand in August 2018 with a proposal for a much less comprehensive (and less effective) plan called the Affordable Clean Energy Plan (ACE). Unlike the CPP, the ACE would not reach "beyond the fenceline," focusing instead on efficiency measures that power plant owners could apply directly at the source. It would also give much broader discretion to the states to come up with their own determinations of "best system of emissions reduction" for particular plants. Finally, it would give power plants more latitude in making "efficiency improvements" without triggering new source review.[12] As of this writing, EPA has not finalized the proposal. When it does become final, it will be challenged in court in a lawsuit that may elicit from the Supreme Court a definitive interpretation of "best system of emission reduction" in section 111(d) of the Clean Air Act.

Pulling Back the MATS

By the outset of the Trump administration, the electric power industry had already adapted to the mercury and air toxics standards (MATS), and nearly all coal-fired units in the country complied with the standards. Between December 2014 and April 2016 when state-granted time extensions expired, companies retired 20 gigawatts of the 299 gigawatts of existing coal-fired capacity, converted 6 gigawatts to natural gas, and received an additional one-year extension for 2.3 gigawatts. Plants representing 87.4 gigawatts installed pollution controls, mostly activated carbon and baghouses, but plants representing 160 gigawatts complied with the standard without having to install any additional controls. The D.C. Circuit in April 2017 granted EPA's request to hold in abeyance the challenge to EPA's April 2016 reaffirmation of its "appropriate and necessary" finding to give the Trump administration an opportunity to reevaluate that finding one more time. In the dead week between Christmas and New Year's Day, with much of the federal government shut down because of Congress's inability to pass an appropriations bill, EPA issued a proposed finding that additional controls on power plants to limit mercury and other air toxics were not "appropriate" because the benefits of the controls would not exceed their cost. In a preamble that resembled a legal brief, EPA dramatically changed its approach to calculating environmental benefits to eliminate consideration of "co-benefits" attributable to reductions in emissions of other pollutants (in this case NOx and particulate matter) that would accompany the installation of controls to reduce mercury emissions. It also concluded that the "unquantified" health benefits, including reduced neurologic, cardiovascular, genotoxic, and immunotoxic effects "were not sufficient to overcome the significant difference between the monetized benefits and costs" of the Obama regulation. Although environmental groups and representatives of many electric power companies urged EPA to retain the MATS, the coal industry demanded a full repeal of the standards. It may be hard for the agency to justify leaving the standard in effect after having reversed the "appropriate and necessary" finding upon which it based its decision to promulgate the standards. As of this writing, EPA has neither finalized the finding nor determined whether the standard should remain in effect.[13]

Pulling Back Regional Haze Plans

At his confirmation hearing, Administrator Pruitt was highly critical of EPA's regional haze program, which he had vigorously challenged as Oklahoma's attorney general. He decided to use Texas as a stalking horse for a radical change in direction. After persuading the Fifth Circuit Court of Appeals to remand the pending challenge to the Obama administration's substitution of its own federal implementation plan (FIP) for the Texas regional haze plan, EPA came up with a new FIP that established an emissions trading program instead of emissions limitations for individual plants representing the "best available retrofit technology." The affected companies were pleased with the flexibility that the Trump administration plan afforded. The Sierra Club complained that EPA's new plan would "do nothing" to bring about additional SO_2 emissions reductions, whereas the Obama administration proposal would have decreased SO_2 emissions by 190,000 tons per year. Scientists at Rice University concluded that the change would eliminate controls that would have prevented three hundred deaths per year.[14]

The Texas action was the first of many EPA retreats from Obama administration positions on regional haze.[15] It also joined the Department of Interior in a campaign to save the massive 2,250-megawatt Navajo plant on the

Navajo reservation in northern Arizona. After agreeing in a 2014 settlement with EPA to reduce regional haze by retiring one unit and installing NO_x controls in the remaining units, the utility companies that operated the plant decided in February 2017 to retire all three units by the end of 2019. The companies' decision was driven by the impossible economics of competing with cheap natural gas and renewables while complying with the settlement. The Institute for Energy Economics and Financial Analysis concluded that the plant's owners would lose around $2 billion if they continued to operate the plant until 2030. Although it was operating only 61 percent of the time, the plant supplied 40 percent of the Navajo Nation's income. The prospect of losing more than eight hundred jobs, most of which were held by members of the tribe, came as a devastating blow to the economically depressed tribe. The U.S. Department of Interior's Bureau of Reclamation, which had a 24.3 percent ownership interest in the plant, opposed the shutdown, but it was not prepared to take over full ownership of the failing plant. The tribe urged the federal government to ease all environmental regulations affecting the plant, but that would require a fresh rulemaking exercise in which EPA would have difficulty demonstrating why the NO_x controls were not necessary to protect downwind national parks and wilderness areas. Instead, EPA in November 2018 proposed to tighten the PM emissions limitation to one-half the previous limitation. In the meantime, two prospective buyers of the plant passed on the opportunity after examining the plant's economics.[16]

Revamping New Source Review

The incoming assistant administrator for air and radiation, William Wehrum, made new source review (NSR) reform one of his top priorities.[17] Rather than the large-scale overhauls that the agency attempted during the George W. Bush administration, however, he planned to take a series of modest steps. The first change was a December 2017 memo replacing EPA's long-standing policy of using the agency's estimates of future emissions attributable to modifications with a policy of relying on the company's subsequent assessment of actual post-modification emissions. The second change came when Administrator Pruitt announced that the agency would no longer be "second-guessing" pre-construction permitting decisions in Title V operating permit proceedings, despite the fact that Title V explicitly required EPA to object to state permitting decisions that were not in compliance with the applicable legal requirements. In a third policy shift, the agency allowed sources to take into account any simultaneous emissions reductions attributable to a project at an existing plant in determining whether the project would result in an overall increase in emissions that exceeded the NSR thresholds. Finally, the ACE proposal would allow power plants to compare hourly emissions instead of annual emissions in deciding whether physical changes triggered new source review. The first two changes were designed to reduce EPA's role in NSR permitting, and the latter two changes would allow sources to avoid NSR altogether. Critics complained that they would enable sources to "skirt the law" and "cook the books" to avoid NSR.[18] At the same time that it was relaxing the NSR requirements, EPA announced that NSR enforcement was no longer a priority.[19]

Revising the NAAQS Process

In early April 2018, President Trump signed a presidential memorandum ordering EPA to change several aspects of its implementation of the Clean Air Act. In response, EPA took two important actions. First, it published a notice of proposed rulemaking establishing requirements for the agency's use of science in rulemaking. Among other things, the proposal required the agency staff to "ensure that dose response data and models underlying pivotal regulatory science are publicly available in a manner sufficient for independent validation." EPA would no longer be able to rely on scientific studies or mathematical models in which all of the underlying data, assumptions, and algorithms were not available to the public. The administrator could, however, grant exemptions from the requirements if it was "not feasible" to comply.[20]

While this appeared on the surface to be a laudable attempt to bring greater transparency to the regulatory process, it was in fact the most recent manifestation of a decades-long effort by the fossil fuel industry to gain access to medical information on individual participants in three epidemiological studies upon which EPA had relied in setting the primary NAAQS for particulate matter for decades. As with most epidemiological studies, the scientists promised the subjects of those studies to keep information about individuals confidential. If strictly enforced, the regulation would preclude EPA from using such studies or studies that contained a company's proprietary information to support its regulations. Critics of the proposal, including environmental groups, scientists, and legal scholars, noted that there were ways to make the underlying data available to scientists who agreed to keep the details confidential for purposes of peer review or replication without making confidential data available to the public. The Chamber of Commerce was pleased with the changes. After the scientific community came out strongly against the proposal and Scott Pruitt resigned as EPA administrator, the proposal quietly faded away.[21]

Second, Administrator Pruitt issued a memorandum that changed the way the staff went about setting NAAQS. The memo directed the CASAC to begin offering its advice on the adverse social, economic, and energy impacts of the standards and to focus more specifically on the relative contributions of natural and anthropogenic activities when it reviewed the NAAQS. The memo further directed the staff to focus more heavily on the possibility of thresholds and background levels of the criteria pollutants "for context." And it told the EPA staff to do a better job of distinguishing between science and policy considerations in all of its documents. Environmental groups and legal scholars criticized the memo as a backdoor way to allow the agency to consider cost in setting NAAQS, despite a unanimous Supreme Court holding that cost was not a relevant consideration at the standard-setting stage. An industry lawyer, however, defended the memo, arguing that even if the administrator could not consider CASAC's input on economic issues, the information would still be relevant in the political process.[22]

Relaxing the Coal Ash Rule

In May 2017, the Utility Solid Waste Activities Group (USWAG) petitioned EPA to reconsider several aspects of its December 2014 coal ash regulations, which had been challenged in the D.C. Circuit but remained in effect with implementation deadlines looming. The group cited President Trump's executive orders and a 2016 amendment to the Resource Conservation and Recovery Act (discussed in Chapter 9) clarifying that the rule could be implemented through EPA-approved state programs as reasons for the reconsideration. In July 2018, EPA finalized changes to the 2015 regulations designed to provide more "flexibility" to comply with alternative standards prescribed by state agencies under the new permitting program. Among other things, they allowed state permitters to: (1) suspend the regulation's requirements for groundwater monitoring if convinced by the operator that CCR constituents would not migrate; (2) use alternative "risk-based" groundwater protection standards for some contaminants instead of technology-based requirements; and (3) allow state officials to issue required certifications instead of professional engineers. The electric power industry was delighted with the changes, which reflected many of their suggestions. Environmental groups claimed that they would "eviscerate even the modest standards put in place in 2015" because they gave far too much authority to the states to waive critical aspects of those standards without EPA oversight. After the D.C. Circuit Court of Appeals in October 2018 found that the Obama Administration regulations were insufficiently stringent, the groups challenged the rollback in the same court.[23]

The Coal Industry's Emergency Petition to DOE

In July 2017, several coal companies petitioned DOE for an emergency order under section 202(c) of the Federal Power Act, which authorizes the secretary of energy to order a power plant to operate when an emergency exists because of a sudden increase in demand for electricity or a sudden reduction in supplies due to the loss of power producing facilities or fuel for such facilities. A 2015 amendment to the statute excuses a company subject to a section 202(c) order from any violations of state or federal environmental laws. The petitioners wanted DOE to declare a two-year moratorium on closures of coal-fired power plants to provide relief from environmental requirements and market stresses due to low electricity prices. According to Robert Murray of Murray Energy, President Trump told Energy Secretary Rick Perry in Murray's presence that he "want[ed] this done." Nevertheless, DOE denied the request, finding that the evidence did not support an emergency finding.[24]

DOE's Grid Resiliency Directive

Energy Secretary Rick Perry attempted to provide more permanent relief to the coal and nuclear industries by invoking his extraordinary powers under section 403 of the Department of Energy Organization Act of 1977 to order FERC to publish a proposed rule (written by DOE staff) allowing companies selling electricity in wholesale markets to fully recover the costs of providing generation from units that retained a ninety-day fuel supply on-site. Perry concluded that the diverse mix of resources necessary to maintain grid "resiliency" had to include traditional baseload generation with on-site fuel storage that could withstand "major fuel supply disruptions caused by natural and man-made disasters." The problem was that competition from natural gas and renewables had made it uneconomical to keep coal and nuclear plants operating so that they could be available during such disrupting events. DOE's proposed rule ordered the system operators that managed wholesale electricity markets to establish "just and reasonable" tariffs ensuring that each eligible unit was "fully compensated for the benefits and services it provid[ed] to grid operations, including reliability, resiliency, and on-site fuel assurance." In effect, system operators would pay operators of idle coal and nuclear plants to keep their plants ready to run, so long as they hoarded ninety days' worth of fuel. The letter specified that eligible units would have to comply with all environmental regulations and could not be subject to cost-of-service regulation by any state or local authority.

In support of his directive, Perry's letter cited a recently completed DOE staff report concluding that 80 percent of the 59,000 megawatts of generation capacity that companies retired between 2002 and 2016 consisted of coal-fired power plants and that an additional 12,700 megawatts of coal-fired capacity was scheduled for retirement by 2020. The letter noted that the electric power industry's response to the 2014 "polar vortex," during which coal-fired units operating at reduced loads were called into full-time service and nuclear units ran at 95 percent capacity, demonstrated the need for fuel-secure resources. Perry also cited "a growing recognition" that FERC-approved wholesale markets did not necessarily "pay generators for all the attributes that they provide[d] to the grid, including resiliency."[25]

The coal industry, the nuclear power industry, and power companies that depended heavily on coal and nuclear power strongly supported the directive, as did the trade associations and think tanks that they supported.[26] The directive, however, encountered a firestorm of opposition from a wide variety of stakeholders, including the natural gas industry, renewable energy companies, independent power producers, vertically integrated utility companies, commercial and industrial electricity consumers, system operators, and environmental groups. Even some coal-dependent power companies were opposed to any interventions that were not fuel-neutral.[27]

Critics denied the proposal's underlying premise that grid resiliency was at risk. Problems with resiliency, they argued, had more to do with downed power lines and other failings of local grids than with the sources of power to those grids. Commenters ranging across the political spectrum from the Sierra Club to the Competitive Enterprise Institute criticized the proposal as a call for a return to regulated wholesale markets that would sacrifice all of the consumer benefits of FERC's restructuring efforts dating back to the Clinton administration. Environmental groups saw the proposal as an attempt "to prop up uneconomic resources that pollute." At the same time, the proposal would not necessarily provide additional resiliency during severely cold weather and floods, because those conditions could compromise coal and nuclear plants as well as natural gas pipelines and plants. During the 2014 polar vortex, for example, around 25 percent of the coal-fired capacity in the PJM Interconnection was forced off-line by low temperatures, while wind power and demand response programs helped fill the void. More recently, floodwaters from Hurricane Harvey forced NRG to take two coal-fired units at its W. A. Parish Generating Station off-line because the coal piles were too wet to burn. Yet the Texas grid, with its heavy reliance on wind and natural gas from West Texas, proved quite resilient during the hurricane.[28]

The statute required FERC to "consider and take final action on any proposal made by the Secretary." After reviewing more than one thousand comments, the commission unanimously rejected Secretary Perry's proposal, finding that DOE had not demonstrated that existing rules governing wholesale electricity pricing were unjust and unreasonable. The grid operators were successfully adapting to their greater reliance on natural gas and renewables. Furthermore, DOE had not demonstrated that its proposed remedy was not unduly discriminatory or preferential. The commission did, however, create a new docket to "examine holistically the resilience of the bulk power system" and to come to some agreement about "what resilience of the bulk power system means and requires."[29]

Invoking National Defense

Undeterred by the FERC setback, DOE used a March 2018 petition for assistance under section 202(c) from the bankrupt merchant subsidiary of FirstEnergy Corporation to put the contribution of "fuel-secure" coal and nuclear power to grid resiliency back on the table. But this time Secretary Perry also invoked the federal government's power to protect national security during emergencies under the Defense Production Act of 1950 (DPAct).

That statute empowers the president to order companies to accept contracts for energy and other "strategic and critical material" under terms and conditions specified by the government, effectively allowing the president to nationalize entire industries. Perry reasoned that 99 percent of the military bases in the United States relied on the grid, but electric power companies were retiring coal and nuclear plants at an alarming rate, leaving grid operators increasingly reliant on natural gas plants, which were more susceptible to cyberattack and weather-related disruption than plants that could store fuel on the premises. In his view, protecting the grid from disruption was worth the increase in electricity prices that would result from the plan. As President Trump explained it to a crowd of supporters in West Virginia: "You bomb a pipeline, that's the end of the pipeline. With coal, that stuff is indestructible."[30]

Once again, DOE had few allies beyond the owners of coal and nuclear plants, the coal industry, and the United Mine Workers union. FirstEnergy, which was virtually unique among large merchant generators in its heavy reliance (88 percent) on coal and nuclear plants, praised the plan for preserving the "irreplaceable role" that coal and nuclear plants played in securing the national defense. West Virginia's governor and congressional delegation supported the plan. But the Heritage Foundation and the editorial page of the *Wall Street Journal* came out against it.[31]

The same coalition that opposed DOE's grid resiliency directive opposed the new DOE national defense initiative. They argued that there was no evidence that either the national grid as a whole or any part of it were experiencing an emergency that warranted invoking either section 202(c) or the DPAct; nor was electricity "scarce and essential" as required by the DPAct. System operators agreed that the grid was reliable and resilient and that existing procedures could handle any foreseeable perturbations. The natural gas industry pointed to many "fail-safes, redundancies and back-ups" that ensured pipelines against cyberthreats. Critics suggested that FirstEnergy's economic problems stemmed from its own misguided decision in 2010 to merge with Allegheny Energy and assume its fleet of coal-fired plants and $3.8 billion of debt.[32]

Conceding that resiliency of the grid could be improved, critics pointed out that stockpiling coal was not the solution. As coal plant shutdowns during the polar vortex and Hurricane Harvey demonstrated, coal was hardly the "indestructible" fuel that President Trump had touted. In fact, most power outages were caused by problems with transmission and distribution that could affect all forms of centralized power delivery. Similarly, electricity produced by coal and nuclear power was equally susceptible to cyberattacks that could shut down those plants or disrupt supplies. DOE's "Soviet-style" solution would put the government in the position of "picking winners," a function that Republican politicians had accused the Obama administration of undertaking in the Clean Power Plan. The plan would transfer billions of dollars from the pockets of consumers to the coal industry and a few privileged uncompetitive power plants. By allowing uneconomic coal-fired plants to continue operating for two more years, the plan would result in more pollution, more human suffering, and more rapid climate change. After receiving heavy criticism from conservative think tanks, the White House placed a hold on the rulemaking process in mid-October 2018.[33]

Tariff on Solar Panels

In April 2017, Suniva Inc., the nation's largest manufacturer of solar panels, sought bankruptcy protection. It then filed a petition with the United States International Trade Commission (ITC) asking the Trump administration to impose a stiff tariff on imported solar panels. Suniva complained that it could not compete with manufacturers of cheap solar panels from China and elsewhere because they were subsidized by their governments. Suniva was an odd entity to bring the request, because its major shareholder was a Chinese corporation and because it imported many of the components of its solar panels from overseas. Over the strong opposition of solar installation companies who depended on low-priced imported panels, President Trump in January 2018 levied a tariff on imported panels of 30 percent in the first year that would gradually fall to 15 percent in four years.[34]

The tariffs had an immediate positive impact on solar manufacturers, who announced plans to spend around $1 billion on new facilities. But they had an even greater negative impact on solar installers, who canceled or froze projects totaling around $2.5 billion in new investment. Having saved 280 jobs in its solar manufacturing business, Sunpower cut 250 jobs in its solar development business because of the tariff. And the tariffs induced some developers to cancel plans to expand their residential solar operations to undeveloped markets. The tariff did not, however, save Suniva, the assets of which were sold off in the bankruptcy proceeding.[35]

THE TRUMP EFFECT ON THE ELECTRIC POWER INDUSTRY

By the time that Donald Trump entered the Oval Office, most of the major electric power companies were well into the transformation from coal to natural gas and renewables described in Chapter 11. As of this writing, two years into the Trump administration, it seems clear that most of its efforts to reverse that transformation are unlikely to succeed. Cheap natural gas, declining prices for renewable energy, and low demand for electricity due to demand response, distributed generation, and improved end-use efficiency have greatly reduced demand for coal. It simply does not pay to maintain an aging coal or nuclear plant that only runs half the time. And the forces that brought about these changes are not likely to change, despite the Trump administration's efforts to save the domestic coal industry.[36]

Many of the nation's largest power generators say that they will continue to invest in natural gas and renewables and include CO_2 reductions as part of their overall corporate strategies, despite EPA's repeal of the CPP, because those options are less expensive than coal. For example, Xcel Energy, one of the nation's largest generators, doubled down on wind in May 2017 with an announcement that it would be adding eleven new wind farms to its portfolio in seven states because it is simply less expensive to build new wind and solar plants than to keep old coal-fired plants running. Even in coal-friendly Texas, inexpensive natural gas and renewable generation have driven companies to retire coal-fired plants ahead of schedule. As Rice University engineering professor Daniel Cohan observed, "Trump can repeal environmental rules, but he can't repeal economics."[37]

Demand from environmentally conscious customers further ensures that electric power companies will continue to expand their renewable offerings. More than 215 of the Fortune 500 companies have established clean energy targets, sometimes as a hedge against uncertain prices for fossil fuel power. Others have done so to improve their images with environmentally conscious consumers. For example, Walmart's "Project Gigaton" has a goal of removing a billion metric tons of GHG from its supply chains by 2030, and it has already installed solar panels on the roofs and parking lots of 350 stores. Other big-box stores, like Home Depot, Target, Ikea, and Kohl's have made similar commitments. Major technology companies like Google, Apple, Amazon, and Microsoft have invested billions of dollars in wind and solar projects to power operations and to offset fossil fuel use. Apple announced in April 2018 that all of its sites were completely powered with renewable energy, by which it meant that it purchased credits for times in which its operations had to rely on backup fossil fuel power. In 2017, nineteen large companies entered into arrangements with power companies to build 2.78 gigawatts of wind and solar capacity. When investor-owned utilities cannot provide sufficient renewable power, the companies purchase it directly from suppliers or build their own facilities. Even in coal-rich states like Kentucky, Virginia, and West Virginia, utility companies have moved away from coal to meet consumer demand for renewable energy. These actions were undertaken in spite of the Trump administration's deregulatory activities.[38]

Environmental groups were discouraged by the 2016 election's outcome, but they remain determined to do everything within their power to facilitate the electric power industry's transformation away from fossil fuels. They played a role in defeating Energy Secretary Perry's grid resiliency rule, and they have fiercely resisted the Trump administration's efforts to repeal Obama administration regulations. With an infusion of another $64 million from billionaire Michael Bloomberg, the Sierra Club continues to sue coal-fired power plants that violate their permits or undertake modifications without undergoing new source review. As in the past, progressive state attorneys general are also playing a role in resisting the Trump administration's rollbacks.[39]

The effect of the Trump administration's pullbacks will be blunted by the commitment of many states and cities to reduce GHG emissions. After President Trump withdrew from the Paris Agreement, twelve governors formed the U.S. Climate Alliance and pledged to make extra efforts to meet their states' shares of the U.S. commitment to reduce the nation's GHG emissions by 26 to 28 percent from 2005 levels by 2025 with or without the support of the federal government. By early 2019, the group had grown to more states, making up more than 40 percent of the country's population and producing half its economic output. California remains on target to meet its goal of reducing GHG emissions by 40 percent from 1990 levels by 2030, and it upped the ante in September 2018 by enacting legislation to join Hawaii in requiring 100 percent carbon-free electricity by 2045.[40]

The nine northeastern states in the Regional Greenhouse Gas Initiative (RGGI), which required power plants to participate in a multistate cap-and-trade program, had reduced CO_2 emissions by 37 percent from 2008 levels by 2016. The price of allowances, however, dropped dramatically with the election of Donald Trump. In August 2017, the group agreed to bring about an additional 30 percent reduction in GHG emissions from power plants by 2030. In addition, New Jersey's newly elected governor Phil Murphy signed an executive order to rejoin the group, and Virginia governor Terry McAuliffe signed an executive order requiring state regulatory agencies to come up with a plan to cap CO_2 emissions from power plants with the goal of joining the RGGI.[41]

The nation's cities are also resisting the Trump administration's initiatives. More than three hundred mayors have pledged their cities to meeting the goals of the Paris Agreement. The United States Conference of Mayors approved a resolution in June 2017 committing its members to running their cities on 100 percent renewable energy by 2035, and more than 80 cities have committed themselves to the pledge. Several cities, including Burlington, Vermont, and Georgetown, Texas, are already purchasing all of their power from renewable sources. Cities that own their own distribution utilities have set ambitious goals for relying on renewable energy.[42]

THE TRUMP EFFECT ON THE COAL INDUSTRY

When President Trump entered office, the coal industry had been on a rather steep decline, as we observed in Chapter 10. As a result, more than one thousand coal mines had closed during the Obama administration. In mid-April 2017, EPA Administrator Pruitt traveled to Pennsylvania to tell coal miners that the federal government's "war on coal" was over. By the end of the year, the Trump administration had completed many of the coal-friendly regulatory actions that Robert Murray had suggested to Vice President Mike Pence in early March. It is, however, highly unlikely that the administration's deregulatory initiatives will bring about the revival of the coal industry that President Trump promised during the 2016 campaign.[43]

First, electric power companies continue to retire coal-fired plants at a rapid pace. Between 2010 and late 2018, they had shuttered more than 200 coal-fired plants. And by early 2017, they had committed to retiring or converting another fifty-one gigawatts (18 percent of operating coal-fired capacity). New England's last coal-fired plant, the Brayton Point Power Station near Somerset, Massachusetts, ceased operations on May 31, 2017. A week later New Jersey's largest public utility PSEG shuttered its two remaining coal-fired plants.[44] Several companies announced that they would soon be retiring several plants that had survived the wave of retirements brought on by EPA's MATS regulations because they were no longer profitable.[45]

Second, there is not likely to be a new "rush to coal" during the Trump administration as there was during the first few years of the George W. Bush administration. At the time of this writing gas remains cheaper than coal, and many companies have made long-term commitments to purchase that fuel. Cost overruns and delays plagued all of the last few coal-burning "mega-projects" like Southern Company's notorious Plant Ratcliffe in Kemper County, Mississippi (discussed in Chapter 10). The Clean Air Act's "best available control technology" and "lowest achievable emissions rate" requirements for emissions of criteria pollutants from new coal-fired power plants and its "maximum available control technology" requirements for emissions of hazardous pollutants will require expensive control technologies. And the BACT requirement for CO_2 emissions will remain in effect for new and modified power plants so long as EPA leaves the endangerment finding in effect. Indeed, the regulatory uncertainty over CO_2 emissions will probably take coal-fired power plants off the planning table until Congress or EPA achieves a lasting regulatory resolution of the issue and banks are once again willing to finance large coal-fired projects.[46]

Third, electric power companies are not likely to fire up mothballed and retired coal-fired plants. In most cases, such "zombie" plants would trigger new source review and its BACT requirement. Efforts to recover the cost of bringing retired plants back to life would encounter stiff resistance in state PUCs. Even if reviving dead plants did not involve large capital expenditures, coal is no longer cheaper than natural gas in most regions.[47]

THE TRUMP EFFECT ON JOBS

Trump's cruelest campaign promise was his guarantee that unemployed coal miners would be going back to work. Most coal miners believed him. For years, their employers, local politicians, and media pundits had told them that EPA regulations were the primary cause for the coal industry's decline and the "economic plague" that had swept through Appalachia. In the real world, the president cannot deliver on that promise. First, as discussed above, companies are not building new coal-fired plants. Second, to the extent that Trump administration rollbacks also help natural gas, demand for coal will continue to suffer. Third, the coal mining industry has become increasingly mechanized to meet the coal industry's long-term business goal of "producing more coal with fewer workers."[48]

The coal industry experienced a 1 percent increase in jobs during Trump's first year in office, but that was mostly attributable to an increase of metallurgical coal production and production of coal for export. If EPA rescinds all of the regulations mentioned in President Trump's Executive Order 13783, the decline in coal mining jobs may be slowed, but there will be no dramatic increase in employment in the coal mining industry. DOE's Energy Information Administration concluded in March 2017 that, at best, rescinding the CPP might save some jobs in the western United States. In its *2017 U.S. Coal Outlook,* the Institute for Energy Economics and Financial Analysis (IEEFA) bluntly concluded that "promises to create more coal jobs will not be kept" and "the industry will continue to cut payrolls."[49]

THE TRUMP EFFECT ON THE NATURAL GAS INDUSTRY

The Trump administration has had little impact on the natural gas industry. The DOE's Energy Information Administration predicts that natural gas consumption will increase at an average of 1.4 percent per year through 2040, with or without the CPP. The president did help broker a deal to bring a Chinese energy company to West Virginia to build an $83.7 billion electric power and chemical complex powered by local shale gas. But the gas industry was not happy about his efforts to subsidize its competitors in the coal industry. Of even greater concern to natural gas producers is the renewed determination of local and national environmental groups to oppose new gas-fired power plants, natural gas pipelines, and other infrastructure.[50]

THE TRUMP EFFECT ON RENEWABLES

President Trump is not a fan of wind energy, having unsuccessfully fought to stop a "monstrous" offshore wind farm within eyesight of his golf course in Aberdeenshire, Scotland. Ignoring the president's animosity, the electric power industry has continued to pour resources into renewable power. During the first year of the Trump administration, companies installed more wind capacity than at any time since the beginning of the Obama administration, with a new wind turbine going up every 2.5 hours on average. Between April and July 2017, American Electric Power, Duke Energy, and PacifiCorp announced plans to install 3.3 gigawatts of wind capacity in Oklahoma, Wyoming, and Indiana. Offshore wind energy has an "enormous potential" to supply energy to East Coast cities that would otherwise require long-distance transport of solar- and wind-generated electricity. Thus far, the Trump administration has done little to discourage offshore wind leasing. In June 2016, the Energy Information Administration predicted that renewable power would increase from less than 300 to more than 400 gigawatts by 2040 with or without the Clean Power Plan.[51]

The 30 percent tariff that President Trump imposed on imported solar cells, however, caused an increase in the price of solar panels, 95 percent of which were imported, just at a time when solar energy was becoming competitive with fossil fuel–fired power. So long as the tariff remains in effect, electric power companies will have a reduced incentive to invest in renewables, and banks may be less inclined to support renewable investments. While this might not have a huge impact on the domestic renewables industry or the overall move away from coal, it could have a devastating effect on some small solar installation companies. During 2018, the first year the tariffs were in effect, 10.6 gigawatts of solar capacity were added to the grid, but that represented a 2 percent decline from the 2017 addition. Jobs in the solar installation industry declined by 3.2 percent in 2018. Experts attributed both declines to the tariff. But solar capacity was still expected to double over the next five years as the tariff declined to 15 percent and perhaps disappeared with the election of a new president. To the extent that Trump's tariff puts renewable energy companies out of business, it will have a negative impact on jobs in an industry that has been one of the largest sources of jobs in the current American economy.[52]

THE TRUMP EFFECT ON THE ENVIRONMENTS

It is too early to measure the effect of the Trump administration's rollbacks on the environment, but they could have a profound effect in future years if they are not reversed by a future administration. While EPA was repealing the Clean Power Plan in 2018, CO_2 emissions in the United States increased 3.4 percent and power plant emissions increased 1.9 percent. Replacing the Clean Power Plan with the proposed Affordable Clean Energy plan will make it very difficult to achieve President Obama's commitment in the Paris Agreement to reduce greenhouse gas emissions by 26 to 28 percent below 2005 levels by 2025, a commitment that President Trump has since abandoned.[53] The Trump administration's changes to the process for setting national ambient air quality standards may pave the way for relaxations of the ozone, SO_2, and particulate matter standards, which could make room for new coal-fired plants and other industrial facilities in nonattainment areas. The administration's relaxation of the modest protections provided by the Obama administration's coal ash impoundment regulations will subject people living near those facilities to greater risks of contaminated groundwater and catastrophic structural failures. And its regional haze reversals will slow down progress toward achieving the Clean Air Act's goals for visibility in national parks and wilderness areas.

CONCLUSIONS

In announcing the CPP repeal in Kentucky with its senior senator Mitch McConnell on the stage, Administrator Pruitt was playing to the Republican base. A February 2017 poll revealed that only 25 percent of Trump voters believed that anthropogenic GHG emissions were causing global warming compared to 65 percent of voters overall. The announcement was more about politics than law, policy, or science. Pruitt pressed ahead with the revocation even as a committee of federal scientists concluded that "it is extremely likely that human activities, especially emissions of greenhouse gases, are the dominant cause of the observed warming since the mid-20th century." The pullback ran counter to overall public opinion. According to an October 2017 poll, 61 percent of the respondents thought that government action was needed to address climate change and only 20 percent thought that EPA should rescind the CPP.[54]

If EPA finalizes its proposal to repeal the Obama administration's "appropriate and necessary" finding as a prelude to revoking the MATS, it will take a long time and a lot of resources. The agency will have to build an elaborate record to justify its changes, and the action will have to survive judicial review. The outcome is unlikely to have any noticeable effect on existing power plants, because the vast majority of affected plants have already complied with the rule, and the owners of many have already recovered their capital expenditures from ratepayers. And it is unlikely to have any impact on new coal-fired power plants, because nobody is interested in building new plants. As Assistant Administrator Wehrum testified, "the bell has been rung" on this one.[55]

Secretary Perry's efforts to revive the coal and nuclear industries by subsidizing coal and nuclear power plants at the expense of ratepayers have failed.[56] Perry said that he was pleased that his resiliency directive had "initiated a national debate on the resiliency of our electric system," and he promised to work with FERC to address "market distortions that are putting the long-term resiliency of our electric grid at risk."[57] But FERC put the resiliency initiative on the back burner. It remains to be seen whether Perry's more recent effort to invoke national defense to justify subsidizing coal and nuclear power will succeed in extending the lives of aging coal and nuclear plants and whether, if it does succeed, it will bring back any coal mining jobs. Thus far, the efforts have had no apparent effect on power company plans to retire coal-fired plants.

When President Trump entered office, the price of coal had nearly doubled over the previous year, and coal company stocks enjoyed a brief "Trump bump." The two largest companies, Peabody Coal and Arch Coal, emerged from bankruptcy, and they were reporting profits for the first time in many years. The number of employed miners increased by 800 to 50,500 during 2017. Unemployment rates were falling in the Powder River Basin, and local businesses were reviving in Gillette, Wyoming. The increase was due entirely to sales of metallurgical coal and coal exports; coal's share of domestic energy production continued to decline.[58]

There is no credible evidence that the Trump administration's efforts to aid the coal industry has had any impact at all on the electric power industry's move away from coal. In a survey of thirty-two utility companies from twenty-six states, most executives reported that President Trump's deregulatory executive order would not influence their future investments. The Energy Information Administration in early 2017 predicted that repealing the CPP would at best stabilize demand for coal through 2030 instead of a steady decline to near zero consumption if the CPP remained in effect. The best that coal advocates can hope for is that the coal-fired plants that are currently operating, many of which were built in the 1980s and have installed second-generation pollution controls, will do what it takes to keep the plants running for as many years as possible.[59]

CHAPTER 13

Toward a Sustainable Energy Future

According to an American Indian proverb, "we do not inherit the earth from our ancestors, we borrow it from our children." If we begin with the moral imperative to preserve the planet for our children's children, we must concern ourselves with the sustainability of our use of electrical energy. The key to sustainability is achieving production and environmental goals simultaneously. In the absence of major technological breakthroughs, that may require the electrical grid to wean itself from fossil fuels. And if we agree with the vast majority of scientists that climate disruption is imminent, we must further conclude that this move from fossil fuels to sustainable resources must take place quite rapidly.[1]

The good news is that completing the transformation from fossil fuels to renewable resources will not be too costly if done correctly. We have reached a point at which we can achieve economic growth and reduce energy consumption, even as population increases. Moreover, we can meet ambitious environmental goals for the electrical power industry without threatening the economy or the reliability of the nation's electrical grid. It will, however, require "some combination of edict and invention." EPA's Clean Power Plan was one attempt at forging such a combination, and, if anything, it underpredicted what the electric power industry was capable of accomplishing in a brief period of time.[2]

The bad news is that the political environment may not be capable of delivering the governmental policies necessary to complete the transformation quickly enough to avoid major damage to human health and the environment. The Trump administration is determined to leave climate disruption to market forces with minimal pressure from the federal government. In the private realm, the electric power industry has been rapidly evolving away from coal, but coal could see a resurgence if natural gas prices increase. And natural gas is still a fossil fuel that adds CO_2 to the atmosphere. Technological innovations like carbon capture, utilization, and sequestration (CCUS) technology can help achieve a sustainable grid, but meeting environmental goals will also require changes in existing power relationships. And the Trump administration has shown no interest at all in modifying those relationships.[3]

END-USE EFFICIENCY

One of the easiest paths to a sustainable grid is to improve the efficiency with which we consume electricity. Residential and commercial consumers can install more efficient heating and cooling systems, use more efficient electrical appliances, switch to LED lighting, and use more efficient electrical devices. Manufacturing companies can use modern digital technologies to reduce overall energy use. Indeed, end-use efficiency has the potential to become the largest electricity resource of all by 2030. By reducing demand for energy from the grid, investments in end-use efficiency increase competition among energy sources, and that in turn drives more expensive sources like coal-fired power plants out of the electricity market, thereby reducing emissions of harmful pollutants. Electric power companies, however, have a natural incentive to encourage greater electricity use to boost electricity prices in the case of deregulated companies or to justify investments in more power plants to public utility commissions (PUCs) in the case of regulated utilities. They will invest in end-use efficiency only if public utility commissions will allow them to recover their investments from consumers plus a reasonable return on those investments.[4]

States can change this incentive by directly subsidizing end-use efficiency or by rewarding distribution utility company investments in end-use efficiency. The state of Maryland's EmPOWER program used $1.3 billion collected from ratepayers to generate an estimated $1.8 billion in energy savings by paying for energy-saving appliances, home energy audit rebates, and monthly bill credits for reduced energy use. Thirty-three states provide for some form of cost recovery for utility-managed energy efficiency programs. All state PUCs should design rate structures that ensure that utility company investment in end-use efficiency is more profitable than investing in new generation and grid infrastructure. And they should resist recent efforts by Tea Party activists to repeal end-use efficiency programs. City councils should continue to write and enforce energy efficiency requirements in local building codes.[5]

Under the Energy Policy and Conservation Act of 1975, DOE must promulgate and update standards for consumer and industrial products at the most efficient level that is "technologically and economically justified." Federal appliance efficiency standards help to overcome the information disadvantage that consumers encounter when they shop for electricity-consuming products, and they eliminate the "split incentives" that cause apartment owners to install cheap, inefficient appliances when more expensive energy-efficient products would decrease their tenants' electric bills. During the Obama administration, the Department of Energy promulgated energy efficiency standards for electrical appliances and other energy consuming devices at the phenomenal pace of about ten per year. The statute's "anti-backsliding" provision will make it hard for the Trump administration to roll back standards that have already gone into effect. DOE could contribute to this salutary pattern by using the social cost of carbon as one component of the benefits side of the cost-benefit balance that it applies when it updates existing energy efficiency standards. But President Trump signed an executive order in March 2017 requiring DOE to stop using the social cost of carbon in regulatory decision-making. Until we have a more energy-conscious occupant of the Oval Office, any significant improvement in end-use efficiency will probably have to come from the states and the private sector.[6]

TRANSFORMING THE GRID

The nation's electricity grid has been called the most complex machine ever built, and it is becoming even more complex as it evolves from a one-way system into a dynamic, multidirectional network that delivers electricity and information to customers and back to the distributor. Modern digital sensing, communication, and control technologies are permitting grid operators to make the three regional grids increasingly efficient, reliable, and resilient. As of 2017 more than seventy million smart meters had been installed in 55 percent of American households. A well-managed smart grid that incorporates demand response (DR), distributed generation (DG), and renewable power has a huge potential to reduce electricity consumption and enhance both the reliability and the sustainability of the grid.[7]

Smart meters and the internet allow two-way communications that permit consumers to know how much electricity appliances and equipment are consuming in real time. Modern "set-it-and-forget-it" thermostats and usage regulators allow customers to specify their energy consumption in advance. Smart meters can facilitate dynamic "time-of-use rates" under which rates are different at different times of the day and highly sophisticated "real-time" marginal cost pricing under which the meter informs the user of the price of the electricity as it varies to reflect the marginal cost of producing it. Using the internet, consumers can remotely control smart appliances in homes to optimize electricity use, or they can program smart appliances to cycle energy use automatically or in response to predetermined price signals.[8]

DR programs have inspired a budding new industry composed of companies that "aggregate" the promises of retail consumers to cut back consumption during peak hours and sell the combined promises to electric distribution companies or grid operators. Some aggregators and distribution companies manage energy consumption for residences and commercial establishments remotely. FERC Order 745, which requires load serving entities to compensate DR providers at the market price for electricity in most circumstances, provides a powerful incentive to reduce demand during high usage periods with a corresponding reduction in emissions as generators rely less on dirty peaker plants.[9]

Smart meters also play a prominent role in DG programs under which consumers generate electricity for their own use and for selling to distributers on the grid. Modern digital technologies allow companies or cooperatives to knit together multiple distributed generators and storage batteries into "microgrids" that depend only to a limited extent on distribution utility companies for electricity. Many cities have established microgrids of providers of critical services that continue to receive power from distributed resources during primary grid outages.[10]

Virtues of the Smart Grid

Digital technologies have played a powerful role in the transformation of the electric power industry, and that role will likely expand in the future because of the many advantages that they bestow. First, they improve operational efficiency by allowing distribution utilities to read meters and connect or disconnect load remotely, to detect tampering and theft, to improve voltage and outage management, and to improve billing and customer support. Second, they reduce the need to build more transmission lines to bring in power from remote locations. Third, by reducing consumption when demand is otherwise high, DR and DG programs increase overall energy efficiency and reduce the need to build additional peaker plants. Fourth, a grid that allows DG and incorporates power from hundreds of small solar generators and batteries that store power from sunlight should be more resilient because it should be less susceptible to outages attributable to weather, malfunctions in equipment, and terrorist attacks. Fifth, the smart grid empowers consumers by allowing them to manage consumption and take advantage of dynamic marginal cost pricing or to rely on the power they produce themselves. Finally, to the extent that distributed resources are not burning fossil fuels and to the extent that DR programs shift generation from fossil fuels to renewable resources, they reduce overall emissions.[11]

Problems with the Smart Grid

Despite their advantages, adoption of smart grid technologies remains uneven across the country. According to the Department of Energy, nearly 75 percent of all smart meter installations have occurred in only ten states and the District of Columbia. An October 2017 survey found that only 31 percent of electric utility customers participated in demand response programs. The uneven adoption of smart grid technologies has many causes. First, fully implementing a smart grid is expensive. In addition to purchasing thousands of smart meters, grid owners must upgrade circuits and transformers, install additional storage capacity, and invest in flow management programs to deal with the variability of DG. Second, DR and DG pose a serious threat to fossil fuel–reliant generators whose profits are dependent on the sale of capacity, rather than electricity. Third, DG poses a threat to the profitability of local distribution utilities who argue that DG customers do not pay their fair share of the fixed costs of the transmission and distribution systems. Fourth, many state legislatures and PUCs have resisted smart grid programs. Fifth, residential consumers tend to be agnostic about DR and DG programs. They do not want to be constantly worrying about how much electricity they are using, and they are not persuaded that the programs will significantly affect their monthly bills. Sixth, energy consumption data are not generally available to consumers, demand response aggregators, energy efficiency advisers, and municipalities in sufficiently granular form to provide the basis for consumer decisions about managing electricity generation and use. Seventh, relying on the smart grid presents many of the same security problems that the internet faces. Finally, to the extent that DG includes heavily polluting diesel engines, it will continue to contribute to emissions of CO_2 and conventional pollutants.[12]

Solutions: A New Business Model

A consensus is emerging among knowledgeable observers and participants in electricity markets that the century-old business model that rewards large capital investments and contains few incentives to conserve energy or protect the environment must migrate to a "services" model in which retail distribution companies sell cleaner electricity, provide customized consumer services, integrate distributed energy resources, and manage DR to reduce peak hour consumption. The first step in this migration is for distributors to extend smart metering to all of their customers and to replace outdated analog equipment with modern digital equipment. These upgrades should, of course, be undertaken with cybersecurity in mind.[13]

The next step is for retail distributors to embrace DR and DG programs and work with their customers to implement them. Companies that have resisted DG in the "net metering wars" should become sellers and installers of rooftop solar and backyard wind technologies, managers of residential and commercial air-conditioning systems and appliances, energy efficiency auditors, and providers of other customized services to their customers at the same time that they are purchasing and distributing power from wholesale markets. Robust DR and DG programs can provide greater protections against outages than aging coal and nuclear power plants, and advances in electricity storage will greatly enhance the resilience of grids that rely heavily on renewable resources.[14]

The final step is for retail distributors to use modern digital communications technology to combine DR and DG resources into "virtual power plants" that dispatch power from renewable resources and reduce demand simultaneously to achieve system balance and eliminate the need to build new fossil fuel–fired plants in the future. San Antonio's municipal utility, CPS Energy, has already used smart meters to assemble a 250-megawatt prototype virtual power plant. And Green Mountain Power has installed batteries in thousands of Vermont homes to allow it to aggregate power from those batteries to meet surges in demand or sell on the wholesale market at a profit.[15] Other distribution utility companies should follow their lead.

Solutions: A New Regulatory Model

It may be overly optimistic to expect that old-line utility companies, which have rarely been agents of change, will voluntarily transform themselves into multidimensional service platforms. Government regulation will be needed to provide incentives to change their business models. As we have seen, FERC greatly encouraged demand response programs by pressing the limits of its authority in Order No. 745 to require system operators to pay the locational market price for electricity to offerers of DR in wholesale power markets. If we are to recognize the full potential of the smart grid, however, changes are also necessary in retail markets. And that will require action by forward-thinking state legislatures and PUCs.[16]

Decoupling. The first task for state legislatures and PUCs is to "decouple" the rates charged for electricity from the volume of electricity sold to reduce utility company incentives to build big power plants and to enhance incentives to invest in end-use efficiency. This will require ratemaking policies that hold utility companies harmless against declines in sales due to their customers' DR and DG activities. Some states now guarantee utility companies a fixed revenue and allow them to increase profits by demonstrating on an annual basis that they have invested in end-use efficiency programs. Others allow utility companies to receive a share of the value of energy reduction benefits achieved by approved DR and DG programs. The remaining states should move forward with decoupling to make utilities and their customers partners in reducing demand for centralized fossil fuel–fired power.[17]

Encouraging Demand Reduction. Before distribution utilities will invest in innovative demand-reducing measures, they need some assurance that state PUCs will allow them to recover the associated costs plus a reasonable rate of return. A majority of states already allow cost recovery for such investments, often in connection with energy efficiency or renewable portfolio standards. State PUCs should follow the lead of Illinois and Massachusetts in allowing distribution companies to obtain up-front approval of grid modernization projects and to recover prudent expenditures while the projects are ongoing. They might even go a step further by providing a slightly higher rate of return for demand reduction than for supply-side investments. This will cause rates to increase, but they should decline again as the efficiency gains result in lower fuel expenditures. State legislatures can reduce demand by enacting legislation mandating variable rates based on marginal cost or time of use to provide strong incentives to consumers with smart meters to manage their use of electricity more efficiently.[18]

Managing Distributed Resources. State PUCs should allow customers or third-party aggregators to sell energy efficiency resources and distributed electricity to the grid and to each other. At the same time, the rate structures should not leave all of the fixed costs of transmission and distribution on consumers who decline to participate in DR and DG programs. To avoid "cross subsidization," PUCs should require sellers of DR and DG to pay their fair share of those prudently incurred and nondiscriminatory fixed costs. In fairness, they should also pay their fair share of "taxes" that PUCs impose on distribution utilities to subsidize low-income customers. Similarly, transactions within microgrids should include small fees to be paid to the distribution utility that provides the necessary backup power. Several state PUCs are already experimenting with pilot rate structures aimed at achieving an appropriate balance between encouraging distributed renewable energy and compensating utility companies for maintaining a reliable and resilient grid.[19]

Making Consumption Data Available. Electricity generators have no incentive to make hourly or sub-hourly electricity consumption data available to consumers, energy aggregators, and municipalities for use in managing consumption and administering energy conservation programs, because better energy management may reduce overall demand. They also worry that making such data available could violate the privacy of individual consumers. State legislatures or public utility commissions should require electricity distributors to make consumption data available to individual consumers who generate the data, to energy aggregators who have received permission from their clients, and to municipalities who provide adequate assurances that they will protect consumer privacy.[20]

Encouraging Distributed Generation. Widespread use of distributed renewable resources will also require changes in local zoning laws and building codes. For example, municipalities will have to amend zoning laws to permit wind turbines in residential neighborhoods and to protect the rights of building owners who install solar panels to continue to have access to the direct rays of the sun. While many model ordinances are available, difficulties will arise in overcoming the opposition of community members who find distributed resources to be unsightly, too noisy, or both. Like any potential nuisance, cities will have to balance the inconvenience of neighbors with the benefits of wind power, but they should not allow aesthetic objections to rooftop solar cells to outweigh the benefits of solar power. Proponents of DG must also reckon with the difficulty of persuading consumers to make the initial investments of time and resources necessary to implement that solution on a wide-scale basis. This may require some degree of subsidization to allow middle- and low-income homeowners to afford the expensive rooftop solar units that will pay for themselves in the long run in lower utility bills.[21]

Discouraging Polluting Distributed Resources. Many advocates of DG include within that term small diesel engines and gas-fired micro-turbines that have the potential in the aggregate to emit large amounts of CO_2 and other pollutants. Small diesel generators are generally less efficient and emit more pollutants per unit of energy than controlled coal-fired power plants. We cannot reach the goal of sustainable energy production by shifting the load from a few large centralized fossil fuel–fired power plants to a great many small fossil fuel–fired engines. Some states have imposed restrictions on operating hours for diesel generators, and New York City has capped the number of diesel generators that can be in operation at any given time. But the most sustainable solution is to eliminate fossil fuel–fired engines from the list of acceptable candidates for distributed generation.[22]

Cybersecurity. The Achilles' heel of the smart grid is the opportunity that it offers to hackers to harm the grid and its users. Cybersecurity experts reported in July 2017 that a group of suspected Russian hackers had gained access to computers at a dozen U.S. power plants. Once hackers control a power plant's computers, they can easily bring down an entire grid. The system operators and the companies that generate and distribute electricity bear the primary responsibility for cybersecurity, but state public utility commissions could play a vital role by monitoring utility company cybersecurity investments to ensure that they are devoting sufficient attention to that critical threat to grid reliability. At the same time, companies, system operators, and regulators must be prepared to respond to successful cyberattacks in much the same way that they respond to weather emergencies to ensure rapid recovery with a minimum of loss to system assets.[23]

Privacy. As consumers interact more frequently with the smart grid, utility companies are gaining access to a great deal of information about how they use electricity that is valuable to advertisers and marketers. Either Congress or state legislatures should enact legislation to address privacy concerns raised by the smart grid. The legislation should provide that data obtained through smart meters belongs to the customer, and it should prohibit use or disclosure of the data for any purpose other than one directly related to serving the customer's energy needs without the customer's consent. Furthermore, utility companies and third parties with access to consumer data should be required to inform consumers immediately after any breach of security that might make their data available to others.[24]

THE FUTURE ROLE FOR RENEWABLE ENERGY

As we saw in Chapter 11, renewable energy has played a powerful role in the ongoing transformation of the electric power industry. Renewable power can be produced on two scales: (1) utility-grade wind and solar farms, large geothermal plants, hydroelectric dams, and biomass-burning power plants, and (2) distributed systems of small rooftop solar panels and backyard wind turbines that provide more than enough electricity to supply a single user.[25]

The Case for Renewables

Renewable power has many advantages over fossil fuels. First, wind, solar, and hydropower generators do not emit SO_2, NO_x, PM, and CO_2 into the air, and burning biomass emits far lower levels of those pollutants (other than NO_x) over time than burning fossil fuels. Hydroelectric dams have been a

highly reliable and flexible source of power for both urban and rural consumers for decades. Second, renewable power is becoming more economical
than fossil fuel power. The capital investment required for solar and wind
farms is generally much lower than the investment required for a fossil fuel–
fired or nuclear plant, and the cost of the fuel is zero. Initial capital investment in new hydroelectric dams can be quite high, but many of the 2,200
existing hydropower plants can be upgraded and optimized at a much lower
cost. Third, even in areas where renewable power is more costly than fossil
fuel–fired power in the short run, it can quickly pay for itself in times of crisis
when wholesale prices spike due to adverse weather conditions or an acute
shortage of natural gas. Fourth, jobs created in the renewable energy sector
more than offset job losses in the coal and electric power industries caused
by retiring coal-fired plants. Finally, utility-scale solar and wind facilities provide lease income to landowners and add to the tax base of rural counties.[26]

The Case against Renewables

Renewable energy also has disadvantages. First, critics of renewable power
maintain that it would not be competitive with fossil fuels but for the subsidies that it receives in the form of tax credits, grants, and loans from the federal government. Wind generators can receive a tax credit of 2.3 cents per
kilowatt-hour; companies that install solar arrays can receive credits equal
to 30 percent of their capital investments; and both can take advantage of ac-
celerated depreciation. In December 2015, Congress extended the tax credits
for wind projects for five years with a gradual phaseout beginning in 2017
and ending in 2020. The tax credits for installing solar power expire in 2022.
Second, the variable nature of some forms of renewable energy, like wind and
solar, makes it harder to integrate into the grid with possible adverse effects
on reliability. Computer-driven smart grid technologies, however, have
greatly reduced this problem, and rapidly developing battery technology and
advanced pumped hydropower storage will allow storage to play a greater role
in balancing the grid. A June 2017 report by R Street Institute concluded that
the move to renewables was not threatening grid reliability.[27]

Third, moving renewable energy from remote utility-scale installations to
the cities requires large-scale electricity transmission lines that can encounter
serious institutional obstacles. State laws governing siting and eminent domain provide many opportunities for opponents of transmission lines to

object to and delay construction. Companies attempting to build such lines often encounter difficulties in siting them in states that do not directly benefit from them but must suffer the inconvenience of construction and unsightliness of large transmission lines. Fourth, proposals for utility-scale wind farms, solar arrays, and hydroelectric dams have frequently attracted opposition and lawsuits from local landowners and environmental groups who worry about their adverse effects on wildlife, their noise, and their negative aesthetic impacts on the landscape. Offshore wind projects have attracted especially vociferous opposition from fishermen and people living on the coast within eyesight of such facilities.[28]

Solutions

With an adequate commitment of resources and political will, renewable power can fill much of the nation's demand for electricity. Indeed, climate experts warn that to limit global warming to the two degrees centigrade that it will take to prevent massive climate disruption, at least 43 percent of the world's energy must come from renewables by 2030 and 77 percent by 2050. The National Renewable Energy Laboratory estimates that 80 percent of the electricity that the nation consumes can come from renewable energy by 2050 with the technologies that are currently available. The move to renewable energy should not be a partisan issue. In 2016, nearly 90 percent of the nation's wind-generated power came from states that voted for Donald Trump.[29]

Eliminating Cost Differentials. Wind and solar power are already competitive with coal and natural gas in many areas. Big companies like Walmart have discovered that rooftop solar generation can pay for itself in five years, but smaller businesses have encountered difficulty obtaining financing for installing rooftop solar arrays. States could help small companies surmount this hurdle by making financing directly available or by facilitating collective efforts by small entities to negotiate with banks for loans at rates that make rooftop solar projects feasible. At least forty-five states permit a strategy called "leased rooftop solar energy" under which a solar developer arranges with the owner of a building to pay a monthly fee for the right to install solar panels and sell the electricity from the panels to the local utility company. In other states, homeowners can lease solar panels for a low monthly payment from a utility company or third-party distributor. All states should implement such arrangements for reducing or eliminating the upfront cost of rooftop solar panel installation.[30]

Future innovation in the production of solar modules and wind turbines should allow them to remain competitive after the federal subsidies disappear in 2020 and 2022. But some observers believe that tax credits will continue to be critical to the continuation of a flourishing renewable energy sector. The production tax credit for wind energy attracted more than $140 billion in investment from 2007 to 2017, and the cost of wind-generated electricity has decreased dramatically as a result of these investments. Proponents of renewable power observe that fossil fuel–fired power receives "far more pervasive subsidies" than renewable energy. An Environmental Law Institute study found that the federal government awarded nearly $6 in subsidies to traditional fossil fuels for every $1 in subsidies to renewable energy. To ensure that wind and solar power compete on a level playing field, Congress should either extend the tax incentives for those technologies or eliminate the tax incentives for fossil fuels.[31]

State and Local Procurement Programs. State and local governments can encourage the development of renewable energy resources by adjusting their energy procurement policies. Responding to the Trump administration's withdrawal from the Paris Agreement, New York governor Andrew Cuomo announced a $1.5 billion procurement program called Clean Climate Careers under which state agencies purchase an additional 2.5 million megawatts of renewable energy annually. The Illinois legislature created a similar procurement program in its Future Energy Jobs Act that will purchase two million megawatt hours of wind and solar energy by 2020, three million by 2025, and four million by 2030. Several cities power government buildings with renewable energy. Other cities and states should adjust their procurement policies to encourage renewable energy.[32]

Municipal Utilities and Renewable Power. Municipally owned utility companies can afford to make sustainability a higher priority than investor-owned utilities and merchant generators that have to answer to shareholders focused on short-term profits. If privately held utilities present a stumbling block, municipalities can turn them into publicly owned entities, either by purchasing their assets or by exercising their power of eminent domain. Even if the effort fails, it can lead to agreements by the incumbent utility company to invest in renewable resources. For example, an effort by the City of Boulder, Colorado, to study the option of taking over the city's power supply from Xcel Energy precipitated several years of negotiations that failed to accomplish the envisioned takeover. But Xcel Energy became a much greener company by retiring several coal-fired power plants and adding seventeen gigawatts of wind and solar power. In November 2017, the citizens of Boulder voted to continue with the city's efforts to municipalize electricity distribution.[33]

A New Business Model. Like the advent of the smart grid, the move to renewable energy has implications for traditional business models. Among other things, electric power companies will have to figure out how to make large capital investments in renewable energy during a period of declining demand due to DR, DG, and end-use efficiency. This will be especially difficult for merchant generators that are not guaranteed reasonable returns on their investments.[34]

Merchant generator NRG Energy's attempted transformation from fossil fuels to renewables offers a cautionary tale. Under an energetic CEO, David Crane, NRG invested heavily in wind and solar power and promised to reduce its CO_2 emissions 50 percent by 2030 and 90 percent by 2050. But a few of its most ambitious projects suffered expensive setbacks and did not yield as much electricity as predicted. When the company's investments failed to generate high profits in the face of stiff competition from existing gas-fired plants, its stock price fell 63 percent, and its shareholders rebelled. The board of directors fired Crane in December 2015 and replaced him with Mauricio Gutierrez, a pragmatist who recognized that he had to balance business imperatives against environmental improvement. He sold off many of the company's renewable energy projects and scaled back others, with a corresponding increase in the price of the company's shares.[35]

The moral of the story, according to the head of Business for Social Responsibility, is that "it's very hard to be a C.E.O. for tomorrow, when the markets only care about being a C.E.O. for today." Crane had challenged investors to broaden their horizons to see that the power company of the future could not rely on a business model that depended heavily on centralized base-load power plants, but he failed. Likewise, it is unrealistic to suggest that the commendable efforts of large consumers of electrical power like Google and Walmart to reduce their carbon footprints will be sufficient to reduce GHGs to sustainable levels. Environmental regulation must provide a firm backdrop to corporate decision-making in the electric power industry and in other sectors.[36]

Regulated utility companies have an advantage over merchant companies in this regard. Several large electric utility companies are purchasing or building wind farms and solar arrays with the expectation of a reasonable rate of return on those capital investments and a corresponding increase in profits. Some have entered into "build-transfer" agreements under which a private developer assumes the risk of securing property, permits, interconnection rights, and project contracts, and the utility company assumes responsibility for the project once it is "shovel ready." The primary drawback to this solution has been the reluctance of some state PUCs to acknowledge that investments in renewable assets are prudent. Utility regulators should recognize that renewable power is not a luxury for elite customers, but an affordable source of power that utility companies can own and deliver at a price that is often lower than that of electricity derived from fossil fuels.[37] Even in areas where that is not the case at the present time, state legislatures and PUCs should recognize that technologies are evolving so quickly that investments in renewable power today will pay off handsomely in the future.

Creating a Robust Transmission Infrastructure. One of the most pressing needs for the development of renewable resources is a robust transmission infrastructure to transport power from remote wind and solar installations to areas where demand for electricity is high. Proposed transmission lines, however, often attract strong opposition from property owners and local environmental groups. The Energy Policy Act of 2005 allows DOE to designate national interest electric transmission corridors in high congestion areas and empowers FERC to override state disapprovals of transmission projects within those corridors. The courts, however, have interpreted this authority so narrowly that it has not been effectively implemented. Congress should enact legislation empowering DOE, FERC, or some other agency to override state siting laws for electric transmission lines that meet prescribed technical and environmental criteria and demonstrate a need to transport power from renewable resources to consumers.[38]

If the federal government fails to act, states should seize the initiative. As we have seen, Texas has created "competitive renewable energy zones" in which the state has built transmission lines to transport renewable energy to the cities that consume it. Originally built with Texas's abundant wind resources in mind, solar developers were able to take advantage of the same infrastructure to launch a massive build-out of solar power in West Texas. States unwilling to socialize transmission projects should allow public utility companies to recover in advance the costs of creating transmission infrastructures for renewable power. States could reduce local landowner opposition to transmission projects by giving them the option of receiving annual payments from transmission line operators instead of accepting judicially determined adequate compensation for forced easements in eminent domain proceedings.[39]

The Storage Solution. Energy storage is the key to integrating renewable power fully into the grid at the much higher rates that will be necessary to make the grid sustainable. Storage makes renewable energy generated during the productive part of the day available for use during the times when solar and wind power are unavailable. Recent advances in battery storage capacity are dramatically increasing the viability of wind and solar energy. Utility companies or third-party storage providers can store energy in "utility-scale" facilities like pumped hydropower and compressed air storage projects, massive flywheels, or fields of lithium-ion batteries. Residences and commercial buildings can store energy in electric cars and powerful on-site batteries. Better batteries loom on the horizon.[40]

The primary disadvantage of storage has been its expense, but the cost of battery storage declined around 40 percent between 2014 and 2017, making it competitive with the cost of building gas-fired peaker plants. FERC's regulations were not designed with storage in mind, and they have in some instances hindered the integration of storage into wholesale markets. Likewise, some state PUCs have been slow to allow utilities to recover the cost of installing storage capacity, and it is unclear in many states whether nonutility companies that store electricity can lawfully sell that energy on retail markets. FERC should revise its rules and the independent system operators should amend their tariffs to provide a level playing field for storage in wholesale markets. State PUCs should allow cost recovery for additions of storage capacity to utility assets. And state legislatures should enact legislation clarifying that large and small energy storage facilities are allowed to sell energy to the grid.[41]

Renewable Portfolio Standards. At least twenty-nine states have enacted renewable portfolio standards (RPSs) that require providers of retail electricity to acquire a prescribed percentage of their power from renewable resources. During the last two decades, RPSs have played a powerful role in technology innovation, cost reductions, and the nation's overall move toward greater consumption of renewable energy. Progress has been uneven, however, for several reasons: standards vary from state to state; municipal utilities and co-ops are exempted from RPSs in many states; some states have not enacted RPSs; and a few states have rolled back RPSs at the behest of fossil fuel companies and Tea Party activists. The recent rollbacks are a step in the wrong direction. Because a sustainable electrical grid will require far more renewable power than is currently available, all states should adopt RPSs with stringent goals, and they should make RPSs mandatory for all utility companies, including municipally owned utilities and co-ops.[42]

Local Renewable Energy Programs. As we saw in Chapter 12, many cities have taken up the challenge of meeting the Paris Agreement goal of reducing GHG emissions by 26 to 28 percent from 2005 levels by 2025. Cities generally have the power to reduce emissions from their own operations and to provide tax incentives to install efficient appliances or solar panels. The most effective tools available to cities are energy codes that specify energy efficiency measures that must be included in new residential and commercial buildings. In California and a few other states, cities and counties have the authority to create community choice aggregation programs under which a city, a group of local governments, or a private entity operating under contract with a city aggregates the electrical loads of participating consumers and purchases electrical power from generators on the wholesale market where the aggregated demand yields more bargaining power. The city can also decide to purchase more renewable power than the privately owned utility company that serves the city is willing to provide. The city becomes responsible for purchasing the power and balancing supply with demand, but the local utility company retains responsibility for distributing the electricity, metering, and billing. The five community choice aggregators that were in existence in California in 2016 were able to supply power that contained a greater proportion of renewables to customers at slightly lower prices than the investor-owned utility companies that previously served those customers. Although it will be difficult for cities to achieve the Paris Agreement goal on their own, they can make a substantial contribution in that direction.[43]

Reducing Legal Impediments to Renewable Development. Some state PUCs have concluded that they lack authority to allow electric utility companies to recover the additional cost of building utility-scale solar and wind farms or purchasing renewable power from the national grid. The New York and Maryland Public Service Commissions, however, interpreted their statutes to allow the developers of two large offshore wind projects to recover their capital costs from ratepayers. Most state public utility statutes are written in broad enough language to allow PUCs to follow the New York and Maryland lead. Failing that, state legislatures should amend their statutes to ensure that prudent purchases of renewable power and reasonable costs of constructing renewable power plants are recoverable. Model statutes and ordinances designed to foster renewable energy development are available for states with the political will to enact them.[44]

Another legal impediment to developing both wind and solar power is the risk that a neighbor will file a nuisance action against the project or that an environmental group will challenge state or federal permits claiming that the permitting agency violated the National Environmental Policy Act, the Endangered Species Act, or a similar statute.[45] While state courts have generally rejected claims that wind farms and solar arrays constitute nuisances based on their aesthetic attributes, at least two states (West Virginia and Nevada) allow such claims. Courts should hesitate to allow nuisance claims based on aesthetic objections to renewable power projects. If they start down that road, virtually any renewable energy project will be at risk, and we will not meet critical environmental goals. Project proponents can attempt to avoid environmental challenges by assisting permitting agencies in preparing adequate environmental impact statements and by locating projects in areas that do not threaten the habitat of endangered species and do not appear out of place.[46]

Expanding Hydroelectric Power. A major 2017 report from the Department of Energy predicts that new hydropower generation capacity in the United States could increase by thirteen gigawatts by 2050 in a cost-beneficial manner. The cost of building new hydroelectric dams, which includes an outlay of up to $50 million to secure all of the necessary state and federal permits, may limit that option, especially in areas where new dams are likely to encounter opposition by environmental groups. But increasing the efficiency of existing hydroelectric facilities and retrofitting existing dams with power generating capacity could significantly increase hydropower's carbon-free contribution to the nation's grid. In October 2018, President Trump signed a water infrastructure bill aimed at expediting the permitting process for installing turbines at existing dams. Owners of existing dams should take advantage of new technologies to upgrade existing plants and add new generating capacity to existing dams. Ironically, a primary impediment to future reliance on hydropower will be global warming, which could cause extended droughts that lower water levels behind existing dams.[47]

Reasonable Use of Biomass. While burning biomass to produce energy is generally consistent with sustainability because the CO_2 emissions will eventually be converted back to biomass, we have seen that some kinds of biomass remain quite controversial. For example, burning whole trees to produce electricity will add a great deal of CO_2 to the atmosphere that will not be removed by new trees for decades or even centuries. In the meantime, the CO_2 in the atmosphere will be contributing to perhaps irreversible changes. Critics of biomass argue that biomass emissions can be 50 percent as CO_2-intensive as coal, and some kinds of biomass have the potential to emit twenty-eight times more dioxin and fourteen times more mercury.[48] Electric power companies should continue to burn biomass that is rapidly replaceable and combustible waste material, like forestry residues and spent sugarcane. And they should get credit for emissions reductions to the extent that the CO_2 emitted will be recaptured by future plantings. They should not, however, be allowed to burn whole trees and other crops that take years to replace.

Are Renewables Enough? In mid-2017, an acrimonious debate broke out among scientists over a 2015 paper by Stanford professor Mark Jacobson. Using sophisticated mathematical models, Jacobson predicted that the United States could obtain all of the energy it needed in all sectors of the economy from renewable resources other than biomass at a reasonable cost. Inspired in part by that paper, the state of California and at least fifty cities have established 100 percent renewable energy goals for themselves. In June 2017, however, a group of twenty-one scientists headed by Christopher Clack published a stinging critique of Professor Jacobson's modeling exercise. They worried that overly optimistic projections would take away from efforts to scale up renewables gradually over time to avoid reliability crises and public backlashes. Jacobson published a rebuttal to the Clack article in a letter to the editor, claiming that it was "riddled with errors." A far less controversial report by the Intergovernmental Panel on Climate Change concluded that renewable resources could supply as much as 77 percent of global energy by 2050. Given the setbacks to renewable power brought on by the Trump administration, the Jacobson paper is probably overly optimistic. As discussed below, nuclear and fossil fuel power will probably be needed to meet demand for electricity for at least two decades and possibly longer.[49]

THE FUTURE ROLE FOR COAL

The ongoing transformation to a more sustainable grid has taken place rather painlessly for consumers as environmental regulations and market forces have pushed the electric power industry toward greater use of natural gas and renewable resources. The move has not, however, been painless for the coal industry, coal miners, and employees of coal-fired power plants. The electric power industry has retired or converted hundreds of aging coal-fired plants, and it is not building new ones. Consequently, the mining industry has suffered as domestic demand for coal has declined, with several large mining companies seeking bankruptcy protection.[50]

The Case for Coal

Many observers believe that coal must continue to play a major role in providing electrical power for the nation. They argue that natural gas and renewables cannot replace coal for several reasons. First, they worry that a rapid conversion from coal to natural gas will subject generators and their customers to the price volatility that has historically affected gas markets. Second, proponents point out that it will cost hundreds of billions of dollars to build out the natural gas pipeline infrastructure to the point at which it can transport gas from regions with large shale deposits to the power plants that serve major urban areas. Third, unlike coal, which can be stockpiled, gas must be delivered on a continuous basis through pipelines, which are subject to accidental breaches and sabotage. A diversity in fossil fuel supplies, they argue, provides a critical resiliency safety valve and a boost to national security. Finally, environmental concerns should not impel us to abandon coal, because "clean coal" technologies on the horizon will allow coal-fired plants to produce electricity with minimal adverse environmental impacts.[51]

The Case against Coal

Most observers agree, however, that the undesirable aspects of coal will ensure that it plays a modest role in the future. First, building a coal-fired plant is a very expensive undertaking, as demonstrated by Southern Company's failed Ratcliffe IGCC plant, which converted to natural gas after costs exceeded $7 billion. Second, coal-fired plants are more expensive to operate than gas-fired plants. In 2017, 48 percent of coal-fired plants in the United States operated at a net loss but were kept afloat by sympathetic PUCs. Third, fracking technology should ensure that natural gas prices remain stable, and renewable energy is competitive with coal in many places. Fourth, coal-fired plants are no longer essential to a reliable and resilient grid. During the two-week "bomb cyclone" of January 2018, the New England region's grid suffered little disruption without needing to fire up any idled coal-fired plants. Finally, the adverse environmental effects of coal-fired power plants are greater than those of natural gas and far greater than those of renewables. Strip mining of coal in Illinois and the Powder River Basin creates giant scars in the land that are not always restored. Mountaintop removal mining in Appalachia inundates valley streams with overburden that often contains toxic metals and leaves behind unsightly plateaus where mountains used to be. Underground mining subjects miners to risks of death and dismemberment, and the acid mine drainage from underground mining can render miles of streams uninhabitable by aquatic species. Finally, growth in coal-fired production is simply inconsistent with meeting planetary goals for GHG reduction.[52]

Solutions

Because the need for large baseline facilities is diminishing with managed energy demand and the increasing ability of renewables to meet that demand, it makes little sense to invest in coal-fired power. At the same time, it is highly likely that existing coal-fired plants will continue to contribute to meeting the nation's electricity needs for several more years. The Obama administration's Clean Power Plan would have forced operators of existing coal-fired plants to rid their fleets of the least efficient coal-fired plants, but the Trump administration has eliminated that culling device. And the administration's Affordable Clean Energy plan will do little to change the status quo, if it goes into effect. To ensure continued progress toward a sustainable grid, state and federal permitters should continue to insist that coal-fired plants employ the best available pollution control technologies when they undergo changes that increase emissions, and the cost of compliance should play a minor role in those determinations. If EPA fails to insist on that, the courts should be receptive to new source review actions by state attorneys general and environmental groups. When Congress enacted the new source review requirements in the 1970s, it assumed that companies would retire power plants at the end of their forty-year lifetimes. It is time to make that assumption a reality.[53]

THE FUTURE ROLE FOR NATURAL GAS

Proponents refer to natural gas as an essential "bridge fuel" to provide reliable power until nuclear power and renewable resources can supply the bulk of our energy needs. From the rate at which companies are building large gas-fired plants, however, it appears that electric power companies are planning to rely heavily on natural gas for at least the next thirty years and probably much longer than that.[54]

The Case for Natural Gas

Natural gas is generally cheaper than coal to burn, and natural gas plants are much less costly to build than coal plants. Gas-fired plants emit far fewer tons of SO_2, mercury, and CO_2 and somewhat fewer tons of NO_x per unit of electricity produced than coal-fired plants. Gas-fired power plants can be turned on and off quickly to provide reliable power when the wind unexpectedly dies down or clouds cover the sun. Until such time as electricity storage can adequately fill the gaps in noncontinuous renewable power, gas-fired plants will probably be necessary to provide power on short notice to keep the grid resilient.[55]

The Case against Natural Gas

For many environmental activists, the dramatic shift from coal to gas that followed the fracking revolution was a desirable onetime phenomenon that has nevertheless left the nation overly dependent on a fossil fuel that contributes to global warming. The gravest environmental threats posed by natural gas stem from methane emissions from production, transportation, storage, and distribution of natural gas and CO_2 emissions from burning it. In 2016, CO_2 emissions from an expanding number of natural gas plants exceeded CO_2 emissions from coal-fired plants. And methane is a more potent contributor to global warming than CO_2 by a factor of twenty. Some scientists question whether burning natural gas in power plants results in fewer overall GHG emissions than burning coal when measured across the entire life cycle of both fuels. To reduce GHG emissions to noncatastrophic levels, some environmental groups believe that all fossil fuel–fired plants must be retired as soon as possible. Gas-powered plants also require pipelines that frequently encounter opposition from local environmental groups and landowners because of their adverse effects on the environment, the risk of leaks and explosions, and their potential to reduce property values. Pipelines are also subject to sabotage and natural disasters that can disrupt supplies to gas-fired plants. To the extent that natural gas supplies are dependent on hydraulic fracturing technologies, the environmental costs of fracking, which include air pollution, water pollution, contaminated groundwater, greenhouse gas emissions, habitat degradation, and minor earthquakes, should weigh against natural gas in the balance.[56]

Solutions

Assuming that natural gas is to serve as a bridge fuel for a time, the industry will have to build out a more robust pipeline and distribution infrastructure in places where greater use of natural gas will be necessary to fill the gap left by retired coal-fired plants. In individual cases, FERC and state authorities should carefully consider the social cost of the CO_2 emissions that will indirectly result from building a new gas pipeline in deciding whether its benefits outweigh its costs, something that FERC has been unwilling to do during the Trump administration. During the Obama administration, EPA promulgated a fairly weak new source performance standard for oil and gas operations to reduce methane emissions, but the Trump administration stayed the standard for two years while it considered whether to rescind or replace it. Although the D.C. Circuit Court of Appeals vacated the stay, the agency proceeded ahead with a rulemaking initiative that resulted in a September 2018 proposal to reduce the number of inspections for leaks, relax the requirements for pumps, and eliminate the need for oversight of changes by a certified engineer. If EPA succeeds in gutting the standards, state and local policy-makers will have to reconsider whether gas has any role at all to play in America's energy future. We should not be adding to a natural gas infra-structure that lacks the best available emissions controls.[57]

The future of natural gas may depend on whether the industry can come up with a commercially viable carbon capture, utilization, and sequestration (CCUS) program for gas-fired plants. DOE has poured billions of dollars into research and demonstration projects employing CCUS technologies since 1997, including the twice-terminated FutureGen program. And Congress has consistently provided tax credits for companies employing CCUS. The natural gas industry itself has spent at least $10 billion on CCUS technology without any significant breakthroughs. The sole CCUS project associated with an existing power plant in the United States—the Petra Nova project at NRG's W. A. Parish plant near Houston—is successfully employing CCUS technology to reduce CO_2 emissions by 90 percent at one of its coal-fired units. Other attempts to capture carbon at new power plants, like the Southern Company's Plant Ratcliffe in Kemper County, Mississippi, and American Electric Power's Mountaineer Plant in West Virginia, have been spectacular failures.[58]

The primary drawback to CCUS is its cost. It adds about 40 percent to the cost of constructing a new plant, and the cost of retrofitting CCUS into an existing plant is even higher. CCUS on a national scale would require a costly pipeline infrastructure larger than the one that currently transports oil. Seg-regating and compressing CO_2 consumes up to 25 percent of the plant's en-tire energy output, far more than the energy penalty of conventional pollutant controls. Sequestered CO_2 can "burp" out of the facility and smother nearby livestock or people. In addition, forcing CO_2 into geologic formations could cause earthquakes of the sort that are plaguing the fracking industry. Per-haps the most significant risk from CCUS stems from the possibility that the pressurized CO_2 will mobilize salts or toxic metals that migrate to aquifers used for drinking water. During the Obama administration, EPA promul-gated regulations to protect the environment from risks posed by CCUS fa-cilities, but a waiver provision allows facilities to sequester CO_2 at shallow depths if they can persuade state permitting agencies that there will be no migration into surrounding aquifers. While it is certainly possible that greater spending on developing CCUS technology will yield new technologies that decrease its expense and increase its environmental acceptability, policy-makers should not assume that technological breakthroughs will be forth-coming in deciding whether to phase out our reliance on fossil fuels to produce electricity.[59]

Even if natural gas is the bridge to the future, it is not clear that we should be building new gas-fired plants. Because they have forty-year lifetimes, their owners will complain of stranded assets (the capital invested in the plants and the associated pipeline infrastructure) if government regulation threatens their viability in the interim. We should instead be building "off-ramps" to guarantee that we do not become dependent on natural gas beyond 2050 and to ensure that plentiful supplies of natural gas do not discourage further de-velopment of renewable resources, more efficient buildings, and other cli-mate mitigation strategies. The Conservation Law Foundation and Footprint Power Company provided a good model in February 2014 when the former agreed not to object to a new $800 million gas-fired power plant and the latter agreed to gradually phase down production in the 2040s and to cease gen-eration altogether by 2050.[60]

THE FUTURE ROLE FOR NUCLEAR POWER

As of 2016, ninety-nine nuclear reactors at sixty-three plants provided about 20 percent of the nation's electricity and 60 percent of the nation's non-fossil fuel related electrical power. During the George W. Bush administration, it looked like nuclear power was on the way to a major revival. Stimulated by several hundred million dollars in DOE grants and loan guarantees, com-panies applied to the Nuclear Regulatory Commission to build twenty-eight new nuclear plants under a new streamlined licensing process. In June 2016, the first new nuclear plant in twenty years came online when TVA connected its Watts Bar Unit 2 to the grid at a cost of $4.7 billion. But by then, inexpen-sive natural gas, less expensive renewable energy, low electricity prices, and the March 2011 Fukushima disaster had dampened enthusiasm for nuclear power. By mid-2017, more than half of the nation's existing nuclear plants (mostly in deregulated markets) were losing money, and five power compa-nies had retired six nuclear reactors. The four units under construction at the Southern Company's Plant Vogtle and the V. C. Summer plant owned by South Carolina Gas & Electric (SCG&E) and Santee Cooper were behind schedule and way over budget.[61]

In March 2017, Westinghouse Electric Company, the primary contractor for all four of the Georgia and South Carolina reactors declared bankruptcy and abandoned the projects. In July 2017, SCG&E and Santee Cooper announced that they were canceling the Summer project, and in late 2018 they put up the plant's assets for sale and agreed to pay its customers $2 billion to offset the higher rates they had been paying to finance the project. The Southern Company reached an agreement with its partners in September 2018 to continue construction on Plant Vogtle with Southern picking up the tab for additional expenses over specified thresholds. President Trump promised the nuclear industry that his administration would "revive and expand our nuclear energy sector," but the "nuclear review" the he ordered bogged down in DOE.[62]

The Case for Nuclear Power

Nuclear power proponents argue that it offers a carbon-free alternative to fossil fuels for meeting the nation's future energy needs without contributing to climate disruption. Its consistency and reliability make it ideal for providing baseload power in a world without fossil fuel–fired power plants. Nuclear power provides 100,000 direct jobs, and nuclear power plants are financial anchors for many rural communities. Proponents claim that as the largest source of zero-carbon-emitting electricity, nuclear power has a critical role to play in meeting GHG reductions, because renewable resources cannot replace both fossil fuel and nuclear power. The thirteen nuclear plants most at risk for retirement in 2016 produced three times as much electricity as all current solar production. All of the recently retired nuclear reactors have resulted in increased emissions from fossil fuels that replaced them, and it is likely that retiring nuclear plants in the near future will likewise result in greater emissions.[63]

The Case against Nuclear Power

Nuclear power can play a major role in achieving the nation's GHG reduction goals only if several problems that have persistently plagued the industry can be resolved. The roughly one-half of the nation's nuclear plants in restructured markets have found it hard to compete with plants running on cheap natural gas and renewables. About one-third of the existing nuclear power

plants will have to be relicensed by the Nuclear Regulatory Commission between 2030 and 2035 at a cost of $1.5 billion to $2.5 billion if their owners
plan to operate them beyond that point. Power companies will find it difficult to justify new nuclear units to state PUCs, given the recent history of bad
investments in nuclear power. Nuclear power still suffers from public concerns that a meltdown like the Chernobyl accident of 1986 could result in the
release of highly radioactive material over a large area. Nuclear power plants
contribute to the risk that nuclear materials capable of creating an atomic
bomb could find their way into the hands of terrorists or rogue states. Perhaps the most serious impediment to nuclear power is the still unsolved
problem of disposing of spent nuclear fuel. The federal government has spent
decades attempting to identify an appropriate location for a radwaste reposi-
tory and to come up with adequate disposal and monitoring technologies,
all to no avail.[64]

Solutions

In the long run, controlled nuclear fusion technology may provide a reasonably safe source of virtually limitless electricity, but we are not likely to locate that holy grail anytime soon. The best hope for nuclear power in the intermediate term is the prospect of a new generation of modular technologies
that operate safely and efficiently, use fuels that are free of proliferation risks,
are less expensive than the current fleet of aging plants, and produce less
radwaste. Unlike existing plants, modular units would have the capacity to
raise and lower output incrementally, allowing them to fill in for variable renewable power. A gravity-controlled cooling system would ensure against
potential meltdowns during storms instead of emergency power systems like
the ones that failed during the Fukushima disaster. Developing a new generation of nuclear power plants, however, will require large public investments
in research and development, and it is not clear that the price of electricity
from such plants will be any lower than that of existing plants. Until the public
can be confident that the government will safely dispose of radwaste over the
centuries-long time span that it will take for it to "cool down," that problem
is likely to be a conversation stopper in future discussions about whether to
build new nuclear plants.[65]

Government action may be needed to preserve the existing nuclear fleet
until renewables can absorb the load that it currently bears. We have seen

how DOE's efforts to rescue the coal and nuclear industries with various subsidy programs have not succeeded. Three states, New York, Illinois, and
New Jersey, have created programs that reward seven nuclear power plants
with "zero-emissions credits," which are essentially payments to the plant
owners for producing electricity without corresponding increases in GHG
emissions, and other states have enacted or are considering similar programs.
Connecticut ordered its distribution utilities to enter into contracts with
owners of two nuclear plants (and nine solar projects) to purchase electricity at higher than market rates. Thus far, challenges to such state programs in court alleging that they are preempted by the statutes that FERC
administers or inconsistent with the dormant Commerce Clause have been
unsuccessful.[66]

Although they do not favor the construction of new nuclear plants, most
environmental groups supported these life-extension programs for nuclear
power plants. The Union of Concerned Scientists concluded that in the absence of such programs, the lost electricity from nuclear units that retired
before the end of their useful lives would be replaced by fossil fuel–fired
plants, which would be worse for the environment. At the same time, the
groups have resisted attempts by the nuclear power industry to reduce safety
standards for nuclear plants. And a few mostly local groups have opposed
any nuclear subsidies. Programs that reward nuclear power plants that meet
stringent safety standards for providing electricity without generating air pollutants may be necessary until research yields safer and less expensive nuclear power and solves the radwaste problem or until the renewable energy
sector is strong enough to avoid a reversion to fossil fuels as large nuclear
plants come off-line.[67]

ENSURING A RELIABLE AND RESILIENT GRID

One weapon that the coal and electric power industries wielded in virtually
every struggle going back to EPA's promulgation of the original NAAQS in
1971 was concern for the reliability of the electricity grid. Because Americans have become highly dependent on electricity in their jobs, their recre-
ation, and their lifestyles, public officials can never put reliability concerns
aside. But we have also seen that predictions that EPA regulations and environmental group lawsuits would cause the lights to go out have consistently
been overly pessimistic. A good example is EPA's MATS regulation, which

precipitated a large number of coal-fired power plant retirements and repowerings. In the Southeast, where some experts had predicted retirements of
coal-fired units would put a severe strain on capacity reserves, electricity markets continued to function smoothly as low natural gas prices and the MATS
rule gradually pushed older, less efficient coal-fired plants off-line ahead of
schedule. Nevertheless, as the electric power industry moves away from fossil
fuels toward renewables and demand management, policymakers must
develop "more robust planning approaches" to ensure that the new mix of
resources remains reliable and resilient.[68]

Shifting National Goals

One serious impediment to a smooth transition to renewables is the fact that
national goals for the electric power industry are constantly shifting with
technological change and the political winds. During the Carter administration, the goal was to diversify fuels by shifting from oil and natural gas to
coal and renewables in the wake of the Arab oil embargoes. During the
Obama administration, the goal was to shift away from coal to natural gas
and renewables to reduce GHG emissions that cause climate disruption. As
of 2019, with President Trump in the Oval Office and a Republican-controlled
Congress, the goal has shifted away from protecting the environment and
future generations to expanding job opportunities in coal-dependent states.
Shifting goals makes it very difficult for electric power companies and PUCs
to plan for the capital investments needed to ensure continued reliability of
the grids they serve. Policymakers can partially address this problem by
devoting resources to predicting how technologies are likely to affect the
grid, but it will be very difficult to prepare in advance for shifting political
fortunes.[69]

Beyond Baseload

The traditional public utility model of electricity generation and distribution
assumed that continuously running "baseload" plants were necessary to provide a steady flow of electricity that could be supplemented by other power
sources operating intermittently to meet peak load demand. In recent years,
that model has incorporated a concern for resilience, which is the capacity
of the grid to avoid or resist shocks to the system, manage disruption, quickly

respond to shocks, and adapt to future shocks. The assumption that large fossil fuel–fired power plants with stockpiled supplies of fuel are essential to maintaining reliability and resiliency, however, may no longer be warranted. Future policymakers should focus on overall system resources and how they can best be integrated to ensure against outages during times of stress on the system. In particular, they should consider greater reliance on distributed energy and microgrids, devoting resources toward hardening transmission and distribution networks, and creating advanced awareness systems to increase grid resiliency. As discussed above, this will require a willingness on the part of public utility commissions to allow utilities to derive a reasonable profit from nonconventional sources of power. Utility companies will have to make full use of computer power, big data, and storage technologies to manage a spectrum of energy resources simultaneously. A good example is the decision by the Electric Reliability Council of Texas (ERCOT), the Texas system operator, to add to its control room a "Reliability Risk" desk that employs digital technologies and predictive models to achieve better coordination of the state's fossil fuel plants with its ample renewable resources.[70]

MAKING ENVIRONMENTAL REGULATION MORE EFFECTIVE

Despite considerable progress during the twenty-first century, it is clear that the electric power industry will have to do more to meet the protective goals of the nation's environmental statutes. This is especially true for CO_2 emissions that contribute to climate disruption, but the industry also needs to accomplish more reductions in NO_x emissions that affect ozone nonattainment areas and emissions of SO_2 and PM from plants that cause regional haze in clean air areas. With more than 90 percent of coal ash impoundments leaking toxic metals into groundwater, the environmental disruption caused by leaking coal ash impoundments still needs attention.[71] And billions of marine and aquatic organisms continue to be destroyed every year at cooling water intake structures for power plants that do not employ closed-cycle cooling systems. But the adamant refusal of the Trump administration to address climate disruption and its strong disinclination to address the other adverse environmental effects of power plants will make it very difficult to accomplish the statutory goals in the near term.

Progress toward a sustainable electric power grid through environmental regulation is possible even in the face of opposition from the White House and EPA's leadership. As we have seen, states, cities, and the companies that use electricity have made impressive gains in reducing CO_2 emissions. Environmental groups continue to file lawsuits to force EPA to perform its statutory duties and to require power plants undertaking modifications to undergo new source review. While these efforts will bring about incremental progress, the kinds of regulatory changes necessary to bring about a sustainable power grid will not be possible until American voters elect a president and a Congress that are committed to that goal. On the assumption that this may happen one day in the future, this chapter suggests changes in the current environmental regulatory regime aimed at making environmental regulation of power plants more effective.

Preventing Backsliding

The immediate task for individuals, companies, and groups committed to a sustainable grid is to prevent backsliding. As we saw in Chapter 11, environmental groups have consistently objected to Trump administration attempts to roll back Obama administration regulatory initiatives. The EPA staff had spent tens of thousands of hours researching, debating, drafting, and refining those rules. They had survived the crucible of interagency review and congressional oversight. And they were undergoing judicial review when EPA pulled them back for reconsideration. If EPA had stretched its legal authority too far or had failed to provide adequate scientific and technical support, the courts would not have been shy about remanding rules to the agency. Nevertheless, the Trump administration was determined to withdraw them and replace them with regulations that reflected its deregulatory policies. There are, however, limits on the extent to which the agency can lawfully substitute policy for science and interpret protective statutes to justify backsliding. The reviewing courts should require EPA to support any rollbacks with the same kind of robust data and analysis that it employed in promulgating those regulations.[72]

Should the Trump administration successfully prevent the Obama administration's power plant regulations from going into effect, some companies may conclude that coal-fired plants are worth a gamble. If so, environmental groups will no doubt redouble their efforts to demonstrate to state PUCs the shortsightedness of that conclusion, and they will continue the Beyond Coal tradition of challenging PSD permits for new sources. In addition, state attorneys general and environmental groups will continue to insist that companies proposing to build new gas-fired plants demonstrate that they are needed to meet genuine energy needs that cannot be met by DR, DG, and renewables.

The most effective vehicle for encouraging companies to retire existing plants is the Clean Air Act's new source review (NSR) program. We saw in Chapter 11 how NSR lawsuits brought about huge reductions in power plant emissions through settlements under which companies agreed to retire old units and to install pollution controls on units undergoing major modifications. Critics argue that NSR is costly, burdensome, and inflexible and produces unconscionable delays in projects needed to improve efficiency, safety, and reliability, and, in the case of retired plants, job losses. On balance, however, the benefits of the NSR program outweigh its cost. If EPA and the Justice Department are no longer willing to enforce the agency's NSR regulations, environmental groups and state attorneys general should fill the enforcement vacuum.[73]

One lesson that should be apparent from EPA's experience with regulating power plants is that voluntary programs rarely work. The George W. Bush administration's "Global Climate Change Initiative" and VISION programs, for example, did not come close to reaching their goals. As EPA's Inspector General found, it is difficult to "convinc[e] companies to spend money on activities that are entirely optional."[74] EPA should not easily be persuaded to allow electric power companies to meet their environmental obligations through programs that are not backed up by sanctions. And when it does acquiesce in ineffective voluntary programs, the courts should insist on mandatory programs that work.

The environmental statues allow states to regulate sources within their borders more stringently than EPA, and many have done just that. As the Trump administration rolls back the Obama administration's protective regulations, states can promulgate them as state requirements. States can also join together in compacts like the regional greenhouse gas initiative (RGGI), a CO_2 cap-and-trade regime that ten New England and mid-Atlantic states launched in 2003 over the opposition of the Bush administration and the electric power industry. The RGGI initiative reduced CO_2 emissions 51 percent between the first auction in 2008 and 2017, while the economies of the participating states have grown. States in other parts of the country should create similar multistate regimes.[75]

Moving Forward

At the heart of the pollution problem is the tragedy of the common. Free access to a resource (air, water, land) held in common for all people will inevitably result in the inefficient depletion of that resource if government does not restrict the use by individuals. The owners of power plants will emit pollutants into the air, kill billions of aquatic organisms in intake structures, and hold coal ash in leaky retention ponds until government provides a strong incentive for them to stop. The problem is especially acute in the case of CO_2 emissions because we will all pay today for the controls necessary to ensure against catastrophic climate disruption, but many of us will not live long enough to experience the benefits of our efforts.[76] Of course, government regulation has its own flaws that can result in production inefficiencies, unnecessarily high prices for electricity, and job losses. The goal should be to achieve necessary reductions in pollution in a cost-effective manner. Although it is unlikely that EPA or Congress will take action to improve environmental regulation during the Trump administration, a future administration or Congress may see fit to remove existing impediments to effective regulation. In anticipation of that day, the following discussion highlights those impediments and offers suggestions for environmental improvement through better environmental regulation.

In its efforts to regulate power plants, EPA has frequently encountered issues of enormous technical complexity. Although EPA has amassed a great deal of expertise over the years, it constantly faces the challenge of explaining its resolution of those issues to the public and to reviewing courts.[77] Some of the scientific issues that EPA has to resolve in writing NAAQS, making "endangerment" findings, determining the impact of a power plant's emissions on visibility, and similar tasks require the agency to resolve questions that are scientific in nature but cannot be resolved definitively with existing scientific knowledge. This is virtually always the case when EPA has to choose among competing air quality, water quality, and economic models, all of which are driven by assumptions about how the world works. It is, however, difficult for the agency to exercise sound policy judgment and apply sound scientific analysis in the politically charged atmosphere of power plant regulation.

Another impediment to effective regulation of power plants is the fragmentation of the existing policymaking framework. Many decisions affecting the future of the industry depend on multiple institutional actors, including EPA, DOE, FERC, system operators, state environmental regulators, state public utility commissions, state siting agencies, and others. Each of these institutional actors has its own policy agenda, which may or may not be consistent with the agendas of the other entities.[78] For example, the Office of Information and Regulatory Affairs (OIRA) in the Office of Management and Budget has played a powerful role in reviewing proposed and final EPA rules and coordinating interagency critiques of those regulations all the way back to the original 1971 state implementation guidelines. We have seen how OIRA has played a largely constraining role in EPA's attempts to increase the stringency of the NAAQS, in administering new source review, in regulating interstate transport of pollutants, in dealing with mercury emissions, in addressing climate disruption, in regulating cooling water intake structures, and in addressing the disposal of coal ash.

Although the transparency of OIRA review has waxed and waned over the years, the process has never been especially open. Procedures worked out in a compromise with Congress in the 1980s require the content of communications between outsiders and OIRA and between OIRA and the agency during the pendency of a rulemaking to be made available to the public after the agency promulgates a rule. Even those modest guidelines, however, are frequently ignored in practice. The content of the discussions that OIRA officials have with representatives of the regulated industries remains undisclosed. The number of contacts between EPA and other White House personnel and the content of those communications are not docketed. Therefore, the extent to which the public can hold OIRA and the other White House offices accountable for their influence on the substance of EPA regulations remains unclear.[79]

In addition to OIRA review, Congress, presidents, and the courts have added to the minimal procedural demands of the Administrative Procedure Act requirements that EPA assemble a comprehensive "rulemaking record," respond to public comments that pass a threshold of materiality, and prepare various analyses of the impact of proposed regulations on the economy and small businesses, most of which were designed to make agency rulemaking more transparent and accountable. These additional requirements have had their intended effect, but they have also made the rulemaking process far more burdensome and expensive for all of the participants in the policymaking process. And EPA has suffered from a perennial lack of resources that has made it difficult for the agency to write and enforce the regulations that its statutes require. The net effect has been long, drawn-out standard-setting exercises that rarely meet statutory deadlines and a willingness on EPA's part to compromise with well-endowed industry groups.[80]

From the advent of the modern environmental statutes, regulated entities have predicted that environmental regulations would produce "regulatory train wrecks" that would kill jobs and devastate the nation's economy. In retrospect, however, these dire threats have never materialized. In virtually every instance, the industry managed to comply without the predicted rolling blackouts, spiking electricity prices, and bankruptcies. Predictions about the employment effects of EPA's regulations fit a similar pattern. There is no doubt that the move away from coal to natural gas and renewable resources has had a profound impact on employment in the coal industry. But industry-funded job impact studies typically ignore jobs created in the pollution control, energy efficiency, and renewable power industries as a result of power plant regulation. The resolution of this debate is greatly hindered by EPA's general disinclination to focus specifically on employment effects in its economic analyses.[81]

Fortunately, many tools exist to improve environmental regulation, ranging from market-based approaches to better technology-based standards. Some solutions can be implemented administratively, but others will require legislation. And that may have to await a changed political climate.

A Pollution Tax. A pollution tax is probably the easiest and most effective way to bring about a reduction in emissions over time. Rather than pay a per-ton tax on emissions, the emitter has the option of installing technology, changing processes, or producing less product. A rational emitter will reduce emissions if the cost per ton of implementing one of the options is lower than the tax. Because the tax will result in higher prices for electricity, consumers will also have an incentive to reduce consumption. The tax discourages polluting power-producing technologies while at the same time rewarding less-polluting technologies. The revenue from the tax can support pollution research, offset inequitable losses to low-income consumers, help restore dislocated communities, or reduce the federal deficit. Most environmental groups, many electric power and petroleum companies, and a few conservative think tanks support a carbon tax as the most effective way to accomplish the nation's GHG reduction goals.[82]

Despite its advantages, the pollution tax idea has been poorly accepted in the United States. President Nixon could not persuade a single Republican member of Congress to sponsor a tax on SO_2 emissions. We saw in Chapter 7 how the Clinton administration tried unsuccessfully to persuade Congress to enact a Btu tax in the early 1990s, and we followed the unsuccessful attempt to enact a carbon tax idea during the Obama administration in Chapter 9. President Trump has firmly rejected any suggestion that he work with Congress to enact a carbon tax to discourage burning fossil fuels, and efforts in 2017 by a group of statespersons and major corporations to persuade Congress to enact a carbon tax fell on deaf ears. Congress is not likely to enact anything that looks like a pollution tax any time in the foreseeable future.[83]

A Cap-and-Trade System. Although it is generally more difficult to administer than a pollution tax, a cap-and-trade regime is another way to overcome many of the impediments to effective regulation. EPA's acid rain program and the multistate RGGI have demonstrated how cap-and-trade regimes can be administered transparently and efficiently to the great benefit of consumers and the environment. The electric power industry generally supports "cap-and-trade" as the least-cost vehicle for achieving pollution reduction goals. Environmental groups remain divided on the cap-and-trade system, but most of the mainstream groups support it. Trading regimes are, however, subject to abuse, as demonstrated by EPA's experience with "open market" trading during the George W. Bush administration. The cap-and-trade system is a viable alternative to technology-based regimes so long as the caps are sufficiently tight, accurate monitoring is available to prevent cheating, penalties are sufficiently high, and steps are taken to ensure transparency, prevent market manipulation, and eliminate hot spots for pollutants that are not evenly dispersed throughout the atmosphere.[84]

Technology-Based Standards. From the outset, technology-based standards have been the workhorses of environmental regulation. Although they are generally less efficient than market-based tools, technology-based standards are unlikely to fade away in the foreseeable future. In jurisdictions that do not adopt market-based approaches, technology-based standards are the most effective way to drive down emissions. For some pollutants like mercury, the threat of hot spots may render market-based approaches ineffective. And even in jurisdictions that adopt market-based regimes, technology-based standards can serve as a useful backstop to ensure that emissions do in fact decrease when the tax or the cap is set too high.[85]

Deadlines and Hammers. Statutory deadlines and hammers can be an effective antidote to the tendency of agencies to push controversial issues down the road without resolving them. Statutory deadlines help agencies set priorities, and they give citizens a firm basis for lawsuits challenging agency inaction for being "unreasonably delayed" under the Administrative Procedure Act. Statutory "hammers" are likewise effective in stimulating agency action. One reason for the success of the Clean Air Act's acid rain program was a hammer providing that if EPA missed a deadline, power plants would automatically become subject to source-specific technology-based standards. In writing future environmental legislation, Congress should include statutory deadlines and hammers to ensure that the jobs that it assigns to EPA get done.[86]

Increasing Transparency in EPA Rulemaking. One way to ensure against excessive influence by outsiders on EPA decision-making is to increase the transparency of its interactions with outside interest groups, OIRA, and the White House. If outside participants who might be tempted to interject illegitimate considerations into the decision-making process know that the content of their overtures will be prominently displayed in the rulemaking docket or the agency's website, they should be less inclined to do so. In the context of OIRA and White House communications, transparency enhances accountability by making the White House's preferences and strategic political maneuvers available for all to see.[87]

Better Estimates of the Impacts of Environmental Regulations. Environmental regulation of power plant emissions could benefit from better estimates of the likely effects of EPA's regulations on affected industries, employment, and the overall economy. We have seen over and over again how industry participants in EPA rulemakings have overestimated the cost of compliance. Since cost estimates depend largely on the assumptions that drive the economic models upon which the estimates are based, it is critical that the analysts who prepare such estimates state those assumptions clearly in language that is accessible to laypersons. EPA and DOE should continue to put resources into economic modeling and to conduct retrospective assessments of the accuracy of past modeling exercises. We have seen how EPA's failure to detail the impact of its greenhouse gas regulations on jobs inspired Murray Energy to embroil the agency in a costly and time-consuming sideshow that could have been avoided by devoting modest resources to a jobs analysis. Even though it prevailed in that litigation, EPA should still devote more resources to estimating the impact of individual regulations on employment.[88]

Finally, EPA should carefully assess the cumulative impact of its regulations on particular industries. At one point in 2013, electric power companies were struggling to adapt to EPA's MATS, CSAPR, cooling water intake structure, and coal ash regulations while complying with state implementation plans addressing the regional haze rule and recent amendments to the ozone, $PM_{2.5}$, NO_x, and SO_2 national ambient air quality standards. When necessary to avoid premature shutdowns of facilities and stranded assets, EPA should adjust the timing of implementation to allow companies to avoid wasteful duplication of effort.

Addressing Coal Ash. The coal ash regulations that EPA promulgated at the end of 2014 after five years of painstaking deliberation were not especially stringent or burdensome. Among other things, they allowed companies to address inactive impoundments by simply dewatering them and capping them in place. Nevertheless, the Trump administration decided to weaken them by giving far too much unreviewable discretion to state permitting agencies. In the meantime, toxic constituents from coal ash continue to leach into groundwater from hundreds of impoundments around the country. While it may put too great a strain on hazardous waste disposal facilities to characterize coal ash as a hazardous waste, a future EPA should tighten the restrictions beyond the Obama administration's regulations and allow EPA to play a larger role in overseeing state implementation efforts.[89]

Judicial Review. Administrative law scholars have written hundreds of law review articles on judicial review of agency rulemaking, and a book on power plants is not an appropriate place to rehearse the arguments for and against stringent judicial oversight. We have seen many instances of deferential judicial review. The *Chevron* case involving EPA's 1981 bubble policy contains the classic articulation of the deferential standard for reviewing agency interpretations of their authorizing statutes, and the *Entergy* case involving EPA's cooling water intake structure regulations is an excellent example of the Supreme Court bending over backwards to uphold EPA's interpretation of its statutes. At the same time, we have seen cases in which the courts have been unwilling to defer to EPA's interpretations that have been clearly off base, as when the D.C. Circuit likened the agency's reasoning process to that of the Queen of Hearts in overturning EPA's ill-conceived mercury rule during the George W. Bush administration. The same court did not defer to the Clinton administration EPA's interpretation of "low-NO_x burner" when it based emissions limitations for NO_x on overfire air, a decision that upset the compliance strategies of several companies and postponed compliance with the statutory requirement for two years.

The D.C. Circuit has occasionally been quite aggressive in setting aside protective EPA regulations under the vague "arbitrary and capricious" standard for judicial review of the substance of agency rules. After upholding all but two aspects of EPA's first attempt to regulate interstate air pollution in its NO_x SIP call during the Clinton administration, the D.C. Circuit rejected EPA's attempts to address interstate pollution in the George W. Bush administration's Clean Air Interstate Rule and the Obama administration's Cross-State rule, only to be overturned by the Supreme Court in the *EME Homer City* case. This kind of arbitrary judicial review can have unanticipated consequences. When the D.C. Circuit set aside the CAIR without any forewarning, the markets for NO_x and SO_2 allowances cratered and did not recover until after the court stayed its ruling.

The lower courts should follow the deferential approach to statutory interpretation that the Supreme Court adopted in the *Chevron* case when EPA is interpreting its statutes in ways that further their protective goals, and they should emulate the pragmatic approach that the Supreme Court adopted in the *EME Homer City* case in reviewing EPA's attempts to navigate uncertainty-laden technical terrain.

Prospects for Future Legislation. Proposals for legislative solutions to environmental problems cannot ignore the fact that one of the two political parties is dominated by politicians who are generally opposed to expanding environmental protections. In the 1970s some of the most vigorous proponents of environmental protection were Republicans from the Northeast like

Senators Jacob Javits and John Chaffee. With the rise of the Tea Party in mid-2009, however, any Republican who was willing to compromise with Democrats on environmental issues risked a challenge in the next primary. Although there are still vigorous Democratic supporters of the coal and electric utility industries in the South and Midwest, their numbers have diminished as voters replaced them with Republicans.[90]

This radical divide on environmental issues apparently fails to reflect the opinions of voters. A January 2015 poll concluded that 88 percent of Democrats, 83 percent of independents, and 71 percent of Republicans believed that climate change "was caused at least in part by human activities." In addition, 91 percent of Democrats, 78 percent of independents, and 51 percent of Republicans thought that the government should be taking action to address climate change. Yet, so long as Republican leaders remain convinced that they will not suffer at the polls for opposing environmental improvements, the gulf between the two parties will render negotiation and compromise virtually impossible.[91]

Effective environmental legislation at the federal level will have to await a change in the political climate that produces a Congress composed of pragmatic representatives of the people who are willing to reason together in pursuit of the common good. And that will take a political movement built from the ground up of people who are committed to making the planet sustainable for the generations that follow.[92]

ASSISTING DISPLACED WORKERS AND REBUILDING DAMAGED COMMUNITIES

Critical to the success of any proposed solution to the power plant emissions problem will be the public perception that its costs and benefits are being distributed fairly. We saw in Chapter 11 how the rapid retirement of coal-fired power plants during the past decade has displaced workers and caused economic disruption in the towns where those plants were located and in coal mining regions in Appalachia, the Midwest, and the Powder River Basin. Environmental groups recognize society's moral obligation to assist workers and communities in adapting to the transition away from coal. The Sierra Club has compromised with community representatives in settling new source review cases on terms that permitted the continued operation of coal-fired power plants for a time as an alternative to destroying local economies.

At the same time, labor unions and community leaders must recognize that the job losses brought on by the transformation of the electric power industry are permanent. Unfortunately, political leaders in coal-dependent states and President Trump have raised false hopes by promising to restore jobs by scaling back environmental regulations. When the promised coal revival fails to materialize, the time may be ripe for a serious dialogue over how best to alleviate the adverse impacts on jobs and local communities of the move away from fossil fuels.[93]

One example of a successful dialogue is the negotiations leading to the retirement of TransAlta's Centralia plant, the state of Washington's only coal-burning facility. Responding to grassroots pressure to rid the state of a plant that imported coal from the Powder River Basin, sold half of its electricity to out-of-state customers, and degraded visibility in Mount Rainier National Park, Governor Christine Gregoire persuaded the company and its unions to negotiate with environmental groups to come up with a compromise that would protect the environment and preserve jobs. The participants hammered out an agreement under which the company would retire one unit in 2020 and the other unit in 2025, create a $55 million fund to diversify the region's job base, provide $30 million for end-use energy efficiency projects, and devote $25 million to supporting new energy-efficient technologies. The phased shutdown allowed 40 percent of the plant's employees to retire before the plant closed.[94]

Environmental groups generally support programs for early retirements with pensions, retraining, and, if necessary, relocating workers in the electric utility and coal industries. But it is often difficult to implement job training and community redevelopment programs in practice. First, many laid-off workers have deep attachments to their locations, and they cannot easily relocate to places thousands of miles from their families and support networks. Second, many coal miners and power plant workers lack the skills to switch fields of work. Even in states like Wyoming, where jobs in the wind industry are available, it takes a lot of patience and training to turn a coal miner into a wind technician. Third, entry-level jobs in the renewables industry and the manufacturing, service, and retail sectors generally pay far less than existing unionized jobs in coal mining and electric power.[95]

Congress has created programs in the Department of Commerce's Economic Development Administration, the Department of Labor's Economic

Training Administration, and the Appalachian Regional Commission (ARC) to assist workers and communities that are hard hit by economic change. The ARC sponsored 662 projects that created or maintained more than 23,000 jobs and trained more than 49,000 residents of Appalachia between October 2015 and January 2017. In 2016, the Obama administration directed $65.8 million in federal funds toward efforts to diversify local economies in Appalachia and assist Native American tribes to begin using existing transmission infrastructure to move electricity from solar and wind farms to the grid.[96]

The Trump administration, however, is moving in exactly the opposite direction. The president's proposed budget for fiscal year 2018 eliminated funding for the ARC. It also eliminated several other programs that were part of President Obama's Partnerships for Opportunity and Workforce and Economic Revitalization (POWER) initiative, which was designed to help communities dependent on coal or coal-fired power plants to diversify their economies and lower unemployment. After Congress refused to defund the ARC in the fiscal year (FY) 2018 appropriation, the Trump administration left it intact in its proposed FY 2019 budget. Rather than denying climate disruption and killing environmental regulations, the Trump administration should be developing new programs for easing the inevitable transition.[97]

In the absence of federal leadership, state legislatures, PUCs, and local governments should seize the initiative. When Southern California Edison and its partners closed the Mohave plant in Nevada in 2005, a coalition of environmental and Native American advocates persuaded the California Public Utilities Commission to require the company to spend the revenue from the sale of the plant's NO_x allowances on renewable energy projects on tribal land. Recognizing that NRG would soon be retiring its Huntly plant in Tonawanda, New York, a coalition of local environmental groups, teachers, union members, and other community activists persuaded the state legislature to appropriate $45 million to provide transitional funding for schools and public services while the town redeveloped the site for use by less polluting industries. And a local community college launched a program to train miners to work as linemen for electric power companies, which has placed 90 percent of its graduates in jobs paying more than their former jobs. Not all "transition assistance" programs, however, have met expectations. For example, the West Virginia legislature devoted a percentage of the state's coal severance tax to a Future Fund for the purpose of economic diversification in its coal-mining regions, but the state has for many years diverted severance revenues to other programs. And in some hard-hit locations, job training programs are undersubscribed because local miners are convinced that President Trump will bring about a coal comeback. Following the California and New York examples, state legislatures and agencies should design innovative "just transition" programs to reduce the pain to workers and communities of the transition to a more sustainable electricity grid.[98]

HELPING LOW-INCOME CONSUMERS

As we transition away from natural gas and the "temporary" tax advantages currently enjoyed by renewable power, the price of electricity may well increase. Price increases should stimulate further efforts to conserve energy, but they will also make it more difficult for low-income consumers to afford electrical power. States that move to time- or marginal cost-based rate-making should make accommodations for low income customers who may not be able to afford higher than average rates. They could accomplish this through special low-income rates or rebates that smooth out the price curve. The difference would, of course, have to be made up through higher rates for well-off customers. Utility companies and solar installers can provide affordable financing to low-income homeowners to allow them to share in the benefits of distributed generation. For example, the San Antonio municipal utility CPS Energy has partnered with a solar installer to allow customers to have solar panels installed at no upfront cost and receive credit on their bills for a portion of the energy savings. But significant penetration of rooftop solar into low income neighborhoods will probably require government assistance.[99]

Congress established the Low Income Home Energy Assistance Program (LIHEAP) in 2009 to help low-income Americans pay their energy bills, deal with energy-related crises, and pay for weatherization of their homes.[100] The program was initially funded at $5.1 billion, but it rapidly dropped to $3.39 billion in 2015. And the Trump administration's proposed budgets for FY 2018 and FY 2019 zeroed out the program. The administration would also defund DOE's Weatherization Assistance Program, which since 1976 has provided funds to low-income families, the elderly, and the disabled to reduce home energy consumption through improvements like additional insulation and more efficient heating and cooling. To ensure the equitable distribution of the added cost of energy conservation incentives, Congress should restore funding to these programs to at least their 2009 levels.[101]

Conclusion

The electric power industry was profoundly affected by the environmental movement of the late 1960s and early 1970s. As the industry struggled to comply with new environmental requirements, it also undertook reorienting structural changes in response to economic deregulation. As we enter the 2020s, neither of those struggles has come to an end. The industry's emissions profile is radically different from that of the 1960s as coal is no longer king, natural gas and renewables are playing much larger roles, and fossil fuel–fired power plants are now equipped with expensive pollution control technologies. Yet progress toward a more efficient and sustainable power grid has come in fits and starts as the battles rage between fossil fuel interests, conservative activists, and upwind states in one camp and renewable energy interests, environmental groups, and downwind states in the other camp, with EPA weighing in on one side or the other as the occupant of the Oval Office changes. All the while, electricity has remained a critical source of the energy powering an increasingly digital economy, and it will become even more essential in the future as consumer demand and government regulations force the electrification of the transportation sector and residential and commercial heating.[1]

A GREAT DEAL OF PROGRESS

Over the past fifty years, we have made a great deal of progress toward achieving a sustainable grid. As we saw in Chapter 11, emissions of power plant pollutants have declined significantly since the enactment of the 1990 Clean Air Act amendments while overall electricity consumption has increased. At the same time, economic growth in the twenty-first century has far outpaced energy consumption. Progress toward reducing regional haze in eastern national parks is proceeding more rapidly than expected. The coal combustion residuals that are not being recycled are being disposed of in more secure facilities than the leaky impoundments of the past. Despite the Trump Administration's aggressive efforts to subsidize the coal industry, this trend has continued unabated. Most of the big dirties are no longer with us. Absent dramatic breakthroughs in carbon capture, utilization, and sequestration (CCUS) technology, coal is on the way out, driven by its inability to compete with natural gas and renewables when forced to internalize the costs that its consumption imposes on others.[2]

As more utility companies recognize that they can make a profit as sophisticated managers of power derived from large renewable generators, demand response, and distributed generation, talk of "death spirals" has faded from the discussion. Electrical appliances are more efficient, and digital technologies are allowing commercial, industrial, and residential consumers to manage electricity consumption far more effectively than in the past. Companies that install rooftop solar panels, build solar arrays, and erect wind turbines are doing well as renewable power plays a larger role in meeting energy needs, and they are employing tens of thousands of workers in well-paying jobs. By mid-2017, installed wind capacity at eighty-four gigawatts exceeded the capacity of hydropower, and renewable resources (including hydropower) produced more electricity than nuclear energy for the first time since 1984. By the end of 2018, thirty-five more gigawatts of capacity were under construction or in advanced development. Even small rural electric co-ops are investing heavily in renewable power or purchasing it from recently constructed wind farms.[3]

STILL A LONG WAY TO GO

Despite this undeniable progress, we still have a long way to go. The transformation has not been especially effective in reducing the impact of power plants on marine and aquatic organisms. Despite the efforts by environmental groups and the Obama administration to force old power plants through new source review, nearly one-quarter of the coal-fired power plants in operation in 2017 still lacked any controls on SO_2 emissions. Those plants supplied only 17 percent of the nation's electrical energy, but they accounted for almost 44 percent of power plant SO_2 emissions. Greenhouse gas reduction presents the heaviest lift ahead for regulators and the electric power industry. CO_2 levels in the atmosphere continued to increase by 3 ppm in both 2015 and 2016, the largest increases since scientists began measuring those levels in the 1950s. At 403 parts per million, the CO_2 level in the atmosphere is 1.45 times its preindustrial level and the highest it has been in 800,000 years. The Energy Information Administration concluded in June 2016 that much larger GHG reductions would be necessary to achieve a long-term goal of an 80 percent reduction in GHG emissions from 2005 levels by 2040, and that prediction assumed that the Clean Power Plan (CPP) and other Obama administration initiatives would go into effect. As we have seen, however, the Trump administration has done everything within its power to roll back all of the Obama administration's climate change initiatives. And CO_2 emissions from power plants increased for the first time since 2013 by 1.9 percent.[4]

REASONS FOR PESSIMISM

There are certainly good reasons to be pessimistic about the prospects for further progress toward sustainable electric power, especially in the near term. It may be true that the transformation of the electric power industry away from coal and toward natural gas and renewables will continue if the Trump administration succeeds in pulling back the Obama Administration initiatives, but it will probably proceed at a slower pace as natural gas becomes the dominant fossil fuel and renewables become less attractive. As of the end of 2018, 255 coal-fired power plants were still operating. The United States may meet its commitment under the Paris Agreement to reduce GHG emissions to 17 percent below 2005 levels by 2020, but it will probably not meet its commitment to reduce GHG emissions to 26–28 percent below 2005 levels by 2025. Because the Obama Administration's regulations served as a critical backstop against regression in states that are disinclined to regulate power plants, it is even possible that the industry will regress toward greater use of coal.[5]

Fossil Fuel Industry Opposition

The coal industry remains adamantly opposed to environmental regulations, and most of the natural gas industry consistently opposes environmental regulation that does not enhance its competitive status relative to coal.[6] Although coal's economic power is diminishing, its political influence remains strong. Coal plays a powerful role in the political culture of regions like central Appalachia, southern Illinois, and the Powder River Basin that are heavily dependent upon that mineral. Politicians from both parties must pay obeisance to coal to get elected. Natural gas dominates the culture of other regions like North Dakota and the Louisiana/Oklahoma/Texas oil patch. As we saw in the battles over the Waxman-Markey climate change legislation at the outset of the Obama administration, politicians from these regions are in a position to erect impenetrable roadblocks to much-needed environmental legislation even when national public opinion is favorable and the Democratic Party is in power.[7]

Resistance from Conservative Idea and Influence Infrastructures

The companies that have in recent years committed themselves to pursuing green goals are discovering that the think tanks, media outlets, and grassroots organizations that conservative foundations and the business community created and nurtured over the past forty years have minds of their own. The U.S. Chamber of Commerce, the largest and most visible representative of the business community, has remained steadfast in its opposition to any form of mandatory climate change legislation, as has the National Association of Manufacturers. Driven by a strong ideological commitment to free markets, they are not likely to temper their opposition to environmental regulation in the foreseeable future, despite the rifts in the electric power industry. Consequently, few Republican senators and representatives are likely to change their positions on those issues.[8]

As we saw in the legislative battles over climate change bills in the first year of the Obama administration, environmental and public health groups are generally outgunned and outclassed in the political battles over environmental legislation. They have not been able to match the highly organized grassroots campaigns of Citizens for a Sound Economy, Americans for Prosperity, and the many ad hoc organizations that the coal and electric power industries have assembled to sway public opinion. Nor do they have media equivalents of Fox News and Rush Limbaugh to spread their message over the airwaves and through social media. Even under a highly motivated future leadership, EPA will encounter strong political resistance that can stymie the most well-supported rulemaking initiatives.[9]

Electric Power Industry Ambivalence

If we are going to meet the goals of the Paris Agreement, GHG emissions nationwide will have to decrease by around 5 percent per year, and that will require halting all new fossil fuel–fired power plants and shuttering many fossil fuel–fired power plants before the end of their planned lifetimes. To make up the shortfall, the electric power industry will have to make massive investments in renewable power that match or exceed the large investments of the past few years. Although most large electric power companies have reduced their reliance on coal and have made major commitments to renewable power, the industry remains deeply ambivalent about moving away from natural gas. Electric power companies are especially effective at lobbying members of Congress, because nearly every congressional district contains one or more power distribution companies with executives and employees who are willing to go to bat for their companies.[10]

While emissions of SO_2, NO_x, and mercury from the one hundred largest power plants have declined dramatically since the enactment of the 1990 Clean Air Act amendments, emissions of CO_2 have increased slightly. The breakaway companies that support strong climate change policies are for the most part public utilities that stand to gain economically from greater reliance on natural gas and nuclear power. The co-ops that serve much of rural America have shown little appetite for shedding fossil fuel assets and little commitment to expanding renewable assets. When it comes to cooling water intake structures, the electric power industry is unified in its opposition to stronger regulation. The industry will no doubt continue to reduce air emissions as companies retire the remaining big dirties and build more solar arrays and wind farms. Visibility in national parks will improve, and mercury levels in fish will drop. But without strong pressure from government to phase out natural gas plants, ozone levels along the Eastern Seaboard will remain high and the nation will not reduce CO_2 emissions to the degree necessary to prevent further climate disruption.[11]

Increased Demand for Electricity

Although greater efficiency, demand response, and distributed generation have greatly decreased demand for electricity during the 2010s, a growing economy could well offset these decreases in the future. Burgeoning data centers and cryptocurrency miners are also driving up demand for electricity.[12] Moreover, meeting greenhouse gas reduction goals will require a transformation of the transportation sector that rivals the transformation in the electric power industry.[13] That transformation from reliance on fossil fuels to reliance on electricity will put enormous pressure on the electric power industry to keep old fossil-fueled power plants online and perhaps to build new ones to meet the increased demand. Under such pressure, the electric power industry's ambivalence may shift, and it may well form alliances with vehicle manufacturers and proponents of fossil fuels to lift governmental restrictions on power plant emissions.

Insufficient State and Local Initiatives

While many states have taken forceful steps to reduce power plant pollution, others have strongly resisted EPA initiatives. In some cases, this resistance to regulation defies logical explanation. States like Texas, Oklahoma, and Kansas that have challenged nearly every EPA regulation affecting coal-fired power plants have few coal reserves and are leading the nation in energy from wind and solar power. Ohio, West Virginia, Wyoming, and many southeastern states remain heavily dependent on coal, and environmental agencies in those states are not likely to push the electric power industry toward a more sustainable power grid. A few state legislatures have even enacted legislation designed to discourage investment in renewable energy. Cities that have made ambitious commitments to renewable energy goals may find it difficult to achieve in states that support fossil fuels, and they may struggle with previous commitments to power companies who want to avoid stranded assets. Given the very real phenomenon of interstate transport of pollutants, restrictions by downwind states on their own sources will not be sufficient to protect their citizens from asthma and lung disease, to reduce mercury in fish to acceptable levels, or to protect their lakes from acid rain. The most effective way to make use of the nation's ample renewable resources is to incorporate them into a nationwide grid that allows a single grid manager to achieve a balance as weather conditions vary across the continent, but that is not something that individual states can accomplish.[14]

Distrust of Government

Proponents of governmental solutions to environmental problems must overcome a deep-seated distrust in government on the part of a large percentage of the U.S. population. Public distrust in government institutions began as far back as the 1970s with government dissembling about the Vietnam War and Watergate. In 1972, William Ruckelshaus, then EPA administrator, complained that "we live in a society in which the mistrust of our institutions, governmental, religious, industrial, the institutions of our society are badly mistrusted by many of the people who live in the society." Distrust in government has only grown more pronounced with the advent of the Tea Party and siloed social media. When a substantial proportion of the public does not trust government to protect them from external threats or even to recognize and define those threats, the government's power to protect is diminished, and the companies that put people at risk are empowered.[15]

Little Progress in Capturing Carbon

The history of carbon capture, utilization, and sequestration technology over the past decade has been the opposite of the history of wind and solar energy and storage technology. While we have seen tremendous strides toward more efficient and cheaper solar panels, wind turbines, and batteries, CCUS technology has developed excruciatingly slowly and has gotten more and more expensive. Despite billions of federal dollars and tax incentives devoted to CCUS research and development, including the ill-fated FutureGen

program, the only large power plant in the United States employing that
technology is the recently completed Petra Nova project at NRG's W. A.
Parish plant near Houston. It is looking more and more like CCUS will not
be the savior that the fossil fuel industry has been anticipating for the past
twenty years.[16]

A Regressive Administration

Most presidential elections feature promises of change from one side or the
other, but the change in environmental policy that the 2016 election brought
about was more radical than most. A cautiously progressive administration
committed to forcing the electric power industry to confront the looming
catastrophe of climate disruption was replaced by an avowedly regressive
administration with no affirmative environmental agenda and a strong commitment to undoing every reform of its predecessor. The 30 percent tariff that
President Trump imposed on imported solar cells has increased the cost of
new solar power installations and reduced incentives to invest in renewables.
While this might not have a huge overall impact on the renewables industry, it
could have a devastating effect on some small solar installation companies.
And to the extent that the tariff puts renewable energy companies out of business, it will have a negative impact on jobs in an industry that has been one of
the largest generators of jobs in the current American economy. At the very
least, the Trump administration's actions have created uncertainty, which is
likely to hamper investment. That by itself could be enough to slow the movement away from coal and the corresponding reductions in GHG emissions.[17]

A Gridlocked Congress

The prospects for legislation providing for a sustainable grid are quite grim
in the short run. Environmental protection has become a highly partisan
issue. For the first two years of the Trump administration, the Republican
leadership in Congress was just as committed to rolling back past environmental protections as was President Trump. In part, this reflects the geographical fact that few Republican members come from states that stand
to benefit from protective legislation. But it also reflects a deep ideological
commitment to a minimalist approach to the role of government in society.

Democratic politicians have not been as monolithic on environmental issues as their Republican counterparts, as demonstrated by the difficulties that sponsors of a climate change bill encountered when the Democrats controlled the House and owned a supermajority in the Senate. Still, a sufficient number of Democrats in Congress are concerned about climate disruption to consistently forestall efforts by Republicans to enact deregulatory legislation.[18]

The result of the partisan divide has been gridlock in Congress. In the best of times, it is very difficult to persuade legislators to force power plants to reduce emissions when the benefits of those reductions flow to nonvoters in other states and to future generations. But in these highly partisan times, serious federal legislation to reduce the adverse environmental effects of power plants is off the table. For example, despite the fact that nearly all of the relevant interest groups agree that a cap-and-trade program or a carbon tax would be the most efficient approach to reducing GHG emissions, gridlock has prevented Congress from adopting either option. The same gridlock prevented the Tea Party wing of the Republican Party from enacting legislation preventing the Obama administration from moving forward with the Clean Power Plan.[19] This could change with a "wave" election in one direction or the other, but for the time being, any action to regulate or deregulate power plant emissions will have to come from the executive branch or the states.

REASONS FOR OPTIMISM

There remains room for optimism that the United States will achieve a sustainable electrical grid in the long term. Former New York mayor Michael Bloomberg believes that "forces beyond Washington" have reached such a "critical mass" that "we should be more optimistic than ever about our ability to lead—and win—the fight against climate change." He was sufficiently convinced that the fight was worth it that he donated another $64 million to the Sierra Club's efforts to reduce reliance on fossil fuels in October 2017 for a total of $168 million since his first donation in 2011.[20] And it is always possible that the political climate will change in ways that bring about a fresh effort in state capitals and Washington to reduce the adverse environmental effects of fossil fuel–fired power plants.

Administrative Law

The Trump administration's efforts to roll back previously promulgated regulations have to comply with the same procedural, analytical, and review requirements that made promulgating those regulations very difficult. And they must be consistent with the relevant statutes. Thus, if EPA attempts to withdraw the endangerment finding, the agency will have a steep hill to climb to overcome the accumulated scientific evidence that GHG emissions cause climate disruption. The Clean Air Act requires EPA to regulate a pollutant if it "may reasonably be anticipated to endanger" public health. Even if scientific uncertainties remain about the nature of the relationship between anthropogenic emissions and global warming, the agency is obliged under this precautionary language to take action to protect people and the environment. And the cost of complying with any regulations that flow from that finding are irrelevant in making the finding. Finally, any changes in the current regulations will be subject to judicial review to determine whether they are consistent with the relevant statute or are arbitrary and capricious.[21]

Strong Statutes

Although the environmental statutes of the 1970s are definitely due for updates, they provide a strong foundation for continued environmental improvement under the guidance of a presidential administration that is committed to that goal. The Resource Conservation and Recovery Act and the Clean Water Act give EPA ample authority to protect rivers and groundwater from leaking coal ash impoundments and to protect marine and aquatic species from impingement and entrainment on cooling water intake structures. The Clean Air Act's margin of safety requirement for primary ambient air quality standards and its insistence that those standards not be compromised by cost considerations ensures that EPA will continue to protect public health, even in times like the present when the president wants to move in the opposite direction. We may never know whether the Obama administration's ambitious attempts to adapt the Clean Air Act to the daunting challenge of climate disruption were lawful, but the fact that CO_2 remains a pollutant subject to regulation under that statute means that plans for every new fossil fuel–fired power plant and every major modification of an existing plant will have to include the best available control technology for reducing greenhouse gas emissions. Because of strong public support for environmental regulation and the gridlock that paralyzes Congress, these powerful statutes are not likely to be weakened in the foreseeable future.

Committed Environmental Groups

The major environmental groups are committed to resisting the Trump administration rollbacks and to challenging new and modified fossil fuel–powered power plants. With major new infusions of resources from wealthy donors like Michael Bloomberg and Tom Steyer, they have the resources to go toe-to-toe with EPA and fossil fuel interests. The Sierra Club also plans to focus on state and local initiatives on the assumption that EPA will not want to tighten power plant emissions standards during the Trump administration. The environmental groups have also launched major grassroots educational efforts to tout the virtues of strong environmental protection regulation to voters in anticipation of future elections.[22]

Improvements in Energy Conservation

Improvements in end-use efficiency and the willingness of state legislatures and regulators to facilitate DR and DG bode well for the future of demand-side management. More people are recognizing that energy conservation, rooftop solar, and backyard wind can save them many dollars in the long run. By one estimate, energy efficiency programs cost about 2.5 cents per kilowatt-hour compared to building a new gas-fired power plant at 6 to 15 cents per kilowatt-hour. As of late 2017, more than 13.6 gigawatts of demand response were enrolled with utility companies, about 78 percent of which were actually used. The time is at hand when residential customers everywhere can allow DR aggregators to control their thermostats and appliance use in return for a payment or credit against electricity bills. As of 2016, forty-four states and the District of Columbia had implemented some form of "net metering" program to allow customers to sell electricity that they generate to the distributor, usually at the full retail price. The electric utility industry expects exponential growth in distributed generation as solar and fuel cell technologies become less expensive and the electricity distribution system evolves into a hybrid system in which customers produce a significant amount of the electricity needed to meet demand. All of these developments greatly reduce the need to build new fossil fuel–fired power plants and make it feasible to retire existing plants with corresponding benefits to the environment.[23]

Bright Prospects for Renewables

There is a huge potential for renewable resources to supply most of the nation's demand for electricity. Google reports that 80 percent of the nation's rooftops are candidates for solar power, a total of around sixty million buildings. According to the research firm Lazard, the "levelized" price (across a twenty-five-year lifetime) of wind power is now lower than the levelized price of electricity from coal-fired plants. A study undertaken by the National Renewable Energy Laboratory concluded that renewable resources could provide 80 percent of the nation's electricity by 2050 if we make the grid infrastructure sufficiently flexible to accommodate new kinds of services as they become available. The California grid regularly uses six gigawatts of wind, ten gigawatts of utility-scale solar, and at least five gigawatts of rooftop solar. Renewable energy companies are locating in coal-rich areas like Wyoming and West Virginia and offering to train coal miners how to install wind turbines and rooftop solar panels. As storage technologies advance, renewable power's capacity to provide all or a substantial portion of the nation's electricity will expand dramatically. There is little evidence thus far that President Trump's tariff has had a large impact on adoption of solar technologies, and it will disappear after its fourth year.[24]

Technological Innovation

Over the years, the companies that design pollution control devices have consistently come up with workable ways to reduce the externalities that power plants imposed on their neighbors and the environment. These solutions have not always been cheap, but they have invariably cost less than predicted, and the costs have usually declined (sometimes dramatically) as companies gained experience with using them in the real world. Scrubbers were once bulky, messy, and prone to intermittent failure. They are still bulky, but they now operate smoothly and are far more effective at removing

SO_2. In early years, the only way to reduce NO_x emissions from power plants was to fine-tune combustion by adjusting burners and air supplies. But engineers soon developed selective catalytic reduction technologies to reduce NO_x emissions dramatically at a reasonable cost. When EPA first took up mercury emissions from power plants, the most effective technologies to address that problem were scrubbers, electrostatic precipitators, and baghouses that could achieve 60 percent removal. But engineers developed and refined dry sorbent injection technologies capable of removing more than 90 percent of mercury from the exhaust at much less cost than scrubbers. The one pollutant that has so far proved intractable is CO_2, which is the most heavily emitted pollutant from power plants. Technologies exist for removing CO_2 from exhaust streams, but they are quite expensive and energy intensive. If the past is any guide, however, we can expect the cost of carbon capture, utilization, and sequestration to decline as engineers gain more experience with that technology, or that engineers will come up with new CO_2 removal technologies that are less expensive and perform more effectively than CCUS.

Some Support in the Electric Power Industry

We have seen that the electric power industry is divided on many of the environmental issues that affect power plants. Companies that are heavily dependent on coal tend to oppose stringent limitations on SO_2 and CO_2, but we no longer hear complaints from power company CEOs about an EPA controlled by "rabid environmentalists" like we did in the mid-1970s. Companies that were on the front lines challenging EPA regulations have been hedging their bets by reducing their consumption of coal and investing in CCUS technology. The Southern Company, historically one of the most vigorous opponents of EPA regulation, reduced coal consumption by 49 percent between 2008 and 2015 for economic reasons largely unrelated to environmental regulation. And its chief operating officer Tom Fanning told shareholders in May 2018 that the company was committed to deeper reductions through such innovations as "smart neighborhoods" and investing $500 million a year in wind projects for five years. Electric power company consumption of fossil fuels in 2017 was at the lowest level since 1994. Companies that rely more on nuclear power, natural gas, and renewables are often strong supporters of stringent air pollution standards.[25]

Private Sector Momentum

Encouraging developments in the private sector suggest that the movement away from fossil fuels and toward energy conservation and renewables will continue as large commercial customers demand more renewable power. Many major American companies have pledged to reduce their carbon footprints through greater energy conservation and by purchasing more renewable power. For example, Walmart's "Project Gigaton" has a goal of removing a billion metric tons of GHG from its supply chains by 2030. Other big-box stores like Home Depot, Target, Ikea, and Kohl's, have made similar commitments. Major technology companies like Google, Apple, Amazon, and Microsoft have committed themselves to relying more heavily on renewable energy. These developments will not ensure that the national ambient air quality standards are achieved or that we will meet the nation's commitments to reduce CO_2 emissions under the still-applicable Paris Agreement. But it is encouraging that so many companies have maintained their commitments to doing something about power plant pollution despite clear signals from the Trump administration that it will not be requiring emissions reductions.[26]

State and Local Efforts

State and local governments have also stepped up to promote sustainable policies for the electric power industry as the Trump administration has tried to roll back environmental protections. Many states regulate power plants more stringently than required by the federal statutes, and even more states have enacted renewable portfolio standards and end-use energy conservation programs that are having a significant impact on how electrical power is generated and used. These clean energy initiatives have been successful in both red and blue states in all regions of the country. Republican governors in some states that voted for Trump, including Nevada, Florida, and Iowa, are strong supporters of renewable energy. Even in coal-dependent states, public utility commissions driven by cost considerations may put pressure on companies to repower or retire old coal-fired plants. At the end of the day, however, state and local regulation cannot solve the national problems of interstate transport and climate disruption.[27]

A Reliable and Resilient Grid

Despite frequently repeated warnings that reducing the role of fossil fuels will threaten the reliability and resiliency of the nation's electric grids, there is very little evidence that the grid is at risk of a breakdown as companies move coal-fired plants off-line. In fact, the grid has become more resilient with the recent addition of efficient natural gas plants, wind farms, and utility-scale and rooftop solar installations. As the nation moves away from natural gas in the future, reliability and resilience may again become a legitimate concern. But that is by no means predetermined. Perhaps the greatest threat to grid reliability is the prospect of a massive move away from fossil fuels in the transportation sector toward electric vehicles. Such a move would put a great deal of pressure on the grid, but it would also result in a 60 percent reduction in CO_2 emissions from vehicles if 80 percent of the electrical power comes from renewable resources. It will be very difficult to achieve an 80 percent reliance on renewables while meeting the increased demand of a transportation sector that relies heavily on electricity, but it should not be impossible, given sufficient time, resources, and attention to managing the grid so that electric vehicles are charging when demand from other uses is low. Should a genuine reliability crisis arise, DOE has the power to issue an order under section 202(c) of the Federal Power Act requiring one or more plants to remain online.[28]

Public Support for Strong Environmental Regulation

The best reason for optimism is the fact that a large majority of the American people support strong environmental regulation. Almost one year into the Trump administration, a nationwide poll found that 61 percent of the respondents thought that government action was needed to address climate change and only 20 percent thought that EPA should rescind the CPP. The election of Donald Trump inspired strong civic engagement on environmental issues, including a March for Science and a People's Climate March in April 2017 in which hundreds of thousands of concerned citizens throughout the country participated. This outpouring of public support for the environment could translate into political change in the not-too-distant future.[29]

THE COST OF DELAY

Environmental groups and the electric power industry can agree that we should be "building a bridge to a clean energy future by using a balanced energy mix," but the length of that bridge depends on how convinced one is that emissions from fossil fuel–fired power plants are causing serious harm to human health and the environment. As we have seen, the scientific evidence on the adverse effects of power plant emissions on human health, the environment, and the global climate strongly suggests that the bridge should be very short and that the future energy mix should be heavily weighted toward efficiency and renewable energy. We are already paying a price for past delays in more severe droughts, heat waves, wildfires, and storms. Insurance rates are increasing to account for the greater risks of catastrophes caused by climate disruption, and federally subsidized disaster assistance payouts are rising dramatically. The White House Council of Economic Advisers issued a report in July 2014 concluding that the cost of reducing GHG emissions will increase by about 40 percent for every decade of delay. In the meantime, the level of CO_2 in the atmosphere reached 410.31 ppm in May 2018, a level that has not been exceeded in 800,000 years, and temperatures in Pakistan reached 122.3°F, the highest ever recorded in human history.[30]

TRANSFORMATIVE POLITICAL MOMENTS

Progress has often come in "transformative moments" when a "confluence of crises" precipitates forceful demands for change from a broad array of deeply affected interest groups and the general public. In such moments, the need for government intervention into the economy is apparent to everyone, and voters elect representatives who promise to create programs aimed at protecting the public by restraining economically powerful actors.[31] We experienced such a moment in the late 1960s and early 1970s when a confluence of environmental crises impelled Congress to enact the Clean Air Act, the Clean Water Act, and the Resource Conservation and Recovery Act. We may have experienced another in 2008 through 2010, when a confluence of economic crises precipitated public demands for governmental action on a number of fronts, including climate change. But well-organized fossil fuel interests withstood demands for change, and Congress failed to enact comprehensive climate legislation.

Those interests are now well positioned to forestall any significant federal legislation aimed at reducing the impact of power plants on health and the environment in the absence of some major crisis. Yet the effects of pollution are usually difficult to tie to particular causes like power plants, and many of those effects occur over long periods of time. Change can come in small bursts like the outpouring of power plant regulation during the Obama administration. Because those changes are not solidified in legislation, however, they are always subject to reexamination and retrenchment, as we have witnessed during the first two years of the Trump administration.

Still, it is possible that the drought, wildfires, flooding, and superstorms that rising global temperatures bring on will cause the kind of dramatic disruption that brings about a transformative moment. If that should happen, a political infrastructure needs to be in place to transform the confluence of crises into political action. Only a strong political movement has a chance of bringing about the kind of significant change in the political climate necessary to overcome the opposition of fossil fuel interests in Congress, state legislatures, EPA, and state agencies. Political scientist Theda Skocpol's profound critique of the 2009–2010 attempt to pass climate disruption legislation concludes that legislative action to prevent climate disruption must begin with local organizing and activism. That kind of activism was on display in the March for Science and the People's Climate March.[32] If environmental activists can sustain the enthusiasm that impelled hundreds of thousands of individuals to participate in those protests, it may be possible to achieve a sustainable grid in time to prevent the worst-case predictions of the climate scientists from becoming a reality.

Notes

INTRODUCTION

1. Robert Walton, "Alliant Plans to Eliminate Coal, Cut Emissions 80% by 2050," *Utility Dive,* August 3, 2018; Hannah Northey, "Trump Touts 'Indestructible' Coal, and Gas Industry Fumes," *E&E News,* July 5, 2018; Travis M. Andrews, "Kentucky Coal Mining Museum in Harlan County Switches to Solar Power," *Washington Post,* April 6, 2017; Katie Valentine, "Obama Administration Becomes the Third to Install Solar Panels on White House Grounds," *ThinkProgress,* August 15, 2013, available at https://thinkprogress.org/obama-administration -becomes-the-third-to-install-solar-panels-on-white-house-grounds-8afd1867cd0f /; Kentucky Coal Museum website, available at http://kycoalmuseum.southeast .kctcs.edu.

1. THE BIG DIRTIES

1. The description of the Gavin plant is drawn from: Bob Matyi, "New Owners to Keep Gavin Coal Plant in Ohio Running: Official," *Platts Coal Trader,* February 7, 2017; Bradley Adams and Constance Senior, "Curbing the Blue Plume: SO_3 Formation and Mitigation," *Power,* May 2006, 39; Katharine Q. Seelye, "Utility Buys Town It Choked, Lock, Stock and Blue Plume," *New York Times*, May 13, 2002, A1; "AEP Struggles to Solve Acid 'Cloud' from Its 2,600-Mw Gavin Coal Plant," *Utility Environment Report,* August 10, 2001, 1; David Stout, "7 Utilities

Sued by U.S. on Charges of Polluting Air," *New York Times,* November 4, 1999, A1; "New York Suit Will Target 17 Coal Plants," *Platts Coal Outlook,* September 20, 1999; "AEP Waits for Word on SO_2 Emissions Credits," *Coal & Synfuels Technology,* March 13, 1995, 5; James Bradshaw, "PUCO, AEP Triumph on Electric Rate Ruling," *Columbus Dispatch,* March 31, 1994, C4; "State Groups Seek to Block Plan by Ohio Utility to Install Scrubbers," *BNA Environmental Law Update,* March 31, 1992; "Tightening Compliance Policy Could Reduce Sulfur Dioxide Emissions but Cost to Midwest Would Be High, State, EPA, Industry Officials Say," *BNA Environment Reporter,* March 16, 1984, 2059.

2. U.S. Government Accountability Office, *Air Emissions and Electricity Generation at U.S. Power Plants* (Washington, DC, April 18, 2012), 3–4; Environmental Integrity Project, *Dirty Kilowatts: America's Most Polluting Power Plants* (Washington, DC, July 2007).

3. The Beckjord description is drawn from Duke Energy, "Batteries Spring to Life at Retired Duke Energy Coal Plant," press release, November 18, 2015, https://news.duke-energy.com/releases/batteries-spring-to-life-at-retired-duke -energy-coal-plant; Rebecca Smith, "Upgrade Costs Doom Older Plants," *Wall Street Journal,* December 23, 2011, B2; "Utility Settles Lawsuit for $1.4 Billion," *Washington Post,* December 22, 2000, A5.

4. The Big Brown, Martin Lake, and Monticello descriptions are drawn from Environmental Integrity Project, *Dirty Kilowatts,* 22, table 8; Vistra Energy, "Luminant to Close Two Texas Power Plants," press release, October 13, 2017, https://www.luminant.com/luminant-close-two-texas-power-plants/ (quote); Luminant, "Luminant Announces Decision to Retire Its Monticello Power Plant," press release, October 6, 2017, https://www.luminant.com/luminant-announces -decision-retire-monticello-power-plant/; Annalee Grant, "Appeals Court Orders EPA to Review Good Neighbor's Smog Limits, Upholds Other Challenges," *SNL Daily Coal Report,* July 31, 2015; Barry Cassell, "Luminant Switches a Second Unit at the Martin Lake Coal Plant into Seasonal Operations," *GenerationHum,* July 24, 2015; Annalee Grant, "Texas Judge Orders Sierra Club to Pay Luminant $6.4M for 'Frivolous' Suit," *SNL FERC Power Report,* September 10, 2014; Kate Galbraith, "Sierra Club Takes Aim at Coal Plants in East Texas," *Texas Tribune,* February 10, 2013; "EPA, Activists Reject Luminant Blaming Transport Rule for Plant Closures," *Inside EPA Weekly Report,* September 16, 2011; S. C. Gwynne, "Coal Hard Facts," *Texas Monthly,* January 2007; John Javetski, "Monticello Steam Electric Station, Mount Pleasant, Texas," *Power,* August 15, 2006; Luminant, "Martin Lake," https://www.luminant.com/wp-content/uploads/2015/02/MartinLake _Facts.pdf.

5. The description of the Bowen plant is drawn from Lisa Evans et al., *State of Failure: How States Fail to Protect Our Health and Drinking Water from Toxic Coal Ash* ([Oakland, CA]: Earthjustice; [Lewisburg, WV]: Appalachian Mountain

Advocates, 2011), 4–5; Environmental Integrity Project, *Dirty Kilowatts*; David
Whitman, "Burning Atlanta," *Washington Monthly,* September 1, 2005, 17;
"Georgia Regulators' NO$_x$ Emissions Plan Will Cost Georgia Power \$840 Million,"
Utility Environment Report, January 28, 2000, 1; Southern Company, "Georgia
Power's Plant Bowen," http://naygn.org/wp-content/uploads/2013/09/Plant-Bowen
-Factsheet.pdf.

    6. The description of the Cumberland plant is drawn from Hancock v.
Train, 426 U.S. 282 (1976); Environmental Integrity Project, *Dirty Kilowatts*;
Andrew M. Ballard, "TVA to Spend up to \$5 Billion to Upgrade Pollution
Controls at Coal-Fired Power Plants," *BNA Environment Reporter,* April 15, 2011,
789; "TVA Adopts Catalytic Technology to Substantially Reduce NO$_x$ Emissions," *BNA Environment Reporter,* August 7, 1998, 761; "TVA Unveils Compliance Plan; Scrubbing for Cumberland, Gallatin Switch," *Coal Week,* October 14,
1991, 1; U.S. Department of the Interior, Bureau of Reclamation, "Navajo
Generating Station," https://www.usbr.gov/ngs/; Tennessee Valley Authority,
"Cumberland Fossil Plant," https://web.archive.org/web/20080916035936
/http://tva.gov/sites/cumberland.htm.

    7. The description of the Navajo plant is drawn from Environmental
Integrity Project, *Dirty Kilowatts*; Bryan Koenig, "Navajo President Signs Off on
New Arizona Coal Plant Lease," *Law360,* July 5, 2017; Benjamin Storrow, "Coal-
Reliant Tribes Ponder a Future without Their Power Plant," *E&E News,* April 3,
2017 (quote); Keith Goldbert, "9th Circ. Rejects Challenges to Navajo Plant Smog
Plan," *Law360,* March 20, 2017; Ryan Randazzo, "Utilities Vote to Close Navajo
Coal Plant at End of 2019," *Arizona Republic,* February 13, 2017; Jeff Stanfield,
"Tribal Environmental Groups Sue EPA over Alleged Special Deal for Navajo
Plant," *SNL Daily Coal Report,* October 30, 2015; Dean Scott, "House Passes
Measure to Block EPA on Greenhouse Gases, Other Rules," *BNA Environment
Reporter,* September 28, 2012, 2486; Dan Lowrey, "Study: Closure of Navajo
Coal-Fired Plant Would Cost Arizona Billions," *SNL Coal Report,* February 27,
2012; Ethan Howland, "Southwestern Coal-Fired Plants Facing Scrutiny This
Week," *Electric Power Daily,* February 15, 2011, 1; "Power Plant Near Grand
Canyon to Get Scrubbers to Curb Haze Problem," *BNA Environment Reporter,*
August 16, 1991, 1053; Margaret L. Knox, "Arizona Cliff Notes: Smoke Gets in
Your Eyes," *Chicago Tribune,* June 26, 1991, A6; Bob Secter and Rudy Abramson,
"Clearing the Air for All to See," *Los Angeles Times,* May 1, 1990, A1; SRP, "Navajo
Generating Station," November 29, 2017, https://www.srpnet.com/about/stations
/ngs/default.aspx.

    8. Simon Romero, "2 Industry Leaders Bet on Coal but Split on Cleaner
Approach," *New York Times,* May 28, 2006, 1.

    9. Margaret Kriz Hobson, "The Sierra Club's Burning Desire," *National
Journal,* September 5, 2009.

10. Taylor Kuykendall, "Judicial Battlefront: Courtroom Battles Targeting Coal Where It Digs," *SNL Generation Markets Week,* September 8, 2015; Molly Christian, "Coal Industry Lashes Out against Another Bloomberg Anti-Coal Donation," *SNL Daily Coal Report,* April 9, 2015; Housley Carr, "Coal Opponents Turn Focus to Second Phase of Campaign: Older, Smaller Plants," *Electric Utility Week,* September 7, 2009, 19.

11. Maya Weber, "Sierra Club Ups Investment in Blocking Natural Gas," *Platts Energy Trader,* October 3, 2016 (quote); Mark Hand, "Sierra Club's Campaigning Extends beyond Coal to 'Dirty Fuels,'" *SNL Daily Gas Report,* April 27, 2016; Jonathan Crawford, "Sierra Club Ramps Up Climate Fight against Gas-Fired Power Plants," *SNL FERC Power Report,* September 11, 2013; Andrew Engblom, "Sierra Club Plans to Oppose 'A Whole Lot More' Gas-Fired Power Plants," *SNL Electric Utility Report,* April 1, 2013; Eric Lipton, "Even in Coal Country, the Fight for an Industry," *New York Times,* May 30, 2012, A1; "Sierra Club Seeks to Move 'Beyond Natural Gas,' Drawing New Criticisms," *Clean Energy Report,* May 14, 2012.

12. Union of Concerned Scientists, *A Dwindling Role for Coal* (Cambridge, MA: Union of Concerned Scientists, 2017), 1.

13. Joseph P. Tomain, "Traditionally-Structured Electric Utilities in a Distributed Generation World," *Nova Law Review* 38 (2014): 479; Naureen S. Malik and Brian Eckhouse, "'Gas Apocalypse' Looms amid Power Plant Construction Boom," *BNA Energy & Climate Report,* May 23, 2017; Timothy Gardner, "'Bring on More Renewables,' U.S. Regulator Says as Grid Study Looms," Reuters, June 27, 2017.

2. ELECTRICITY, POWER PLANTS, AND THE ENVIRONMENT

1. U.S. Energy Information Administration, "Use of Electricity," https://www .eia.gov/energyexplained/index.php?page=electricity_use.

2. Daniel Yergin, *The Quest: Energy, Security, and Remaking of the Modern World* (New York: Penguin, 2011), 350–52; Hari M. Osofsky and Hannah J. Wiseman, "Dynamic Energy Federalism," *Maryland Law Review* 72 (2013): 787.

3. Ari Peskoe, "A Challenge for Federalism: Achieving National Goals in the Electricity Industry," *Missouri Environmental Law and Policy Review* 18 (2011): 212–14; Sidney A. Shapiro and Joseph P. Tomain, "Rethinking Reform of Electricity Markets," *Wake Forest Law Review* 40 (2005): 505–6.

4. U.S. Energy Information Administration, "Electric Power Industry Overview 2007," http://www.eia.gov/electricity/archive/primer/; Uma Outka, "Cities and the Low-Carbon Grid," *Environmental Law* 46 (2016): 112–13, 119.

5. Ted Nace, *Climate Hope: On the Front Lines of the Fight against Coal* (San Francisco: CoalSwarm, 2010), 66; America's Electric Cooperatives, "Powering America," http://electric.coop/our-mission/powering-america/.

6. U.S. Energy Information Administration, "Electric Power Industry Overview."

7. International Energy Agency, "2008 Energy Balance for United States," http://www.iea.org/stats/balancetable.asp?COUNTRY_CODE=US; Osofsky and Wiseman, "Dynamic Energy Federalism," 783–84.

8. Richard J. Campbell, "Increasing the Efficiency of Existing Coal-Fired Power Plants," Congressional Research Service, December 20, 2013, 3–4; World Coal Association, "What Is Coal?," http://www.worldcoal.org/coal/what-is-coal/.

9. U.S. Congress, Office of Technology Assessment, *The Direct Use of Coal* (Washington, DC: U.S. Government Printing Office, 1979), 62; U.S. Geological Survey, "Coal Fields of the Contiguous United States, 1996," http://pubs.usgs.gov /of/1996/of96-092/other_files/us_coal.pdf.

10. U.S. Congress, Office of Technology Assessment, *Direct Use of Coal*, 7–8, 53; U.S. Geological Survey, "Mercury in U.S. Coal—Abundance, Distribution, and Modes of Occurrence," September 2001, http://pubs.usgs.gov/fs/fs095-01/fs095-01 .html; Osofsky and Wiseman, "Dynamic Energy Federalism," 782.

11. U.S. Energy Information Administration, "What Is U.S. Electricity Generation by Energy Source?," http://www.eia.gov/tools/faqs/faq.cfm?id=427&t =3; Robert Swanekamp, "Fuel Management: Natural Gas / Fuel Oil," *Power,* January 1996, 11 (quote).

12. U.S. Government Accountability Office, *Information on Shale Resources, Development, and Environmental and Public Health Risks* (Washington, DC, September 2012), 5, 25; Union of Concerned Scientists, "How Natural Gas Works," http://www.ucsusa.org/clean_energy/our-energy-choices/coal-and-other-fossil -fuels/how-natural-gas-works.html; John M. Golden and Hannah J. Wiseman, "The Fracking Revolution: Shale Gas as a Case Study in Innovation Policy," *Emory Law Journal* 64 (2015): 964–68; Swanekamp, "Fuel Management," 11.

13. U.S. Department of Energy, *More Economical Sulfur Removal for Fuel Processing Plants* (Washington, DC, 2010); "Natural Gas Prices—Historical Chart," Macrotrends, https://www.macrotrends.net/2478/natural-gas-prices-historical -chart; Union of Concerned Scientists, "How Natural Gas Works."

14. Nuclear Energy Institute, "U.S. Nuclear Power Plants," https://www.nei .org/Knowledge-Center/Nuclear-Statistics/US-Nuclear-Power-Plants.

15. U.S. Department of Energy, *Hydropower Vision* (Washington, DC, March 2017), 2; U.S. Environmental Protection Agency, "Carbon Pollution Emission Guidelines for Existing Stationary Sources: Electric Utility Generating Units," 80 Fed. Reg. 64662, 64803 (October 23, 2015); Tiffany Hsu, "Powering Up

with Landfill Methane," *Los Angeles Times,* August 14, 2010, B1; Craig D. Reber, "Making the Switch," *Dubuque (IA) Telegraph Herald,* September 28, 2008, A10.

16. Campbell, "Increasing the Efficiency," 3–6; Steven E. Kuehn, "Power for the Industrial Age: A Brief History of Boilers," *Power Engineering,* February 1, 1996, 15; Melissa Powers, "The Cost of Coal: Climate Change and the End of Coal as a Source of 'Cheap' Electricity," *University of Pennsylvania Journal of Business Law* 12 (2010): 421.

17. Campbell, "Increasing the Efficiency," 6; Kuehn, "Power for the Industrial Age," 15; Powers, "Cost of Coal," 422.

18. U.S. Energy Information Administration, "Electric Grid Operators Forecast Load Shapes to Plan Electricity Supply," *Today in Energy,* July 22, 2016, http://www.eia.gov/todayinenergy/detail.php?id=27192.

19. Shapiro and Tomain, "Rethinking Reform," 508; U.S. Energy Information Administration, "Electric Power Industry Overview."

20. Richard A. Drom and Christian D. McMurray, "Maintaining System Reliability: Responding to the Retirement of Coal-Fueled Electric Generation Resources," *Energy Law Journal* 34 (2013): 591–92; Powers, "Cost of Coal," 413–14.

21. 16 U.S.C.§ 824e; U.S. Energy Information Administration, "Electric Power Industry Overview"; Alexandra B. Klass and Elizabeth J. Wilson, "Interstate Transmission Challenges for Renewable Energy: A Federalism Mismatch," *Vanderbilt Law Review* 65 (2012): 1814–15; Osofsky and Wiseman, "Dynamic Energy Federalism," 803–4; Jim Rossi and Thomas Hutton, "Federal Preemption and Clean Energy Floors," *North Carolina Law Review* 91 (2013): 1317.

22. Timothy P. Duane, "Regulation's Rationale: Learning from the California Energy Crisis," *Yale Journal on Regulation* 19 (2002): 479–81; Jim Rossi, "The Electric Deregulation Fiasco: Looking to Regulatory Federalism to Promote a Balance between Markets and the Provision of Public Goods," *Michigan Law Review* 100 (2002): 1775–76; Shapiro and Tomain, "Rethinking Reform," 509, 513.

23. Public Utility Regulatory Policies Act of 1978, §§ 202(c), 205, 207, Pub. L. 95–617, 91 Stat. 3117 (1978); U.S. Energy Information Administration, "Electric Power Industry Overview"; Shapiro and Tomain, "Rethinking Reform," 510; Jacqueline Lang Weaver, "Can Energy Markets Be Trusted? The Effect of the Rise and Fall of Enron on Energy Markets," *Houston Business and Tax Law Journal* 4 (2004): 14–15; Osofsky and Wiseman, "Dynamic Energy Federalism," 788.

24. Federal Energy Regulatory Commission, "Regional Transmission Organizations (RTO) / Independent System Operators (ISO)," available at https://www.ferc.gov/industries/electric/indus-act/rto.asp; Klass and Wilson, "Interstate Transmission Challenges," 1843–44; Drom and McMurray, "Maintaining System Reliability," 592–93, 600; Marc B. Mihaly, "Recovery of a Lost Decade (or Is it Three?): Developing the Capacity in Government Necessary to

Reduce Carbon Emissions and Administer Energy Markets," *Oregon Law Review* 88 (2009): 453–54.

25. Larry B. Parker and John E. Blodgett, "Air Quality and Electricity: Enforcing New Source Review," Congressional Research Service, January 31, 2000, 12; William Boyd and Ann E. Carlson, "Accidents of Federalism: Ratemaking and Policy Innovation in Public Utility Law," *UCLA Law Review* 63 (2016): 836–38; Klass and Wilson, "Interstate Transmission Challenges," 1807; Peskoe, "Challenge for Federalism," 242.

26. U.S. Energy Information Administration, "Electric Power Industry Overview."

27. Elesha Simeonov, "Just Not Reasonable: What the FERC's Order on Demand Response Compensation Reveals about the Current Shortfalls in 'Just and Reasonable' Rulemaking," *Temple Journal of Science, Technology and Environmental Law* 31 (2012): 314–15; Weaver, "Can Energy Markets Be Trusted?," 123–24.

28. Henry Lee, "Assessing the Challenges Confronting Distributive Electric Generation," *Electricity Journal,* June 2003; Richard L. Revesz and Burcin Unel, "Managing the Future of the Electricity Grid: Distributed Generation and Net Metering," *Harvard Environmental Law Review* 41 (2017): 45; Shapiro and Tomain, "Rethinking Reform," 518.

29. U.S. Energy Information Administration, "Electric Power Industry Overview."

30. Peskoe, "Challenge for Federalism," 245–46, 264; Hydropower Reform Coalition, "Renewable Portfolio Standards (RPS)," https://www.hydroreform.org /policy/rps.

31. Marc K. Landy, Marc J. Roberts, and Stephen R. Thomas, *The Environmental Protection Agency: Asking the Wrong Questions* (New York: Oxford University Press, 1990), 24.

32. The North American Electric Reliability Corporation defines a "reliable bulk power system" as "one that is able to meet the electricity needs of end-use customers even when unexpected equipment failures or other factors reduce the amount of available electricity." How Does NERC Define Reliability?," NERC FAQs, November 2012, https://www.nerc.com/AboutNERC/Documents/FAQs _DEC12.pdf).

33. Christopher E. Van Atten et al., *Benchmarking Air Emissions of the 100 Largest Electric Power Producers in the United States, July 2016* (Concord, MA: M. J. Bradley & Associates, 2016), 3, available at https://www.nrdc.org/sites/default /files/benchmarking-air-emissions-2016.pdf; Oliver Millman, "Vehicles Are Now America's Biggest CO_2 Source but EPA Is Tearing Up Regulations," *Guardian,* January 1, 2018.

34. U.S. Environmental Protection Agency, "National Ambient Air Quality Standards for Ozone; Proposed Rule," 72 Fed. Reg. 37818, 37823 (2007).

35. James J. Schlesselman, "'Proof' of Cause and Effect in Epidemiologic Studies: Criteria for Judgment," *Preventive Medicine* 16 (1987): 197, 200–201.

36. U.S. Environmental Protection Agency, *Integrated Science Assessment for Particulate Matter* (December 2009).

37. U.S. Environmental Protection Agency, *Integrated Science Assessment for Sulfur Oxides—Health Criteria* (September 2008).

38. U.S. Environmental Protection Agency, *Integrated Science Assessment for Oxides of Nitrogen—Health Criteria* (July 2008); Byron Swift, "How Environmental Laws Work: An Analysis of the Utility Sector's Response to Regulation of Nitrogen Oxides and Sulfur Dioxide under the Clean Air Act," *Tulane Environmental Law Journal* 14 (2001): 353.

39. U.S. Environmental Protection Agency, *Air Quality Criteria for Ozone and Related Photochemical Oxidants* (February 2006).

40. U.S. Environmental Protection Agency, *Integrated Science Assessment for Oxides of Nitrogen and Sulfur—Ecological Criteria* (December 2008).

41. U.S. Environmental Protection Agency, *Integrated Science Assessment for Particulate Matter*, 2–27; William C. Malm, *Introduction to Visibility* (Fort Collins, CO: Cooperative Institute for Research in the Atmosphere, NPS Visibility Program, Colorado State University, 1999), 10–18.

42. *Review of Mercury Pollution's Impacts to Public Health and the Environment: Hearing Before the Subcommittee on Clean Air and Nuclear Safety and the Committee on Environmental and Public Works,* U.S. Senate, 112th Cong., 2nd sess., April 17, 2012 (statement of Jerome Paulson, American Academy of Pediatrics); Jonathan Skinner, "Myths of Coal's Clean Future: The Story of Methylmercury," *Virginia Environmental Law Journal* 29 (2011): 176–78.

43. Intergovernmental Panel on Climate Change, *Climate Change 2007: The Physical Science Basis; Summary for Policymakers* (Geneva: IPCC, 2007); Ross Gelbspan, *The Heat Is On* (Reading, MA: Addison-Wesley, 1997); Jeremy Knee, "Rational Electricity Regulation: Environmental Impacts and the 'Public Interest,'" *West Virginia Law Review* 113 (2011): 741–42; Amy Sinden, "Climate Change and Human Rights," *Journal of Land, Resources and Environmental Law* 27 (2007): 255; "Methane Emitted from Natural Gas Activities Might Offset Climate Benefits, Group Says," *BNA Environment Reporter,* June 7, 2013, 1674.

44. U.S. Environmental Protection Agency, "Hazardous and Solid Waste Management System; Identification and Listing of Special Wastes; Disposal of Coal Combustion Residuals from Electric Utilities; Proposed Rule," 75 Fed. Reg. 35128, 35133–39 (2010); U.S. Environmental Protection Agency, *Steam Electric Power Generating Point Source Category: Final Detailed Study Report* (October 2009); Deborah Elcock and Nancy L. Ranek, *Coal Combustion Waste Management at Landfills and Surface Impoundments, 1994–2004* (Washington, DC: U.S. Department of Energy; U.S. Environmental Protection Agency, 2006).

45. Ellen Baum, *Wounded Waters: The Hidden Side of Power Plant Pollution* (Boston: Clean Air Task Force, 2004), 6.

46. U.S. Congress, Office of Technology Assessment, *Direct Use of Coal*, 93.

47. Kennecott Copper Corp. v. Train, 526 F.2d 1149 (9th Cir. 1975).

48. Ibid.

49. "Many Factors Must Be Evaluated When Considering a Switch or Blend of Western Coals," *Power Engineering,* August 1, 1999, 18.

50. U.S. Department of Energy, "Cleaning Up Coal," August 13, 2010, https://www.energy.gov/articles/cleaning-coal.

51. Clean Air Task Force, IGCC, http://www.fossiltransition.org/pages /gasification_carbon_capture/19.php.

52. U.S. Congress, Office of Technology Assessment, *Direct Use of Coal,* 12, 94–95.

53. Sierra Club v. Costle, 657 F.2d 298, 323, n. 69 (D.C. Cir. 1981); "Scrubber Myths and Realities," *Power Engineering,* January 1, 1995, 35; National Lime Association, "Flue Gas Desulfurization," https://www.lime.org/lime-basics/uses-of -lime/environmental/flue-gas-desulfurization/.

54. Filip Johnsson, "Fluidized Bed Combustion for Clean Energy," The 12th International Conference on Fluidization—New Horizons in Fluidization Engineering (2007), http://dc.engconfintl.org/cgi/viewcontent.cgi?article =1134&context=fluidization_xii.

55. U.S. Environmental Protection Agency, *Nitrogen Oxides (NO$_x$), Why and How They Are Controlled* (November 1999), 15; U.S. Environmental Protection Agency, Air Pollution Control Technology Fact Sheet, Selective Catalytic Reduc- tion (SCR), http://www.epa.gov/ttncatc1/dir1/fscr.pdf; John Macphail and Les King, "New Laws Prompt Focus on Low NO$_x$ Options," *Modern Power Systems,* November 1999, 29.

56. U.S. Environmental Protection Agency, *Mercury Study: Report to Congress* (December 1997), vol. 8, pp. 2.26, 2.46–49; Frank B. Meserole et al., Modeling Mercury Removal by Sorbent Injection," *Journal of the Air and Waste Management Association* 49, no. 6 (1999): 694–704.

57. U.S. Environmental Protection Agency, "Frequent Questions about the 2015 Coal Ash Disposal Rule," https://www.epa.gov/coalash/frequent-questions -about-coal-ash-disposal-rule#3.

58. U.S. Environmental Protection Agency, "Carbon Pollution Emission Guidelines," 64,689.

59. Peter Folger, *Carbon Capture and Sequestration (CCS) in the United States,* Congressional Research Service, August 9, 2018; Powers, "Cost of Coal," 422–23.

3. FIRST STEPS

1. Devra Davis, *When Smoke Ran Like Water: Tales of Environmental Deception and the Battle Against Pollution* (New York: Basic Books, 2002), chap. 1; Charles O. Jones, *Clean Air: The Policies and Politics of Pollution Control* (Pittsburgh: University of Pittsburgh Press, 1975), 25–27.

2. Congressional Research Service, "Environmental Impacts of the President's Energy Plan," September 13, 1977, 19; J. Clarence Davies III, *The Politics of Pollution* (New York: Pegasus, 1970), 167; John C. Esposito, *Vanishing Air* (New York: Grossman, 1970), 95–96, 104–8; Jones, *Clean Air,* 79, 128.

3. Alfred A. Marcus, *Promise and Performance: Choosing and Implementing an Environmental Policy* (Westport, CT: Greenwood Press), 85–87; John C. Whitaker, *Striking a Balance: Environment and Natural Resources Policy in the Nixon-Ford Years* (Washington, DC: American Enterprise Institute for Public Policy Research, 1976), 13; "William D. Ruckelshaus: Oral History Interview," EPA Web Archive, January 1993, https://archive.epa.gov/epa/aboutepa/william-d -ruckelshaus-oral-history-interview.html; Richard Corrigan, "EPA Ending Year-Long Shakedown Cruise; Ruckelshaus Cast as Embattled Spokesman," *National Journal,* October 9, 1971, 2039.

4. Jones, *Clean Air,* 181; Marc K. Landy, Marc J. Roberts, and Stephen R. Thomas, *The Environmental Protection Agency: Asking the Wrong Questions* (New York: Oxford University Press, 1990), 30; Richard J. Lazarus, *The Making of Environmental Law* (Chicago: University of Chicago Press, 2004), 75–77; John Quarles, *Cleaning Up America: An Insider's View of the Environmental Protection Agency* (Boston: Houghton Mifflin, 1976), 78; Whitaker, *Striking a Balance,* 93.

5. Jones, *Clean Air,* 189–93, 207–8; Whitaker, *Striking a Balance,* 93; Richard Corrigan, "Muskie Plays Dominant Role in Writing Tough New Air Pollution Law," *National Journal,* January 2, 1971, 25.

6. 42 U.S.C. §§ 7408–12, 7604; S. Rept. No. 91–1196 10 (1970).

7. Marcus, *Promise and Performance,* 87–90.

8. Philip Shabecoff, *A Fierce Green Fire: The American Environmental Movement* (New York: Hill and Wang, 1993), 116–17; Robert E. Taylor, "Group's Influence on U.S. Environmental Laws, Policies Earns It a Reputation as a Shadow EPA," *Wall Street Journal,* January 13, 1986; James R. Wagner, 'Washington Pressures / Environment Groups Shift Tactics from Demonstrations to Politics, Local Action," *National Journal,* July 24, 1971, 1557.

9. U.S. Environmental Protection Agency, "National Ambient Air Quality Standards," 35 Fed. Reg. 8186 (1971); "EPA Proposes Ambient Air Standards, Publishes Nitrogen Oxides Criteria," *BNA Environment Reporter,* May 8, 1971, 1065.

10. Thomas O. McGarity, *Reinventing Rationality: The Role of Regulatory Analysis in the Federal Bureaucracy* (Cambridge: Cambridge University Press, 1991); "Memorandum for the Heads of Departments and Agencies," from George P. Shultz, Director, Office of Management and Budget, October 5, 1971, reprinted in *Clean Air Act Oversight: Hearings before the Subcommittee on Public Health and Environment of the House Committee on Interstate and Foreign Commerce,* 92nd Cong. 99 (1971); "EPA Receives Attack of Agencies of Ambient Air Quality Standards," *BNA Environment Reporter,* June 11, 1971, 157.

11. Kennecott Copper Corp. v. EPA, 462 F.2d 846, 847–49 (D.C. Cir. 1972); "EPA Revokes Secondary Sulfur Dioxide, Ambient Air Quality Criteria," *Environmental Law Reporter* 3 (1973): 10159.

12. "EPA Receives Critical Comments on Proposed Implementation Plans," *BNA Environment Reporter,* July 16, 1971, 306; "Ruckelshaus Calls for Every Effort to Comply with Clean Air Standards," *BNA Environment Reporter,* July 2, 1971, 239; "Proposed Rules for State Plans Upheld by Defense Counsel, Opposed by Others," *BNA Environment Reporter,* May 21, 1971, 65.

13. *Implementation of the Clean Air Act Amendments of 1970—Part 1: Hearings before the Subcommittee on Air and Water Pollution of the Senate Committee on Public Works,* 92nd Cong. (1972) (testimony of Richard Ayres); interview with Richard Ayres, O'Melveny & Myers, Washington, DC, April 11, 1994 (quote).

14. Buckeye Power, Inc. v. EPA, 481 F.2d 163 (8th Cir. 1973); *Implementation of the Clean Air Act . . . Hearings,* 28, 31 (testimony of Richard Ayres), 171 (testimony of Frank Partee), 197 (cover letter to the Ohio SIP); James A. Noone, "Doubts about 'Clean' Fuels Fail to Deter EPA, States on Air Pollution Battle Plans," *National Journal,* June 24, 1972, 1050 (quoting Don Sorrels, Colorado Department of Health); "Ruckelshaus Announces EPA Approval of State Plans to Meet Air Standards," *BNA Environment Reporter,* June 2, 1972, 123; "Implementation Plans to EPA," *BNA Environment Reporter,* February 4, 1972, 1205.

15. Union Electric Co. v. EPA, 427 U.S. 246 (1976).

16. Big Rivers Electric Corp. v. EPA, 523 F.2d 16 (6th Cir. 1975); Natural Resources Defense Council, Inc. v. EPA, 489 F.2d 390 (5th Cir. 1974); "Tightening Compliance Policy Could Reduce Sulfur Dioxide Emissions but Cost to Midwest Would Be High, State, EPA, Industry Officials Say," *BNA Environment Reporter,* March 16, 1984, 2059 (Gavin plant); James A. Noone, "Great Scrubber Debate Pits EPA Against Electric Utilities," *National Journal,* July 27, 1974, 1103.

17. Dick Kirschten, "The New Clean Air Regs—More at Stake Than Breathing," *National Journal,* September 2, 1978, 1392; "New Process Makes Coal Come Clean," *Chemical Week,* January 7, 1976, 33; Noone, "Doubts."

18. Robert L. Sansom, *The New American Dream Machine* (Garden City, NY: Anchor Press, 1976), 80, 83; U.S. Congress, Office of Technology Assessment, *The*

Direct Use of Coal (Washington, DC: U.S. Government Printing Office, 1979), 170–72; Noone, "Great Scrubber Debate."

19. 42 U.S.C. § 7401(b); Fri v. Sierra Club, 412 U.S. 541 (1973).

20. 42 U.S.C. § 1857c-6(a)(1), (b) (c) (1971) [repealed].

21. United States v. Ohio Edison, 276 F. Supp. 2d 829, 836 (S.D. Ohio 2003); U.S. Environmental Protection Agency, "Rules and Regulations," 36 Fed. Reg. 24876, 24877 (1971).

22. Sierra Club v. Costle, 657 F.2d 298, 312 n. 6 (D.C. Cir. 1981).

23. Bruce A. Ackerman and William T. Hassler, "Beyond the New Deal: Coal and the Clean Air Act," *Yale Law Journal* 89 (1980): 1483–84; "Coming Government Moves in War against Pollution: Interview with William D. Ruckelshaus, Administrator, Environmental Protection Agency," *U.S. News & World Report,* March 29, 1971, 74.

24. U.S. Environmental Protection Agency, "Standards of Performance for New Stationary Sources: Proposed Standards for Five Categories," 36 Fed. Reg. 15704 (August 17, 1971).

25. Oljato Chapter of the Navajo Tribe v. Train, 515 F.2d 654, 656 (D.C. Cir. 1975) (lawyers' explanation); U.S. Environmental Protection Agency, "Standards of Performance," 15706; Ackerman and Hassler, "Beyond the New Deal," 1486–87.

26. U.S. Environmental Protection Agency, "Standards of Performance for New Stationary Sources," 36 Fed. Reg. 24876 (December 23, 1971).

27. Ackerman and Hassler, "Beyond the New Deal," 1482; Elizabeth H. Haskell, *The Politics of Clean Air* (New York: Praeger, 1982), 9.

28. Essex Chemical Corp. v. Ruckelshaus 486 F.2d 427, 440–42 (D.C. Cir. 1973).

29. Yanek Mieczkowski, *Gerald Ford and the Challenges of the 1970s* (Lexington: University of Kentucky Press, 2005), 199–200; John F. Burby, "EPA's New Team Girds for Stubborn Fight to Fulfill Mandate," *National Journal,* April 13, 1974, 533 (quoting John J. Coffey Jr., U.S. Chamber of Commerce); Sansom, *Dream Machine,* 81–82; "Clean Air Act Could Affect Economy, Military Security, NCA Tells Senate," *BNA Environment Reporter,* April 14, 1972, 1512 (quoting Carl E. Bagge, president, National Coal Association); "Environmental Regulations Crippling Utility Industry, EEI President Says," *BNA Environment Reporter,* January 21, 1972, 1150.

30. John F. Burby, "White House, Activists Debate Form of Sulfur Tax; Industry Shuns Both," *National Journal,* October 21, 1972, 1643 (quote); "Administration Proposes Bill to Tax Sulfur at 15 Cents per Pound Emitted," *BNA Environment Reporter,* February 11, 1972, 1226.

31. Landy, Roberts, and Thomas, *Environmental Protection Agency,* 38; Quarles, *Cleaning Up America,* 199.

32. Landy, Roberts, and Thomas, *Environmental Protection Agency,* 38; Quarles, *Cleaning Up America,* 198–99; Sansom, *Dream Machine,* 44; Juan

Williams, "Ruckelshaus Sworn in as EPA Head as President Hails a 'New Chapter,'" *Washington Post*, May 19, 1983, A3.

33. Mieczkowski, *Gerald Ford*, 203; Sansom, *Dream Machine*, 46; Gladwin Hill, "Chief of E.P.A. Defines Policy on Coal Emissions," *New York Times*, November 13, 1974, 85; John F. Burby, "Sulfates Present Major New Problem in Growing Debate over Clean Air Act," *National Journal*, September 22, 1973, 1412.

34. Jones, *Clean Air*, 311; Lazarus, *Making of Environmental Law*, 78; Quarles, *Cleaning Up America*, 141; Sansom, *Dream Machine*, 48–49; Whitaker, *Striking a Balance*, 106.

35. Whitaker, *Striking a Balance*, 26, 107–10; E. W. Kenworthy, "Donald Cook vs. E.P.A.," *New York Times*, November 24, 1974, 169.

36. Energy Supply and Environmental Coordination Act, Pub. L. 93–319, 88 Stat. 246 (1974); Reginald Stuart, "Coal's Uncertain Role in Energy Shortage," *New York Times*, September 1, 1974, 113; Noone, "Great Scrubber Debate," 1103.

37. "Tall Stacks versus Scrubbers: $3.5 Million Publicity Campaign Fails to Discredit Emission Reduction Technology," *Environmental Law Reporter* 5 (January 1975): 10009; Edward Cowan, "Way to Remove Gases Is Backed," *New York Times*, September 26, 1974, 39; Noone, "Great Scrubber Debate."

38. "Tall Stacks versus Scrubbers"; Edward Cowan, "Paying for Cleaner Air," *New York Times*, July 21, 1974, 124; James G. Phillips, "Environmentalists, Utilities Argue over East, West Mining," *National Journal*, July 6, 1974, 1014.

39. U.S. Environmental Protection Agency, "Requirements for Preparation, Adaption, and Submittal of Implementation Plans," 41 Fed. Reg. 55524 (December 20, 1976); "EPA Lists States with Inadequate Implementation Plans," *Environmental Law Reporter* 6 (August 1976): 10181; "A Turn in the Tide— Pollution Battle Being Won?," *U.S. News & World Report*, August 4, 1975, 56; Gladwin Hill, "Air Pollution Drive Lags, but Some Gains Are Made," *New York Times*, May 31, 1975, 1; E. W. Kenworthy, "E.P.A. Head Assays Gains in Clean Air," *New York Times*, May 31, 1975, 15; Reginald Stuart, "Utilities' Compliance with Clean-Air Standards Proceeds Slowly," *New York Times*, April 14, 1975, 62.

40. "What's Causing Those Smog Blankets," *U.S. News & World Report*, August 18, 1975, 21.

41. Bayard Webster, "City Ozone Found to Move Up Coast," *New York Times*, January 15, 1976, 14.

42. U.S. Congress, Office of Technology Assessment, *Direct Use of Coal*, 7–8.

43. "The EPA Is in a Bind in West Virginia," *Business Week*, September 6, 1976, 30; "Where Murky Air Means Prosperity," *Business Week*, February 2, 1976, B10.

44. Quarles, *Cleaning Up America*, 172, 174.

45. Robert D. Hershey Jr., "Fuel Conversion Stirs Conflicts," *New York Times*, November 20, 1980, D1.

4. MINOR ADJUSTMENTS, MAJOR CONTROVERSY

1. Marc K. Landy, Marc J. Roberts, and Stephen R. Thomas, *The Environmental Protection Agency: Asking the Wrong Questions* (New York: Oxford University Press, 1990), 39; Philip Shabecoff, *A Fierce Green Fire: The American Environmental Movement* (New York: Hill and Wang, 1993), 203–4; David Vogel, *Fluctuating Fortunes: The Political Power of Business in American* (New York: Basic Books, 1989), 148, 229; Bruce A. Ackerman and William T. Hassler, "Beyond the New Deal: Coal and the Clean Air Act," *Yale Law Journal* 89 (1980): 1466, 1537; Lynn Langway, The EPA's New Man, *Newsweek,* February 21, 1977, 80; Gladwin Hill, "Conservationists Expecting Carter to Open New Era for Environment," *New York Times,* November 5, 1976, 15; History Central, "1976 Elections: Carter vs. Ford," http://www.historycentral.com/elections/1976 .html.

2. Jimmy Carter, "The Environment—The President's Message to the Congress," *Environmental Law Reporter* 7 (1977): 50057.

3. Richard A. Harris and Sidney M. Milkis, *The Politics of Regulatory Change: A Tale of Two Agencies* (New York: Oxford University Press, 1989), 250; Landy, Roberts, and Thomas, *Environmental Protection Agency,* 40.

4. U.S. Congress, Office of Technology Assessment, *The Direct Use of Coal* (Washington, DC: U.S. Government Printing Office, 1979), 190, 192; Lawrence Mosher, "Acid Rain Fallout Threatens Subsidies for Utilities That Convert to Coal," *National Journal,* May 3, 1980, 716; Philip Shabecoff, "Effort to Soften Coal Seems Likely," *New York Times,* November 18, 1980, D1.

5. Executive Office of the President, Energy Policy and Planning, *The National Energy Plan* (Washington, DC: U.S. Government Printing Office, 1977), xix–xx; *Coal Conversion Legislation: Hearings before the Subcommittee on Energy Production and Supply of the Senate Committee on Energy and Natural Resources,* 95th Cong. (March 21 and 28, 1977) (testimony of John F. O'Leary, Administrator, Federal Energy Administration) [hereinafter *Senate Coal Conversion Legislation Hearings*]; Jimmy Carter, "The President's Proposed Energy Policy" (address delivered April 18, 1977), *Vital Speeches* 43, no. 14 (May 1977), 418; Jack McWethy, "Carter's Bid: Both Energy and Clean Air," *U.S. News & World Report,* May 2, 1977, 18.

6. Pub. L. 95–95, 91 Stat. 685 (1977); on the poll, see *Clean Air Act Amendments of 1977: Hearings before the Subcommittee on Health and the Environment of the House Committee on Interstate and Foreign Commerce,* pt. 2, 95th Cong. 1341 (1977) [hereinafter *House Clean Air Act Amendments of 1977 Hearings*] (testimony of Richard E. Ayers, NRDC).

7. *Clean Air Act Amendments of 1977: Hearing before the Subcommittee on Environmental Pollution of the Senate Committee on Environment and Public*

Works, pt. 2, 95th Cong. (1977) [hereinafter *Senate Clean Air Act Amendments of 1977 Hearings*] 26 (testimony of Donald G. Allen, Edison Electric Institute), 462 (testimony of Otes Bennett Jr., National Coal Association), 151 (testimony of Carl M. Shy, American Lung Association) and 113 (testimony of Charlie Grimm, United Mine Workers); *House Clean Air Act Amendments of 1977 Hearings,* 1372–73 (testimony of A. Joseph Dowd, Edison Electric Institute).

 8. 42 U.S.C. § 7607(d)(3).

 9. *Senate Clean Air Act Amendments of 1977 Hearings,* 201 (testimony of Richard E. Ayres, NRDC); *House Clean Air Act Amendments of 1977 Hearings,* 1403 (testimony of A. Joseph Dowd, Edison Electric Institute); "Carter Takes Compromise Stand on Clean Air Act," *National Journal,* April 23, 1977, 641.

 10. Pub. L. 95–95, 91 Stat. 685 (1977); "The Clean Air Act Amendments of 1977: Expedient Revisions, Noteworthy New Provisions," *Environmental Law Reporter* 7 (October 1977): 10182.

 11. *Senate Clean Air Act Amendments of 1977 Hearings,* 1118 (testimony of J. William Haun, National Association of Manufacturers), 113 (testimony of Charlie Grimm, United Mine Workers), 198–99 (testimony of Richard E. Ayres, NRDC); *House Clean Air Act Amendments of 1977 Hearings,* 1034 (testimony of Edward F. Weber Jr., U.S. Chamber of Commerce), 1536 (answers to supplemental questions by Joseph Dowd).

 12. Pub. L. 95–95, 91 Stat. 685 (1977) § 123.

 13. *House Clean Air Act Amendments of 1977 Hearings,* 1584 (testimony of Sid Oren, Industrial Gas Cleaning Institute); Ackerman and Hassler, "Beyond the New Deal," 1488; Elizabeth H. Haskell, *The Politics of Clean Air* (New York: Praeger, 1982), 12; "Carter Takes Compromise Stand on Clean Air Act," *National Journal,* April 23, 1977, 641.

 14. *Senate Clean Air Act Amendments of 1977 Hearings,* 226 (testimony of Donald G. Allen, Edison Electric Institute), 472 (testimony of Otes Bennett Jr., National Coal Association).

 15. *Senate Coal Conversion Legislation Hearings,* 3 (statement of Senator Howard Metzenbaum, D-OH); J. Dicken Kirschten, "Converting to Coal—Can It Be Done Cleanly?," *National Journal,* May 21, 1997, 781.

 16. Pub. L. 95–95, 91 Stat. 685 (1977) § 109; H.R. Rep. No. 564, 95th Cong. 130 (1977); Ackerman and Hassler, "Beyond the New Deal," 1507.

 17. Maria H. Grimes, *Summary of Testimony on Provisions of the Proposed Clean Air Act Dealing with Attainment of Ambient Air Quality Standards, Control of Emissions from Stationary Sources, and Other Related Subjects* (Washington, DC: Congressional Research Service, 1977), v, xviii; *Senate Clean Air Act Amendments of 1977 Hearings,* 1123 (testimony of J. William Haun, National Association of Manufacturers).

18. *Senate Clean Air Act Amendments of 1977 Hearings,* 476–77 (testimony of Otes Bennett Jr., National Coal Association); *House Clean Air Act Amendments of 1977 Hearings,* 485 (testimony of Nita Molyneaux, National Parks and Conservation Association), 1679 (testimony of Douglas W. Costle, Administrator, EPA), 1087–88 (testimony of Frank R. Millikan, Business Roundtable).

19. Pub. L. 95–95, 91 Stat. 685 (1977) § 127.

20. Ibid. § 128.

21. Ibid.

22. Ibid. § 107.

23. *Senate Clean Air Act Amendments of 1977 Hearings,* 40–41 (testimony of Donald G. Allen, Edison Electric Institute).

24. Congressional Research Service, "Environmental Impacts of the President's Energy Plan," September 13, 1977, 30–31.

25. Haskell, *Politics of Clean Air,* 21; Warren Brown, "Pollution's Cost Tops Controls, Costle Asserts," *Washington Post,* April 28, 1978, F2; Ward Sinclair, "Environmentalists Angry at Strauss," *Washington Post,* April 21, 1978, A2; Timothy B. Clark, "Carter's Assault on the Costs of Regulation," *National Journal,* August 12, 1978, 1281.

26. Library of Congress, *Summary of Department of Energy Organization Act of 1977,* July 27, 1977.

27. Natural Gas Policy Act, Pub. L. 95–621, 95th Cong. (1978); Powerplant and Industrial Fuel Use Act, Pub. L. 95–620, 95th Cong. (1978); Public Utility Regulatory Policies Act, Pub. L. 95–617, 95th Cong. (1978).

28. Janet Gellici, "The Road *Recently* Less Traveled: Public Policy Influences on Coal's Path Forward," *Electricity Journal,* March 2013, 32.

29. U.S. Environmental Protection Agency, "1977 Clean Air Act Amendments to Prevent Significant Deterioration, Final Rule," 43 Fed. Reg. 26388 (June 19, 1978).

30. Alabama Power Co. v. Costle, 636 F.2d 323, 399–402 (D.C. Cir. 1979).

31. U.S. Environmental Protection Agency, "Requirements for Preparation, Adoption, and Submittal of Implementation Plans; Approval and Promulgation of Implementation Plans," 45 Fed. Reg. 52676 (August 7, 1980).

32. U.S. Environmental Protection Agency, "Visibility Protection for Federal Class I Areas, Final Rulemaking," 45 Fed. Reg. 80084 (December 2, 1980); U.S. Environmental Protection Agency, "Visibility Protection for Federal Class I Areas, Proposed Rulemaking," 45 Fed. Reg. 34762 (May 22, 1980); "Protecting Visibility under the Clean Air Act: EPA Establishes Modest 'Phase I' Program," *Environmental Law Reporter* 11 (February 1981): 10053.

33. 42 U.S.C. § 7411(b)(2); Sierra Club v. Costle, 657 F.2d 298, 313 (D.C. Cir. 1981); Ackerman and Hassler, "Beyond the New Deal," 1509; William C. Banks, "EPA Bends to Industry Pressures on Coal NSPS—and Breaks," *Ecology Law*

Quarterly 9 (1980): 81; Dick Kirschten, "The New Clean Air Regs—More at Stake than Breathing," *National Journal,* September 2, 1978, 1392.

34. U.S. Congress, Office of Technology Assessment, *Direct Use of Coal,* 167, 172.

35. Haskell, *Politics of Clean Air,* 25–28; Ackerman and Hassler, "Beyond the New Deal," 1537.

36. Haskell, *Politics of Clean Air,* 29–31; Ackerman and Hassler, "Beyond the New Deal," 1540; Charles Mohr, "Billions at Stake as U.S. Weighs Clean-Air Rules," *New York Times,* August 2, 1978, A1.

37. *Oversight: Effect of the Clean Air Act Amendment on New Energy Technologies and Resources: Hearing before the Subcommittee on Fossil and Nuclear Energy Research, Development and Demonstration of the House Committee on Science and Technology,* 95th Cong. 7 (April 19, 26, 1978) (testimony of George Freeman, Hunton & Williams); Haskell, *Politics of Clean Air,* 37–38; Kirschten, "New Clean Air Regs."

38. Ackerman and Hassler, "Beyond the New Deal," 1547–49; Mohr, "Billions at Stake"; "Inflation Fighters Draw a Bead on the Regulators," *U.S. News & World Report,* September 18, 1978, at 27; Kirschten, "New Clean Air Regs."

39. U.S. Environmental Protection Agency, "Standards of Performance for New Stationary Sources; Electric Utility Steam Generating Units," 43 Fed. Reg. 42154, 42156 (September 19, 1978); Mohr, "Billions at Stake"; Kirschten, "New Clean Air Regs."

40. U.S. Environmental Protection Agency, "Standards of Performance for New Stationary Sources; Electric Utility Steam Generating Units."

41. Dick Kirschten, "Politics at the Heart of the Clean Air Debate," *National Journal,* May 19, 1979, 812; Dick Kirschten, "EPA Proposes Rules for Cleaner Coal Burning," *National Journal,* September 16, 1978, 1482; Kirschten, "New Clean Air Regs."

42. Richard F. Ayres and David D. Doniger, "New Source Standard for Power Plants II: Consider the Law," *Harvard Environmental Law Review* 3 (1979): 64, 70–71, 78–80.

43. Haskell, *Politics of Clean Air,* 75–76.

44. Sierra Club v. Costle, 385; Haskell, *Politics of Clean Air,* 79–80; Ackerman and Hassler, "Beyond the New Deal," 1544–45.

45. Sierra Club v. Costle, 388; Ackerman and Hassler, "Beyond the New Deal," 1550–51; Dick Kirschten, "Politics at the Heart."

46. Haskell, *Politics of Clean Air,* 89.

47. Sierra Club v. Costle, 350 n. 184 (dry scrubbing cheaper); Haskell, *Politics of Clean Air,* 91–92; Ackerman and Hassler, "Beyond the New Deal," 1553–54.

48. Haskell, *Politics of Clean Air,* 98; Margot Hornblower, "EPA Will Relax Pollution Rules for Coal Power," *Washington Post,* May 5, 1979, A1. But see

B. Drummond Ayres Jr., "Byrd Denies Tying Arms Treaty Vote to Coal Rule," *New York Times,* May 6, 1979, A27.

49. U.S. Environmental Protection Agency, "New Stationary Sources Performance Standards; Electric Utility Steam Generating Units, Final Rule," 44 Fed. Reg. 33580 (June 11, 1979); Sierra Club v. Costle, 316; Margot Hornblower, "EPA Issues Rules on Power Plant Coal Emissions," *Washington Post,* May 26, 1979, A1.

50. Sierra Club v. Costle, 347–52, 384, 408.

51. Dick Kirschten, "EPA's Ozone Standard Faces a Hazy Future," *National Journal,* December 16, 1978, p. 2015.

52. U.S. Environmental Protection Agency, "Revisions to the National Ambient Air Quality Standards for Photochemical Oxidants," 44 Fed. Reg. 8202, 8204 (February 8, 1979); Landy, Roberts, and Thomas, *Environmental Protection Agency,* 63.

53. U.S. Environmental Protection Agency, "Photochemical Oxidants: Proposed Revisions to the National Ambient Air Quality Standard," 43 Fed. Reg. 26962 (June 22, 1978); Landy, Roberts, and Thomas, *Environmental Protection Agency,* 63–64.

54. Margot Hornblower, "EPA Set to Ease Smog Standards for Urban Areas," *Washington Post,* January 21, 1979, A1; Robert Crandall and Lester Lave, "The Cost of Controlling Smog," *Washington Post,* January 6, 1979, A17; Kirschten, "Hazy Future"; "EPA Proposes to Ease Smog Standards," *Oil & Gas Journal,* June 19, 1978, 82.

55. Kirschten, "Hazy Future."

56. Landy, Roberts, and Thomas, *Environmental Protection Agency,* 67–69; Kirschten, "Hazy Future."

57. U.S. Environmental Protection Agency, Revisions; Landy, Roberts, and Thomas, *Environmental Protection Agency,* 67–68, 72–73 (cribbed quote); Christopher C. DeMuth, "The White House Review Programs," *Regulation,* January / February 1980, 13, 20 (Carter quote).

58. American Petroleum Institute v. Costle, 665 F.2d 1176, 1184–87 (D.C. Cir. 1981).

59. U.S. Congress, Office of Technology Assessment, *Direct Use of Coal,* 169; James L. Rowe Jr., "In Once-Sooty Pittsburgh, Duquesne Light Has Scrubbed Up Its Act—For $300 Million," *Washington Post,* July 16, 1978, F1.

60. Cleveland Electric Illuminating Co. v. EPA, 572 F.2d 1150, 1157 (6th Cir. 1978); U.S. Environmental Protection Agency, "Approval and Promulgation of Implementation Plans: Ohio-Sulfur Dioxide Plan," 41 Fed. Reg. 36324, 36324–25 (August 27, 1976); Gregory Wetstone, "Air Pollution Control Laws in North America and the Problem of Acid Rain and Snow," *Environmental Law Reporter* 10 (January 1980): 50001; Philip Shabecoff, "Ohio Told to Meet Clean

Air Deadline," *New York Times,* October 18, 1979, A16; Dick Kirschten, "Coal War in the East—Putting a Wall Around Ohio," *National Journal,* January 13, 1979, 50.

61. Cincinnati Gas & Electric Co. v. EPA, 578 F.2d 660 (6th Cir. 1978); U.S. Environmental Protection Agency, "Approval and Promulgation of Implementation Plans: Ohio-Sulfur Dioxide Plan," 36,324.

62. Dick Kirschten, "Coal Politics with an Eastern Tilt May Boost Carter Stock in Key States," *National Journal,* September 13, 1980, 1519; "Ohio Air Pollution Rules Are Relaxed to Save Jobs," *New York Times,* June 7, 1979, A18.

63. Emily Bamforth, "Residents Protest Avon Lake Power Plant Wastewater Permit," *Cleveland.com,* May 2, 2018, https://www.cleveland.com/metro/2018/05 /avon_lake_power_plant_wastewat.html; Kirschten, "Coal Politics"; Joanne Omang, "2 Plants Told to Cut Emissions Causing Acid Rain," *Washington Post,* June 18, 1980, A14.

64. Hancock v. Train, 426 U.S. 282 (1976); "EPA's 'Bubble Concept' Could Cut Clean Air Costs," *National Journal,* December 30, 1978, 2082; Merrill Sheils, "Updating TVA," *Newsweek,* May 22, 1978, 55; Wayne King, "T.V.A., a Major Polluter, Faces Suit to Cut Sulfur Dioxide Fumes," *New York Times,* July 4, 1977, 1; "TVA Clears the Air," *Chemical Week,* March 30, 1977, 16.

65. Pub. L. 95–95, 91 Stat. 685 (1977) §§ 108(a)(4), 123.

66. Air Pollution Control District of Jefferson County, Kentucky v. EPA, 739 F.2d 1071, 1077 (6th Cir. 1984).

67. 42 U.S.C. § 7415; Gregory Wetstone, "Air Pollution Control Laws; Air Act Provisions on Interstate Pollution Triggered by Violation of NAAQS, Court Rules," *BNA Environment Reporter,* July 20, 1984, 440; Shabecoff, "Ohio Told to Meet Clean Air Deadline."

68. "Utility Signs $500-Million Cleanup," *Engineering News-Record,* January 29, 1981, 16; "Pollution Still Plagues West Virginia, Ohio Towns," *Washington Post,* December 25, 1978, E2.

69. Congressional Research Service, "Summary of Testimony," reprinted in *Environmental Implications of the New Energy Plan: Hearings before the Subcommittee on the Environment and the Atmosphere of the House Committee on Science and Technology,* 95th Cong. 421, 425 (June 8, 9, July 21, 26, 27, and September 27, 28, 29, 1977).

70. Ibid.; Energy and Policy Institute, *Utilities Knew: Documenting Electric Utilities' Early Knowledge and Ongoing Deception on Climate Change from 1968–2017,* July 2017, 12 (quoting presidential science adviser Donald F. Hornig), https://www.energyandpolicy.org/utilities-knew-about-climate-change/; U.S. Congress, Office of Technology Assessment, *Direct Use of Coal,* 10; "The Pitfalls of Counting Too Much on Coal," *U.S. News & World Report,* April 25, 1977, 75.

71. Landy, Roberts, and Thomas, *Environmental Protection Agency,* 245; Douglas E. Kneeland, "Carter Softens His Criticism of Foe," *New York Times,* October 10, 1980, A1; Joanne Omang, "Reagan Criticizes Clean Air Laws and EPA as Obstacles to Growth," *Washington Post,* October 9, 1980, A2; Dick Kirschten, "Environmentalists Tell Carter Thanks But No Thanks," *National Journal,* June 23, 1979, 1036; History Central, "1980 Elections Carter vs. Reagan," https://www .historycentral.com/elections/1980.html.

72. Ronald A. Taylor, "In Clean-Air Drive, a Pause to Help Industry," *U.S. News & World Report,* December 15, 1980, 62; Ben Franklin, "Coal Outlook Troubled Despite High Hopes," *New York Times,* November 16, 1980, A1.

5. THE LOST DECADE

1. Marc K. Landy, Marc J. Roberts, and Stephen R. Thomas, *The Environmental Protection Agency: Asking the Wrong Questions* (New York: Oxford University Press, 1990), 245; Daniel Yergin, *The Quest: Energy, Security, and Remaking of the Modern World* (New York: Penguin, 2011), 381; Ari Peskoe, "A Challenge for Federalism: Achieving National Goals in the Electricity Industry, *Missouri Environmental Law and Policy Review* 18 (2011): 230–31 (wind generation); Robert Swanekamp, "Fuel Management: Natural Gas / Fuel Oil," *Power,* January 1996, 11 ("PURPA machines"); Philip Shabecoff, "Effort to Soften Coal Seems Likely," *New York Times,* November 18, 1980, 17.

2. Janet Gellici, "The Road *Recently* Less Traveled: Public Policy Influences on Coal's Path Forward," *Electricity Journal,* March 2013, 32; "Amendment Proposals Would Make Air Act More Difficult to Carry Out, Barber Says," *BNA Environment Reporter,* May 21, 1982, 54; "Coal Conversion Boom May Pose Clean Air Threat," *Engineering News-Record,* December 10, 1981, 13; Bruce Horovitz, "Coal's Soiled Promise," *Industry Week,* April 6, 1981, 61.

3. Horovitz, "Coal's Soiled Promise"; Lawrence Mosher, "Environmentalists Question Whether to Retreat or Stay on the Offensive," *National Journal,* December 13, 1980, 2116; Shabecoff, "Effort to Soften Coal."

4. Anne M. Burford, *Are You Tough Enough?,* with John Greenya (New York: McGraw-Hill, 1986), 68, 98; Jonathan Lash, Katherine Gillman, and David Sheridan, *A Season of Spoils: The Reagan Administration's Attack on the Environment* (New York: Pantheon Books, 1984), 5; Joanne Omang, "Nominee for EPA Voices a Credo of 'Non-Confrontation,'" *Washington Post,* May 2, 1981, A10.

5. Landy, Roberts, and Thomas, *Environmental Protection Agency,* 247–50; Thomas O. McGarity, *Reinventing Rationality: The Role of Regulatory Analysis in the Federal Bureaucracy* (Cambridge: Cambridge University Press, 1991), chap. 16.

6. Executive Order 12291, 3 C.F.R. 127 (1982); Donald J. Devine, *Reagan's Terrible Swift Sword: Reforming and Controlling the Federal Bureaucracy* (Ottawa, IL: Jameson Books, 1991), 2; George C. Eads and Michael Fix, *Relief or Reform? Reagan's Regulatory Dilemma* (Washington, DC: Urban Institute Press, 1984), 9; Cass Peterson, "Rare Criticism of OMB Draws Some Attention," *Washington Post,* July 18, 1985, A21.

7. "High Court Ruling on Utility's Emissions Comes after Settlement Already Reached," *Coal Week,* December 3, 1984, 3; "Tightening Compliance Policy Could Reduce Sulfur Dioxide Emissions but Cost to Midwest Would Be High, State, EPA, Industry Officials Say," *BNA Environment Reporter,* March 16, 1984, 2059; "Coalition Says 30-Day Averaging Proposal Would Mean Increased Sulfur Oxide Emissions," *BNA Environment Reporter,* November 18, 1983, 1314.

8. U.S. Environmental Protection Agency, "Requirements for Preparation, Adoption, and Submittal of Implementation Plans, Final Rule," 46 Fed. Reg. 50,766 (October 14, 1981); "Utility Observers Are Generally Pleased with EPA's Proposal to Redefine 'Source,'" *Electrical Week,* March 23, 1981, 6.

9. Chevron USA Inc. v. Natural Resources Defense Council Inc., 467 U.S. 837, 842–44, 860, 864 (1984).

10. Rochelle L. Stanfield, "Environmentalists Try the Backdoor Approach to Tackling Acid Rain," *National Journal,* October 19, 1985, 2365; "Effects of Schemes to Control Acid Rain Given to Senate Panel by EPA, OTA, Others," *BNA Environment Reporter,* June 11, 1982, 158; Ralph Blumenthal, "Study Finds That Acid Rain Is Intensifying in Northeast," *New York Times,* January 25, 1981, A30.

11. 42 U.S.C. § 7415; Carol Garland, "Acid Rain over the United States and Canada: The D.C. Circuit Fails to Provide Shelter under Section 115 of the Clean Air Act While State Action Provides a Temporary Umbrella," *Boston College Environmental Affairs Law Review* 16 (1988): 18; "Air Act Provisions on Interstate Pollution Triggered by Violation of NAAQS, Court Rules," *BNA Environment Reporter,* July 20, 1984, 440; Lawrence Mosher, "Acid Test for Reagan," *National Journal,* April 4, 1981, 579.

12. Irvin Molotsky, "Canadian-Films Ruling Assailed in Washington," *New York Times,* February 26, 1983, A16; Robert D. McFadden, "3 Canadian Films Called 'Propaganda' by U.S.," *New York Times,* February 25, 1983, C4; "Congress to Weigh Curb on 'Acid Rain,'" *New York Times,* January 17, 1982, Wk13.

13. Burford, *Tough Enough,* 89; Richard A. Harris and Sidney M. Milkis, *The Politics of Regulatory Change: A Tale of Two Agencies* (New York: Oxford University Press, 1989), 255; "Tightening Compliance Policy Could Reduce Sulfur Dioxide Emissions"; Howard Kurtz, "Since Reagan Took Office, EPA Enforcement Actions Have Fallen," *Washington Post,* March 1, 1983, A6; Sandra Sugawara, "EPA Enforcement Team Is Off or Running," *Washington Post,* June 23, 1982, A25.

14. Michael Barone, "Tactics of an Ace in the Congressional Air Wars," *Washington Post*, December 14, 1982, A27; Greg Johnson, "Efforts to Strip Bill of Its Major Goals Will Ignite Capitol Hill Fight," *Industry Week*, June 1, 1981, 19; "The Great Clean Air Act Debate of 1981: Environmentalists, Industry, Air Quality Commission Take Positions," *Environmental Law Reporter* 11 (January 1981): 10027; Shabecoff, "Effort to Soften Coal Seems Likely."

15. Ronald A. Taylor, "On Tap: Another Bitter Battle Over Clean Air," *U.S. News & World Report*, October 19, 1981, 46; "EEI Wants Fewer Restrictions, Larger State Role in New Clean Air Act," *Electrical Week*, April 20, 1981, 3; "Coal State Groups Launch Separate Lobbying Efforts," *Coal Week*, February 9, 1981, 6; Ben Franklin, "Coal Outlook Troubled Despite High Hopes," *New York Times*, November 16, 1980, A1.

16. "Dingell Adjourns Air Act Markups amid Signs of Eroding Republican Support," *BNA Environment Reporter*, August 27, 1982, 548; Sandra Sugawara and Philip J. Hilts, "Clean Air Act Rewrite Tangled in Thicket of Conflicting Interests," *Washington Post*, August 2, 1982, A6; Lawrence Mosher, "Courting Trouble," *National Journal*, November 7, 1981, 1996 (poll results); Lawrence Mosher, "The Clean Air That You're Breathing May Cost Hundreds of Billions of Dollars," *National Journal*, October 10, 1981, 1816; Joanne Omang, "Firms Urged to Back Air Act Changes," *Washington Post*, August 23, 1981, A10; "Leaked Report Rallies Clean Air Act Defenders," *Oil & Gas Journal*, June 29, 1981, 56.

17. Lash, Gillman, and Sheridan, *A Season of Spoils*, 77; Richard J. Lazarus, *The Making of Environmental Law* (Chicago: University of Chicago Press, 2004), 102.

18. Burford, *Tough Enough*, 149, 157, 163; Lazarus, *Making of Environmental Law*, 102; Martin H. Belsky, "Environmental Policy Law in the 1980s: Shifting Back the Burden of Proof," 12 *Ecology Law Quarterly* 12 (1984): 79–80; Dale Russakoff, "Gorsuch: A Loyal Soldier in the Center of a Constitutional Storm," *Washington Post*, December 19, 1982, A2; Lawrence Mosher, "EPA, Critics Agree Agency under Gorsuch Hasn't Changed Its Spots," *National Journal*, November 13, 1982, p. 1941; Joanne Omang, "Senate Panel Decides to Take Look at Troubled EPA," *Washington Post*, October 1, 1981, A5.

19. Landy, Roberts, and Thomas, *Environmental Protection Agency*, 251; Dale Russakoff, "Ruckelshaus Given an Emotional Welcome by 1,000 Employees of Embattled EPA," *Washington Post*, March 23, 1983, A3.

20. Rochelle L. Stanfield, "Regional Tensions Complicate Search for an Acid Rain Remedy," *National Journal*, May 5, 1984, 860; "New Focus on Acid Rain at EPA, Stress Shifts from Research to Control Strategies," *BNA Environment Reporter*, May 27, 1983, 125.

21. "Clean Air Issue Reveals Crack in Coal Industry Lobby Front," *Coal Week*, August 22, 1983, 1.

22. David Maraniss, "Hard Questions in Acid-Rain Control Are Who Benefits and Who Must Pay," *Washington Post,* January 29, 1984, A1; Philip Shabecoff, "Ruckelshaus Puts Off Plan to Curb Acid Rain," *New York Times,* October 23, 1983, A27; "Ruckelshaus to Begin Narrowing 11 Options for Acid Rain Control in Task Force Report," *BNA Environment Reporter,* August 5, 1983, 572.

23. Landy, Roberts, and Thomas, *Environmental Protection Agency,* 255; Dale Russakoff and Lou Cannon, "Thomas Nominated as EPA Chief," *Washington Post,* November 30, 1984, A1.

24. Patrick Charles McGinley and Charles H. Haden II, "Climate Change and the War on Coal: Exploring the Dark Side," *Vermont Journal of Environmental Law* 13 (2011): 274; Philip Shabecoff, "Acidic Pollution May Be on the Rise," *New York Times,* December 12, 1985, A23; "Old King Coal Comes Back from the Depths," *Chicago Tribune,* June 30, 1985, A1 (chaos quote); "Clean Coal Steps to the Front," *Engineering News-Record,* November 17, 1983, 28.

25. "TVA Steps Out over Acid Rain," *Engineering News-Record,* October 4, 1984, 54; "Public Power Group Endorses Acid Rain Plan," *BNA Environment Reporter,* February 17, 1984, 1789.

26. "Acid Deposition Documented in Southeastern Region, but Scientists, State Officials Uncertain about Its Effects," *BNA Environment Reporter,* October 10, 1986, 874; "NAS Study Confirms Past Reports Linking Environmental Damage with Sulfur Dioxide," *BNA Environment Reporter,* March 21, 1986, 2085; Nelson Bryant, "In the West, Concern over Acid Rain," *New York Times,* January 12, 1986, E10; "Reagan Has Decided Not to Act on Acid Rain, Mitchell Claims at Senate Oversight Hearing," *BNA Environment Reporter,* December 20, 1985, 1626 (Thomas quote); Jane Perlez, "Bush Reassures Miners on Acid Rain," *New York Times,* September 25, 1984, A21; Stanfield, "Regional Tensions."

27. Michael Weisskopf, "Industry's Covert War on an Acid Rain Bill," *Washington Post,* September 25, 1986, A23.

28. Thomas v. New York, 802 F.2d 1443 (D.C. Cir. 1986).

29. New York v. EPA, 852 F.2d 574 (D.C. Cir. 1988); Felicity Barringer, "Drifting Air Pollution Beginning to Pit States against Their Neighbors," *Washington Post,* October 24, 1983, A9.

30. "Air-Quality Upgrade Impacts Coal Plant," *Engineering News-Record,* April 28, 1983, 51; Guide to the Seattle City Light Creston Coal Plant Project Records, 1980–1984, http://nwda.orbiscascade.org/ark:/80444/xv20019 (plant never built).

31. Cass Peterson, "EPA Announces New Standards on Soot, Dust," *Washington Post,* March 10, 1984, A1; "EPA Staff Wants Lower Particulate Standards," *Coal Week,* June 22, 1981, 6.

32. McGarity, *Reinventing Rationality,* 50; "Ruckelshaus Said to Want Standard Proposal at Low End of Science Board Recommendation," *BNA Environment Reporter,* July 15, 1983, 398.

33. U.S. Environmental Protection Agency, "Proposed Revisions to the National Ambient Air Quality Standards for Particulate Matter," 49 Fed. Reg. 10,408 (1984); McGarity, *Reinventing Rationality,* 51 (conversion); "Proposed Revision to Particulate Standard Too Lenient, Too Strict, EPA Told at Hearing," *BNA Environment Reporter,* May 11, 1984, 43.

34. U.S. Environmental Protection Agency, "Revisions to the National Ambient Air Quality Standards for Particulate Matter," 52 Fed. Reg. 24634 (1987); Natural Resources Defense Council, Inc. v. EPA, 902 F.2d 962 (D.C. Cir. 1990).

35. Duquesne Light Co. v. EPA, 791 F.2d 959, 963–64 (D.C. Cir. 1986).

36. Michael Weisskopf, "U.S. Contribution to 'Greenhouse Effect' Rises," *Washington Post,* September 16, 1989, A1; "Senate Panel Warned of Politics Delaying Action on Greenhouse Effect," *Platts Oilgram News,* August 15, 1988, 4 (EPA quote); "Senators, Scientists Urge Policy Shifts to Avert Problems from Atmospheric Changes," *BNA Environment Reporter,* June 13, 1986, 164; "Climate Changes Foreseen in NAS Study but No Energy Policy Change Said Warranted," *BNA Environment Reporter,* October 28, 1983, 1143.

37. Abramowitz v. EPA, 832 F.2d 1071 (9th Cir. 1988); Michael Weisskopf, "EPA Threatens 14 Cities with Construction Ban," *Washington Post,* June 30, 1987, A17.

38. Gerald F. Seib and Bruce Ingersoll, "This Year, Events Force Topics of Environment, Energy, Agriculture to Fore in Presidential Race," *Wall Street Journal,* October 7, 1988, A1; "Candidates Outline Environmental Issues, Seek Air, Hazardous Waste, Wetlands Reforms," *BNA Environment Reporter,* September 16, 1988, 982; Philip Shabecoff, "Environmentalists Say Either Bush or Dukakis Will Be an Improvement," *New York Times,* September 1, 1988, B9; "1988 Presidential General Election Results," http://uselectionatlas.org/RESULTS/national.php?year =1988.

39. Joan Claybrook, *Retreat from Safety: Reagan's Attack on America's Health* (New York: Pantheon Books, 1984), xvi (stalemate); Steven J. Marcus, "Acid Rain and Pollution Curbs," *New York Times,* November 7, 1983, D1; Greg Johnson, "Acid-Rain Debate Heats Up," *Industry Week,* August 8, 1983, 61; Philip Shabecoff, "Pollution Control Industry in Peril," *New York Times,* July 4, 1982, C4.

40. Rochelle L. Stanfield, "Punching at the Smog," *National Journal,* March 5, 1988, 600; Robert E. Taylor, "Group's Influence on U.S. Environmental Laws, Policies Earns It a Reputation as a Shadow EPA," *Wall Street Journal,* January 13, 1986; Russakoff and Cannon, "Thomas Nominated as EPA Chief."

6. MAJOR ADJUSTMENTS

1. Barbara Rosewicz, "The War Within: Environmental Chief Clashes with New Foe: Deregulation Troops," *Wall Street Journal,* March 27, 1992, A1; Casey Bukro, "The True Greenhouse Effect," *Chicago Tribune,* December 31, 1989, A1; "Reilly's Selection as EPA Head Draws Warm Reaction from Environmentalists," *BNA Toxics Law Reporter,* January 11, 1989, 983; Josh Getlin, "Reilly an Activist Who Seeks Consensus," *Los Angeles Times,* December 23, 1988, A14; Philip Shabecoff, "Bush Tells Environmentalists He'll Listen to Them," *New York Times,* December 1, 1988, B15.

2. Philip Shabecoff, *A Fierce Green Fire: The American Environmental Movement* (New York: Hill and Wang, 1993), 252; Kirk Victor, "Quayle's Quiet Coup," *National Journal,* July 6, 1991, 1676; "White House 'Bullies' EPA into Weakening Clean Air Act Regulations, Waxman Charges," *BNA Environment Reporter,* March 29, 1991, 2124; Maureen Dowd, "Who's Environment Czar, E.P.A.'s Chief or Sununu?" *New York Times,* February 15, 1990, A1.

3. Michael Weisskopf, "A Changed Equation on Pollution," *Washington Post,* June 7, 1989, A1; Philip Shabecoff, "E.P.A. Nominee Says He Will Urge Law to Cut Acid Rain," *New York Times,* February 1, 1989, A1; Philip Shabecoff, "Bolstering Clean Air Act: Fingers Point in All Directions after Latest Failure," *New York Times,* October 7, 1988, B7.

4. Trip Gabriel, "Greening the White House," *New York Times Magazine,* August 13, 1989, 25.

5. Ann McDaniel, "The EPA's Man in the Middle," *Newsweek,* April 16, 1990, 26; Gabriel, "Greening the White House."

6. Michael Weisskopf and Ann Devroy, "Bush Presents Clean Air Package," *Washington Post,* July 22, 1989, A5.

7. "Report on State Acid Rain Program Formed Basis of Bush Proposal, EPA Says," *BNA Environment Reporter,* August 25, 1989; "Bush Spells Out Clean Air Proposals," *Oil & Gas Journal,* July 31, 1989, 36; Rose Gutfeld, "Debate Begins on Bush's Clean-Air Plan, Which Could Cost $18 Billion a Year," *Wall Street Journal,* July 24, 1989; "Bush Acid Rain Bill May Allow Lower Cuts Than 10 Million Ton SO_2, Analysis Shows," *Coal Week,* July 24, 1989, 2.

8. Michael Weisskopf, "Bush Weakens Clean Air Proposal," *Washington Post,* July 21, 1989, A7; "President Proposes to Amend Clean Air Act to Limit Acid Rain, Ozone, Toxic Emissions," *BNA Environment Reporter,* June 16, 1989, 427.

9. "Air Bills' Industrial Permitting System Hit as Regulatory 'Behemoth' by Trade Group," *BNA Environment Reporter,* January 19, 1990, 1620.

10. "Air Toxics Bill Could Hurt Coal in Utilities, Steel; Sets End Run on Acid Rain," *Coal Week,* November 27, 1989, 8; Rose Gutfeld, "Pure Plays: For Each Dollar Spent on Clean Air Someone Stands to Make a Buck," *Wall Street*

Journal, October 29, 1990, A1; "SO$_2$ Emission Trading Provisions Attacked at Hearing on Administration's Clean Air Bill," *BNA Environment Reporter,* October 13, 1989, 1049; "Watkins Appearance on Air Bill Uncertain," *Inside Energy with Federal Lands,* September 11, 1989, 1; "The Cost of Acid Rain Measures in the President's Clean Air Bill," *Inside Energy with Federal Lands,* September 4, 1989, 7.

11. Margaret Kriz, "Emission Control," *National Journal,* July 3, 1993, 1696; Gabriel, "Greening the White House"; Patrice Apodaca, "Firm Hopes Clean-Air Laws Will Boost Sales," *Los Angeles Times,* October 30, 1990, D9; Lois Ember, "Bush's Atmospheric Ambitions," *Chemistry & Industry,* July 17, 1989, 437.

12. "Conferees Agree on Permits, Non-Attainment, Put Aside Dispute on Small Business Permits," *BNA Environment Reporter,* September 21, 1990, 1047; "Clean Air Conference Gets Under Way as Senate Offers CFC, Permitting Plans," *BNA Environment Reporter,* July 20, 1990, 499; Richard L. Berke, "House, 401–21, Votes to Widen Clean Air Curbs," *New York Times,* May 24, 1990, A1.

13. 42 U.S.C. §§ 181–89; "Conference Agreement on Clean Air Comprehensively Revises U.S. Air Law," *BNA Environment Reporter,* November 2, 1990, 1247.

14. "Clean Air Conference Staff Agree on Bill's Toxic Air Emissions Title," *BNA Environment Reporter,* October 19, 1990, 1171; Michael Weisskopf, "Panel Reaches Tentative Pact on Industry Emission Controls," *Washington Post,* October 18, 1990, A3.

15. Byron Swift, "How Environmental Laws Work: An Analysis of the Utility Sector's Response to Regulation of Nitrogen Oxides and Sulfur Dioxide under the Clean Air Act," *Tulane Environmental Law Journal* 14 (2001): 321, 355; "Conference Agreement"; Michael Ross, "Negotiators Reach Accord on Acid Rain," *Los Angeles Times,* October 22, 1990, A1.

16. "Bush Signs Clean Air Act Amendments, Predicts Benefits for All U.S. Citizens," *BNA Environment Reporter,* November 23, 1990, 1387.

17. *Clean Air Act Implementation (Part 1): Hearings before the Subcommittee on Health and the Environment of the House Committee on Energy and Commerce,* 102d Cong. 129 (1991) (statement of Rep. Henry Waxman); Rosewicz, "The War Within"; "Agency Working with 'Evangelic Fervor' to Meet Air Act Deadlines, Analyst Says," *BNA Environment Reporter,* April 5, 1991, 2141; William H. Miller, "Congress Cleans the Air," *Industry Week,* November 19, 1990, 58.

18. "Operational Flexibility Issue Continues to Delay Permits Proposal, EPA Official Says," *BNA Environment Reporter,* April 19, 1991, 2255; "Draft EPA Rules Afford Broad Protection to Sources that Hold, Apply for Air Permits," *BNA Environment Reporter,* March 29, 1991, 2123; "Permitting Rules Should Be Drafted to Be Flexible, Industry Group Says," *BNA Environment Reporter,* January 18, 1991, 1677.

19. *Clean Air Act Implementation (Part 1): Hearings,* 187 (testimony of E. Donald Elliott, EPA); "EPA Hits Snags with White House on CAA Permitting Reg, Delaying Proposal," *Inside EPA,* April 12, 1991, 8; "Competitiveness Council Begins Informational Meetings on CAA Permit Reg," *Inside EPA,* March 22, 1991, 9.

20. U.S. Environmental Protection Agency, "Operating Permit Program; Proposed Rule," 56 Fed. Reg. 21712 (1992); "Permits Proposal Draws Fire at Hearing," *BNA Environment Reporter,* June 7, 1991, 302; "Industry Prefers Proposed EPA Clean Air Permit Rule to Earlier Drafts," *Utility Environment Report,* May 3, 1991, 8; Michael Ross, "Proposed Clean Air Rules Draw Criticism," *Los Angeles Times,* April 25, 1991, A9.

21. David Sive, "Sparks Fly over Clean Air Act Permits," *National Law Journal,* June 29, 1992, 21; "White House Competitiveness Council to Craft Options for CAA Permit Rules," *Inside EPA,* April 4, 1992, 1 (quote); Michael Ross, "Proposed Clean Air Rules Changes Spark Battle," *Los Angeles Times,* November 21, 1991, A20.

22. "Bush Sides with Quayle Council on Dispute with EPA over Clean Air Permit Notification," *BNA Environment Reporter,* May 22, 1992, 395; Ann Devroy, "Bush between Push and Pull on Clean Air," *Washington Post,* April 26, 1992, A18; "DOJ Memo Suggests No Public Participation Needed for Minor CAA Permit Changes," *Inside EPA,* April 24, 1992, 15; "EPA, White House Air Act Permitting Talks Halt as Justice Dept. Reviews Issue," *Inside the White House,* November 28, 1991; Philip J. Hilts, "Questions on Role of Quayle Council," *New York Times,* November 19, 1991, A11.

23. U.S. Environmental Protection Agency, "Operating Permit Program; Final Rule," 57 Fed. Reg. 32250 (July 21, 1992); "NRDC, Sierra Club, EDF Sue Agency, Call New Air Permitting Rule Illegal," *BNA Environment Reporter,* August 14, 1992, 1188; Keith Schneider, "Industries Gaining Broad Flexibility on Air Pollution," *New York Times,* June 26, 1992, A1.

24. "EPA Issues Final Acid Rain Rules, Sets Official Program in Motion," *Utility Environment Report,* October 30, 1992, 3; "Allowance Market at Risk from Poor Public Relations, Proponents Fear," *Utility Environment Report,* May 29, 1992, 1; "WP&L Makes First Announced Sales of Allowances; TVA, Duquesne Buy," *Utility Environment Report,* May 15, 1992, 1; Matthew L. Wald, "Utility Is Selling Right to Pollute," *New York Times,* May 12, 1992, A1; "Chicago Board of Trade Proposes to Run Allowance Trading Program," *Utility Environment Report,* July 12, 1991, 1.

25. "Control of Nitrogen Oxide Emissions Seen as Critical in Meeting Ozone Standard," *BNA Environment Reporter,* January 17, 1992, 2139.

26. "Utility Limits on NO_x to Curb Acid Rain, Guidance for Ozone Control Announced by EPA," *BNA Environment Reporter,* October 30, 1992, 1668; "NO_x

Control Proposals Could Weaken Ability to Reduce Ozone, State Group Says," *BNA Environment Reporter,* July 17, 1992, 885; "Administration Wrestling with NOx Control Regulations for Utilities," *Utility Environment Report,* January 24, 1992, 6; "Control of Nitrogen Oxide Emissions Seen as Critical in Meeting Ozone Standard," *BNA Environment Reporter,* January 17, 1992, 2139.

27. *New Source Review Policy, Regulations and Enforcement Activities: Hearings before the Senate Committee on Environment and Public Works and the Senate Committee on the Judiciary,* 107th Cong. 114 (July 16, 2002) (testimony of Jeffrey Holmstead, EPA).

28. Bruce Buckheit, testimony before the Senate Democratic Policy Committee, in "Clearing the Air: An Oversight Hearing on the Administration's Clean Air Enforcement Program," February 6, 2004, https://www.dpc.senate.gov /hearings/hearing11/buckheit.pdf.

29. *New Source Review Policy . . . Hearings,* 114 (testimony of Jeffrey Holmstead, EPA).

30. *Clean Air Act: New Source Review Regulatory Program: Hearings before the Subcommittee on Clean Air, Wetlands, Private Property and Nuclear Safety of the Senate Committee on Environment and Public Works,* 106th Cong. 99 (February 28, 2000) (testimony of David Hawkins, NRDC).

31. Ibid., 100.

32. See, e.g., United States v. Louisiana-Pacific Corp., 682 F. Supp. 1141, 1162–63 (D. Colo. 1988).

33. Cassie N. Aw-yang, "EPA's Changes to the Routine Maintenance, Repair and Replacement Rule of the New Source Review Program: An Unlawful Threat to Public Health and Welfare?," *Environs* 27 (2004): 336; David B. Spence, "Coal-Fired Power in a Restructured Electricity Market," *Duke Environmental Law & Policy Forum* 15 (2005): 187, 198.

34. Wisconsin Electric Power Co. v. Reilly, 893 F.2d 901, 910 (7th Cir. 1990) (quoting memorandum from Don R. Clay, EPA Acting Assistant Administrator for Air and Radiation, to David A. Kee, Director of Air and Radiation Division, EPA Region V [September 9, 1988]).

35. Larry B. Parker and John E. Blodgett, "Air Quality and Electricity: Enforcing New Source Review," Congressional Research Service, January 31, 2000, 8.

36. Wisconsin Electric Power, 893 F.2d at 905–6.

37. Ibid., 910–12, 916–18.

38. "An Investigation of EPA's Clean Air 'WEPCO' Rule," Staff Report of the Subcommittee on Health and the Environment of the House Committee on Energy and Commerce, undated, reprinted in *Clean Air Act Implementation (Part 1): Hearings,* 314–15.

39. U.S. Environmental Protection Agency, "Requirements for Preparation, Adoption, and Submittal of Implementation Plans; Approval and Promulgation of

Implementation Plans; Standards of Performance for New Stationary Sources," 56 Fed. Reg. 27630, 27632–37 (June 14, 1991).

40. "Utility Pollution Control Projects Exempt from New Source Review under Air Act, EPA Says," *BNA Environment Reporter,* May 29, 1992, 422; "NRDC Faults EPA Rule on Source Modification, Says 'One-Sided' Approach Allows More Emissions," *BNA Environment Reporter,* July 26, 1991, 780.

41. U.S. Environmental Protection Agency, "Requirements for Preparation, Adoption and Submittal of Implementation Plans; Approval and Promulgation of Implementation Plans; Standards of Performance for New Stationary Sources," 57 Fed. Reg. 32314, 32325–32 (July 21, 1992).

42. Keith Schneider, "Utilities to Take Steps to Cut Haze at Grand Canyon," *New York Times,* August 9, 1991, A1; "Arizona Power Plant Responsible for Haze in Grand Canyon Park, EPA Says in Proposal," *BNA Environment Reporter,* September 8, 1989, 794; Michael Weisskopf, "Air Pollution at Grand Canyon May Pose Test for Bush Administration," *Washington Post,* August 28, 1989, A4.

43. "Grand Canyon Haze-Tracing Experiment Identifies Source, Not Quantity, NRC Says," *BNA Environment Reporter,* October 19, 1990, 1182; "Plant Said to Foul Grand Canyon Air," *New York Times,* October 16, 1990, A26; "Salt River Challenges Park Service Methods," *Coal & Synfuels Technology,* October 30, 1989, 5; Paul Houston, "Haze at Grand Canyon Blamed on Power Plant," *Los Angeles Times,* August 30, 1989, B3.

44. Kirk Victor, "Quayle's Quiet Coup," *National Journal,* July 6, 1991, 1676 (quote); "Emission Reductions Proposed at Power Plant Near Grand Canyon," *BNA Environment Reporter,* February 8, 1991, 1798; Michael Weisskopf and John Lancaster, "Reilly Invoking Clean Air Law to Clear Vista at Grand Canyon," *Washington Post,* February 1, 1991, A1; Bob Secter and Rudy Abramson, "Clearing the Air for All to See," *Los Angeles Times,* May 1, 1990, A1.

45. Central Arizona Water Conservation District v. EPA, 990 F.2d 1531 (9th Cir. 1993); "Remarks at an Environmental Agreement Signing Ceremony at the Grand Canyon, Arizona," *Weekly Compilation of Presidential Documents,* September 18, 1991, 1292; "Salt River Project and Its Partners Have Completed a $420 Million Cleanup," *U.S. Coal Review,* October 11, 1999; "Navajo Agreement Requires 90% SO_2 Reduction, but Includes Concessions," *Utility Environment Report,* August 23, 1991, 3; Schneider, "Utilities to Take Steps."

46. "D.C. Circuit Decides Acid Rain Case: No EPA Duty Yet on International Pollution," *BNA Environment Reporter,* September 7, 1990, 878; Michael Weisskopf, "U.S., in Court, Doubts Theory of Acid Rain," *Washington Post,* November 28, 1989, A23 (quote); "Federal Government to Join in Ontario Acid Rain Lawsuit," *BNA International Environment Reporter,* January 11, 1989, 15.

47. Energy Policy Act of 1992, Pub. L. 102–486, §2, 102 Stat. 2776 (1992); Rudy Abramson, "U.S. Flexes Its Muscles before Earth Summit," *Los Angeles Times,*

May 30, 1992, A19; Douglas Jehl, "Study Finds Surge in U.S. Emissions of Carbon Dioxide," *Los Angeles Times,* September 16, 1989, A1.

48. Robert A. Reinstein, "Climate Negotiations," *Washington Quarterly* 16 (1993): 79–95; Abramson, "U.S. Flexes Its Muscles."

49. William Drozdiak, "U.S. Refuses to Pledge Limit on Greenhouse Gases Emissions," *Washington Post,* November 8, 1990, A22; Rudy Abramson, "Bush Urges More Study of Climate Change Data," *Los Angeles Times,* April 18, 1990, A1; Glenn Frankel, "Nations Pass Resolution on Environment," *Washington Post,* November 8, 1989, A33.

50. Sheldon Rampton and John Stauber, *Trust Us, We're Experts!* (New York: Jeremy P. Tarcher; Putnam, 2001), 272 (quote); Keith Schneider, "Bush Plans to Join Other Leaders at Earth Summit in Brazil in June," *New York Times,* May 13, 1992, A8; Margaret E. Kriz, "Warm-Button Issue," *National Journal,* February 8, 1992.

51. Abramson, "U.S. Flexes Its Muscles"; Rose Gutfeld, "Earth Summitry: How Bush Achieved Global Warming Pact with Modest Goals," *Wall Street Journal,* May 27, 1992, A1.

52. Larry B. Stammer and Michael Parrish, "Ex-Environment Activist Will Take Helm at Edison," *Los Angeles Times,* October 1, 1990, A1 (quote); Margaret E. Kriz, "Balancing Act," *National Journal,* January 20, 1990, 125.

53. "Colorado Utility Agrees to Pay $600,000, Spend $66 Million to Settle Air Act Charges," *BNA Environment Reporter,* August 28, 1992, 1277; "Puerto Rico Electric Power Authority Cited for Emitting Pollutants Four Times Federal Limit," *BNA Environment Reporter,* August 14, 1992, 1191; "Ariz. Co-Op Snared by EPA in Crackdown on Polluters along U.S. / Mexican Border," *Utility Environment Report,* June 12, 1992, 1; "The Environmental Protection Agency Has Levied Fines against Two Illinois Companies," *U.S. Coal Review,* May 27, 1992; "EPA Fines Nevada Power over Clean Air Act Violations," *U.S. Coal Review,* December 23, 1991.

54. Michael Levy, "United States Presidential Election of 1992," June 7, 2018, https://www.britannica.com/event/United-States-presidential-election-of-1992; "Clinton Defends Environmental Record; Reilly Defends Gore against White House," *BNA Environment Reporter,* August 14, 1992, 1194 (quote); Michael Weisskopf, "Environment Fades as Political Issue," *Washington Post,* February 1, 1992, A14.

7. CAUTIOUS IMPLEMENTATION AND VIGOROUS ENFORCEMENT

1. "White House Announces Six EPA Nominees, Including Choices for Water, Air Offices," *BNA Environment Reporter,* July 23, 1993, 509; Editorial, "Bill Clinton, Environmentalist?," *New York Times,* January 5, 1993, A14; "Browner Named for EPA, O'Leary for DOE; Both Boast Environmental Credentials," *Utility*

Environment Report, December 25, 1992, 4; "Browner Portrayed as Hard-Working, Results-Oriented," *BNA Environment Reporter,* December 18, 1992, 2086.

2. Executive Order 12,866, 58 Fed. Reg. 190 (September 30, 1993); "Development of Regulations Will Quicken, EPA Chief, OMB Official Tell Senate Panel," *BNA Environment Reporter,* October 1, 1993, 1005; "Clinton Administration Orders Retraction of Dozens of Last-Minute Bush Regulations," 23 *BNA Environment Reporter,* January 29, 1993, 2571.

3. "Switching to Beat Out FGDs for Clean Air," *Coal & Synfuels Technology,* May 9, 1994, 5; "AEP Submits Clean Air Compliance," *Coal & Synfuels Technology,* February 22, 1993, 4; "Utilities Struggle with Acid Rain Control Compliance Decisions," *Power Engineering,* August 1, 1991, 17.

4. E. A. Bretz, "Gavin Scrubber Key to AEP's Clean-Air Compliance Plan," *Electrical World,* December, 1993, 12; Elizabeth A. Bretz, "The Clean Air Act Amendments: How Utilities Will Reduce Pollutant Emissions," *Electrical World,* July, 1993, 49; "AEP Submits Clean Air Compliance," *Coal & Synfuels Technology,* February 22, 1993, 4; "Illinois Power Works Scrub Deal for Baldwin," *Coal & Synfuels Technology,* February 17, 1992, 5.

5. Wyoming v. Oklahoma, 502 U.S. 437 (1992); Jason Makansi, "PSI Gibson Turns Compliance into a Vision for the Future," *Power,* December 1993, 37; "Two Senators Urge Agency to Oppose Protectionist Legislation on Coal Use," *BNA Environment Reporter,* December 3, 1993, 1475; "Acid Rain Program Brings Clash in States over Jobs, Trading Rights, Committee Told," *BNA National Environment Daily,* October 25, 1993; Bretz, "Clean Air Act Amendments."

6. James Bradshaw, "PUCO, AEP Triumph on Electric Rate Ruling," *Columbus Dispatch,* March 31, 1994, C4; Alan Johnson, Ohio Top Court Asked to Scrub Scrubber Plan," *Columbus Dispatch,* March 9, 1993, C2; "Ohio PUC Approves $815-Million Gavin Scrubber Plan for Clean Air Compliance," *Utility Environment Report,* December 11, 1992, 1; "Ohio Consumers' Counsel Slams AEP for Stalling on Compliance Decision," *Utility Environment Report,* April 19, 1991, 5.

7. Jason Makansi, "Phase I of the Clean Air Act Amendments of 1990," *Power,* September 1993, 9 (quote); "ComEd Opts for Mid-Sulfur Coal / Credit Compliance Combo," *U.S. Coal Review,* February 2, 1993; Matthew L. Wald, "Risk-Shy Utilities Avoid Trading Emission Credits," *New York Times,* January 25, 1993, D2.

8. Byron Swift, "How Environmental Laws Work: An Analysis of the Utility Sector's Response to Regulation of Nitrogen Oxides and Sulfur Dioxide under the Clean Air Act," *Tulane Environmental Law Journal* 14 (2001): 322; "Utilities' Paperwork, Auction Cited as Causing Price Drop to $80 Range," *Utility Environment Report,* February 16, 1996, 10; Jeff Bailey, "Utilities Overcomply with Clean Air Act, Are Stockpiling Pollution Allowances," *Wall Street Journal,* November 15,

1995, B8; "Phase-1 Bank Is Blooming with Credits," *Coal & Synfuels Technology,* October 2, 1995, 3; Jeffrey Taylor, "CBOT Plan for Pollution-Rights Market Is Encountering Plenty of Competition," *Wall Street Journal,* August 24, 1993, C1; Casey Bukro, "Allowing for Pollution," *Chicago Tribune,* March 31, 1993, B1.

9. Swift, "How Environmental Laws Work," 325, 332–36; "Phase-One Affected Units Emitted 5.3 Million Tons of SO_2 Last Year," *Utility Environment Report,* March 29, 1996, 5.

10. 42 U.S.C. § 7651f(b); Jason Makansi, "SO_2 / NO_x Control Fine-Tuning for Phase 1 Compliance," *Power,* March 1994, 15; "Environmental Group Challenges New EPA to Tighten NO_x Controls," *Utility Environment Report,* February 19, 1993, 7; "NCA, EEI Critical of Proposed NO_x Rules," *Coal Week,* February 15, 1993, 8; "Utility Limits on NO_x to Curb Acid Rain, Guidance for Ozone Control Announced by EPA," *BNA Environment Reporter,* October 30, 1992, 1668.

11. Alabama Power Co. v. EPA, 40 F.3d 450 (D.C. Cir. 1994); "EPA Reissues Acid Rain NO_x Rules, Extends Compliance Date to 1996," *Utility Environment Report,* April 14, 1995, 3; "Court Ruling Scrambles NO_x Plans," *Coal & Synfuels Technology,* January 9, 1995, 4; "Industry Likely to Seek Speedy Court Review of EPA NO_x Rules," *Utility Environment Report,* March 4, 1994, 1.

12. Swift, "How Environmental Laws Work," 357–67.

13. "Many Factors Must Be Evaluated When Considering a Switch or Blend of Western Coals," *Power Engineering,* August 1, 1999, 18; "Utilities Aren't in a Rush to Scrub Plants," *Energy Report,* August 17, 1998.

14. Joel Achenbach, "See You in September," *Washington Post,* August 14, 1995, D1; Margaret Kriz, "Cough. Wheeze. There Goes Party Unity," *National Journal,* May 27, 1995, 1289; Jerry Gray, "Grading G.O.P. Freshmen in House: High in Ambition, Low in Humility," *New York Times,* April 11, 1995, A20; "GOP Launching Broad Legislative Bid to Reduce Environmental Regulation," *Utility Environment Report,* February 3, 1995, 6.

15. "Legislative 'Rollback' of Regulations Would Undermine Law, EPA Air Chief Says," *BNA Environment Reporter,* May 19, 1995, 218; "GOP Launching Broad Legislative Bid"; Cindy Skrzycki, "Hill Republicans Promise a Regulatory Revolution," *Washington Post,* January 4, 1995, A1.

16. Swift, "How Environmental Laws Work," 370–71; Michael J. Meagher and Craig Jakubowics, "Eastern States Convene over Ozone Compliance," *National Law Journal,* October 14, 1996, 7; "Environmentalists Holding Utilities' Feet to the Fire on 1995 NO_x Deadlines," *Utility Environment Report,* March 19, 1993, 11.

17. "Availability, Price Uncertainty of NO_x Offsets May Hinder Independent Projects," *Utility Environment Report,* October 29, 1993, 10; "Mass. Unveils Final Emissions Trading Program for NO_x, VOC, CO Compliance," *Utility Environment Report,* October 1, 1993, 1; "Gade Delineates Illinois' NO_x Trading Scheme," *Coal & Synfuels Technology,* September 27, 1993, 3.

18. "SCAQMD's Annual 'Reclaim' Audit Reveals $20-Million in Credits Were Traded in '96," *Utility Environment Report,* March 14, 1997, 8; Amy Gahran, "Users Aim to Cash in on Conservation through Emissions Credits," *Energy User News,* August 1, 1993, 1; Marie Leone, "State Launches Emissions-Trading Program," *Electrical World,* April, 1992, 21; Judy Pasternak, "QMD Officials Propose Pollution Rights Market," *Los Angeles Times,* January 30, 1992, A1.

19. 42 U.S.C. §§ 7410(a)(2)(D), 7410(k), 7426; "EPA Forges Ahead with Trading Plan in Face of Criticism, Uncertainties," *BNA Environment Reporter,* May 15, 1998, 194; "FERC, EPA Likely to Face More Battles over Utility Deregulation, Official Predicts," *BNA Environment Reporter,* August 9, 1996, 813.

20. "MOG: States, EPA Colluded on NO_x Rule," *Electricity Daily,* February 5, 1998 (cover quote); "Midwest Ozone Group Asks EPA to Reject Northeast States' Section 126 Petitions," *Utility Environment Report,* September 12, 1997, 1; "Eight Northeastern States Petition EPA for Direct Action on Transported Pollution," *BNA Environment Reporter,* August 15, 1997, 709.

21. "Proposal Calls on 22 States to Cut NO_x Emissions by 1.6 Million Tons per Year," *BNA Environment Reporter,* October 17, 1997, 1198; John H. Cushman Jr., "E.P.A. Acts to Require Big Cut in Air Emissions by 22 States," *New York Times,* October 11, 1997, A9.

22. "EPA NO_x Rule Will Harm Reliability of Midwest Transmission Grid: Study," *Utility Environment Report,* August 14, 1998, 5 (Applied Economics Research Study for UARG); "Will EPA Ozone Rules Crash the Grid?," *Electricity Daily,* August 6, 1998 (H. Zinder & Associates study); "EPA Forges Ahead"; "Comments on EPA Proposed SIP Call Reflect Regional Biases on Ozone," *Utility Environment Report,* March 13, 1998, 6; "Coal Interests Retaliate: Cite EPA's 'War on Coal,'" *Coal Week,* February 16, 1998, 3 (war on coal quote); "Midwest, Southern Utilities Blast EPA on Ozone Transport Proposal," *Utility Environment Report,* February 13, 1998, 1; Cushman, "E.P.A. Acts."

23. U.S. Environmental Protection Agency, "Finding of Significant Contribution and Rulemaking for Certain States in the Ozone Transport Assessment Group Region for Purposes of Reducing Regional Transport of Ozone," 62 Fed. Reg. 57356 (August 1997); John Wile, Mark Berkman, and Jonathan Falk, "Complying with New Rules for Controlling Nitrogen Oxides Emissions," *Electricity Journal,* January / February 2000, 18; "EPA's Regional Emissions Plan Leaves Utilities Breathless," *Engineering News-Record,* October 5, 1998, 12; "NO_x Cuts of 1.1 Million Tons Annually Sought in Rule on Transported Pollution," *BNA Environment Reporter,* October 22, 1998, 1093; "Pool of Credits for States Could Help Utilities Meet Low Emission Goals, EPA Says," *BNA Environment Reporter,* October 2, 1998, 1097; "States' Backlash against NO_x Rules May Spawn Lawsuits," *Megawatt Daily,* September 30, 1998.

24. Michigan v. EPA, 213 F.3d 663 (D.C. Cir. 2000); "Meeting October Deadline for NO$_x$ Plans Called Impossible for Some Southeast States," *BNA Environment Reporter,* January 5, 2001, 8; "Utilities to Spend Hundreds of Millions in Three States to Reduce NO$_x$ Emissions," *BNA Environment Reporter,* November 10, 2000, 2357; "New Pa. Rules Cut NO$_x$ 75% by 2003, Gives Bonuses for Early Compliance," *Utility Environment Report,* July 28, 2000, 5; "Ohio, Michigan Devise NO$_x$ Control Strategies in Challenge to U.S. EPA," *Utility Environment Report,* January 29, 1999, 10; "TVA Adopts Catalytic Technology to Substantially Reduce NO$_x$ Emissions," *BNA Environment Reporter,* August 7, 1998, 761.

25. 42 U.S.C. § 7509; "EPA Triggers Sanction Clock on 11 States, D.C. for Failing to Submit Plans to Cut NO$_x$," *BNA Environment Reporter,* January 5, 2001, 7.

26. 42 U.S.C. §§ 7409, 7410; Thomas O. McGarity, "Science and Policy in Setting National Ambient Air Quality Standards: Resolving the Ozone Enigma," *Texas Law Review* 93 (2015): 1783–1809.

27. "NESCAUM Pushes EPA to Impose New Daily Standard for Fine Particles," *Utility Environment Report,* September 13, 1996, 11; "Standard-Setting Process for Ozone, PM to Be Coordinated under New EPA Process," *BNA Environment Reporter,* May 31, 1996, 372; "EPA Staff Proposes Daily, Annual Standards for Fine Particulates," *Utility Environment Report,* November 10, 1995, 1.

28. U.S. Environmental Protection Agency, "National Ambient Air Quality Standards for Ozone: Proposed Decision," 61 Fed. Reg. 65716, 65733 (December 13, 1996); John H. Cushman Jr., "Administration Issues Its Proposal for Tightening of Air Standards," *New York Times,* November 28, 1996, A1.

29. Joby Warrick, "A Dust-Up over Air Pollution Standards," *Washington Post,* June 17, 1997, A1; Joby Warrick, "Panel Seeks Cease-Fire on Air Quality but Gets a War," *Washington Post,* February 6, 1997, A24; Joby Warrick, "Clean Air Standards Opponents Circle the Backyard Barbecues," *Washington Post,* January 24, 1997, A1; "Forces Gather against EPA Particulates Plan," *Coal & Synfuels Technology,* November 15, 1996, 5; Alec C. Zacaroli, "Revising the PM, Ozone Standards: Major Questions Remain On Major Projects," *BNA Environment Reporter,* August 23, 1996, 929; "National Assn. of Manufacturers Forms Coalition to Fight PM-10 Rule," *Utility Environment Report,* April 12, 1996, 7.

30. Claudia H. Deutsch, "Cooling Down the Heated Talk," *New York Times,* May 27, 1997, D1; Peter Fairley, "EPA Wading Through Comments on Tougher Standards," *Chemical Week,* March 19, 1997, 20; "Environmental Group Launches Ads Backing EPA Air Standards," *National Journal's Congress Daily,* February 11, 1997; "Environmental Groups Support Basis for Ozone, Particulate Matter Proposals," *BNA Environment Reporter,* February 7, 1997, p. 2024.

31. Michael D. Lemonick, "The Queen of Clean Air," *Time,* July 7 1997, 32; James Gerstenzang, "White House Urges EPA to Ease Up," *Los Angeles Times,*

June 14, 1997, A22; John J. Fialka, "EPA Plan to Tighten Pollution Rules Continues to Divide White House Aides," *Wall Street Journal*, June 6, 1997, A16; "Generators Lobby OMB for Compromise on Ozone, PM," *Utility Environment Report*, June 6, 1997, 1; "Agency Chief Answers Bipartisan Attacks on Air Proposals, Says No Decision Made," *BNA Environment Reporter*, May 23, 1997, 150.

32. U.S. Environmental Protection Agency, "National Ambient Air Quality Standards for Ozone," 62 Fed. Reg. 38856 (1997); "EPA to Focus on Power Plants to Meet New Air Standards," *Megawatt Daily*, June 30, 1997; "Clinton Announces Support for Finalizing Tougher Restrictions on Ozone, Particulates," *BNA Environment Reporter*, June 27, 1997, 397; "Clinton Backs Modification of EPA Clean Air Regulations," *National Journal's Congress Daily*, June 25, 1997.

33. "Attempts to Delay Clean Air Regulations Postponed," *National Journal's Congress Daily*, November 10, 1997; "Commerce Committee Seeks Answers on Implementation Plan for New Air Rules," *BNA Environment Reporter*, September 12, 1997, 829; "Bipartisan Group Vows to Attempt Override of New Air Standards," *BNA Environment Reporter*, June 27, 1997, 397.

34. American Trucking Ass'ns v. EPA, 175 F.3d 1027 (D.C. Cir. 1999).

35. Whitman v. American Trucking Ass'ns, 531 U.S. 457 (2001).

36. American Trucking Ass'ns v. EPA, 283 F.3d 355 (D.C. Cir. 2002).

37. U.S. Environmental Protection Agency, "Proposed Decision Not to Revise the National Ambient Air Quality Standards for Sulfur Oxides (Sulfur Dioxide)," 53 Fed. Reg. 14926 (1988); "Lung Association, EPA Agreement Sets Deadline for Assessing SO_2 Standards," *Utility Environment Report*, May 28, 1993, 1; "Lung Association, Two States Sue EPA Seeking Review of Health-Based SO_2 Standard," *BNA Environment Reporter*, November 20, 1992, 1871; "24-Hour Sulfur Dioxide Limit Not Surrogate for One-Hour Standard, EPA Tells Scientists," *BNA Environment Reporter*, July 16, 1982, 372.

38. American Lung Ass'n v. EPA, 134 F.3d 388, 390–91 (D.C. Cir. 1998).

39. Ibid.

40. "GAO Says Most Air Pollution Sources Near National Parks Fall Outside PSD Rules," *BNA Environment Reporter*, March 16, 1990, 1861; Jim Robbins, "Pollution Shrouding National Parks," *New York Times*, December 3, 1989, E19.

41. American Corn Growers Ass'n v. EPA, 291 F.3d 1, 3 (D.C. Cir. 2002); Eric L. Hiser, Rolf R. von Oppenfeld, and Dan P. Kravet, "Regional Haze and Visibility: Potential Impacts for Industry," *BNA Environment Reporter*, April 30, 1999, 2597; "U.S. Regional Haze Rule Finalized; Western Governors Praise Amended Plan," *BNA Environment Reporter*, April 30, 1999, 2558.

42. *American Corn Growers Ass'n*, 291 F.3d 1.

43. Lisa Heinzerling and Rena Steinzor, "A Perfect Storm: Mercury and the Bush Administration," *Environmental Law Reporter* 34 (2004): 10306.

44. "EPA Report on Utility Air Toxics Finds Mercury Is Biggest Potential Threat," *Utility Environment Report,* February 27, 1998, 1; "Utility Air Toxics Report Highlights Mercury as Pollutant of Concern, Lists Other HAPs," *BNA Environment Reporter,* February 27, 1998, 2285; "EPA Says Mercury Emissions from Coal-Fired Plants Are a Concern," *Coal Week,* March 2, 1998, 8; James Gerstenzang, "EPA Report Raises New Concerns about Mercury," *Los Angeles Times,* December 19, 1997, A37.

45. "EEI Challenges EPA-NRDC Mercury Pact to Study Strategies for Reducing CO_2," *Utility Environment Report,* July 17, 1998, 5; "Utilities Lobby against Emission Regs," *National Journal's Daily Energy Briefing,* June 17, 1998; David H. Festa and Stacey E. Davis, "Moving on Mercury: First Steps for Electric Utilities," *Electricity Journal,* August 1997.

46. "Mercury Emissions Likely to Be Regulated as NAS Study Affirms Health Concerns," *Utility Environment Report,* July 14, 2000, 1; "EPA to Require Coal-Fired Utilities to Report Data on Mercury Emissions," *Utility Environment Report,* November 20, 1998, 5.

47. U.S. Environmental Protection Agency, "Regulatory Finding on Emissions of Hazardous Air Pollutants from Electric Utility Steam-Generating Units," 65 Fed. Reg. 79825 (December 20, 2000); John J. Fialka, "EPA Plans Curbs on Mercury Emissions from Coal-Fired Power Plants by 2004," *Wall Street Journal,* December 15, 2000, B2.

48. Ross Gelbspan, *The Heat Is On* (Reading, MA: Addison-Wesley, 1997), 33; Al Gore, *Earth in the Balance: Ecology and the Human Spirit* (Boston: Houghton Mifflin, 1992), chap. 6; Marc K. Landy, Marc J. Roberts, and Stephen R. Thomas, *The Environmental Protection Agency: Asking the Wrong Questions* (New York: Oxford University Press, 1990), 258; "EPA Releases Report on Greenhouse Gas Emission Status," *Coal Week,* November 14, 1994, 8; "Greenhouse Effect Gases Biggest Threat to Environment, Gore Says at Conference," *BNA Environment Reporter,* April 29, 1994, 2227; "U.S. Greenhouse Gas Inventory for 1985–1990 Outlined in DOE Report," *BNA National Environment Daily,* October 27, 1993; "Global Warming Remains Low Priority for Electric Utilities, Survey Finds," *Utility Environment Report,* January 8, 1993, 3.

49. Thomas W. Lippman, "Energy Tax Would Touch All," *Washington Post,* February 18, 1993, A1; "Clinton Plan: Btu's Bearing the Brunt," *Platts Oilgram News,* February 18, 1993, 1.

50. Craig S. Cano, "Gas Groups Gain Some Ground in Btu-Tax Debate; Discussions Continue," *Inside FERC,* March 15, 1993; "Environmental Groups Flex for Industry Opposition to Btu Tax," *BNA National Environment Daily,* February 22, 1993 (false choice quote); "Wide Range of Lobbies Pull Together on Taxes," *Platts Oilgram News,* February 8, 1993; David Wessel and Rick Wartzman,

"Energy Interests Mobilize against a New Tax, but Real Fight May Focus on What Form It Takes," *Wall Street Journal,* January 26, 1993, A16.

51. "Lobbyists Boast Btu Tax Beaten in the House," *Platts Oilgram News,* May 24, 1993; Patrick Crow, "BTU Tax Battle Hits Capitol Hill," *Oil & Gas Journal,* May 10, 1993, 25; "Congress Checks Impact of Btu Tax," *Coal Week,* March 1, 1993; "Environmental Groups Flex for Industry Opposition to Btu Tax," *BNA National Environment Daily,* February 22, 1993.

52. "In Close Vote, House Sends Energy Tax and Budget Plan on to Senate," *Inside FERC,* May 31, 1993; "House Ways and Means Committee Passes Higher Btu Tax Rate but Accommodates Gas Industry Demand for Collection Directly from End Users," *Foster Natural Gas Report,* May 13, 1993; Jackie Calmes and David Wessel, "Clinton Changes Course on Part of Energy Tax," *Wall Street Journal,* May 11, 1993.

53. Michael Weisskopf, "Fanning a Prairie Fire," *Washington Post,* May 21, 1993, A1; "Gas Lobby Wins Btu-Tax Concessions as Ways and Means Reports Bill," *Inside FERC,* May 17, 1993; Patrick Crow, "BTU Tax Battle Hits Capitol Hill," *Oil & Gas Journal,* May 10, 1993, 25.

54. Paul Sonali, "The Big Btu Victory," *Platts Oilgram News,* August 9, 1993; David E. Rosenbaum, "Clinton Backs Off Plan for New Tax on Heat in Fuels," *New York Times,* June 9, 1993, A1.

55. Paul Sonali, "Death of Energy Tax Makes Carbon Levy, Other Environmental Taxes Less Likely," *Utility Environment Report,* August 6, 1993, 12.

56. "The Electric Utility Industry Withheld a Blanket Endorsement," *Inside Energy with Federal Lands,* May 16, 1994, 4; Timothy Noah, "Environmentalists Fault Clinton Plan on Emissions Tied to Global Warming," *Wall Street Journal,* April 19, 1994, A22; John H. Cushman Jr., "Clinton Urging Voluntary Goals on Air Pollution," *New York Times,* October 19, 1993, A23.

57. Gelbspan, *Heat Is On,* 64–70; Dick Thompson, "Capitol Hill Meltdown," *Time,* August 9, 1999, 56; Robert Wright, "Some Like It Hot," *New Republic,* October 9, 1995, 6 (Rohrabacher quote).

58. "No Plans to Regulate Carbon Dioxide, Although Legal Authority Exists, EPA Says," *BNA Environment Reporter,* October 15, 1999, 1085; "EPA Moving to Define CO_2 as a Criteria Pollutant," *Electricity Daily,* July 13, 1998.

59. "Industry Groups Opposing Clean Air Act CO_2 Rules," *Electricity Daily,* July 31, 2000; Gerald Karey, "'Green' Groups Want Us to Regulate CO_2," *Platts Oilgram News,* October 21, 1999, 3.U.S.

60. 42 U.S.C. §§ 6921(b)(3)(A)(i), 6921(b)(3)(C), 6982(n); U.S. Environmental Protection Agency, "Final Regulatory Determination on Four Large-Volume Wastes from the Combustion of Coal by Electric Utility Power Plants," 58 Fed. Reg. 42466, 42467–68 (1993).

61. U.S. Environmental Protection Agency, "Final Regulatory Determination," 42472–73.

62. David J. Tenenbaum, "Questioning Coal Ash Deregulation," *Environmental Health Perspectives* 108 (2000): 496; "EPA Decides against Regulating Coal-Burning Plants' Waste," *Washington Post*, April 26, 2000, 7; Judith Jacobs, "Senator Suggests EPA Ignoring Science in Fossil Fuel Waste Regulation Effort," *BNA Environment Reporter*, April 7, 2000, 630; "22 Groups Ask Clinton to Intercede on EPA Coal Waste Disposal Rules," *Utility Environment Report*, March 27, 1998.

63. "EPA Reaches Settlements with ComEd, Consumers on Air, Water Violations," *Utility Environment Report*, December 20, 1996, 4; John H. Cushman Jr., "E.P.A. Is Canceling Pollution Testing across the Nation," *New York Times*, November 25, 1995, A1; Gary Lee, "Browner Strengthens Enforcement Office," *Washington Post*, October 14, 1993, A29.

64. Congressional Research Service, "Air Quality and Electricity: Enforcing New Source Review," January 31, 2000, 5–6; Bruce Buckheit, testimony before the Senate Democratic Policy Committee, in "Clearing the Air: An Oversight Hearing on the Administration's Clean Air Enforcement Program," February 6, 2004, https://www.dpc.senate.gov/hearings/hearing11/buckheit.pdf; Matthew L. Wald, "Old Plants with New Parts Present a Problem to E.P.A.," *New York Times*, December 26, 1999, A22.

65. Government Accountability Office, "Air Pollution: EPA Needs Better Information on New Source Review Permits," June 2012, 18.

66. Department of Justice, Office of Legal Policy, "New Source Review: An Analysis of the Consistency of Enforcement Actions with the Clean Air Act and Implementing Regulations," January 2002, 109; John H. Stam, "Clinton's Big Enforcement Initiative Poses Major Challenge for Bush Team, Lawyer Says," *BNA Environment Reporter*, February 16, 2001, 293; Andrew M. Ballard, "Duke Power Claims EPA Is Changing Definition of 'Routine Maintenance' Unfairly," *BNA Environment Reporter*, January 5, 2001, 11; Pamela Najor, "Government Sues Electric Companies over New Source Review at 17 Power Plants," *BNA Environment Reporter*, November 12, 1999, 1269; David Stout, "7 Utilities Sued by U.S. on Charges of Polluting Air," *New York Times*, November 4, 1999, A1.

67. *New Source Review Policy, Regulations and Enforcement Activities: Hearings before the Senate Committee on Environment and Public Works and the Senate Committee on the Judiciary*, 107th Cong. 608 (July 16, 2002) (testimony of John Walke, NRDC); Bruce Barcott, "Changing All the Rules," *New York Times Magazine*, April 4, 2004, 38.

68. Wisconsin Electric Power Co. v. Reilly, 893 F.2d 901 (7th Cir. 1990); *Clean Air Act: New Source Review Regulatory Program: Hearings before the Subcommittee on Clean Air, Wetlands, Private Property and Nuclear Safety of the Senate Com-*

mittee on Environment and Public Works, 106th Cong. 74 (February 28, 2000) (testimony of Eric Schaeffer, Environmental Integrity Project); Christopher W. Armstrong, "EPA's New Source Review Enforcement Initiatives," *Natural Resources & Environment* 14 (2000): 204; Ballard, "Duke Power Claims"; Barcott, "Changing All the Rules" (quoting Dan Riedinger, EEI).

69. *Clean Air Act: New Source Review Regulatory Program: Hearings,* 626 (testimony of Joseph Bast, Heartland Institute); Laura Mahoney, "California, Colorado Regulators Disagree on Need to Change Controversial Program," *BNA Environment Reporter,* July 20, 2001, 1399 (quoting Dave Ouimette, Colorado Department of Public Health and the Environment).

70. Armstrong, "New Source Review Enforcement," 203; Buckheit, testimony; *Public Health and Natural Resources: A Review of the Implementation of Our Environmental Laws—Parts I and II: Hearings before the Senate Committee on Governmental Affairs,* 107th Cong. 35 (March 7, 2003) (testimony of Eric Schaeffer).

71. Makram B. Jaber, "Utility Settlements in New Source Review Lawsuits," *Natural Resources & Environment* 18 (2004): 23; *Clean Air Act: New Source Review Regulatory Program: Hearings,* 101 (testimony of William F. Tyndall, Cinergy Services Inc.); Conrad Schneider, *Power to Kill: Death and Disease from Power Plants Charged with Violating the Clean Air Act* (Boston: Clean Air Task Force, July 2001), 8; Larry B. Parker and John E. Blodgett, "Air Quality and Electricity: Enforcing New Source Review," Congressional Research Service, January 31, 2000, 12.

72. Jonathan Remy North and Richard L. Revesz, "Grandfathering and Environmental Protection," *Northwestern University Law Review* 101 (2007): 1696; "Ohio-Based Utility to Reduce Emissions at 10 Coal-Fired Plants in Air Act Settlement," *BNA Environment Reporter,* January 5, 2001, 10; "Utility Settles Lawsuit for \$1.4 Billion," *Washington Post,* December 22, 2000, A5; "Virginia Firm Agrees to Major Reductions in NO_x, SO_2, Conversion of Two Coal Plants," *BNA Environment Reporter,* November 24, 2000, 2453.

73. Marc B. Mihaly, "Recovery of a Lost Decade (or Is It Three?): Developing the Capacity in Government Necessary to Reduce Carbon Emissions and Administer Energy Markets," *Oregon Law Review* 88 (2009): 475–77.

74. Steven D. Czajkowski, "Notes: Focusing on Demand Side Management in the Future of the Electric Grid," *Pittsburgh Journal of Environmental and Public Health Law* 4 (2010): 117–18, 132; Kay L. Gehring, "Can Yesterday's Demand-Side Management Lessons Become Tomorrow's Market Solutions?," *Electricity Journal,* June 2002, 15; "Environmentalists Urge Aggressive DSM in Florida to Reduce Pollution," *Utility Environment Report,* July 22, 1994, 10.

75. Mihaly, "Lost Decade," 472; David B. Spence, "Regulation, 'Republican Moments,' and Energy Policy Reform," *BYU Law Review* 2011 (2011): 1564–65.

76. "East Ky. Co-op Gets Approval for SCR at Spurlock, Cooper Coal-Fired Plants," *Utility Environment Report*, December 1, 2000, 11; "Kentucky Power Seeks PSC Reversal to Recoup Low-NO$_x$ Burner Expense," *Utility Environment Report*, July 4, 1997, 9; "Indiana URC Approves 1.9% Psi Energy Rate Hike to Cover Clean Air Expenses," *Utility Environment Report*, March 17, 1995, 9.

77. Alexandra B. Klass and Elizabeth J. Wilson, "Interstate Transmission Challenges for Renewable Energy: A Federalism Mismatch," *Vanderbilt Law Review* 65 (2012): 1821–22; Mihaly, "Lost Decade," 477; David B. Spence, "Coal-Fired Power in a Restructured Electricity Market," *Duke Environmental Law & Policy Forum* 15 (2005): 202.

78. "Moler: FERC Taking Another Look at EIS but Is 'Not Environmental Regulator,'" *Utility Environment Report*, March 1, 1996, 1; "NO$_x$ Emissions Prompted FERC EIS," *Coal & Synfuels Technology*, August 7, 1995, 4.

79. "EPA Tells FERC to Go After Power Producers," *Coal & Synfuels Technology*, February 26, 1996, 1; "FERC Calls Open-Access Rule Benign, but EPA Official Questions Analysis," *Utility Environment Report*, November 24, 1995, 1; "DOE, Justice Urge FERC to Stretch Envelope on Corporate Divestiture," *Inside FERC*, August 14, 1995, 3.

80. "Letter to Clinton, House Resolution Raise Profile of Protest on FERC Rule," *Utility Environment Report*, May 10, 1996, 6; "EPA Refers Order No. 888 to Council on Environmental Quality," *Foster Natural Gas Report*, May 16, 1996, 1; "FERC Determines Only Slight Environmental Impact from Proposed Rule to Require Open Access Transmission by Electric Utilities," *Foster Natural Gas Report*, April 18, 1996, 1.

81. "Claiming Victory, EPA Says White House Backs Drive for Pollution Mitigation," *Utility Environment Report*, June 21, 1996, 1; Gary Lee, "New Curbs on Pollution Set for Electric Utilities; Deregulation Impact Feared for Northeast," *Washington Post*, June 19, 1996, A17.

82. Alan Weisman, "Power Trip," *Harper's*, October 1, 2000, 76; Andrew C. Revkin, "Power Plants Lure Investors, and Critics," *New York Times*, April 27, 1999, B1; John Richardson, "Restructuring Review," *Power Economics*, December 31, 1998, 24; Martha M. Hamilton, Deregulation Recharges an Industry, *Washington Post*, December 21, 1998, F10; Richard Munson and Tina Kaarsberg, "Unleashing Innovation in Electricity Generation," *Issues in Science and Technology*, March 22, 1998, 51.

83. Taylor Moore et al., "Merchant Plants Drive Market Competition," *EPRI Journal* 24 (Summer 1999): 8–17; Revkin, "Power Plants"; Agis Salpukas, "GPU Sells 23 Power Plants to a French-Linked Company," *New York Times*, November 10, 1998, C6; Munson and Kaarsberg, "Unleashing Innovation"; Beth Belton, "Utility Mergers Raise Question: Is Bigger Better?," *USA Today*, August 13, 1996, B4.

84. Moore et al., "Merchant Plants"; Munson and Kaarsberg, "Unleashing Innovation."

85. Chris Deisinger, "The Backlash against Merchant Plants and the Need for a New Regulatory Model," *Electricity Journal,* December, 2000, 13; Casey Bukro, "Perfect Views of Exhaust Stacks," *Chicago Tribune,* April 9, 2000, C1; "Power Developers Flood Illinois Market with Plans for 17,000 Mw in New Capacity," *Power Markets Week,* January 31, 2000, 12; Tom Ragan, "Another Power-Plant Bid Gathers Steam," *Chicago Tribune,* December 8, 1999, MC1.

86. Robert Swanekamp, "Rise of the Merchant Class," *Power,* September / October 1999, 48; Mark R. Madler, "Proposed Power Plant Generating Opposition," *Chicago Tribune,* March 9, 1999, MC1.

87. Illinois Environmental Protection Agency, "Electric Power Plant Construction Projects since 1998" (2003), http://www.epa.state.il.us/air/permits /electric/2003-july.pdf; Hal Dardick, "Peaker Plant Beats Protest, Gets Approval," *Chicago Tribune,* February 25, 2000, W3; Madler, "Proposed Power Plant."

88. Jacqueline Lang Weaver, "Can Energy Markets Be Trusted? The Effect of the Rise and Fall of Enron on Energy Markets," *Houston Business and Tax Law Journal* 4 (2004): 15–16, 114; Sheila Slocum Hollis, "Marketing Hydrocarbons and Electricity," *Oil & Gas Journal,* June 29, 1998, 56.

89. "Vepco Shifts from Coal to Gas at Possum Point Unit," *Coal Week,* April 17, 2000, 4; "Pa. Generators Question Content of Pro-Green Ads," *Megawatt Daily,* July 28, 1999; Jason Makansi, "From the Fuel Use Act to the 'Use This Fuel' Act," *Power,* May / June 1998, 4; Marie Leone, "Independent Power: State of the Market," *Power,* February 1994, 45.

90. "Louisiana Community Group Files Suit to Block Restart of Energy Plant," *Utility Environment Report,* July 16, 1999, 9; "Environmentalists, EPA Challenge Plan to Restart Conners Creek Plant," *Utility Environment Report,* June 19, 1998, 4; "Report Indicates Utility Deregulation May Be Increasing Air Pollution," *BNA Environment Reporter,* January 23, 1998, 1789; Bruce Biewald, "Competition and Clean Air: The Operating Economics of Electricity Generation," *Electricity Journal,* January 1997; "Centerior Move to Restart Lake Shore Coal-Fired Unit Raises Air Concerns," *Utility Environment Report,* June 21, 1996, 3.

91. Deisinger, "Backlash"; Wallace Roberts, "Power Play," *American Prospect,* January / February 1999, 71.

92. The description of the California crisis is drawn from Timothy P. Duane, "Regulation's Rationale: Learning from the California Energy Crisis," *Yale Journal on Regulation* 19 (2002): 503–8; Steven Ferrey, "Soft Paths, Hard Choices: Environmental Lessons in the Aftermath of California's Electric Deregulation Debacle," *Virginia Environmental Law Journal* 23 (2004): 255–62, 286–88; Jim Rossi, "The Electric Deregulation Fiasco: Looking to Regulatory Federalism to Promote a Balance between Markets and the Provision of Public Goods," *Michigan Law*

Review 100 (2002): 1778; Weaver, "Can Energy Markets Be Trusted?," 31–34, 83–85; Rich Heidorn Jr., "Electric Competition Isn't Happening in Deregulated States," *Philadelphia Inquirer,* June 14, 1998, D1; Terence Monmaney, "Rush for Power Plant in Chino Raises Concerns," *Los Angeles Times,* July 24, 2001, B1; Gary Polakovic, "Agreement Lets Power Station Triple Output," *Los Angeles Times,* August 5, 2000, B1.

93. Barton Gellman, *Angler: The Cheney Vice Presidency* (New York: Penguin, 2008); Margaret Kriz, "Power Struggle," *National Journal,* March 31, 2001, 942; Yuval Rosenberg, "Testing the Air in Texas," *Newsweek,* October 19, 2000; Margaret Kriz, "Presidential Positions on the Environment," *National Journal,* April 1, 2000, 1032.

94. Sonali, "Big Btu Victory."

95. "Prices Tumble on Va. Power Accord, Rumors about Cinergy EPA Settlement," *Utility Environment Report,* December 1, 2000, 4; Mark Golden, "Dirty Dealings," *Wall Street Journal,* September 13, 1999, R13; "Utilities Continue Full Compliance with Program to Cut SO_2, Agency Says," *BNA Environment Reporter,* August 6, 1999, 696.

96. "Acid Rain Control: Success on the Cheap," *Science,* November 6, 1998, 1024; Dallas Burtraw, "Call It 'Pollution Rights,' but It Works," *Washington Post,* March 31, 1996, C3.

97. Kathy Dhanda Kanwalroop, "A Market-Based Solution to Acid Rain: The Case of the Sulfur Dioxide (SO_2) Trading Program," *Journal of Public Policy & Marketing* 18 (Fall 1999): 258; "Major Study Finds Acid Rain Program Cutting Sulfur Deposition in Northeast," *BNA Environment Reporter,* August 14, 1998, 781; "Study Shows No Improvement in Adirondacks," *National Journal's Daily Energy Briefing,* August 6, 1998.

98. Daniel Yergin, *The Quest: Energy, Security, and Remaking of the Modern World* (New York: Penguin, 2011), 384; Daniel Gabaldon, Eric Spiegel, and Joe Van den Berg, "Betting Big on Baseload," *Electricity Journal,* August / September 2006, 74; David Haarmeyer and Benjamin Tait, "The Race to Competitiveness: US Power Market In Transition," *Power Economics,* February 29, 2000, 15.

99. "FERC Underestimated Air Effects of Competition, Study Says," *Megawatt Daily,* November 27, 2001, 1; "Greenpeace Asks Federal Trade Commission to Bar AGA Environmental Ads as 'Deceptive' and Require Corrective Information," *Foster Natural Gas Report,* February 10, 1994, 21.

100. "Coal Industry Goes from Bad to Worse as Doom and Gloom Usher in New Century," *Energy Report,* January 10, 2000; "EPA's Clean Air Rules Have Wide Impact on Utilities," *Energy Report,* October 5, 1998; "Electric Industry Report Sees Dire Effects of EPA Regulations on Power Cost, Supply," *BNA Environment Reporter,* May 29, 1998, 257.

8. RETRENCHMENT

1. Bruce Barcott, "Changing All the Rules," *New York Times Magazine,* April 4, 2004, 38; "Older Electric Power Plants Produce High Percentages of Pollutants, GAO Says," *BNA Environment Reporter,* June 21, 2002, 1375; "Study Fingers Plants Burning Coal as Key Sources of Smog Formation," *Coal Week,* May 7, 2001, 8; "Coal Scores with Wager on Bush," *Washington Post,* March 25, 2001, A5; "Report Says Major Switch from Coal to Gas Would Achieve Reductions in Key Pollutants," *BNA Environment Reporter,* November 24, 2000, 2459.

2. Robert Swanekamp, "High Gas Prices and Technology Brighten King Coal's Star," *Global Energy Business,* November / December, 2001, 43; "Coal Scores with Wager on Bush"; Eric Pianin, "As Coal's Fortunes Climb, Mountains Tremble in W.Va.," *Washington Post,* February 25, 2001, A3.

3. Shankar Vedantam, "Scientist Named to Head the EPA," *Washington Post,* March 5, 2005, A1; Barcott, "Changing All the Rules"; Katharine Q. Seelye, "Bush Nominates Utah Governor to Lead Environmental Agency," *New York Times,* August 12, 2003, A1; "Whitman Announces Resignation in Letter to Bush, Citing Wish to Return to New Jersey," *BNA Environment Reporter,* May 23, 2003, 1157.

4. Steven Ferrey, "Soft Paths, Hard Choices: Environmental Lessons in the Aftermath of California's Electric Deregulation Debacle," *Virginia Environmental Law Journal* 23 (2004): 267; Peter Behr and Rene Sanchez, "Bush Seeks Speedy Review of Calif. Power Plants," *Washington Post,* February 17, 2001, A8; "EPA Extends Guidance to California on Use of Emergency Electric Generators," *BNA Environment Reporter,* February 16, 2001, 314; Peter Behr and Walter Pincus, "Power Crisis Awaits New Administration," *Washington Post,* January 19, 2001, at A16.

5. Ferrey, "Soft Paths," 263, 313; "California Emergency Peaker Placement Overdue," *Generation Week,* April 11, 2001; "White House Weighing Davis Request to Expedite Federal Reviews of Power Plants," *BNA Environment Reporter,* February 16, 2001, 313.

6. Timothy P. Duane, "Regulation's Rationale: Learning from the California Energy Crisis," *Yale Journal on Regulation* 19 (2002): 517; Ferrey, "Soft Paths," 305; "Market Assesses New Calif. Role, Sees Supply Crunch, High Prices Continuing," *Power Markets Week,* February 5, 2001, 1.

7. Michael Janofsky, "In the Race to Produce More Power, States Are Faced with Environmental Tradeoffs," *New York Times,* March 26, 2001, A17; "Abraham Hits Environmental Regulations in Drive to Increase National Energy Supply," *BNA Environment Reporter,* March 23, 2001, 532; Jeanne Cummings, "Power Politics: Energy Crisis Offers Clues to the Workings of Bush Administration," *Wall Street Journal,* February 16, 2001, A1; Behr and Sanchez, "Speedy Review."

8. Duane, "Regulation's Rationale," 519–20; Ferrey, "Soft Paths," 283; Margaret Kriz, "Power Struggle," *National Journal,* March 31, 2001, 942; "California's Power Authority Puts State in Infrastructure Business," *Power Markets Week,* May 21, 2001, 3; Edwin Chen and Richard Simon, "Bush Offers No Answers for California," *Los Angeles Times,* May 9, 2001, 1.

9. Gary Polakovic, "AQMD Moves to Overhaul Power Plant Emission Rules," *Los Angeles Times,* January 20, 2001, 23.

10. Ferrey, "Soft Paths," 294; Jacqueline Lang Weaver, "Can Energy Markets Be Trusted? The Effect of the Rise and Fall of Enron on Energy Markets," *Houston Business and Tax Law Journal* 4 (2004): 86; Gary Polakovic, "State Losing Ground in War on Dirty Air," *Los Angeles Times,* July 17, 2001, A1; "California AG to Convene Grand Jury on Allegations of Price Manipulation," *Power Markets Week,* June 18, 2001, 3.

11. Duane, "Regulation's Rationale," 498–99; Ferrey, "Soft Paths," 271, 276–86.

12. Duane, "Regulation's Rationale," 509–10; Weaver, "Can Energy Markets Be Trusted?," 20, 38–39, 42–45, 65; Peter H. King, Nancy Vogel, and Nancy Rivera Brooks, "Paper Trail Points to Roots of Energy Crisis," *Los Angeles Times,* June 16, 2002, 1.

13. Duane, "Regulation's Rationale," 512–13; Ferrey, "Soft Paths," 334–36; Weaver, "Can Energy Markets Be Trusted?," 47–49.

14. Duane, "Regulation's Rationale," 507, 516, 525; Ferrey, "Soft Paths," 335–36; Weaver, "Can Energy Markets Be Trusted?," 79–81.

15. *Clean Air Act: New Source Review Regulatory Program: Hearings before the Subcommittee on Clean Air, Wetlands, Private Property and Nuclear Safety of the Senate Committee on Environment and Public Works,* 106th Cong. 611 (February 28, 2000) (testimony of John Walke, NRDC); Michael Abramowitz and Steven Mufson, "Papers Detail Industry's Role in Cheney's Energy Report," *Washington Post,* July 18, 2007, A1; Christopher Drew and Richard A. Oppel Jr., "How Power Lobby Won Battle of Pollution Control at E.P.A.," *New York Times,* March 6, 2004, A1; Cathy Landry, "Cheney to Head US Task Force on Energy," *Platts Oilgram News,* January 30, 2001, 3; "Bush Official Suggests Clean-Air Rules Will Loosen," *Platts Coal Outlook,* January 29, 2001.

16. Drew and Oppel, "How Power Lobby Won" (memo quote); Neela Banerjee, "Files Detail Debate in E.P.A. on Clean Air," *New York Times,* March 21, 2002, A32; Michael Isikoff, "A Plot to Foil the Greens," *Newsweek,* June 4, 2001, 36.

17. National Energy Policy Development Group, *National Energy Policy* (Washington, DC: U.S. Government Printing Office, 2001), viii, xii; "Bush's Energy Plan Bares Industry Clout," *Los Angeles Times,* August 26, 2001, A1.

18. National Energy Policy Development Group, *National Energy Policy,* 7.18, app. 1; Jeff Goodell, "Blasts from the Past," *New York Times Magazine,* July 22, 2001, 31.

19. Executive Order 13212, 66 Fed. Reg. 28357 (May 22, 2001).

20. Goodell, "Blasts from the Past"; U.S. Department of Energy, National Energy Technology Laboratory, *Tracking New Coal-Fired Power Plants,* January 24, 2007; "Illinois Initiative Sparked Interest, Proposed Power Projects Advance," *U.S. Coal Review,* June 10, 2002; Geraldine Baum, "Suddenly, Dirty Old Coal Is the Fossil Fuel of the Future," *Los Angeles Times,* May 27, 2001, 24.

21. Barry Cassell, "New Coal Unit at Santee Cooper Goes Commercial," *SNL Generation Markets Week,* October 7, 2008; "Hawthorn Unit 5: Harbinger of Coal's Comeback," *Power,* May / June 2001, 19.

22. Daniel Gabaldon, Eric Spiegel, and Joe Van den Berg, "Betting Big on Baseload," *Electricity Journal,* August / September 2006, 74; Jeff Ryser, "With 50,000 MW Seen as Key to Merchant Sector, Analysts See Consolidation on Horizon," *Power Markets Week,* March 14, 2005; "Future of Gas-Based Generation Up for Grabs," *Natural Gas Week,* July 18, 2003.

23. Jeff Goodell, *Big Coal: The Dirty Secret behind America's Energy Future* (Boston: Houghton Mifflin, 2006), 193; Goodell, "Blasts from the Past"; Dan Morgan and Peter Behr, "Developing Energy Bill Ignites Power Scramble," *Washington Post,* May 20, 2001, A1; "Multi-Pollutant Strategy Would Disrupt Economy, Analyst Says," *Coal Week,* April 23, 2001, 8; Andrew C. Revkin, "New Alliance Forms to Cut Plants' CO_2 Emissions," *Chicago Tribune,* March 11, 2001, C9; "Senate Republicans Preparing Broad Energy Bill for Quick Action," *Inside FERC,* January 29, 2001, 1.

24. "EPA May Scale Back Multi-Pollutant Bill as Bush Administration Focuses on Terrorism," *BNA Environment Reporter,* September 28, 2001, 1862; Katharine Q. Seelye, "White House Rejected a Stricter E.P.A. Alternative to the President's Clear Skies Plan," *New York Times,* April 28, 2002, A26; Eric Pianin, "EPA Mulls Limits for Power Plant Emissions," *Washington Post,* February 28, 2001, A13.

25. White House, "Fact Sheet: President Bush Announces Clear Skies & Global Climate Change Initiatives," February 14, 2002, http://georgewbush -whitehouse.archives.gov/news/releases/2002/02/20020214.html; "EPA: Bush Proposal Eliminates Need for NSR," *Megawatt Daily,* February 20, 2002, 1; Elizabeth Shogren and Gary Polakovic, "Bush Seeks to Curb Power Plant Emissions, Sets Climate Goals," *Los Angeles Times,* February 15, 2002, A18.

26. "Power Plant Pollution, Climate Change Top Priorities for Jeffords in Senate Panel," *BNA Environment Reporter,* July 20, 2001, 1425; "Sen. Jeffords' Jump from Republican Party Derails Coal Industry's Agenda in Congress," *Coal Week,* May 28, 2001, 6.

27. Juliet Eilperin, "Standoff in Congress Blocks Action on Environmental Bills," *Washington Post,* October 18, 2004, A2; "With 'Clear Skies' Bogged Down in Congress, EPA Offers Rules as Alternative to Initiative," *BNA Environment*

Reporter, August 27, 2004, 1840; "Power Companies Seeking Changes in Administration's 'Clear Skies' Measure," *BNA Environment Reporter,* January 31, 2003, 239; Eric Pianin, "Fight Ahead on Emissions," *Washington Post,* January 7, 2003, A4; Eric Pianin and Helen Dewar, "Oil, Air, Energy Laws in Play," *Washington Post,* November 18, 2002, A1.

28. U.S. Environmental Protection Agency, *New Source Review: Report to the President, June 2002* (Washington, DC: U.S Environmental Protection Agency, 2002), 1–3, 11; Kara Sissell, "Industry Asks EPA to Overhaul NSR Program," *Chemical Week,* July 25, 2001.

29. U.S. General Accounting Office, *Clean Air Act: New Source Review Revisions Could Affect Utility Enforcement Cases and Public Access to Emissions Data* (Washington, DC: GAO, October 2003), 17; Steve Cook, "EPA Officials Were in Conflict over Bush Plan to Change New Source Review, Records Show," *BNA Environment Reporter,* March 22, 2002, 622; Drew and Oppel, "How Power Lobby Won" ; Barcott, "Changing All the Rules"; Eric Pianin, "EPA Veteran Resigns over Pollution Policy," *Washington Post,* March 1, 2002, A5.

30. U.S. Environmental Protection Agency, "Prevention of Significant Deterioration (PSD) and Nonattainment New Source Review (NSR); Routine Maintenance, Repair and Replacement," 67 Fed. Reg. 80290, 80293–95 (2002).

31. Cassie N. Aw-yang, "EPA's Changes to the Routine Maintenance, Repair and Replacement Rule of the New Source Review Program: An Unlawful Threat to Public Health and Welfare?," *Environs* 27 (2004): 342, 345; Jonathan Remy North and Richard L. Revesz, "Grandfathering and Environmental Regulation," *Northwestern University Law Review* 101 (2007): 1703–4; Steve Cook, "Legality of 20 Percent Cost Threshold in Equipment Replacement Rule Disputed," *BNA Environment Reporter,* September 10, 2004, 1859.

32. *New Source Review Policy, Regulations and Enforcement Activities: Hearings before the Senate Committee on Environment and Public Works and the Senate Committee on the Judiciary,* 107th Cong. 619 (July 16, 2002) (testimony of E. Donald Elliott, of Paul, Hastings, Janofsky & Walker); Tripp Baltz, "Industry Says Rule Will Bring Predictability; Others Fear Rollback of Air Quality Progress," *BNA Environment Reporter,* April 4, 2003, 779; Kenneth W. Betz, "New Source Review Revisions Fail to Satisfy," *Energy User News,* January 1, 2003, 1.

33. Barton Gellman, *Angler: The Cheney Vice Presidency* (New York: Penguin, 2008), 208; U.S. Environmental Protection Agency, Office of Inspector General, *New Source Review Rule Change Harms EPA's Ability to Enforce against Coal-Fired Electric Utilities,* Report No. 2004-P-00034, September 30, 2004, 10–11; Barcott, "Changing All the Rules"; "Moderate Christie Whitman Resigns as Administrator of EPA," *Oil Daily,* May 21, 2003.

34. U.S. Environmental Protection Agency, "Prevention of Significant Deterioration (PSD) and Non-Attainment New Source Review (NSR): Equipment

Replacement Provision of the Routine Maintenance, Repair and Replacement Exclusion," 68 Fed. Reg. 61248, 61250 (October 27, 2003); Barcott, "Changing All the Rules"; Al Kamen, "Not Wavering on Policy Changes," *Washington Post,* November 3, 2003, A17.

35. New York v. EPA, 443 F.3d 880, 885, 888 (D.C. Cir. 2006).

36. *Clean Air Act: New Source Review Regulatory Program: Hearings before the Subcommittee on Clean Air, Wetlands, Private Property and Nuclear Safety of the Senate Committee on Environment and Public Works,* 106th Cong. 97 (February 28, 2000) (testimony of David Hawkins, NRDC); U.S. Environmental Protection Agency, "NSR 90-Day Review Background Paper," June 22, 2001, 10; John J. Fialka, "Part of Clean Air Act Is Termed 'Widespread Failure' in Study," *Wall Street Journal,* April 22, 2003, A21; "Old Fossil Plants Much Dirtier, Pose Tough Problem: GAO," *Inside Energy with Federal Lands,* June 24, 2002, 7; John J. Fialka and David S. Cloud, "White House Review Freezes EPA Inquiry," *Wall Street Journal,* June 28, 2001, A3.

37. Richard Perez-Pena, "Possible Federal Pullout Clouds Northeast States' Pollution Suits," *New York Times,* August 20, 2001, A1; Elizabeth Shogren, "Decision Near on Air Rule Review," *Los Angeles Times,* August 13, 2001, A1; Fialka and Cloud, "White House Review."

38. Bruce Buckheit, testimony before the Senate Democratic Policy Committee, in "Clearing the Air: An Oversight Hearing on the Administration's Clean Air Enforcement Program," February 5, 2004, https://www.dpc.senate.gov /hearings/hearing11/buckheit.pdf; Mike Ferullo, "EPA Pursuing New Source Review Violations amidst Efforts to Revise Rules, Official Says," *BNA Environment Reporter,* October 18, 2002, 2261; Gerald B. Silverman and Steve Cook, "DOJ Announces New Source Review Cases May Proceed, Asserting Law Supports Them," *BNA Environment Reporter,* January 18, 2002, 117; Christopher Marquis, "E.P.A. Power Plant Cases to Proceed, Ashcroft Says," *New York Times,* January 16, 2002, A5.

39. Susan Bruninga, "Ohio Edison Co. Agrees to Pay $1.1 Billion to Cut Emissions 212,000 Tons from Plants," *BNA Environment Reporter,* March 25, 2005, 573; Eric Pianin, "Clean-Air Ruling Puts Blame on Ohio Utility," *Washington Post,* August 8, 2003, E1.

40. United States v. Southern Indiana Gas & Electric Co., 245 F. Supp. 2d 944 (S.D. Ind. 2003); Barcott, "Changing All the Rules"; Christopher Brown, "Federal Trial Opens in Clean Air Lawsuit Alleging Violations by Illinois Electric Utility," *BNA Environment Reporter,* June 6, 2003, 1261.

41. See, e.g., Environmental Defense v. Duke Energy Corp., 549 U.S. 561 (2007); United States v. Ohio Edison Co., 276 F. Supp. 2d 829 (S.D. Ohio 2003).

42. Matthew Bandyk, "Duke Energy Gets Victory on Appeal in Wabash River NSR Case," *SNL FERC Power Report,* October 20, 2010; Bob Matyi, "Jurors Rule in

Duke Energy's Favor on Four of Six Projects in Federal Clean Air Act Suit,"
Electric Utility Week, May 25, 2009, 8; Andrew Childers, "Cinergy Cleared of 10 of
14 Charges That It Violated New Source Review," *BNA Environment Reporter,*
May 30, 2008, 1048; Steven D. Cook, "U.S. Court Adopts Electric Utilities' View on
Definition of 'Routine Maintenance,'" *BNA Environment Reporter,* September 9,
2005, 1814; Steven D. Cook, "Federal Court in Alabama Deals Setback to EPA in
Case against Power Company," *BNA Environment Reporter,* June 10, 2005, 1165.

43. Buckheit testimony; Steve Cook, "EPA Asked to Provide Enforcement Files
on Cases Dropped because of New Rules," *BNA Environment Reporter,* November 14, 2003, 2478; Eric Pianin, "White House to End Power Plant Probes,"
Washington Post, November 6, 2003, A31; Eric Pianin, "EPA Rule Revisions
Roil U.S. Case against Power Plant," *Washington Post,* October 6, 2003, A8.

44. Steve Cook, "EPA Cites Indiana Company for Violations of Clean Air Act
at Three Coal-Fired Plants," *BNA Environment Reporter,* October 1, 2004, 2050;
Juliet Eilperin, "Justice Gets 14 Air Pollution Cases," *Washington Post,* July 15,
2004, E1; Patricia Ware, "Kentucky Coal-Fired Plant Agrees to Pay Nearly $650
Million to Reduce Emissions," *BNA Environment Reporter,* July 6, 2007, 1496; Eric
Pianin, "Justice Dept. Sues Ky. Utility for Breach of Clean Air Act," *Washington
Post,* January 29, 2004, A2; Steve Cook, "EPA to File New Lawsuits for Violations
of New Source Review, EPA Chief Says," *BNA Environment Reporter,* January 23,
2004, 158.

45. Steven D. Cook, "EPA Says Violations of New Source Review by Power
Plants Top Enforcement Priority," *BNA Environment Reporter,* February 15, 2008,
303; Tom Tiernan, "New-Source Ruling Leaves Utilities Seeking New Law, Other
Side Hoping for Enforcement," *Electric Utility Week,* March 27, 2006, 1; Steven D.
Cook, "Power Company Seeks Stay of Lawsuit Pending Separate Ruling, Proposed
Rule," *BNA Environment Reporter,* November 11, 2005, 2288.

46. U.S. Government Accountability Office, *Air Pollution: EPA Needs Better
Information on New Source Review Permits* (Washington, DC, June 2012), 30, app.
3; Steven D. Cook, "AEP Settles Lawsuit Alleging Violations, Will Spend $4.6
Billion on Emissions Cuts," *BNA Environment Reporter,* October 12, 2007, 2165;
Steve Cook, "Government, Cinergy Unlikely to Reach Settlement of Litigation,
Company Says," *BNA Environment Reporter,* March 12, 2004, 529; Mike Ferullo
and Steve Cook, "Virginia Electric Utility Will Spend $1.2 Billion to Cut Air
Pollution from Eight Power Plants," *BNA Environment Reporter,* April 25, 2003,
925; Christopher Lee, "Effort to Ease Air Rules Decried," *Washington Post,*
October 19, 2002, A19; ; Perez-Pena, "Possible Federal Pullout."

47. U.S. Environmental Protection Agency, "National Emission Standards for
Hazardous Air Pollutants from Coal- and Oil-Fired Electric Utility Steam
Generating Units and Standards of Performance for Fossil-Fuel-Fired Electric
Utility, Industrial-Commercial-Institutional, and Small Industrial-Commercial-

Institutional Steam Generating Units; Final Rule," 77 Fed. Reg. 9304, 9310 (February 16, 2012); Barcott, "Changing All the Rules"; "Almost 10 Percent of Women Have Levels of Mercury Near Hazardous Level, CDC Says," *BNA Environment Reporter*, March 9, 2001, 442.

48. Lisa Heinzerling and Rena Steinzor, "A Perfect Storm: Mercury and the Bush Administration," *Environmental Law Reporter* 34 (2004): 10306; Catherine A. O'Neill, "Mercury, Risk and Justice," *Environmental Law Reporter* 34 (2004): 11081; "TXU Official: New Mercury Regulations Will Be Onerous," *Platts Coal Outlook*, October 21, 2002, 1.

49. Goodell, *Big Coal*, 144; Margaret Kriz, "Mercury Uprising," *National Journal* 36 (2006): 33; "Leavitt Signs Emissions Trading Proposal to Cut Mercury Emissions at Power Plants," *BNA Environment Reporter*, December 19, 2003, 2741; Eric Pianin, "EPA Announces 'Cap and Trade' Plan to Cut Mercury Pollution," *Washington Post*, December 16, 2003, A35.

50. Tom Hamburger and Alan C. Miller, "Emissions Rule Geared to Benefit Industry, Staffers Say," *Los Angeles Times*, March 16, 2004, A1; Eric Pianin, "Proposed Mercury Rules Bear Industry Mark," *Washington Post*, January 31, 2004, A4; Jennifer 8. Lee, "New Policy on Mercury Pollution Was Rejected by Clinton E.P.A.," *New York Times*, December 16, 2003, A31.

51. Goodell, *Big Coal*, 143; Michael Schmidt, "EPA Seeks More Mercury Comments, Plans to Analyze Data on Non-Energy Sources," *Inside Energy with Federal Lands*, December 6, 2004, 7; "NMA Says Poor Data Flaws EPA's Mercury Proposals," *Platts Coal Outlook*, May 24, 2004, 11; "Mercury Proposal, Emissions Banking Targeted during Hearing in Philadelphia," *BNA Environment Reporter*, February 27, 2004, 413.

52. O'Neill, "Mercury, Risk and Justice," 11071; Kriz, "Mercury Uprising"; "Witnesses in North Carolina Tell EPA to Scrap Mercury, Interstate Air Rules," *BNA Environment Reporter*, February 27, 2004, 415; Margaret Kriz, "The Next Arsenic," *National Journal*, February 14, 2004.

53. U.S. Environmental Protection Agency, "Standards of Performance for New and Existing Stationary Sources: Electric Utility Steam Generating Units," 70 Fed. Reg. 28606 (2005); U.S. Environmental Protection Agency, "Revision of December 2000 Regulatory Finding," 70 Fed. Reg. 15994 (2005); "EPA Announces First-Ever Regulation to Limit Power Plant Mercury Emissions," *BNA Environment Reporter*, March 18, 2005, 525.

54. New Jersey v. EPA, 517 F.3d 574, 582–84 (D.C. Cir. 2008).

55. "EPA Cites Continuing Progress on Reducing Nitrogen Oxide Emissions in Eastern States," *BNA Environment Reporter*, September 28, 2007, 2061; "EPA Estimates Rule for Lower NO_x Emissions Improved Eastern States' Air Quality in 2004," *BNA Environment Reporter*, August 26, 2005, 1759; "FirstEnergy Agrees to Invest in Polluting Power Plant," *Coal Americas*, July 12, 2004; "EPA Begins

Enforcing Multistate Program to Cut Nitrogen Oxides from Power Plants," *BNA Environment Reporter,* June 4, 2004, 1199; William J. Angelo, "Powerplant Retrofits Require Muscular Cranes with Long Arms," *Engineering News-Record,* January 21, 2002, 23; "Natsource, Evolution Claim 'First' NO_x Deals under SIP Call," *Coal Americas,* June 18, 2001.

56. "Bush to Spare NO_x Rules," *U.S. Coal Review,* April 9, 2001; "New Acid Rain Study Says Problem Is More Serious Than Once Believed," *Utility Environment Report,* April 6, 2001, 3.

57. "EPA Releases Major Plan to Cut SO_2, NO_x Emissions," *Megawatt Daily,* December 19, 2003, 8; "EPA Offers Plan to Reduce Sulfur Dioxide, Nitrogen Oxides from Plants in 29 States," *BNA Environment Reporter,* December 19, 2003, 2742.

58. "Eastern States Need Stronger CAIR, Says Environmental Defense," *Platts Coal Outlook,* August 23, 2004, 12; "Utilities Wary of Costs of EPA Emissions Rule; Environmental Groups Urge Early Compliance," *Electric Utility Week,* August 2, 2004, 12; Felicity Barringer, "Critics Say Clean-Air Plan May Be a Setback for Parks," *New York Times,* May 31, 2004, A12.

59. U.S. Environmental Protection Agency, "Rule to Reduce Interstate Transport of Fine Particulate Matter and Ozone (Clean Air Interstate Rule); Revisions to Acid Rain Program; Revisions to the NO_x SIP Call," 70 Fed. Reg. 25162 (May 12, 2005); "EPA Issues Final Rule to Reduce Emissions from Power Plants in 28 States by 2015," *BNA Environment Reporter,* March 11, 2005, 461.

60. Barry Cassell, "SCANA Works on SO_2 Scrubbers at Wateree and Williams Coal Plants," *SNL Coal Report,* March 24, 2008; "AEP Settles Lawsuit Alleging Violations, Will Spend $4.6 Billion on Emissions Cuts," *BNA Environment Reporter,* October 12, 2007, 2165; Marcin Skomial, "Utilities to Continue with Coal, Emissions under CAIR," *Platts Coal Outlook,* July 23, 2007, 1; Wayne Barber, "East Kentucky Power to Invest $656 Million in Air Controls to End NSR Litigation," *SNL Electric Utility Report,* July 9, 2007; "Alabama Power Sees Partial Clean-Air Deal with Government as a Win, but Issues Remain," *Electric Utility Week,* May 1, 2006, 12; "Two North Dakota Co-Ops Make $100 Million Deal with Justice to Clean Up Young Coal Plant," *Electric Utility Week,* May 1, 2006, 30; Cathy Cash, "EPA Plan for Powerplants Will Push Pollution Control Work," *Engineering News-Record,* March 21, 2005, 12; "Cinergy Begins Its Biggest Ever $2B Emissions-Reduction Project," *Platts Coal Outlook,* September 6, 2004, 11.

61. North Carolina v. EPA, 531 F.3d 896, 907 (D.C. Cir. 2008).

62. Ibid.; Barry Cassell, "Pennsylvania Says It Will Impose CAIR Restrictions as of Jan. 1," *SNL Power Daily Northeast,* December 31, 2008; Michael Niven, "CAIR Ruling Sparks Immediate Surge in Emissions Allowance Prices," *SNL Power Daily Northeast,* December 24, 2008; Amena Saiyid, "Investment Banks, Hedge Funds Leave Emissions as Market Dries Up," *Platts Coal Outlook,* October

13, 2008, 10; Michael Niven, "CAIR's Demise Throws US Emissions Trading Market into Turmoil," *SNL Coal Report,* July 21, 2008; "CAIR Decisions Leaves Turmoil in Its Wake for EPA, Environmentalists, Power Plants," *BNA Environment Reporter,* July 18, 2008, 1429.

63. For an extensive analysis of this standard-setting exercise, see Thomas O. McGarity, "Science and Policy in Setting National Ambient Air Quality Standards: Resolving the Ozone Enigma," *Texas Law Review* 93 (2015): 1783–1809.

64. U.S. Environmental Protection Agency, "National Ambient Air Quality Standards for Ozone; Proposed Rule," 72 Fed. Reg. 37818, 37822 (2007).

65. U.S. Environmental Protection Agency, *Air Quality Criteria for Ozone and Related Photochemical Oxidants,* vol. 1 (February 2006), i–iii, 6.10, 7.85, 7.110, 8.44, E-13, E-28; National Academies of Sciences, National Research Council, *Estimating Mortality Risk Reduction and Economic Benefits from Controlling Ozone Air Pollution* (Washington, DC: National Academies Press, 2008); "OMB and EPA Eyeing Outside Scientific Review of Ozone Risk Estimates," *Risk Policy Report,* September 27, 2005.

66. U.S. Environmental Protection Agency, *Review of the National Ambient Air Quality Standards for Ozone: Policy Assessment of Scientific and Technical Information* (January 2007), 3–8, 3–81, 6–76, 8–13, 8–25.

67. U.S. Environmental Protection Agency, "National Ambient Air Quality Standards for Ozone; Proposed Rule," 72 Fed. Reg. 37818, 37824, 37827–28, 37837, 37840, 37883 (2007).

68. U.S. Environmental Protection Agency, *Responses to Significant Comments on the 2007 Proposed Rule on the National Ambient Air Quality Standards for Ozone* (March 2008), 109, 160; Lorraine McCarthy, "EPA's Proposed Standard for Ozone Draws Fire at Philadelphia Hearing," *BNA Environment Reporter,* September 7, 2007, 1881; Barney Tumey, "Industry, Environmentalists Far Apart on Ozone Proposal at Atlanta Hearing," *BNA Environment Reporter,* September 7, 2007, 1876.

69. Stephen Power, "Industry Groups Seek to Kill Ozone Rules," *Wall Street Journal,* March 7, 2008, A3; Michael Bologna, "Witnesses at Chicago Hearing Urge EPA to Retool Proposal for Ozone Standard," *BNA Environment Reporter,* September 7, 2007, 1877; Susanne Pagano, "Witnesses in Houston Offer Differing Views on Proposed Changes to Ozone Standard," *BNA Environment Reporter,* September 7, 2007, 1879.

70. *Federal Rulemaking and the Unitary Executive Principle: Hearings before the Subcomm. on Commercial and Administrative Law, of the House Comm. on the Judiciary,* 110th Cong. 126 (2008) (statement of Curtis W. Copeland, Congressional Research Service) (Barbour letter; Dudley letter); Steven D. Cook, "EPA Sets Stricter Standards for Ozone, but at Level Weaker Than Advisers Sought," *BNA Environment Reporter,* March 14, 2008, 493); Juliet Eilperin, "EPA Tightens

Pollution Standards," *Washington Post,* March 13, 2008, A1; "Industry Launches Late OMB Lobbying Bid to Retain Current Ozone NAAQS," *Inside EPA,* March 7, 2008, 14; Power, "Industry Groups Seek to Kill Ozone Rules."

71. *EPA's New Ozone Standards: Hearing before the House Committee on Oversight and Governmental Reform,* 110th Cong. 233 (2008) (testimony of Rogene Henderson); Erik Stokstad, "EPA Adjusts a Smog Standard to White House Preference," *Science* 319 (2008): 1602; John Sullivan, Tom Avril, and John Shiffman, "Politics Choke Clean-Air Efforts," *Philadelphia Inquirer,* December 10, 2008, A1; Juliet Eilperin, "Ozone Rules Weakened at Bush's Behest," *Washington Post,* March 14, 2008, A1.

72. U.S. Environmental Protection Agency, "National Ambient Air Quality Standards for Ozone; Final Rule," 73 Fed. Reg. 16436, 16483, 16496–500 (2008).

73. Mississippi v. EPA, 744 F.3d 1334, 1342–44, 1347–48, 1350–54 1358–62 (D.C. Cir. 2013).

74. "EPA Inspector General Calls on Agency to Improve Oversight of State Programs," *BNA Environment Reporter,* October 4, 2002, 2142; Michael Grunwald and Eric Pianin, "Bush EPA Expands Emission Trading," *Washington Post,* February 15, 2001, A2.

75. Tim Dickinson, "The Secret Campaign of President Bush's Administration to Deny Global Warming," *Rolling Stone,* June 20, 2007, 20; "Bush Administration Rolls Out Voluntary Plan to Cut Pollution," *Oil Daily,* February 12, 2003; "EPA Announces Voluntary Partnerships under Policy to Limit Greenhouse Gases," *BNA Environment Reporter,* February 22, 2002, 387; Katharine Q. Seelye and Andrew C. Revkin, "Panel Tells Bush Global Warming Is Getting Worse, *New York Times,* June 7, 2001, A1.

76. "Agency's Voluntary Programs Insufficient for Cutting Emissions, Internal Report Finds," *BNA Environment Reporter,* August 1, 2008, 1532; Guy Gugliotta and Eric Pianin, "Bush Plans on Global Warming Alter Little," *Washington Post,* January 1, 2004, A10; Elizabeth Shogren, "13 Industries Set Emissions Targets as Part of Bush Initiative," *Los Angeles Times,* February 13, 2003, 30; Jennifer 8. Lee, "Voluntary Pacts to Curb Greenhouse Gases," *New York Times,* February 13, 2003, A22.

77. U.S. Environmental Protection Agency, "Control of Emissions from New Highway Vehicles and Engines," 68 Fed. Reg. 52922 (2003); "EPA Says It Does Not Have Authority to Regulate Carbon Dioxide from Vehicles," *BNA Environment Reporter,* September 5, 2003, 1925; "Lawsuit Filed to Force EPA to Regulate Greenhouse Gases from Mobile Sources," *BNA Environment Reporter,* December 13, 2002, 2681.

78. David Greising, "Big Business Sweats Climate Change Laws," *Chicago Tribune,* March 2, 2007, C1; Jeffrey Ball, "In Climate Controversy, Industry Cedes

Ground," *Wall Street Journal,* January 23, 2007, A1; Miguel Bustillo, "A Shift to Green," *Los Angeles Times,* June 12, 2005, 1.

79. Peter Haldis, "Lieberman-Warner Climate Change Bill Fails to Overcome Filibuster," *World Refining and Fuels Today,* June 9, 2008; "Bush Administration Still Opposes CO_2 Mandate," *Electric Utility Week,* December 18, 2006, 21; "Despite Election Gains, Environmentalists Face Limits in New Congress," *Risk Policy Report,* November 14, 2006.

80. Massachusetts v. EPA, 549 U.S. 497, 534 (2007).

81. "EPA, NHTSA to Align GHG Auto Rules to Ensure Joint Compliance," *Inside EPA Weekly Report,* August 31, 2007; Maura Reynolds and Richard Simon, "Bush Moves to Regulate Auto Emissions," *Los Angeles Times,* May 15, 2007, 9; Felicity Barringer and William Yardley, "Bush Splits on Greenhouse Gases with Congress and State Officials," *New York Times,* April 4, 2007, A1.

82. John Shiffman and John Sullivan, "An Eroding Mission at EPA," *Philadelphia Inquirer,* December 7, 2008, A1; Margaret Kriz, "Vanishing Act," *National Journal* 38 (2008): 22; Juliet Eilperin and R. Jeffrey Smith, "EPA Won't Act on Emissions This Year," *Washington Post,* July 11, 2008, A1; Felicity Barringer, "White House Refused to Open E-Mail on Pollutants," *New York Times,* June 25, 2008, A15.

83. Robert V. Percival, "Who's in Charge? Does the President Have Directive Authority over Agency Regulatory Decisions?," *Fordham Law Review* 79 (2011): 2520; Juliet Eilperin, "EPA Seeks Comment on Emissions Rules, Then Discredits Effort," *Washington Post,* July 12, 2008, A4; Shiffman and Sullivan, "Eroding Mission"; Ian Talley and Siobhan Hughes, "White House Blocks EPA Emissions Draft," *Wall Street Journal,* June 30, 2008, A3.

84. Felicity Barringer, "States' Battles over Energy Grow Fiercer with U.S. in a Policy Gridlock," *New York Times,* March 20, 2008, A18; Andrew Engblom, "Joined by Schwarzenegger, Florida Governor Orders Emissions Reductions," *SNL Electric Utility Report,* July 23, 2007; "Energy Northwest to Slow IGCC Development While Law Takes Shape," *Platts Coal Outlook,* May 21, 2007, 5; Janet Wilson and Marla Cone, "What Bill Would Do, Who's Affected," *Los Angeles Times,* September 1, 2006, 6; "Mass. Becomes First State to Impose CO_2 Controls under Tough New Standards," *Utility Environment Report,* May 4, 2001, 1; Jennifer 8. Lee, "The Warming Is Global but the Legislating, in the U.S., Is All Local," *New York Times,* October 29, 2003, A20; John J. Fialka, "States Are Stepping in to Reduce Levels of Carbon Dioxide," *Wall Street Journal,* September 11, 2001, A28.

85. Stephen C. Jones and Paul R. McIntyre, "Filling the Vacuum: State and Regional Climate Change Initiatives," *BNA Environment Reporter,* July 27, 2007, 1640; Brian Hansen, "Power Groups Air Concerns over Plan to Cap Carbon Emissions in Northeast," *Inside Energy with Federal Lands,* April 3, 2006, 8.

86. Gail Roberts, "Sequestration 'Overblown' as a Technology to Solve Greenhouse Gas Problem, Top Utility Execs Say," *Electric Utility Week,* June 23, 2008, 1; Wayne Barber, "Greenpeace Questions Carbon Capture as a Solution to Climate Change Concerns," *SNL Electric Utility Report,* May 12, 2008; Kathleen Hart, "Carbon Price Key to Developing Capture and Storage Technology, Bingaman Says," *SNL Coal Report,* March 10, 2008; Suzanne McElligott, "U.S. DOE Tells Senate: CCS to Be Common Practice by 2045," *Gasification News,* April 18, 2007.

87. U.S. Department of Energy, "Notice of Intent to Prepare and Environmental Impact Statement for Implementation of the FutureGen Project," 71 Fed. Reg. 42840, 42841 (July 28, 2006); Elizabeth McGowan, "FutureGen Pullout Spurs Carbon Cap Debate," *Waste News,* February 18, 2008, 12; Wayne Barber, "DOE Raises Cost Concerns as FutureGen Picks Illinois Site," *SNL Electric Utility Report,* December 24, 2007.

88. 42 U.S.C. § 1326(b); U.S. Environmental Protection Agency, "National Pollutant Discharge Elimination System—Regulations Addressing Cooling Water Intake Structures for New Facilities," 65 Fed. Reg. 49060, 49072 (August 10, 2000); "EPA Looking at Power Plant Water Intake Rules," *Electricity Daily,* August 10, 1998.

89. Riverkeeper Inc. v. EPA, 358 F.3d 174, 183, 187 (2d Cir. 2004); U.S. General Accounting Office, *Rulemaking: OMB's Role in Reviews of Agencies' Draft Rules and the Transparency of Those Reviews* (Washington, DC: U.S. General Accounting Office, September 2003), 194.

90. *Riverkeeper Inc.,* 358 F.3d at 188–89, 196–97.

91. E. P. Taft, J. L. Black, and L. R. Tuttle, "Clean Water Act Compliance," *Power Magazine,* March 2007, 48; Thomas F. Armistead, "New Cooling-Water Intake Is Low-Cost and Fish-Friendly," *Engineering News-Record,* April 4, 2005, 17; Teresa Hansen, "Innovative Intake Screen Protects Fish at Newington Energy," *Power Engineering,* November 1, 2004, 190.

92. Entergy Corp. v. Riverkeeper Inc., 129 S. Ct. 1498, 1504 (2009); U.S. General Accounting Office, *OMB's Role,* 198; "Power Plants Will Have to Cut Number of Organisms Affected by Intake Systems," *BNA Environment Reporter,* February 20, 2004, 374.

93. *Entergy Corp.,* 129 S. Ct. at 1504; U.S. Environmental Protection Agency, "National Pollutant Discharge Elimination System—Final Regulations to Establish Requirements for Cooling Water Intake Structures at Phase II Existing Facilities," 69 Fed. Reg. 41576, 41592 (July 9, 2004).

94. *Entergy Corp.,* 129 S. Ct. 1498; "Cooling Water Rule May Be Early Test for Obama EPA on Cost-Benefit," *Inside EPA Weekly Report,* November 6, 2009.

95. Steven D. Czajkowski, "Note: Focusing on Demand Side Management in the Future of the Electric Grid," *Pittsburgh Journal of Environmental and Public*

Health Law 4 (2010): 121; Sidney A. Shapiro and Joseph P. Tomain, "Rethinking Reform of Electricity Markets," *Wake Forest Law Review* 40 (2005): 510–11; Weaver, "Can Energy Markets Be Trusted?," 113–14.

96. Weaver, "Can Energy Markets Be Trusted?," 124–25; Kay L. Gehring, "Can Yesterday's Demand-Side Management Lessons Become Tomorrow's Market Solutions?," *Electricity Journal,* June 2002, 15.

97. Marilyn A. Brown, Dan York, and Martin Kushler, "Reduced Emissions and Lower Costs: Combining Renewable Energy and Energy Efficiency into a Sustainable Energy Portfolio Standard," *Electricity Journal,* May 2007, 62; Susan Joy Hassol, "A Change of Climate," *Issues in Science and Technology,* March 22, 2003, 39; Benjamin K. Sovacool and Christopher Cooper, "Big *Is* Beautiful: The Case for Federal Leadership on a National Renewable Portfolio Standard," *Electricity Journal,* May 2007, 48; David B. Spence, "Coal-Fired Power in a Restructured Electricity Market," *Duke Environmental Law & Policy Forum* 15 (2005): 213; "Growth of Green Power Market Accelerating," *Megawatt Daily,* October 25, 2005, 5; Congressional Research Service, "The Renewable Electricity Production Tax Credit: In Brief" (July 26, 2017).

98. Nidhi Thakar, "The Urge to Merge: A Look at the Repeal of the Public Utility Holding Company Act of 1935," *Lewis & Clark Law Review* 12 (2008): 920–21; Mike Boyer, "Duke Merger Nearly Done," *Cincinnati Enquirer,* March 25, 2006, D1; "Analysts: 'Size Envy' to Inspire Power Mergers," *Megawatt Daily,* March 7, 2006; Mike Boyer, "Cinergy / Duke Gets Another OK," *Cincinnati Enquirer,* December 16, 2005, D1; "FERC Approves Mergers, Grid Coordinators, *Megawatt Daily,* December 16, 2005, 7; Jad Mouawad, "Duke Energy Will Acquire Cinergy for $9 Billion in Stock," *New York Times,* May 10, 2005, C1; "Cinergy's CEO: Deal Took 10 Years to Arrive," *Megawatt Daily,* May 10, 2005, 9.

99. "Adviser Says McCain Would Not Use Air Act to Cut Emissions, in Contrast with Obama," *BNA Environment Reporter,* October 31, 2008, 2168; Andrew C. Revkin, "On Global Warming, McCain and Obama Agree: Urgent Action Is Needed," *New York Times,* October 19, 2008, A22; Stephen Power, "In Energy Policy, McCain, Obama Differ on Role of Government," *Wall Street Journal,* June 9, 2008, A2; Steven Mufson, "Coal Industry Plugs into the Campaign," *Washington Post,* January 18, 2008, D1.

100. Weaver, "Can Energy Markets Be Trusted?," 112–13.

101. Ted Nace, *Climate Hope: On the Front Lines of the Fight against Coal* (San Francisco: CoalSwarm, 2010), 97; Lynn Doan, "LS Power Group Cancels Plans for Texas Gas Plant," *SNL Electric Utility Report,* September 28, 2009; Elisa Wood, "Coal versus the Politicians," *Platts Energy Economist,* September 1, 2007, 4; Robert Peltier, "Repowering Breathes New Life into Old Plants," *Power,* June 2003, 30.

102. Simon Romero, "2 Industry Leaders Bet on Coal but Split on Cleaner Approach," *New York Times,* May 28, 2006, 1.

103. *North Carolina,* 531 F.3d 896; Felicity Barringer, "Decisions Shut Door on Bush Clean-Air Steps," *New York Times,* July 12, 2008, A17 (quoting Jim Owen, Edison Electric Institute).

9. THE OBAMA ADMINISTRATION TAKES A NEW APPROACH

1. Alexander Duncan, "Obama Takes Charge, Citing Energy as Priority and Halting Bush Rules," *Inside Energy with Federal Lands,* January 26, 2009, 3.

2. Coral Davenport, "EPA: The World in Microcosm," *National Journal,* September 22, 2010, 20; "Obama Pick for EPA Air Chief May Signal Quick Action on Climate Rules," *Risk Policy Report,* March 24, 2009; "Obama Team Seen Bringing Cohesion to Administration Response on Warming," *BNA Environment Reporter,* December 19, 2008, 2516; "Industry Backs New Jersey's Jackson as Top Pick to Head Obama EPA," *Risk Policy Report,* December 9, 2008.

3. Mark Peters, "Natural Gas Is Threat to Fast Development of Coal-Fired Power," *Wall Street Journal,* March 31, 2010, A5; Bob Matyi et al., "Coal Is Taking Its Lumps From Environmentalists, but It Counts Some Successes in Friendly States," *Electric Utility Week,* November 16, 2009, 1; Stephen Power and Siobhan Hughes, "Coal Industry Digs Itself Out of a Hole in the Capitol," *Wall Street Journal,* January 15, 2009, A4; Kris Maher, "Coal Producers Struggle to Meet Demand," *Wall Street Journal,* June 24, 2008.

4. Patrick Charles McGinley, "Climate Change and the War on Coal: Exploring the Dark Side," *Vermont Journal of Environmental Law* 13 (2011): 322; "Industry Fractures on Climate Policy," *Electricity Journal,* December 2009, 1.

5. For a more extensive description of efforts to pass climate change legislation during President Obama's first term, see Thomas O. McGarity, "The Disruptive Politics of Climate Disruption," *Nova Law Review* 38 (2014): 393–472.

6. Charles Homans, "Marathon Man," *Washington Monthly,* May 1, 2009, 10; Margaret Kriz, "Changed Climate," *National Journal,* February 7, 2009, 40; "Greenhouse Gas Levels Hit Record High in 2008, Meteorological Organization Says," *BNA Environment Reporter,* December 26, 2008, 2711; Cathy Cash, "Shift from Dingell to Waxman Signals New Era for Energy Interests," *Electric Utility Week,* November 24, 2008, 1.

7. Cathy Cash, Christine Cordner, and Alexander Duncan, "Obama Sets Pace for Congress on Carbon," *Electric Utility Week,* March 2, 2009, 1; "Hearings Draw Out the Climate Change, Clean Coal, Practical Sides of Appointees," *Platts Coal Outlook,* January 19, 2009, 1; Kara Sissell, "USCAP Offers Congress a Blueprint for Climate Action," *Chemical Week,* January 19, 2009, 6; Steven Mufson, "Coalition Agrees on Emissions Cuts," *Washington Post,* January 15, 2009, D1.

8. "Small Businesses, Farmers Ask Congress to Mitigate Costs of Climate Legislation," *BNA Environment Reporter,* May 1, 2009, 981; Cassandra Sweet,

Coal-Burning Utilities Want Time on CO_2 Rules," *Wall Street Journal*, March 11, 2009.

9. "Waxman-Markey, Characterized as 'A Pile of (Bleep)' by One Lawmaker," *U.S. Coal Review*, July 6, 2009; John M. Broder, "Adding Something for Everyone, House Leaders Gained a Climate Bill," *New York Times*, July 1, 2009, 20; "House Leaders Make Dozens of Deals to Draw Votes for Groundbreaking Climate-Energy Bill," *Electric Utility Week*, June 29, 2009, 1; "Utility Industry Leaders Declare Waxman-Markey Bill 'Greatly Improved.'" *SNL Electric Utility Report*, June 29, 2009; "Climate Bill Slated for House Floor Vote; Waxman, Other Chairmen Reach Agreements," *BNA Environment Reporter*, June 26, 2009, 1489; "To Move House Climate Bill, Activists Soften Push for GHG Standards," *Inside EPA Weekly Report*, June 26, 2009; Paul West, "Obama Lobbies for Climate Bill," *Los Angeles Times*, June 25, 2009, A16; Ian Talley and Siobhan Hughes, "Climate Bill Set for Vote after Deal Is Reached," *Wall Street Journal*, June 24, 2009, A6.

10. "How Offset Credits Will Work under Waxman-Markey Climate Bill," *Oil Daily*, July 22, 2009; Jennifer Zajac, "Federal RECs in Waxman-Markey Bill Raise Questions on Fate of State RECs," *SNL Electric Utility Report*, July 13, 2009; Tom Tiernan, "Deep in the Weeds of Allowance Allocations, No Clear Path for Impact on Utility Customers," *Electric Utility Week*, July 6, 2009, 3; Broder, "Adding Something for Everyone"; Greg Hitt and Naftali Bendavid, "Obama Wary of Tariff Provision," *Wall Street Journal*, June 29, 2009, A3.

11. Jane Mayer, "Covert Operations," *New Yorker*, August 30, 2010; Juliet Eilperin, "Climate Bill Faces Hurdles in Senate," *Washington Post*, November 2, 2009, A1; David A. Fahrenthold, "Environmentalists Slow to Adjust in Climate Debate," *Washington Post*, August 31, 2009, A1 (quote).

12. Darren Goode, "Climate Bill Backers Unveil Large-Scale Effort for 28 States," *National Journal's CongressDaily*, September 9, 2009; Cathy Cash, "Unions, Enviros Pressure Congress on Climate Bill," *Platts Coal Outlook*, August 24, 2009, 1; Fahrenthold, "Environmentalists Slow to Adjust."

13. Cathy Cash, "Bargaining on EPA Rule, Nuclear Provisions, Allowances and More as Carbon Bill Advances," *Electric Utility Week*, November 2, 2009; "Boxer's Revisions Match House Allocations; Permits Would Go Largely to Electric Utilities," *BNA Environment Reporter*, October 30, 2009, 2499; Darren Goode, "Dem Divisions Show during Climate Hearing," *National Journal's CongressDaily*, October 27, 2009; Steven Mufson and Juliet Eilperin, "Senate's Climate Bill a Bit More Ambitious," *Washington Post*, October 25, 2009, A3.

14. Eilperin, "Climate Bill Faces Hurdles"; Darren Goode, "Baucus Pleased by Compromise Coal Provision," *National Journal's CongressDaily*, October 29, 2009; Siobhan Hughes and Ian Talley, "Key Democrat Cites Concerns on Climate Bill," *Wall Street Journal*, October 28, 2009, A2; Goode, "Dem Divisions Show."

15. Juliet Eilperin, "EEI, Three Oil Companies to Back Climate Bill," *Washington Post,* April 23, 2010; Juliet Eilperin, "Merkel Urges Congress to Act on Climate," *Washington Post,* November 4, 2009, A4; Eilperin, "Climate Bill Faces Hurdles" (quotes).

16. Stephen Power, "Senate Halts Effort to Cap Emissions," *Wall Street Journal,* July 23, 2010, A3; Cathy Cash, "Electricity Rates on List of Considerations as Senators Contemplate GHG Bill," *Platts Coal Outlook,* May 24, 2010, 1; Jim Tankersley and Richard Simon, "Sen. Lindsey Graham's Bipartisan Efforts Bog Down," *Los Angeles Times,* April 28, 2010, A9; Darren Goode, "Graham Wants Immigration Off Table for Year or He Bolts," *National Journal's CongressDaily,* April 27, 2010; Laura Meckler, "Democrats Revive Immigration Push," *Wall Street Journal,* April 22, 2010, A5.

17. David A. Fahrenthold and Juliet Eilperin, "EPA to Pick Up Climate Change Where Congress Left Off," *Washington Post,* August 4, 2010, A3; Kathleen Hart, "US Government Report Predicts Dire Impacts from Climate Change across America," *SNL Power Daily Northeast,* June 17, 2009.

18. U.S. Environmental Protection Agency, "Proposed Endangerment and Cause or Contribute Findings for Greenhouse Gases under Section 202(a) of the Clean Air Act: Proposed Rule," 74 Fed. Reg. 18886 (2009); U.S. Environmental Protection Agency, *Technical Support Document for Endangerment and Cause or Contribute Findings for Greenhouse Gases under Section 202(a) of the Clean Air Act* 4–5 (Washington, DC: U.S. Environmental Protection Agency, December 7, 2009); "GOP Plays 'Cow Card' at Climate Hearing," *Inside Energy with Federal Lands,* April 27, 2009, 16.

19. Lisa Heinzerling, "Introduction: Climate Change at EPA," *Florida Law Review* 64 (2012): 7–8; Juliet Eilperin, "New Emissions Standards Set for Cars and Light Trucks," *Washington Post,* April 2, 2010, A1; John M. Broder, "Obama Proposes Rules for Automakers on Greenhouse Gas Emissions and Mileage," *New York Times,* September 16, 2009, B4.

20. 42 U.S.C. § 7475, 7479(1); U.S. Environmental Protection Agency, "Prevention of Significant Deterioration and Title V Greenhouse Gas Tailoring Rule," 74 Fed. Reg. 55292 (2009); *H.R. ___, the Energy Tax Prevention Act of 2011: Hearing before the Subcommittee on Energy and Power of the House Committee on Energy and Commerce,* 112th Cong. 204 (2011) (testimony of Peter Glaser, Troutman Sanders LLP); "EPA Proposal Would Require Large Sources to Control Their Greenhouse Gas Emissions," *BNA Environment Reporter,* October 2, 2009, 2281.

21. Coalition for Responsible Regulation Inc. v. EPA, 684 F.3d 102 (D.C. Cir. 2012); U.S. Environmental Protection Agency, "Prevention of Significant Deterioration and Title V Greenhouse Gas Tailoring Rule," 75 Fed. Reg. 31514 (2010); U.S. Environmental Protection Agency, "Endangerment and Cause or Contribute

Findings for Greenhouse Gases under Section 202(a) of the Clean Air Act," 74 Fed. Reg. 66,496 (2009).

22. Utility Air Regulatory Group v. EPA, 134 S. Ct. 2427, 2438–39, 2442–43, 2447 (2014); EPA memorandum from Janet G. McCabe, Acting Assistant Administrator, Office of Air and Radiation, and Cynthia Giles, Assistant Administrator, Office of Enforcement and Compliance Assurance, to Regional Administrators re: "Next Steps and Preliminary Views on the Application of Clean Air Act Permitting Programs to Greenhouse Gases Following the Supreme Court's Decision in *Utility Air Regulatory Group v. Environmental Protection Agency,*" dated July 24, 2014, 4 (EPA retains 75,000 tpy threshold for anyway sources), https://www.epa.gov/sites/production/files/2015-07/documents/2014scotus.pdf.

23. Annalee Grant and Taylor Kuykendall, "Future of CO_2 Capture Called into Question after Government Pulls Project Funds," *SNL Electric Utility Report,* February 9, 2015; "Sierra Club Asks Illinois Court to Overturn Minor Source FutureGen Permit," *InsideEPA / climate,* December 15, 2014; "EPA's CCS Permits for FutureGen Draw Mixed Reviews from Industry, Advocates," *Clean Energy Report,* September 21, 2014; Jeffrey Tomich, "FutureGen Brings New Hope to Old Ameren Plant," *St. Louis Post-Dispatch,* November 7, 2010, E1; U.S. Department of Energy, "Department of Energy Formally Commits $1 Billion in Recovery Act Funding to FutureGen 2.0," press release, September 28, 2010.

24. Coral Davenport, "Heads in the Sand," *National Journal,* December 1, 2011; Margaret Kriz Hobson, "Political Tidal Wave Turns EPA Strategy," *Congressional Quarterly Weekly,* February 14, 2011, 335; Stephen Power, "Not on Ballot, but EPA Chief a Campaign Issue," *Wall Street Journal,* October 8, 2010, A4; Kimberley A. Strassel, "The Cap-and-Trade Crackup," *Wall Street Journal,* October 8, 2010, A17 (laser quote); Davenport, "EPA: The World in Microcosm" (train wreck quote); "Election 2010 House Map," http://elections.nytimes.com/2010/results/house; "Election 2010 Senate Map," http://elections.nytimes.com/2010/results/senate.

25. H.R. 910, 112th Cong., 1st Sess. (2011); "House Appropriations Committee Votes to Cut EPA Funding, Restrict Regulatory Authority," *BNA Environment Reporter,* June 29, 2012, 1695; "Senate Democrats Delay FY12 Bill over Riders Blocking Key EPA Policies," *Inside EPA Weekly Report,* December 16, 2011; John M. Broder, "House Votes to Bar E.P.A. from Regulating Industrial Emissions," *New York Times,* April 7, 2011, A17; Tennille Tracy, "Senate Rejects Proposals to Block EPA Emissions Rules," *Wall Street Journal,* April 6, 2011; "GOP Sharpens Three-Pronged Legislative Strategy to Overturn EPA Rules," *Inside EPA,* January 7, 2011, 1.

26. Andrew Childers, "Industries, Regulators Report Few Problems with Greenhouse Gas Permitting Program," *BNA Environment Reporter,* December 9, 2011, 2790; Cathy Cash and Beth Ward, "Coal States on Learning Curve for

Greenhouse BACT," *Electric Utility Week,* July 4, 2011, 1; "Industry Gives Muted Reaction to First Non-'Disruptive' GHG Coal Permits," *Inside EPA Weekly Report,* June 24, 2011; Jennifer Zajac, "EPA Posts Permitting Guidelines for Greenhouse Gas Emissions," *SNL Electric Utility Report,* November 29, 2010.

27. Arnold W. Reitze Jr., "Federal Regulation of Coal-Fired Electric Power Plants to Reduce Greenhouse Gas Emissions," *Utah Environmental Law Review* 32 (2012): 404 ; Andrew Childers, "EPA Publishes Proposed Emissions Limits on Carbon Dioxide for New Power Plants," *BNA Environment Reporter,* April 20, 2012, 1008; Juliet Eilperin, "EPA's Greenhouse Gas Limits Affect Only New Power Plants," *Washington Post,* March 28, 2012, A16; Amy Harder, "EPA Proposes First-Ever Climate Rules," *National Journal Daily,* March 27, 2012.

28. *The American Energy Initiative: A Focus on H.R.6172: Hearings before the Subcommittee on Energy and Power of the House Committee on Energy and Commerce,* 112th Cong. 203 (September 20, 2012) (testimony of John Thompson, Clean Air Task Force); "Pessimism Grows over Large-Scale Carbon Capture Deployment," *Clean Energy Report,* July 11, 2011 (quote); Andrew Childers, "Single Carbon Standard for Power Plants Breaks EPA Precedent, Commenters Say," *BNA Environment Reporter,* June 29, 2012, 1675; "UMWA's Roberts Has a Clue: It Was Lisa Jackson, in the Capitol, with a Candlestick," *U.S. Coal Review,* April 9, 2012.

29. "EPA Positioned to Stay under Radar through 2012 Election Season," *Inside EPA Weekly Report,* July 20, 2012; Jonathan Crawford, "Environmentalists, Industry Agree: Election-Year Politics Behind EPA Rule Delays," *SNL FERC Power Report,* February 29, 2012; Juliet Eilperin, "Environmental Rules Left Hanging in the Political Balance," *Washington Post,* February 13, 2012, A4.

30. "EPA Rules Will Cut Utility Fuel Options, Raise Electricity Prices, Republicans Say," *BNA Environment Reporter,* March 8, 2013, 635; Dean Scott, "Re-Election Greenlights Host of EPA Rules, Solidifies Regulation of Greenhouse Gases," *BNA Environment Reporter,* November 9, 2012, 2829; Amy Harder, "MIA on Climate Change," *National Journal,* August 1, 2012; Bill Reilly, "Coal Is Orphan Fuel in Obama's State of the Union Address," *SNL Coal Report,* January 30, 2012.

31. Coral Davenport, "Republicans to Begin New Assault on Obama's Climate Plans at Hearing for EPA Nominee," *National Journal,* April 10, 2013 (quoting Senator David Vitter [R-LA]); Jonathan Crawford, "Obama Nominates 'Straight Shooter' McCarthy to Lead EPA," *SNL Electric Utility Report,* March 11, 2013.

32. Barack Obama, "Presidential Memorandum—Power Sector Carbon Pollution Standards, White House, Office of the Press Secretary, June 25, 2013; Barack Obama, "Remarks by the President on Climate Change, Georgetown University, Washington, D.C.," White House, Office of the Press Secretary, June 25, 2013.

33. "Environmental Groups Seek to Rally Support for Obama's Climate Action Plan," *BNA Environment Reporter,* August 16, 2013, 2441; Virgil Dickson, "Coal

Industry Works to Repulse Obama Climate Plan," *PR Week,* July 1, 2013; Juliet Eilperin, "Obama Unveils Climate Agenda," *Washington Post,* June 26, 2013, A1.

34. Coral Davenport, "As Listener and Saleswoman, E.P.A. Chief Takes to the Road for Climate Rules," *New York Times,* March 22, 2014, A11; "EPA Proposal to Limit Greenhouse Gases Allows Options for Power Plant Compliance," *BNA Environment Reporter,* September 27, 2013, 2852; Tennille Tracy, "EPA Turns Up Heat on New Coal Plants," *Wall Street Journal,* September 21, 2013, A3; Juliet Eilperin and Philip Rucker, "EPA Considers Delaying Power Plant Rules," *Washington Post,* March 16, 2013, A1.

35. *EPA Power Plant Regulations: Is the Technology Ready: Hearings before the Subcommittee on Energy and Environment of the House Committee on Science, Space, and Technology,* 113th Cong. (October 29, 2013); "Senate Democrats Aim to Shift Politics of Climate Change," *Congressional Quarterly News,* January 14, 2014; Coral Davenport, "Republicans Pounce on Obama's Global-Warming Regulations for Political Fodder," *National Journal Daily,* September 22, 2013 (email quote).

36. 42 U.S.C. § 7621(a); Mark Latham, Victor E. Schwartz, and Christopher E. Appel, "EPA Ignoring Clean Air Act Mandate to Analyze Impact of Regulations on Jobs?" *Legal Backgrounder,* June 6, 2014 (EPA response); "Suit Seeking Jobs Review of EPA Rules Shows Split on 'Economy' Modeling," *Clean Energy Report,* March 30, 2014.

37. Murray Energy Corp. v. EPA, 861 F.3d 529 (4th Cir. 2017).

38. 42 U.S.C. § 7411(a), (d); "Upcoming Power Plant Rule to Pose Test of EPA's Clean Air Act 111(d) Authority," *BNA Environment Reporter,* September 6, 2013, 2671.

39. Coral Davenport and Julie Hirschfeld Davis, "Move to Fight Climate Plan Started Early," *New York Times,* August 4, 2015; Catherine Ho, "Energy Companies, Utilities Lead EPA Lobbying," *Washington Post,* June 9, 2014, A10; Taylor Kuykendall, "Organizers Expect More than 3,500 to Rally for Coal in Washington, D.C.," *SNL Daily Coal Report,* October 29, 2013 (sign quote).

40. "EPA Is Engaging State Regulators in Writing Emissions Rules," *Congressional Quarterly News,* April 7, 2014; Juliet Eilperin, "Is This Man Obama's Legacy-Builder?," *Washington Post,* March 5, 2014, A4 ("laser focused" quote); Coral Davenport, "E.P.A. Staff Struggling to Create Pollution Rule," *New York Times,* February 5, 2014, A12.

41. U.S. Environmental Protection Agency, "Carbon Pollution Standards for Modified and Reconstructed Stationary Sources: Electric Utility Generating Units; Proposed Rules," 79 Fed. Reg. 34830, 34830, 34835–38; 34851 (June 18, 2014).

42. "Fearing Jobs Impacts, Unions Urge EPA to Ease GHG Rule for Existing Utilities," *InsideEPA / climate,* November 21, 2014; "NRG's Carbon Pledge Highlights Utility Support for EPA's Climate Rules," *InsideEPA / climate,* November 21, 2014 (EEI quote); Dan Testa, "Utility CEOs' Views toward EPA Carbon Rule Fall

along Familiar Fault Lines," *SNL Renewable Energy Weekly,* August 8, 2014; Amy Harder and Cassandra Sweet, "Utilities Size Up Proposed Carbon Limits," *Wall Street Journal,* June 3, 2014, A4; Jeff McMahon, "5 Dene Carbon Polluters in EPA Crosshairs," *Forbes,* June 1, 2014 (rural co-ops).

43. "Environmentalists Urge EPA to Tighten ESPS but Split on Costs, Gas Use," *InsideEPA/climate,* December 3, 2014; Frances Beinecke, "Don't Buy the Smear of the EPA," *Los Angeles Times,* June 4, 2014, A13.

44. Senator Mitch McConnell, "Top Senate Democrat, Harry Reid, Blocks McConnell's 'Coal Country Protection Act,'" press release, Congressional Documents and Publications, June 4, 2014.

45. "Energy Issues Still Divide Congress along Party Lines," *Foster Natural Gas/Oil Report,* November 25, 2015, 46; Ben Geman, "Obama: 'No Challenge Poses a Greater Threat to Future Generations Than Climate Change,'" *National Journal Daily,* January 20, 2015 (Obama quote); Eric Wolff, "Committee on Energy and Commerce to Target EPA CO_2 Rules in Oversight," *SNL Daily Coal Report,* January 15, 2015; "Republican Majority Will Take Obama 'From Tough to Tougher,'" *Congressional Quarterly News,* November 5, 2014; "Senate Election Results," http://elections.nytimes.com/2014/results/senate; "House Election Results," http://elections.nytimes.com/2014/results/house.

46. U.S. Environmental Protection Agency, "Standards of Performance for Greenhouse Gas Emissions from New, Modified, and Reconstructed Stationary Sources: Electric Utility Generating Units," 80 Fed. Reg. 64510, 64512–13 (October 23, 2015); "EPA Sticks to Carbon Capture Requirement for New Coal-Fired Plants," *Congressional Quarterly News,* January 27, 2014.

47. U.S. Environmental Protection Agency, "Carbon Pollution Emission Guidelines for Existing Stationary Sources: Electric Utility Generating Units," 80 Fed. Reg. 64662, 64761–63, 64666, 64677–79, 64867–68, 64881–83, 64901–4, 64928 (October 23, 2015).

48. "Obama Vetoes GOP Measures to Block Power Plant Carbon Rules," *Congressional Quarterly News,* December 19, 2015; Esther Whieldon, "McCabe: EPA Not Using Clean Power Plan to Create Cap-And-Trade Program," *SNL Electric Utility Report,* October 12, 2015.

49. Molly Christian, "GOP Vows to Block Supreme Court Nominee as Energy-Related Legal Disputes Intensify," *SNL Electric Utility Report,* February 29, 2016; Annalee Grant, "After Passing of Scalia, Clean Power Plan's Future More Uncertain Than Ever," *SNL Generation Markets Week,* February 23, 2016; Andrew C. Revkin, "Justice Scalia's Irreplaceable Views on CO2 and Climate," *Dot Earth* (*New York Times* blog), February 13, 2016; Adam Liptak and Coral Davenport, "Justices Deal Blow to Obama Effort on Emissions," *New York Times,* February 10, 2016, A1; "EPA Says Utilities Cite Bogus Plant Closures in High Court ESPS Stay Bid," *Clean Energy Report,* February 7, 2016; "Court Rejects a Bid to Block Coal

Plant Regulations," *New York Times,* January 22, 2016, A13; "At Deadline, Critics File Host of New Challenges to Power Plant GHG Rules," *Inside EPA,* December 23, 2015.

50. Eric Wolff, "Oklahoma Just Says No to CO_2 Rule," *SNL Electric Utility Report,* May 11, 2015; "State Bills Show Passion Stirred by EPA Rules," *Congressional Quarterly News,* April 8, 2015; "Unable to Craft ESPS Plan, Kentucky May Be Early Test of EPA 'FIP' Power," *InsideEPA / climate,* October 31, 2014; Eric Wolff, "Despite Political Opposition, State Clean Air Agencies Grappling with EPA CO_2 Rule," *SNL FERC Power Report,* August 13, 2014; Jeff Stanfield, "Mont. Governor Sees No Need to Shut Coal Plants to Meet EPA's Draft CO_2 Rule," *SNL Electric Utility Report,* September 29, 2014; Coral Davenport, "States Fight Obama's Climate Plan, but Quietly Prepare to Comply," *New York Times,* July 20, 2016, A14; Amanda Luhavalja, "RGGI, Calif. Carbon Markets Likely to Meet New EPA Standards with Minor Program Tweaks," *SNL Renewable Energy Weekly,* June 6, 2014.

51. U.S. Environmental Protection Agency, "Federal Implementation Plans to Reduce Interstate Transport of Fine Particulate Matter and Ozone," 75 Fed. Reg. 45210, 45214–15 (2010).

52. "EPA Proposes GHG Permit Requirements for States, and Federal Implementation Plan," *Electric Utility Week,* August 16, 2010; Catherine Cash, "EPA Proposal to Slash Power-Plant Emissions Triggers Industry Concern," *Inside Energy with Federal Lands,* July 12, 2010, 8; "Health, Environmental Groups Ask EPA to Set Stricter Emissions Limits in Transport Rule," *BNA Environment Reporter* 41 (2010): 1930.

53. Matthew Tresaugue, "Study: Smog in Texas Fouls Neighbors' Air," *Houston Chronicle,* November 17, 2010, B1; "Utilities Question Legality of Tight Deadline for EPA Transport Rule Plans," *Inside EPA Weekly Report,* October 29, 2010.

54. Daniel J. Weiss, "Poor Little Big Coal Says EPA Smog Standards Too Expensive," *Grist,* November 17, 2011 (ad quote); "Energy Wars," *Politico,* August 9, 2011.

55. U.S. Environmental Protection Agency, "Federal Implementation Plans: Interstate Transport of Fine Particulate Matter and Ozone and Correction of SIP Approvals," 76 Fed. Reg. 48,208, 48209–10, 48271 (August 8, 2011); U.S. Environmental Protection Agency, "EPA Reduces Smokestack Pollution, Protecting Americans' Health from Soot and Smog," press release, Environmental Protection Agency Documents and Publications, July 7, 2011.

56. *Lights Out: How EPA Regulations Threaten Affordable Power and Job Creation: Hearings before the Subcommittee on Regulatory Affairs, Stimulus and Oversight of the House Oversight and Government Affairs Committee,* 112th Cong. (2011); "Senate May Target EPA Cross-State Rule with Goal of Forcing Obama to

Accept Delay," *BNA Environment Reporter,* November 21, 2011, 2231; "Poll Finds Support for EPA Power Plant Rules among Both Democratic, Republican Voters," *BNA Environment Reporter,* November 28, 2011, 2289; Amy Harder, "Senate Nixes Resolution Voiding EPA Rule," *National Journal Daily,* November 10, 2011; James Inhofe, "Obama EPA's Transport Rule Latest Impediment to Job Growth," press release, Congressional Documents and Publications, July 7, 2011.

    57. EPA v. EME Homer City Generation, 134 S. Ct. 1584, 1600, 1604–7 (2014).

    58. Annalee Grant, "Appeals Court Orders EPA to Review 'Good Neighbor' Smog Limits, Upholds Other Challenges," *SNL Daily Coal Report,* July 31, 2015; "DOJ Highlights Legal Uncertainty for EPA after Reinstatement of CSAPR," *Clean Energy Report,* May 11, 2014.

    59. "Mercury Emissions from Power Plants Down 6.5 Percent Since 2000, Report Finds," *BNA Environment Reporter,* March 19, 2010, 597; Andrew Childers, "Stronger Air Rules Predicted as EPA Drops Defense of Regulations from Bush Years," *BNA Environment Reporter,* November 13, 2009, 2632 (emissions); "EPA Plans Mercury Rules for Power Plants, Moves to Withdraw Supreme Court Petition," *BNA Environment Reporter,* February 13, 2009, 317.

    60. U.S. Environmental Protection Agency, "National Emission Standards for Hazardous Air Pollutants from Coal- and Oil-Fired Electric Utility Steam Generating Units and Standards of Performance for Fossil-Fuel-Fired Electric Utility, Industrial-Commercial-Institutional, and Small Industrial-Commercial-Institutional Steam Generating Units," 76 Fed. Reg. 24976, 25027, 24977–78 (May 3, 2011); U.S. Environmental Protection Agency, *Reducing Toxic Pollution from Power Plants: Final Mercury and Toxics Standards (MATS),* December 2011, p. 14, https://www.epa.gov/sites/production/files/2015-11/documents /20111216matspresentation.pdf.

    61. American Legislative Exchange Council, *EPA's Regulatory Train Wreck: Strategies for State Legislators* (Washington, DC: American Legislative Exchange Council, 2011), 44; Matthew Bandyk, "GDF SUEZ Sees Its Generation in Strong Position to Comply with New EPA Rules," *SNL Coal Report,* September 26, 2011; "Activists Raise Concerns over Utility MACT Limits, Compliance Extensions," *Inside EPA,* August 19, 2011, 13; Nick Juliano, "Power Sector Split over EPA's Mercury Rule May Be Proxy Fight for Vision of Coal's Future," *Electric Utility Week,* August 15, 2011, 1; Nick Juliano, "Fix It or Kill It? Utility Groups Split over Strategy on Major EPA Air Reg," *Inside Energy with Federal Lands,* August 8, 2011 (EEI position).

    62. Jessica Coomes, "Electricity Reliability Can Be Maintained with EPA Air Rules, Clean Energy Report Says," *BNA Environment Reporter,* December 2, 2011, 2678; Ryan Tracy, "EPA Rules Spark Power-Plant Fray," *Wall Street Journal,* November 10, 2011, A1; "Electric Power Industry Says Proposal Allows Too Little Time to Install Controls," *BNA Environment Reporter,* August 22, 2011, 1806;

"Utilities Challenge Legality of EPA Air Toxics Rule Favored by Gas Sector," *Clean Energy Report,* August 22, 2011.

63. *The American Energy Initiative: Hearings before the Subcommittee on Energy and Power of the House Committee on Energy and Commerce,* 112th Cong., 1st Sess. (2011) (testimony of Susan F. Tierney, Analysis Group Inc.); *Lights Out II: Another Look at EPA's Utility MACT Rule: Hearing before the House Committee on Oversight and Government Reform,* 112th Cong. (2011) (testimony of Josh Bivens, Economic Policy Institute); "Policy Institute Projects Net Job Gains under Proposed Air Toxics Standards," *BNA Environment Reporter* 42 (2011): 1332; Matthew L. Wald, "New Rules and Old Plants May Strain Summer Energy Supplies," *New York Times,* August 12, 2011, B3; Sandy Bauers, "EPA Looks at Crackdown on Smokestack Emissions," *Philadelphia Inquirer,* July 8, 2011, A1 (job losses).

64. "ACCCE Urges Administration to Avoid 'Severe Economic Impacts,' Rethink MACT," *U.S. Coal Review,* December 19, 2011; Anna Palmer and Robin Bravender, "Odd Couple: Gas Firm, Lung Group," *Politico,* December 14, 2011.

65. U.S. Environmental Protection Agency, "National Emission Standards for Hazardous Air Pollutants from Coal- and Oil-Fired Electric Utility Steam Generating Units and Standards of Performance for Fossil-Fuel-Fired Electric Utility, Industrial-Commercial-Institutional, and Small Industrial-Commercial-Institutional Steam Generating Units; Final Rule," 77 Fed. Reg. 9304, 9305–6, 9310, 9326–27, 9362, 9367–68 (February 16, 2012); "EPA Finalizes Rule to Reduce Mercury, Air Toxics Emissions from Power Plants," *BNA Environment Reporter,* December 23, 2011, 2877 (Jackson quote).

66. Cathy Cash, "Mercury Rule Survives Inhofe Attack," *Platts Coal Outlook,* June 25, 2012, 1; Bill Reilly, "Sen. Rockefeller: Some US Coal Operators Cannot 'Face Reality,'" *SNL Coal Report,* June 25, 2012 (Rockefeller quote); "Senators Reject Bid to Scrap EPA Utility MACT as Reliability Bill Advances," *Inside EPA Weekly Report,* June 22, 2012; Darren Epps, "Pro-Coal Group Ad Campaign Pressures Politicians to Overturn EPA's Mercury Rule," *SNL Daily Coal Report,* June 11, 2012; Kathleen Hart, "Whitfield Charges EPA with 'War On Coal' at House Hearing on Agency's $8.3B Budget," *SNL Energy Finance Daily,* February 29, 2012; William H. Carlile, "Power Industry Worries about 'Train Wreck' of Converging Air Pollution Regulations," *BNA Environment Reporter,* February 10, 2012, 319; Amy Harder, "Inhofe Launches Rocket at EPA Mercury Rule," *National Journal Daily,* February 16, 2012.

67. *Oversight: Review of the Environmental Protection Agency's Mercury and Air Toxics Standards (MATS) for Power Plants: Hearings before the Subcommittee on Clean Air and Nuclear Safety of the Senate Committee on Environment and Public Works,* 112th Cong. (March 20, 2012) (testimony of Vickie Patton, Environmental Defense Fund); M. J. Bradley & Associates, "MATS Compliance Extension

Status Update," June 24, 2015, http://www.mjbradley.com/sites/default/files /MATS%20Compliance%20Extension%20Update.pdf (extensions, FERC recommendation); "States Largely Granting Power Plant Requests for One-Year Mercury Compliance Extensions," *BNA Environment Reporter,* July 5, 2013, 1973; Housley Carr, "NRG Latest IPP to Plan for Compliance with EPA's MATS Rule," *Electric Power Daily,* November 26, 2012; "EPA Says Further Guidance on Utility MACT Deadline Extension Unlikely," *Inside EPA Weekly Report,* November 23, 2012 (dry sorbent); Jonathan Crawford, "FirstEnergy Sees Costs Fall, Markets Improve, but Economy Weighs Heavily," *SNL Electric Transmission Week,* August 13, 2012; Cathy Cash, "No Firestorm Appears to Be Raging about EPA Mercury Rule as Court Challenges Start Slowly," *Electric Utility Week,* March 26, 2012, 1.

68. Chris Mooney, "One Statistic Shows Just How Dramatically Our Energy System Is Changing," *Washington Post,* February 4, 2016 (no new coal plants); "EPA Air Rules, Gas Prices Further Weaken Prospects for New Coal Power," *Inside EPA,* February 3, 2012, 1.

69. Michigan v. EPA, 135 S. Ct. 2699, 2707 (2015).

70. U.S. Environmental Protection Agency, "Supplemental Finding That It Is Appropriate and Necessary to Regulate Hazardous Air Pollutants from Coal- and Oil-Fired Electric Utility Steam Generating Units," 81 Fed. Reg. 24420, 24420–21 (April 25, 2016); Patrick Ambrosio, "Trump Has Options in Utility Emissions Battle," *BNA Environment Reporter,* November 21, 2016, 4171; Taylor Kuykendall, "Appeals Court Allows MATS Rule to Stand While EPA Fixes Cost Estimate," *SNL Daily Coal Report,* December 16, 2015.

71. For a detailed description and analysis of this exercise, see Thomas O. McGarity, "Science and Policy in Setting National Ambient Air Quality Standards: Resolving the Ozone Enigma," *Texas Law Review* 93 (2015): 1783–1809.

72. Andrew Childers, "EPA Will Reconsider Air Quality Standards for Ozone Set during Bush Administration," *BNA Environment Reporter,* September 18, 2009, 2173.

73. U.S. Environmental Protection Agency, "National Ambient Air Quality Standards for Ozone," 75 Fed. Reg. 2938 (January 19, 2010); Andrew Childers, "Number of Counties Not Attaining Standards for Ozone Could Double under EPA Proposal," *BNA Environment Reporter,* January 15, 2010, 101; Andrew Childers, "EPA Review Plan Outlines Process Used to Reconsider Ozone Standards," *BNA Environment Reporter,* October 23, 2009, 2440.

74. Andrew Childers, "NRDC Says Industry Estimates of Job Losses from EPA Boiler, Ozone Rules Are 'Flawed,'" *BNA Environment Reporter,* November 5, 2010, 2447; Andrew Childers, "Health Groups Tell EPA to Set Standards for Ground-Level Ozone at Strictest Level," *BNA Environment Reporter,* February 5, 2010, 253; Andrew Childers, "Scientific Advisers Endorse EPA Proposal on Ozone Despite Implementation Concerns," *BNA Environment Reporter,* January 29, 2010, 199

(Arch Coal quote); Juliet Eilperin, "EPA Proposes Stricter Smog Pollutant Limits," *Washington Post,* January 8, 2010, A1 (scientific justification quote).

75. The description of the White House role and associated lobbying is drawn from John M. Broder, "Re-election Strategy Is Tied to a Shift on Smog," *New York Times,* November 17, 2011, A1; Jessica Coomes, "White House Chief of Staff Hears Arguments by Industry, Advocacy Groups on Ozone Rule," *BNA Environment Reporter,* August 26, 2011, 1919; Stephen Power, "Business Blasts Ozone Limits— Trade Groups Warn White House That New EPA Curbs Would Choke Off Growth," *Wall Street Journal,* July 21, 2011, A3 (ads).

76. Barack Obama, "Statement by the President on the Ozone National Ambient Air Quality Standards," White House, Office of the Press Secretary, September 2, 2011; Broder, "Re-election Strategy" (Jackson quote); Jessica Coomes, "Court Dismisses Lawsuit Challenging EPA on Decision Not to Toughen Ozone Standards," *BNA Environment Reporter,* February 24, 2012, 441.

77. U.S. Environmental Protection Agency, "National Ambient Air Quality Standards for Ozone; Final Rule," 80 Fed. Reg. 65292, 65294 (October 26, 2015); Annalee Grant and Molly Christian, "Ozone Rule Expected to Go Easy on Coal as EPA Modeling Shows Only 5 Units May Need Changes," *SNL Generation Markets Week,* October 13, 2015; Eric Wolff, "Court Orders EPA to Propose New Ozone Standards by Dec. 1," *SNL FERC Power Report,* May 7, 2014.

78. National Environmental Development Ass'n v. EPA, 686 F.3d 803 (D.C. Cir. 2012); U.S. Environmental Protection Agency, "Primary National Ambient Air Quality Standard for Sulfur Dioxide; Final Rule," 75 Fed. Reg. 35520 (June 22, 2010); John M. Broder, "E.P.A. Tightens Rule on Sulfur Dioxide," *New York Times,* June 4, 2010, A18; "Energy, Utility Groups Challenge Novel EPA Short-Term NO_x Standard," *Inside EPA Weekly Report,* April 16, 2010; "State, Local Agencies Resist EPA Proposal to Scrap 24-Hour SO_2 NAAQS," *Risk Policy Report,* March 9, 2010 (EEI position); Juliet Eilperin, "Official Questions Pollution Proposal," *Washington Post,* December 4, 2009, A4; Amena Saiyid, "EPA's Proposed New Hourly Standard for SO_2 Has Power Utilities Scrambling," *Electric Utility Week,* November 23, 2009; "Risk Report Recommends EPA Establish One-Hour Air Standard for Sulfur Dioxide," *BNA Environment Reporter,* August 14, 2009, 1937.

79. For an extended analysis of the Kingston spill and the failures to take precautions that precipitated the spill, see Thomas O. McGarity and Rena Steinzor, "The End Game of Regulation: Myopic Risk Management and the Next Catastrophe," *Duke Environmental Law & Policy Forum* 23 (2012): 93–149.

80. Lynn L. Bergeson, "EPA Responds to Coal Ash Release," *Pollution Engineering,* May 1. 2009, 19; Editorial, "Pool of Trouble," *Washington Post,* January 12, 2009, A12; Richard Fausset, "Ash Spill Leaves Future Hazy," *Chicago Tribune,* January 1, 2009, A14.

81. Charlotte E. Tucker, "EPA to Propose Coal-Ash Rule by Year's End, Asks Utilities for Data on Ash Impoundments," *BNA Environment Reporter,* March 13, 2009, 552; Juliet Eilperin, "Disposal of Coal Ash Rises as Environmental Issue," *Washington Post,* January 16, 2009, A4.

82. U.S. Environmental Protection Agency, "Hazardous and Solid Waste Management System; Identification and Listing of Special Wastes; Disposal of Coal Combustion Residuals from Electric Utilities; Proposed Rule," 75 Fed. Reg. 35128, 35143, 35151 (2010); Charlotte E. Tucker, "Advocacy Groups Ask EPA to Take Lead on Regulating Coal-Combustion Waste," *BNA Environment Reporter,* March 6, 2009, 494.

83. Bebe Raupe, "Strict Coal Ash Rule Will Destroy Market for Reuse, Industry Representatives Tell EPA," *BNA Environment Reporter,* October 1, 2010, 2190; Andrew M. Ballard, "Declaring Coal Ash Hazardous Would Strain Landfill Capacity, Industry Officials Tell EPA," *BNA Environment Reporter,* September 17, 2010, 2072; Tripp Baltz, "Industry Decries EPA Coal Ash Rule at Hearing; Others Urge Stricter Option," *BNA Environment Reporter,* September 10, 2010, 1999; "Utility Industry to EPA: Please Regulate Coal Ash," *Pollution Engineering,* June 1, 2009, 9.

84. Jeffrey Tomich, "Ameren Coal Ash Used as Mine Fill Near Ste. Genevieve," *St. Louis Post-Dispatch,* March 30, 2013 (quarries); Asher Price, "Few Rules for Coal Ash Waste," *Austin American-Statesman,* January 11, 2009, B01 (Texas); Shaila Dewan, "Hundreds of Coal Ash Dumps, with Virtually No Regulation," *New York Times,* January 7, 2009, A1; "Economic Benefits of Coal Ash Reuse Exaggerated In EPA Analysis, Groups Charge," *BNA Environment Reporter,* January 7, 2011, 16.

85. Rena Steinzor, "The Case for Abolishing Centralized White House Regulatory Review," 1 *Michigan Journal of Environmental & Administrative Law* 1 (2012): 260–62; Charlotte E. Tucker, "Original Draft Shows Coal-Ash Proposal Substantially Revised during OMB Review," *BNA Environment Reporter,* May 14, 2010, 1061.

86. U.S. Environmental Protection Agency, "Hazardous and Solid Waste Management System; Identification and Listing of Special Wastes; Disposal of Coal Combustion Residuals from Electric Utilities; Proposed Rule," 75 Fed. Reg. 35128, 35133–34, 35152–53 (2010).

87. Darren Sweeney, "Prosecutors Document 'Failure' at Duke Energy as Company Still Faces Civil Action," *SNL Electric Utility Report,* May 25, 2015; Trip Gabriel, "Ash Spill Shows How Watchdog Was Defanged," *New York Times,* March 1, 2014, A1; Sandy Smith, "Federal Prosecutors Launch Criminal Investigation of NC Environmental Regulators," *EHS Today,* February 20, 2014 (EPA technical report); Michael Wines and Timothy Williams, "Huge Leak of Coal Ash Slows at North Carolina Power Plant," *New York Times,* February 7, 2014, A11.

88. U.S. Environmental Protection Agency, "Hazardous and Solid Waste Management System: Disposal of Coal Combustion Residuals from Electric Utilities," 80 Fed. Reg. 21302, 21310 (April 17, 2015); Joby Warrick, "EPA Imposes Coal-Ash Curbs but Stops Short of 'Hazardous' Designation," *Washington Post*, December 20, 2014, A3.

89. Utility Solid Waste Activities Group v. EPA, 901 F.3d 414 (D.C. Cir. 2018); U.S. Environmental Protection Agency, "Hazardous and Solid Waste Management System: Disposal of Coal Combustion Residuals from Electric Utilities; Amendments to the National Minimum Criteria (Phase One); Proposed Rule," 83 Fed. Reg. 11584, 11586 (March 15, 2018); "EPA Coal Ash Permit Guide Limits States' Flexibility but Avoids New Rule," *Inside EPA*, August 18, 2017, 17.

90. U.S. Environmental Protection Agency, "National Pollutant Discharge Elimination System—Cooling Water Intake Structures at Existing Facilities and Phase I Facilities," 76 Fed. Reg. 22174, 22204–5 (April 20, 2011); Jennifer Zajac, "EPA's Proposed Cooling Water Intake Standards Met with Skepticism," *SNL Daily Gas Report*, April 6, 2011.

91. Nick Juliano, "Power Sector Blasts Fish-Kill Standard in EPA's Proposed Cooling-Water Rule," *Inside Energy with Federal Lands*, August 22, 2011, 3; Jennifer Zajac, "Exelon 'Encouraged' by EPA Cooling Water Intake Proposal; Environmentalists Unhappy," *SNL Generation Markets Week*, April 5, 2011; Cathy Cash, "EPA Cooling Water Intake Rule Prompts Concern among Stakeholders," *Platts Coal Outlook*, April 4, 2011, 1.

92. U.S. Environmental Protection Agency, "National Pollutant Discharge Elimination System—Final Regulations to Establish Requirements for Cooling Water Intake Structures at Existing Facilities and Amend Requirements at Phase I Facilities," 79 Fed. Reg. 48300, 48304, 48329–30 (August 15, 2014).

93. Cooling Water Intake Structure Coalition v. EPA, 898 F.3d 173 (2d Cir. 2018); Eric Wolff, "Environmental Groups to Sue EPA over Water Cooling Rule," *SNL Generation Markets Week*, August 26, 2014 (quote).

94. Hilary Costa, "Lost Generation Not Only Result If Navajo Closes," *Megawatt Daily*, November 29, 2010; "EPA's Proposed Haze Rule for Four Corners Plant Too Costly: Navajos," *Platts Coal Outlook*, October 18, 2010, 8; Barry Cassell, "Colorado Officials Turn Up Heat on Desert Rock Plant, Four Corners Cleanup," *SNL Coal Report*, March 23, 2009.

95. Taylor Kuykendall, "Court Denies Environmentalists Bid to Reverse Approval of Four Corners Plant," *SNL Daily Coal Report*, July 25, 2014; Ethan Howland, "Southwestern Coal-Fired Plants Facing Scrutiny This Week," *Electric Power Daily*, February 15, 2011, 1; "Arizona Utility Plans Partial Shutdown of Four Corners Plant to Reduce Emissions," *BNA Environment Reporter*, November 19, 2010, 2561; Barry Cassell, "EPA Proposes Sharp NO_x, Particulate Cuts at Four Corners Coal Plant," *SNL Generation Markets Week*, October 26, 2010; "EPA

Proposes Stringent Pollution Controls for Coal-Fired Power Plant in New Mexico," *BNA Environment Reporter,* October 15, 2010, 2295.

96. Jeff Stanfield, "EPA Approves Colo. 'Novel' Air Quality Plan for Regional Haze Reduction," *SNL Daily Gas Report,* September 13, 2012; Matthew Bandyk, "Oregon Approves Plan to Close Boardman Coal Plant by 2020," *SNL Electric Utility Report,* December 20, 2010; Jeff Stanfield, "Colo. PUC Allows Switch to Gas at Xcel Energy Plant, Closing of 5 More Coal Units," *SNL Electric Utility Report,* December 20, 2010; Jeff Stanfield, "Colo. Governor Signs Bill to Reduce Emissions from Coal-Fired Power Plants," *SNL Power Week Midwest,* April 27, 2010; Barry Cassell, "Xcel Energy Eyes Potential Coal Unit Shutdowns in Favor of Natural Gas," *SNL Power Week Midwest,* March 16, 2010.

97. Oklahoma v. EPA, 723 F.3d 1201 (10th Cir. 2013); *America's Energy Future, Part I: A Review of Unnecessary and Burdensome Regulations: Hearings before the House Committee on Oversight and Government Reform,* 112th Cong. (July 13, 2012) (testimony of Patricia Horn, OG&E Energy Corp.); Jonathan Crawford, "EPA Finalizes Stricter SO_2 Limits for Okla. Coal Plants," *SNL Generation Markets Week,* December 20, 2011; "EPA Rejection of Oklahoma Haze Strategy Could Prompt Federal Air Plan," *Inside EPA Weekly Report,* August 20, 2010; Barry Cassell, "Oklahoma Gas and Electric Signs Coal Deals, Battles over Air Emissions," *SNL Coal Report,* March 1, 2010.

98. Christine Cordner, "DOJ Files Enforcement Action Against 2 Luminant Coal-Fired Plants in Texas," *SNL Electric Utility Report,* August 26, 2013; Brian Hansen, "For Coal Plants, EPA Data Requests Often Spell Trouble," *Inside Energy with Federal Lands,* May 6, 2013, 1; Kathleen Hart, "GAO Finds 'Substantial' Number of Fossil Plants Not Complying with New Source Review," *SNL Coal Report,* July 30, 2012; Andrew Childers, "New Source Review Lawsuit Signals Focus on Enforcement at Obama EPA," *BNA Environment Reporter,* September 11, 2009, 2095; Wayne Barber, "EPA Sues NRG Energy over Operation of Big Cajun 2 Plant," *SNL Electric Utility Report,* March 2, 2009.

99. U.S. Government Accountability Office, *Air Pollution: EPA Needs Better Information on New Source Review Permits* (Washington, DC, June 2012), 30, app. 3; Matthew Bandyk, "Alliant's Iowa Utility Reaches Settlement with EPA on Coal Plants," *SNL Electric Utility Report,* July 20, 2015; Mary Powers, "TVA's 'Historic' Agreement with EPA to Retire 2,700 MW of Coal Capacity Is Praised and Panned," *Electric Utility Week,* April 18, 2011, 1.

100. 28 U.S.C. § 2462; United States v. Luminant Generation Co., 905 F.3d 874, 882 (5th Cir. 2018); "EPA Argues Courts Divided on Whether NSR Violations Are Single Events," *Inside EPA Weekly Report,* January 29, 2016; "Third Circuit Becomes Latest Court to Cite Statute of Limitations in Power Plant Case," *BNA Environment Reporter,* August 23, 2013, 2514; "Appellate Rulings May Shrink Time-Frame for EPA to Pursue NSR Cases," *Clean Energy Report,* August 5, 2013;

"EPA Appeals Ruling Finding New Owners Not Liable for Past NSR Violations," *Inside EPA Weekly Report,* January 20, 2012; Bob Matyi, "DTE Energy Prevails in Federal Government Lawsuit over Michigan's Largest Power Plant," *Electric Utility Week,* September 19, 2011.

101. FERC v. Electric Power Supply Ass'n, 136 S. Ct. 760, 771 (2016); Glen Boshart, "Despite Strong Opposition by Some, FERC Allows Demand Response to Receive Full Market Price," *SNL Electric Utility Report,* March 21, 2011; Glen Boshart, "Industry Divides Sharply over FERC's Proposed Demand Response Compensation Rules," *SNL Electric Utility Report,* May 24, 2010; Glen Boshart, "ELCON, EPSA and Harvard's Hogan Face Off over Demand Response Compensation Rules," *SNL Electric Utility Report,* May 24, 2010.

102. Julia E. Sullivan et al., "Why End Users Are Investing (Big) in DG," *Electricity Journal,* March 2014, 23; Grace Hsu, "Net Metering Wars: What Should We Pay for DG?," *Berkeley Energy & Resources Collaborative,* February 24, 2014.

103. Richard L. Revesz and Burcin Unel, "Managing the Future of the Electricity Grid: Distributed Generation and Net Metering," *Harvard Environmental Law Review* 41 (2017): 47–49; Paula Mints, "Notes from the Solar Underground: The U.S. Utility War against Net Metering," *Renewable Energy World,* February 23, 2016; Tim Dickinson, "The Koch Brothers' Dirty War on Solar Power," *Rolling Stone,* February 11, 2016; Sullivan, et al., Why End Users Are Investing; Hsu, Net Metering Wars.

104. Revesz and Unel, "Managing the Future," 64–69; Amy L. Stein, Distributed Reliability, *University of Colorado Law Review* 87 (2016): 889; Amy Poszywak, "EEI Lays Out Key Objectives for 2014, with DG a Top Priority," *SNL Energy Finance Daily,* February 14, 2014; "Obama Push to Add Distributed Power, CHP Sparks Utility Concerns," *Clean Energy Report,* April 8, 2013.

105. Elisabeth Graffy and Steven Kihm, "Does Disruptive Competition Mean a Death Spiral for Electric Utilities?," *Energy Law Journal* 35 (2014): 32; David Raskin, "Getting DG Right: A Response to 'Does Disruptive Competition Mean a Death Spiral for Electric Utilities?,'" *Energy Law Journal* 35 (2014): 267; Edward Humes, "Throwing Shade: How the Nation's Investor-Owned Utilities Are Moving to Blot Out the Solar Revolution," *Sierra,* May / June, 2014; Peter Maloney, "Brattle: Wider Electrification Key to Averting Both Climate Change and Utility Death Spiral," *Utility Dive,* May 24, 2017.

106. William Boyd and Ann E. Carlson, "Accidents of Federalism: Ratemaking and Policy Innovation in Public Utility Law," *UCLA Law Review* 63 (2016): 869; Steven Ferrey, "Ring-Fencing the Power Envelope of History's Second Most Important Invention of All Time," *William & Mary Environmental Law and Policy Review* 40 (2015): 60; Edward Klump, "Concerns Linger for Solar as Regulators Back Changes," *E&E News,* December 15, 2017 (Texas); Mints, "Notes from the

Solar Underground" (Hawaii); Carl Pope, "Rooftop Solar Wars Continue," *EcoWatch,* January 29, 2016 (other states).

107. Delaware Dept. of Natural Resources and Environmental Control v. EPA, 785 F.3d 1, 15 (D.C. Cir. 2015); Martin Coyne, "Diesel Generators Cleaner as Demand Response Than Conventional Turbines, Wellinghoff Says," *Electric Utility Week,* July 16, 2012; "State, Industry Fear EPA Engine Rule May Deter 'Clean' Power Investment," *Inside EPA Weekly Report,* July 6, 2012 (virtual power plants).

108. William Yardley, "Candidates and the Climate," *Los Angeles Times,* December 30, 2015; Molly Christian, "Hillary Clinton Unveils $30B Plan to Ease Coal Industry Decline," *SNL Daily Coal Report,* November 13, 2015; Coral Davenport, "Beijing Puts Ball Back in Washington's Court in Fight to Curb Climate Change," *New York Times,* September 26, 2015, A8; Alan Neuhauser, "The 2016 Election Is Critical for Stopping Climate Change," USNEWS.com, August 14, 2015; Ben Geman, "Al Gore: Climate Skepticism Will Haunt Republicans in 2016," *National Journal Daily,* September 17, 2014; "Presidential Election Results: Donald J. Trump Wins," https://www.nytimes.com/elections/results/president.

109. Hendrick Hertzberg, "Cooling on Warming," *New Yorker,* February 7, 2011, 21; Jean Chemnick, "Experts: Words Matter in Climate Debate," *Inside Energy with Federal Lands,* August 30, 2010; Paul Burka, "Cap and Tirade," *Texas Monthly,* November 2009, 14.

110. U.S. Environmental Protection Agency, "8-Hour Ozone (2008) Designated Area / State Information," May 31, 2018, https://www3.epa.gov/airquality/greenbook /hbtc.html; Barbara Vergetis Lundin, "CSAPR Impact on Power Plant Emissions Questioned," *FierceEnergy,* May 16, 2014.

111. Eric Wolff, "Supreme Court's Eventual MATS Ruling Will Be (Mostly) Moot," *SNL Daily Coal Report,* May 15, 2015.

112. Annalee Grant, "Utilities Expect Little Change to Coal Fleet Decisions with Carbon Rule on Hold," *SNL Daily Coal Report,* February 23, 2016; Lauren Bellero, "TVA on Track with Plant Retirements Despite Clean Power Plan Stay," *SNL Daily Coal Report,* February 15, 2016 (load shifting); Jeffrey Ryser, "Generators Reshape Portfolios, Despite CPP Stay," *Platts Energy Trader,* February 11, 2016, 1.

113. Juliet Eilperin and Peter Wallsten, "Obama's Strategy Confounds Allies, Foes," *Washington Post,* September 4, 2011, A1.

114. "Power Companies Retreat from 'Train Wreck' Claim in Fighting EPA Rules," *Clean Energy Report,* August 13, 2012; "EPA Rejects Studies Claiming Reliability 'Train Wreck' from Utility Rules," *Inside EPA Weekly Report,* May 20, 2011.

115. Molly Christian, "Gas Overtakes Coal Again for US Power Generation," *SNL Daily Coal Report,* September 28, 2015; Molly Christian, "Coal Regains Top US Power Market Share in May," *SNL Daily Coal Report,* July 29, 2015; "NatGas

Responsible for Dive in Carbon Dioxide Emissions for US," *Natural Gas Week,* August 6, 2012; Abby Gruen, "Commodity, Economic Woes Behind Dearth of Fossil-Fired Power Deals, JP Morgan Says," *SNL FERC Power Report,* May 30, 2012.

116. Everett Wheeler, "How Did We Get Here? The Causes and Consequences of Today's Coal Market," *SNL Daily Coal Report,* December 2, 2015; Glen Boshart, "Coal Still 'Very Relevant,' State Regulators Told," *SNL Daily Coal Report,* November 12, 2015; Erica Martinson, "The Fall of Coal," *Politico,* April 16, 2015 (Carter quote).

10. THE WAR ON FOSSIL FUELS

1. Ted Nace, *Climate Hope: On the Front Lines of the Fight against Coal* (San Francisco: CoalSwarm, 2010), 16–17, 90–93,162–63; Matthew Bandyk, "In Its Environmental Battles, Sierra Club Fights Coal Plants with Concessions," *SNL Electric Utility Report,* February 6, 2012; Margaret Kriz Hobson, "The Sierra Club's Burning Desire," *National Journal,* September 5, 2009; Judy Pasternak, "Coal at Heart of Climate Battle," *Los Angeles Times,* April 14, 2008, A1.

2. Pasternak, "Coal at Heart."

3. Juliet Eilperin and Steven Mufson, "An Uneasy Alliance on Natural Gas Fractures," *Washington Post,* February 20, 2012; Eric Lipton, "Even in Coal Country, the Fight for an Industry," *New York Times,* May 30, 2012, A1.

4. Sierra Club v. EPA, 499 F.3d 653 (7th Cir. 2007).

5. "IGCC and CCS—From Polk I to Edwardsport: Two Decades of IGCC Evolution," *Modern Power System,* September 8, 2010, 20; "Reports Point to IGCC as Less Costly for CO_2 Control Than Other Technologies," *Electric Utility Week,* July 17, 2006, 12; R. C. Balaban, "Impact from Power Plant Would Be Widespread," *Waterloo (IA) Courier,* July 2, 2006.

6. Sierra Club v. EPA, 499 F.3d 653 (7th Cir. 2007); Jeff Goodell, *Big Coal: The Dirty Secret behind America's Energy Future* (Boston: Houghton Mifflin, 2006), 216; Wayne Barber, "Wyo. Supreme Court Upholds Dry Fork Plant Permit," *SNL Power Week Midwest,* March 16, 2010; "IGCC and CCS"; Steve Raabe, "'Clean Coal' Plant Setbacks Mount in U.S.," *Denver Post,* November 1, 2007, C1; "EPA Official Rejects Gasification as Standard for New Coal-Fired Electric Power Plants," *BNA Environment Reporter,* December 23, 2005, 2625.

7. Suzanne McElligott, "Sierra Club Appeals Taylorville IGCC Project Permit," *Gasification News,* July 11, 2007; Steve Blankinship, "Illinois IGCC Gets Air Permit," *Power Engineering,* July 1, 2007, 12; Suzanne McElligott, "Illinois EPA OK's Permit for IGCC Plant," *Gasification News,* June 13, 2007; "Illinois Governor Unveils Landmark Permit for Coal Gasification Plant," *U.S. Coal Review,* June 11, 2007.

8. Bob Matyi, "Tenaska Scraps Coal Gasification Plants," *Megawatt Daily,* June 24, 2013, 1; "EPA Board Rejects Administrative Review of Challenge to Coal-Fired Power Plant," *BNA Environment Reporter,* February 8, 2008, 258; "Sierra Club Opposes IGCC Plant, but Another Enviro Group Favors It," *Platts Coal Outlook,* July 16, 2007, 6; McElligott, "Sierra Club Appeals."

9. Nace, *Climate Hope,* 68–70; Christine Powell, "Kansas' Top Court Affirms Green Light for $2.8B Coal Plant," *Law360,* March 17, 2017; "State Court Reverses Environment Agency on Air Permit for Proposed Coal-Fired Plant," *BNA Environment Reporter,* October 11, 2013, 3052; Housley Carr, "Insider Emails Cited in Push to Invalidate Air Permit for Sunflower Plant," *Platts Coal Outlook,* June 27, 2011, 4; "Activists' Suit Seeks to Reinstate GHG Limits in Controversial Kansas Permit," *Inside EPA Weekly Report,* January 21, 2011; Tim Carpenter, "State's Top Environmental Regulator Ousted," *Topeka Capital-Journal,* November 3, 2010, 10; Housley Carr and Gail Roberts, "New Kansas Governor, Sunflower Co-Op Cut Deal to Build 895-MW Coal Plant," *Electric Utility Week,* May 11, 2009, 15; Housley Carr, "Sunflower Sues Kansas Officials, Saying They Discriminated in Denial of Permit for Project," *Electric Utility Week,* November 24, 2008, 17; Alan Greenblatt, "Guarding the Greenhouse," *Governing Magazine,* July 2008, 21; Stephen Power, "Kansan Stokes Energy Squabble with Coal Ruling," *Wall Street Journal,* March 19, 2008, A6; Housley Carr, "Sunflower Calls Kansas Denial of Air Permit for Coal Units 'Arbitrary' and Vows to Fight It," *Electric Utility Week,* October 22, 2007, 1; Matthew L. Wald, "Citing Global Warming, Kansas Denies Plant Permit," *New York Times,* October 20, 2007, 4; "Enviros Want CO$_2$ Limits, New Technology on Kansas Units," *Platts Coal Outlook,* May 28, 2007, 11; Sunflower Electric Power Corporation, "Holcomb Station Expansion Project," https://www.sunflower.net /holcomb-station-expansion-project%E2%80%8B/.

10. Nace, *Climate Hope,* 139–41; Laura Paskus, "The Life and Death of Desert Rock," *High Country News,* August 16, 2010; Kari Lydersen, "Dirty Smoke Signals," *In These Times,* May 2008, 26; Wayne Barber, "New Mexico Governor Seeks More Hearings on Sithe Plant Proposal," *SNL Electric Utility Report,* August 6, 2007 (quote); "Texas Company, New Mexico Tribe Sign Agreement to Cut Power Plant Emissions," *BNA Environment Reporter,* May 25, 2007, 1167; "EPA Proposes Permit for Power Plant on Navajo Nation Site in New Mexico," *BNA Environment Reporter,* July 27, 2006, 1559.

11. "Private Investors Take the Longview in West Virginia," *Modern Power System,* June 12, 2007, 43; "W.Va. DEP Makes Changes to Longview Permit, Seeks Comment," *Platts Coal Outlook,* January 5, 2004, 9; "Pressure Mounts to Scuttle Proposed WV Plant Project," *U.S. Coal Review,* October 6, 2003.

12. Sierra Club v. Sandy Creek Energy Associates, 627 F.3d 134, 137 (5th Cir. 2010); Barry Cassell, "Sierra Club Lawsuit Seeks to Block Operation of EKPC's New Spurlock Coal Unit," *SNL Electric Utility Report,* May 4, 2009; Wayne Barber,

"Sierra Club Files Notice to Sue Coal Plants over Mercury Emissions," *SNL Power Week West*, May 12, 2008.

13. Tony Bartelme, "Years after Santee Cooper Coal Plant Is Canceled, Millions of Dollars of Equipment Sits Unused," *Charleston Post and Courier*, April 11, 2008, A1; Doug Pardue, "Pulling the Plug on Pee Dee Plant," *Charleston Post and Courier*, September 26, 2009, A1 (win-win); Tony Bartelme, "Health Board Upholds Pee Dee Coal Plant Permit," *Charleston Post and Courier*, February 13, 2009, B1; "Santee Cooper Defends Coal Generation Plan," *Electric Power Daily*, July 23, 2008, 5 (environmental groups' response); Santee Cooper, "Santee Cooper Announces Generation Plan, Submits Mercury Emissions Analysis for Pee Dee Energy Campus," press release, Targeted News Service, July 1, 2008; Tony Bartelme, "Environmental Groups Critical of Proposed Power Plant," *Charleston Post and Courier*, January 23, 2008, B3; Jamie Durant, "Petition Presented in Support of Proposed Coal-Buring Plant," *Florence (SC) Morning News*, November 9, 2007; Jamie Durant, "Power Plant Hearing Draws Crowd," *Florence Morning News*, November 8, 2007; Barry Cassell, "South Carolina Issues Draft Air Permit for Coal-Fired Pee Dee Plant," *SNL Power Daily with Market Report*, October 12, 2007; Kyle Stock, "Burning Issue," *Charleston Post and Courier*, July 16, 2007, E22 (Pee Dee tribe); "Santee Cooper Plan for Coal Generation Units Under Fire," *Electric Power Daily*, July 3, 2007, 1; "Santee Cooper Decides to Self-Build 600-MW Coal Plant in South Carolina," *Global Power Report*, April 27, 2006, 16.

14. Steve Hooks, "Sunflower's 895-MW Kansas Plant Needs Extensive Environment Review, Judge Rules," *Electric Utility Week*, April 4, 2011; "East Kentucky Power Cooperative Admitted to Some Problems but Insisted That It Is Not Near a Default," *SNL Electric Utility Report*, May 3, 2010.

15. Wayne Barber, "Groups Claim Efficiency Programs More Cost-Effective Than Dominion Coal Plant," *SNL Power Week West*, January 12, 2009; Bill Gerhard, "Coal-Based Power to Fuel Expansion," *Des Moines Register*, May 19, 2008, A9; Ryan Self, "Oklahoma Law Judge Recommends Approval of 950-MW Red Rock Coal Plant," *SNL Power Week Southeast*, August 28, 2007.

16. Mick Hinton, "Power Plant Decries Official's Activism," *Tulsa World*, September 6, 2007, A13; Janice Francis-Smith, "Red Rock Media War Draws Fire," *Oklahoma City Journal Record Legislative Report*, September 6, 2007.

17. Nace, *Climate Hope*, 75 (quote); Melissa Powers, "The Cost of Coal: Climate Change and the End of Coal as a Source of 'Cheap' Electricity," *University of Pennsylvania Journal of Business Law* 12 (2010): 424; Anita Weier, "Activists Protest Proposed Coal Plant," *Capital Times* (Madison, WI), May 15, 2008.

18. Ian Urbina, "A Model for 'Clean Coal' Goes Awry," *New York Times*, July 5, 2016, A1; Jennifer Jacob Brown, "Anthony Topazi: Cleaner Coal in Kemper

County," *Meridian Star,* December 22, 2008; Laura Hipp, "Utility Weighs $1.8B Plant," *Jackson Clarion-Ledger,* December 13, 2006, A1.

19. Barry Cassell, "Mississippi Power Doing Feasibility Work on Coal-Fueled IGCC," *SNL Electric Utility Report,* August 18, 2008.

20. Jennifer Jacob Brown, "Mississippi Power Awaits PSC Decision on Kemper Plant," *Meridian Star,* October 3, 2009; Housley Carr, "Sierra Club Opposes Mississippi Power Plan to Build $2.2 Bil, 582-MW IGCC CCS Plant," *Electric Utility Week,* March 2, 2009, 22.

21. Jennifer Jacob Brown, "Protest over Proposed Power Plant," *Meridian Star,* March 2, 2010; "Miss. PSC Eyes Utility Plans for IGCC Plant," *Electric Power Daily,* October 7, 2009, 9 (AARP opposition); "Merchants Press Mississippi to Suspend IGCC Plant Review," *Electric Power Daily,* March 19, 2009, 1; "Mississippi Power IGCC Plant Draws Opponents," *Electric Power Daily,* April 23, 2009, 7.

22. "Mississippi PSC Finds Southern Unit Needs Capacity," *Electric Power Daily,* November 11, 2009, 1; Andrew Engblom, "Mississippi Sets Schedule to Evaluate Kemper IGCC Project," *SNL Electric Utility Report,* June 15, 2009; Housley Carr, "Mississippi Power Seeks Permit to Build 582-MW, $2.2 Billion IGCC Project," *Electric Utility Week,* January 19, 2009, 28; Jeff Ayres, "Mississippi Power Working on Clean-Coal Plant," December 19, 2008, *Jackson Clarion-Ledger,* B7.

23. "PSC Ruling Casts Doubt on Future of 582-MW IGCC Project in Miss.," *Electric Power Daily,* May 3, 2010, 1 (commission quote); Clay Chandler, "Mississippi Public Service Commission Members: Politics Played No Part in Kemper Plant Decision," *Mississippi Business Journal,* May 9, 2010.

24. Housley Carr, "Sierra Club Says Mississippi Governor Helped IGCC Project Win PSC Approval," *Global Power Report,* June 24, 2010, 13; Housley Carr, "Coal Power for Mississippi, but $2.9-Billion Cost Cap Set," *Engineering News-Record,* June 14, 2010, 17; "PSC Approves Miss. Power Plant," *Meridian Star,* May 27, 2010; "Mississippi Power Seeks Rehearing on IGCC Order," *Electric Power Daily,* May 12, 2010, 1; "PSC Ruling."

25. Housley Carr, "Mississippi Power Breaks Ground on 582-Mw IGCC Plant Despite Challenges to Approvals," *Electric Utility Week,* December 20, 2010; Jennifer Jacob Brown, "Mississippi Power Breaks Ground on Kemper County IGCC Power Plant," *Meridian Star,* December 17, 2010.

26. "Groups Try to Stop Construction at New Coal Plant," *Delta Democrat Times,* March 23, 2012; Housley Carr, "Unanimous Mississippi Supreme Court Backs Sierra Club over Coal Gasification Project," *Electric Utility Week,* March 19, 2012.

27. Jeff Barber, "Moody's Downgrades Mississippi Power," *Electric Power Daily,* August 7, 2012; Housley Carr, "Costs Crop Up in Mississippi Power's IGCC Plans," *Electric Power Daily,* July 11, 2012; Wyatt Emmerich, "Lignite Gasification

Plant in Trouble," editorial, *McComb Enterprise-Journal,* July 3, 2012; Housley Carr, "Southern Unit Proceeding with IGCC Build Despite PSC Ruling," *Platts Coal Outlook,* July 2, 2012; Housley Carr, "Miss. PSC OKs New Certificate for IGCC Plant," *Electric Power Daily,* April 25, 2012.

28. Wyatt Emmerich, "Southern Should Pay the Kemper Cost Overruns," editorial, *Jackson Northside Sun,* January 30, 2014 ("special purpose entity"); Matthew Bandyk, "Challenges Remain for Mississippi Power's Plant Ratcliffe Following Settlement," *SNL Electric Utility Report,* February 4, 2013; Housley Carr, "Mississippi PSC Approves $2.4 Billion Settlement for Mississippi Power's Kemper Project," *Electric Utility Week,* January 28, 2013.

29. Mark Drajem, "$5.2 Billion Plant May Help Elevate Fossil Fuel's Profile," *Chicago Tribune,* April 15, 2014, B2; Robert Kunzig, "Clean Coal Test: Power Plants Prepare to Capture Carbon," *National Geographic,* March 2014 (quote); Dan Testa, "Mississippi Power to Forfeit $133M as IGCC Project to Miss In-Service Date," *SNL Electric Utility Report,* October 14, 2013; Jeff Amy, "Kemper Delay Could Force Repayment of $133 Million," *Commercial Dispatch,* October 1, 2013.

30. Clay Chandler, "Utility Faces Court-Ordered Refunds," *Jackson Clarion-Ledger,* February 13, 2015, A1; Housley Carr, "Combined-Cycle Units at Kemper Started, Utility Says," *Platts Megawatt Daily,* August 15, 2014, 13; Sierra Club, "Agreement Brings Clean Energy Investments, Cleaner Air to Mississippi," press release, Targeted News Service, August 4, 2014; "Troubles Mount for Kemper," editorial, *McComb Enterprise-Journal,* May 12, 2014.

31. Matthew Bandyk, "Mississippi Power Refund for Kemper Project Could Top $250M," *SNL Electric Utility Report,* February 23, 2015; Housley Carr, "Miss. IGCC Project Hit by State Supreme Court," *Platts Megawatt Daily,* February 17, 2015, 1.

32. Kristi Swartz, "Investors Sue Southern Co. over Kemper Disclosures," *E&E News,* January 25, 2017; Urbina, "Model for 'Clean Coal.'"

33. Mark Chediak, "Southern Strikes a Deal on Costs of Failed 'Clean Coal' Plant," *BNA Daily Environment Report,* December 1, 2017; Steven Mufson, "This Isn't the 'Clean Coal' News Trump Wanted during 'Energy Week,'" *Washington Post,* June 28, 2017; Kristi E. Swartz, "Southern Co.'s Clean Coal Plant Hits a Dead End," *E&E News,* June 22, 2017.

34. "Seven Utilities to Build 600-MW Coal Plant," *Electric Power Daily,* June 30, 2005, 5; Tom Meersman, "Coal-Fired S.D. Power Plant's Price Escalates," *Minneapolis Star Tribune,* July 27, 2006, A1; Ben Shouse, "Companies Agree to Second Plant," *Sioux Falls Argus Leader,* July 1, 2005, B1.

35. "Backers of S.D. Coal Plant Win Court Victory," *Electric Power Daily,* January 22, 2008, 5; Dennis Lien, "Permit for Coal Plant Appealed," *St. Paul Pioneer Press,* October 3, 2006, B3; Steve Young, "Big Stone Plant Endorsed," *Sioux*

Falls Argus Leader, July 15, 2006, A1; Ben Shouse, "Plant Hinges on PUC," *Sioux Falls Argus Leader,* July 3, 2006, A1.

36. Tom Meersman, "Recommendation Is Blow Against Big Stone II Plans," *Minneapolis Star Tribune,* May 10, 2008, A1; Wayne Barber, "Big Stone II Partners Lay Out Plan for Smaller Coal Plant," *SNL Electric Utility Report,* November 19, 2007; Leslie Brooks Suzukamo, "Two Utilities Pull Out of Big Stone II Power Plant," *St. Paul Pioneer Press,* September 18, 2007; Wayne Barber, "Minnesota Law Judges Rule for Transmission Upgrades Needed for Big Stone II Plant," *SNL Electric Utility Report,* August 27, 2007; "Environmentalists Tout Power Project as Climate Change Test Case," *Clean Air Report,* December 28, 2006.

37. Thom Gabrukiewicz, "Environmental Group Appeals N.D. Ruling to Allow Investment in Big Stone II," *Sioux Falls Argus Leader,* October 2, 2008, A3; Paul Carlsen, "North Dakota PSC Finds Coal Plant 'Prudent' for OTP and MDU, along with More DSM," *Electric Utility Week,* September 1, 2008, 19.

38. Wayne Barber, "With Transmission Line Approval, Big Stone Eyes 2010 Construction, *SNL Electric Utility Report,* March 23, 2009; Ethan Howland, "Minnesota Regulators Approve Power Line Needed for Big Stone II Coal-Fired Project," *Global Power Report,* January 22, 2009, 12; Leslie Brooks Suzukamo, "Big Stone II Pushes Clean-Energy Alternative," *St. Paul Pioneer Press,* January 14, 2009.

39. Wayne Barber, "North Dakota Allows Utilities to Recover Costs of Canceled Power Plant," *SNL Electric Utility Report,* July 12, 2010; "End of Big Stone II Could Be Trouble for S.D. Wind," *Bismarck Tribune,* November 19, 2009, B1; Leslie Brooks Suzukamo, "Utilities Kill Plans for Big Stone II Power Plant," *St. Paul Pioneer Press,* November 2, 2009; Jeff Martin, "Sierra Club Rep Cheers Investor's Withdrawal from Big Stone Plant," *Sioux Falls Argus Leader,* September 30, 2009.

40. 42 U.S.C. § 7503(a)(3); Matthew L. Wald, "Utility and Sierra Club Deal Aims to Cut Carbon Dioxide," *New York Times,* March 20, 2007, C5; Rick LaFrombois, "Power Company Reveals Violations," *Wausau (WI) Daily Herald,* March 4, 2006, A1; Steve Hooks, "In Latest of Long Series of Skirmishes, KCPL Taking Sierra Club to Court over Iatan," *Platts Coal Trader,* March 2, 2007, 1.

41. Nace, *Climate Hope,* 56; Jeff Ryser, "TXU's New Managers Face Daunting Task: Build Revenue Flows to Service Big Debt," *Electric Utility Week,* October 15, 2007, 1; "TXU Deal to Launch National Drive toward IGCC, Carbon Capture: Experts," *Inside Energy with Federal Lands,* March 5, 2007, 10; "TXU-KKR-Texas Pacific Merger Announcement Reveals Practical, Financial, Regulatory and Environmental Components," *Foster Natural Gas Report,* March 2, 2007, 1; Steven Mufson and David Cho, "Energy Firm Accepts $45 Billion Takeover," *Washington Post,* February 26, 2007, A4; Rebecca Smith, Dennis K. Berman, and Henny Sender, "Power Play," *Wall Street Journal,* February 26, 2007, A1 (quote); Andrew Ross

Sorkin, "A $45 Billion Buyout Deal with Many Shades of Green," *New York Times,* February 26, 2007, A19; Felicity Barringer and Andrew Ross Sorkin, "Utility to Limit New Coal Plants in Big Buyout," *New York Times,* February 25, 2007, A1; Regina Johnson, "Texas Legislator Seeks Moratorium on PC Permits," *Platts Coal Outlook,* January 29, 2007, 11.

42. Barry Cassell, "Michigan Governor Seeks Strict Review of New Coal-Fired Projects," *SNL Daily Coal Report,* February 5, 2009; Ryan Self, "Michigan Environmental Groups Mobilizing to Block New Coal Plant Efforts," *SNL Electric Utility Report,* December 10, 2007.

43. "In Uncertain Times, LS Power Cancels 750-MW Coal Plant," *Electric Power Daily,* May 4, 2009; Bob Matyi, "Michigan Public Power Agency Signs Letter of Intent to Participate in Michigan Coal Project," *Electric Utility Week,* September 17, 2007, 18.

44. Consumers Energy Co., "Consumers Energy Announces Cancellation of Proposed New Coal Plant, Continued Substantial Investments in Major Coal Units, Anticipated Suspension of Operation of Smaller Units in 2015," press release, PR Newswire, December 2, 2011; "Consumers Energy Seeks Support for Mich. Plant," *Platts Coal Trader,* March 3, 2010, 3; Bob Matyi, "Environmentalists Say Michigan Regulators Failed 'Leadership' Test with Coal Plant Permit," *Electric Utility Week,* January 4, 2010, 9; Kerry Bleskan, "Michigan: Coal Plants Not the Best Option for Consumers Energy, Wolverine," *SNL Electric Utility Report,* September 14, 2009; Bob Matyi, "Mich. PSC Weighs NRDC Report Shunning Coal," *Electric Power Daily,* August 12, 2009, 6; Barry Cassell, "Michigan DEQ Takes Comment on Consumers Energy Coal Plant Analysis," *SNL Generation Markets Week,* June 23, 2009; "Mich. AG Decides Governor Cannot Ban Coal Plants," *Platts Coal Trader,* February 24, 2009, 1.

45. Bob Matyi, "Co-op Scraps Plans for 600-MW Mich. Plant," *Platts Megawatt Daily,* December 18, 2013; Bob Matyi, "Michigan Court Ruling Upholds Permits for Two Coal Plants," *Electric Power Daily,* March 27, 2013; Bob Matyi, "Wolverine Stops Bid Process on 600-MW Project," *Platts Coal Trader,* May 23, 2012; Bob Matyi, "Michigan Regulators Approve Final Air Permit for Wolverine Co-Op's 600-MW Coal-Fired Plant," *Electric Utility Week,* July 4, 2011 (Snyder quote); Bob Matyi, "With GOP at Helm, Mich. Changes Course, Will Permit Coal Units," *Platts Coal Trader,* February 7, 2011, 1; Bob Matyi, "Mich. House Democrats Support 'Clean Coal,'" *Electric Power Daily,* January 19, 2011; "Michigan Regulators, Calling the Project 'Way Too Expensive,' on Friday Rejected Wolverine Power Cooperative's Request for an Air Quality Permit," *Electric Utility Week,* May 24, 2010, 13; "Michigan PSC Staff Questions Need for Two Coal Plants," *Electric Power Daily,* September 9, 2009; Michael Niven, "Wolverine Submits Analysis Supporting Planned Michigan Coal Plant," *SNL Electric Utility Report,* June 15, 2009.

46. See, e.g., Matthew Bandyk, "Wolverine Cancels Mich. Coal-Fired Power Plant Project," *SNL Electric Utility Report,* December 23, 2013; Tom Henry, "Electric Supplier Pulls Its Plan for $1.2B Power Plant," *Toledo Blade,* November 26, 2009.

47. Pat Kinney, "Railroad Job: Power Plant Traffic Examined," *Waterloo Courier,* July 16, 2006; "Iowa Residents Oppose Elk Run Coal-Fired Plant," *Platts Coal Trader,* July 10, 2006, 4; Balaban, "Impact from Power Plant."

48. Tim Jamison, "Power Plant Annexation Hearing Draws Overflow Crowd," *Waterloo Courier,* September 13, 2007; "Iowa Council Approves Elk Run Plant," *Platts Coal Trader,* May 9, 2007, 2; "LS Power 750-MW Iowa Coal Plant Clears Hurdle," *Electric Power Daily,* April 5, 2007, 3.

49. Charlotte Eby, "State Panel Rejects Waterloo Annexation Plan," *Sioux City Journal,* October 12, 2007; Jamison, "Power Plant Annexation"; Beeman Perry, "Opponents of Waterloo Coal Plant Plan Rally," *Des Moines Register,* September 8, 2007, A1.

50. Pat Kinney, "LS Power Re-Files Power Plant Annexation, Rezoning Request," *Waterloo Courier,* March 18, 2008.

51. Beeman Perry, "Proposal for Waterloo Coal Plant Is Scrapped," *Des Moines Register,* January 7, 2009, A1; Tim Jamison, "LS Power Pulls Plug on Waterloo Plant," *Sioux City Journal,* January 6, 2009 (LS Power quote); Amena Saiyid, "Tight Credit, Regulatory Hurdles Prompt Dynegy to Pull Plug on LS Power," *Platts Coal Trader,* January 5, 2009, 1; Kerry Bleskan, "Waterloo, Iowa, Council Approves LS Power's Land Use Requests," *SNL Electric Utility Report,* May 5, 2008; Drew Andersen, "Waterloo Zoning Commission Again Backs LS Power Annexation and Rezoning Request," *Waterloo Courier,* April 2, 2008.

52. Bandyk, "Sierra Club Fights Coal Plants"; Ethan Howland, "MidAmerican in Deal with Sierra Club to Stop Burning Coal at Six Units in Iowa," *Electric Utility Week,* January 28, 2013; Housley Carr, "Duke in Agreement with Environmental Groups to Retire 1,667 MW of Unscrubbed Coal Capacity," *Electric Utility Week,* January 23, 2012; Michael Hawthorne, "Foes Clear the Air over Power Plant," *Chicago Tribune,* January 6, 2007, C1; "Springfield Council Reverses Vote, Approves Controversial Plant Agreement with Sierra Club," *Electric Utility Week,* August 14, 2006, 16.

53. Joshua Learn, "Coalition Blasts DOE's Funding of Texas Clean Energy Project," *SNL Daily Coal Report,* May 27, 2016; Matthew Bandyk, "AEP Exec: Federal Policy Needed to Address 'Clean Coal' Challenges," *SNL Daily Coal Report,* October 12, 2012; Keith Norman, "Startup Delay: Economy Pushes Opening of GRE Plant Operations to 2012," *Jamestown (ND) Sun,* May 28, 2010; Barry Cassell, "West Virginia IGCC Receives General State Support, but Rate-making Remains a Problem," *SNL Power Week West,* November 26, 2007; Mark Ballard, "Blanco Lauds Cleco's Coal-Fired Plant Plan," *Baton Rouge Advocate,*

July 13, 2005, A1; "Cleco to Self-Build 600 MW, Signs Two PPAs," *Megawatt Daily*, July 13, 2005, 7.

54. Molly Christian, "Coal Industry Lashes Out against Another Bloomberg Anti-Coal Donation," *SNL Daily Coal Report*, April 9, 2015; "Activists' Suits Seek Air Controls on Texas Emitters Ahead of EPA Rules," *Clean Air Report*, May 10, 2012; Bandyk, "Sierra Club Fights Coal Plants"; Marty Toohey, "Power Plant's Fate at Center of Debate," *Austin American-Statesman*, January 2, 2012, A1; Housley Carr, "Coal Opponents Turn Focus to Second Phase of Campaign: Older, Smaller Plants," *Electric Utility Week*, September 7, 2009, 19; Dean Scott, "Older Coal-Fired Units Said to Produce Largest Share of Power Plant Pollution," *BNA Environment Reporter*, May 25, 2012, 1334.

55. "Uncertainty over NSR Spurs Activists' Novel Bid to Cut Utilities' Emissions," *Clean Air Report*, March 1, 2012.

56. Jeff Stanfield, "Deseret Power Signs Agreement with Environmental Groups on Bonanza Plant Emissions," *SNL Generation Markets Week*, October 13, 2015; "Court Affirms EPA's Denial of Petition to Oppose Permit for Coal-Fired Plant," *BNA Environment Reporter*, July 26, 2013, 2193; "Activists See New EPA Hurdles to Pursuing Citizen Enforcement of Air Act," *Inside EPA Weekly Report*, August 12, 2011; "EPA, Activists Turn Focus to 'Opacity' Violations in Utility Air Permits," *Clean Air Report*, July 8, 2010; "Activists' Challenges Aim to Set Precedents on Key Air Permit Provisions," *Inside EPA Weekly Report*, June 25, 2010; "EPA Poised to Address Heat Input Limit Enforceability at Coal Plants," *Inside EPA Weekly Report*, November 27, 2009.

57. See, e.g., Matthew Bandyk, "Alliant's Iowa Utility Reaches Settlement with EPA on Coal Plants," *SNL Electric Utility Report*, July 20, 2015; Christine Cordner, "Xcel Energy Settles Emissions Lawsuit over Cherokee Coal Plant in Colorado," *SNL Daily Coal Report*, March 11, 2013; "EPA, States Sue Penn. Power Plant," *Inside Energy with Federal Lands*, January 10, 2011, 14; Michael Bologna, "Groups Threaten to Sue State Line Energy, Dominion over Air Act Violations in Indiana," *BNA Environment Reporter*, September 17, 2010, 2056; Matthew Bandyk, "New York, Pennsylvania to Sue Homer City Coal Plant over Emissions," *SNL Generation Markets Week*, July 27, 2010; Michael Bologna, "Government Brings Clean Air Act Charges against Midwest Generation in Illinois Case," *BNA Environment Reporter*, September 11, 2009, 2094.

58. "Judge Sides with TVA in NSR Case," *E&E News*, April 5, 2010; Joel Davisof, "Environmentalists, TVA Spar in Court," *Maryville (TN) Daily Times*, June 2, 2009.

59. "Advocates Look Beyond GHG Rules to Shutter High-Emitting Coal Plants," *Inside EPA*, November 26, 2014.

60. U.S. Environmental Protection Agency, "State Implementation Plans: Response to Petition for Rulemaking; Restatement and Update of EPA's SSM Policy

Applicable to SIPs; Findings of Substantial Inadequacy; and SIP Calls to Amend Provisions Applying to Excess Emissions during Periods of Startup, Shutdown and Malfunction," 80 Fed. Reg. 33840 (June 12, 2015); Annalee Grant, "Luminant, Sierra Club to Settle Case Ordering Legal Fee Reimbursement," *SNL Electric Utility Report,* December 8, 2014; Annalee Grant, "Texas Judge Orders Sierra Club to Pay Luminant $6.4M for 'Frivolous' Suit," *SNL FERC Power Report,* September 10, 2014; J. B. Smith, "Power Giant's Bankruptcy Raises Questions of Coal Plants' Future," *Waco Tribune-Herald,* April 30, 2014, 1; Jonathan Crawford, "Luminant Declares Victory in Court Battle over Texas Coal Plant's Emissions," *SNL Generation Markets Week,* March 4, 2014; Kate Galbraith, "Sierra Club Takes Aim at Coal Plants in East Texas," *Texas Tribune,* February 10, 2013; Earthjustice, "TXU-Luminant Violated Clean Air Act over 38,000 Times," press release, October 27, 2011; "EPA, Activists Turn Focus."

61. "Five Owners of Colorado Power Plant to Pay $100 Million to Resolve Citizen Suit," *BNA Environment Reporter,* January 19, 2001, 121; "Northwest Group Sues to Cut Emissions at Centralia Unit to Protect Mt. Rainier," *Utility Environment Report,* August 30, 1996, 5.

62. Marc Lifsher, "Mohave Facility Won't Be Reopened," *Los Angeles Times,* June 20, 2006, C1; Rebecca Smith, "Utilities Settle Grand Canyon Pollution Case," *Wall Street Journal,* October 6, 1999, C13; "Nevada Power Plant Not a Major Cause of Grand Canyon Haze Problem, Study Finds," *BNA Environment Reporter,* March 26, 1999, 2312; "Groups Sue Nevada Power Company, Claim CAA Violations, Visibility Impairment," *BNA Environment Reporter,* February 27, 1998, 2292; Frank Clifford, "Mohave Power Plant's Future a Thorny Dilemma," *Los Angeles Times,* February 15, 1998, A1; Marla Cone, "Historic Plan to Curb Grand Canyon Smog Approved," *Los Angeles Times,* June 11, 1996, A3.

63. Jeff Stanfield, "NM Regulators Approve Stipulation over Future of PNM's San Juan Plant through 2022," *SNL Generation Markets Week,* December 22, 2015; Bob Matyi, "Enviros Using IRP, Rate Cases to Oppose Retrofits," *Platts Megawatt Daily,* March 11, 2015, 1; Ethan Howland, "Regulators Question Puget on Colstrip Plant," *Platts Megawatt Daily,* February 10, 2014; Matthew Bandyk, "Kentucky Power Reaches Agreement to Convert Big Sandy Coal Unit to Gas," *SNL Generation Markets Week,* June 11, 2013; Cathy Cash, "US Could Close an Additional 59,000 MW of Coal-Fired Generation, Group Says," *Inside Energy with Federal Lands,* November 19, 2012.

64. See, e.g., Bob Matyi, "Indiana Regulators Approve Vectren Retrofits," *Platts Megawatt Daily,* June 24, 2016; Jeffrey McDonald, "We Energies Granted Wisconsin Approval to Modify Elm Road for More PRB Coal," *Platts Coal Trader,* May 14, 2015, 2; Bob Matyi, "Missouri Regulators Reject Challenge Seeking to Close Labadie Coal Plant," *Platts Coal Trader,* April 30, 2015.

65. Matthew Bandyk, "Okla. Regulators Reject OG&E Environmental Plan," *SNL Generation Markets Week,* December 8, 2015.

66. Bob Matyi, "LG&E, KU Get OK on $2.3 Bil in Control Projects," *Platts Coal Trader,* December 16, 2011, 5; Bob Matyi, "LG&E/KU Could Recover from Ratepayers $2.25 Billion to Clean Up Coal-Fired Plants," *Electric Utility Week,* November 14, 2011 (quote).

67. Jeremy P. Jacobs, "Utility Accelerates Schedule for Closing 2 Chicago Power Plants," *Greenwire,* May 3, 2012; Michael Hawthorne, "2 Coal Plants to Shut Early," *Chicago Tribune,* March 1, 2012, C1; Dan Lowrey, "Sierra Club Targets Chicago Coal Plants," *SNL Generation Markets Week,* September 20, 2011.

68. "Plant Briefs," *Electric Utility Week,* June 1, 2009; Bob Matyi, "Sierra Club Says Michigan Should Reject All Coal Plants, Files Suit against Holland Board," *Electric Utility Week,* December 22, 2008, 20.

69. Marty Toohey, "Council: Break from Coal Plant Too Costly," *Austin American-Statesman,* February 5, 2014, A1.

70. See, e.g., Sierra Club v. Virginia Elec. & Power Co., 145 F. Supp. 3d 601, 607 (E.D. Va. 2015) (monitoring impoundments); "Suit Poses Early Test for Enforcing Mandates of EPA's Ash Disposal Rule," *Inside EPA,* June 23, 2017, 1 (enforcing EPA regulations); "In Bid for EPA Rules, Environmentalists Target State Coal Ash Programs," *Clean Energy Report,* October 29, 2012 (administrative complaints); John Myers, "Minnesota Power Ash Plans Panned," *Duluth News-Tribune,* May 31, 2012 (objections); Barry Cassell, "NM Plant Operator Tries to Negotiate Coal Combustion Waste Deal," *SNL Coal Report,* November 15, 2010 (section 7003 lawsuits).

71. Andrew M. Ballard, "TVA Ordered to Dig Up, Move Coal Ash at Gallatin Plant," *BNA Daily Environment Report,* August 7, 2017; Sierra Club, "Hidden Camera Operation Exposes Kentucky Utility LH&E for Dumping Dangerous Coal Ash Pollution into Ohio River," press release, Targeted News Service, March 17, 2014; Timothy B. Wheeler, "Coal-Ash Pollution at Three Maryland Landfills to Be Cleaned Up," *Baltimore Sun,* January 13, 2013; Wijdan Khaliq, "PNM, BHP Settle San Juan Coal Ash Pollution Case for $10M," *SNL Daily Coal Report,* April 3, 2012.

72. Kevin Robinson-Avila, "PNM Considers Coal-Mine Purchase for San Juan Fuel," *Albuquerque Journal,* November 18, 2014, B1; Kathy Helms, "Ratepayers Rally for Renewables at PNM Meeting," *Gallup Independent,* May 16, 2014, 2; Michael Hartranft, "Environmentalists Slam Haze Plan," *Albuquerque Journal,* April 16, 2011, B1; Electa Draper, "Power-Plant Paradox in N.M.," *Denver Post,* August 5, 2004, B1.

73. Sierra Club, "Groups Appeal Water Pollution Permit for Waukegan Coal Plant on Lake Michigan," press release, Targeted News Service, April 30, 2015; Sierra Club, "Lake County Community Members Call on Illinois Environmental Protection Agency to Protect Lake Michigan," press release, July 31, 2013; Jason

Fordney, "EPA Needs More Time to Study 'Complex' Coal Ash Issues, It Tells Court in Lawsuit Case," *Electric Utility Week,* October 29, 2012; Michael Hawthorne, "Complaint Alleges Chemicals Leaking from Coal Ash Ponds," *Chicago Tribune,* October 5, 2012, C4; Hawthorne, "2 Coal Plants to Shut Early."

74. NRG, About the (Closed) Potomac River Generating Station, http://www.prgsonline.org; Patricia Sullivan, "Accidental Activists Close to Seeing Coal Plant Shut," *Washington Post,* September 4, 2011, C1; "Virginia Board to Weigh Fate of Mirant Coal Plant," *Megawatt Daily,* May 22, 2007, 7; David A. Fahrenthold, "Power Plant Still Battling to Stay Open," *Washington Post,* September 13, 2007, B2; "EPA Eases Restrictions on Emissions for Mirant's Potomac River Power Plant," *BNA Environment Reporter,* June 9, 2006, 1202; "Bodman Orders Mirant to Resume Some Potomac River Production," *U.S. Coal Review,* December 26, 2005; "Potomac River Runs Dry for Now as Mirant Halts Power Production," *U.S. Coal Review,* August 29, 2005; "Shutting Virginia Plant Could Strain PJM Grid," *Megawatt Daily,* August 24, 2005, 1.

75. Maya Weber, "Sierra Club Ups Investment in Blocking Natural Gas," *Platts Energy Trader,* October 3, 2016; Andrew Engblom, "Sierra Club Plans to Oppose 'A Whole Lot More' Gas-Fired Power Plants," *SNL Electric Utility Report,* April 1, 2013; Jonathan Crawford, "Sierra Club Ramps Up Climate Fight against Gas-Fired Power Plants," *SNL FERC Power Report,* September 11, 2013; Darren Epps, "Beyond Coal Senior Campaign Director Bruce Nilles on the Push to Eliminate Coal," *SNL Daily Coal Report,* January 11, 2013.

76. Ethan Howland, "Calif. PUC OKs SDG&E Power Deal," *Platts Megawatt Daily,* February 6, 2014; "EAB Sets Precedent Backing EPA Discretion to Determine BACT for GHGs," *Clean Energy Report,* August 26, 2013; Matthew Bandyk, "Developers Pick New Site in San Diego County for 300-MW Gas Plant," *SNL Electric Utility Report,* February 21, 2011.

77. In re La Paloma Energy Center, Order Denying Review, PSD Appeal No. 13-10, Environmental Protection Agency Environmental Appeals Board, March 14, 2014; Morgan Lee, "Coastal Power Plant at Carlsbad Approved PUC Oks New Carlsbad Plant," *San Diego Union-Tribune,* May 22, 2015, A1; Matthew Bandyk, "Mass. Gas Plant Settlement Approved, but Called a 'Dangerous Precedent' for Generators," *SNL Electric Utility Report,* March 3, 2014.

78. Sean D. Hamill, "Zion Peaker Set to Power Up," *Chicago Tribune,* June 17, 2002, A3; "Salt River Accepts Tough Limits on New Plant," *Megawatt Daily,* April 30, 2001.

79. Vic Kolenc, "Agreement Ends Montana Vista Residents' Fight over El Paso Electric's Power Plant," *El Paso Times,* December 11, 2013; Julian Aguilar, "Community Determined to Fight Power Plant," *New York Times,* April 5, 2013, A21; Aileen B. Flores, "Planned El Paso Electric Plant Concerns Far East Side Residents," *El Paso Times,* December 3, 2012.

80. Mark Hand, "Sierra Club's Campaigning Extends Beyond Coal to 'Dirty Fuels,'" *SNL Daily Gas Report,* April 27, 2016; "Sierra Club Seeks to Move 'Beyond Natural Gas,' Drawing New Criticism," *Clean Energy Report,* May 14, 2012; Jonathan Crawford, "NRDC's Dave Hawkins: New Coal Plants Not Dead," *SNL Daily Coal Report,* April 30, 2012.

81. David Spence, "The New Politics of (Energy) Market Entry," *Notre Dame Law Review* (forthcoming), table 3; Lorne Stockman, *A Bridge Too Far: How Appalachian Basin Gas Pipeline Expansion Will Undermine U.S. Climate Goals* (Washington, DC: Oil Change International, July 2016), 6, 20, 30; Weber, "Sierra Club Ups Investment."

82. Stockman, *A Bridge Too Far,* 15; Pamela King and Jenny Mandel, "Pipeline Tug of War: Where Are We Now?," *E&E News,* November 26, 2018; Sibyl Layat and Ximena Mosqueda-Fernandez, "Va. Agency Votes to Keep Permits for Mountain Valley, Atlantic Coast Pipelines," *SNL Daily Gas Report,* August 29, 2018; Miguel Angel Cordon, "FERC OK's Construction on Atlantic Coast Pipeline over Environmentalist Challenge," *SNL Daily Gas Report,* July 26, 2018; Ximena Mosqueda-Fernandez, "Court Stays Permit for Mountain Valley Pipeline River Crossings," *SNL Gas Week,* July 2, 2018; Ximena Mosqueda-Fernandez, "FERC Denies Attempts to Recall Approval of 2-Bcf/d Mountain Valley Pipeline," *SNL Daily Gas Report,* June 18, 2018; Sierra Club, "Virginia Supreme Court Hears Arguments over Sierra Club's Affiliates Act Challenge to Atlantic Coast Deal," press release, Target News Service, June 6, 2018; Patrick Wilson, "Anti-Terrorism Agencies Aided Police," *Roanoke Times,* June 2, 2018; Sierra Club, "Atlantic Coast Pipeline Opponents Rally outside Bank of America Shareholder Meeting in Charlotte," press release, States News Service, April 25, 2018; "Dominion-Led Atlantic Coast Pipeline Wins Key Forest Service Permit," *SNL Daily Gas Report,* November 22, 2017; Ken Ward Jr., "DEP, Its Cabinet Secretary Challenged on Gas Pipeline Approval," *Charleston Gazette-Mail,* June 10, 2017.

83. Sierra Club v. Department of Interior, 899 F.3d 260 (4th Cir. 2018) (Fish and Wildlife Service's incidental take permit and National Park Services right-of-way overturned); Sierra Club v. State Water Control Bd., 898 F.3d 383 (4th Cir. 2018) (Virginia certification upheld). See also City of Boston Delegation v. FERC, 897 F.3d 241 (D.C. Cir. 2018) (FERC grant of certificate for pipeline upgrade upheld); Atchafalaya Basinkeeper v. Corps of Engineers, 894 F.3d 692 (5th Cir.2018) (Corps of Engineers permit to construct pipeline through Louisiana basin upheld).

84. Pamela King, "Atlantic Coast Permit Pileup: Where Things Stand," *E&E News,* February 7, 2019; Ken Ward Jr., "Another Court Ruling against a West Virginia Pipeline, Then Another Effort to Change the Rules," *Charleston Gazette-Mail,* October 4, 2018; Ken Ward Jr., "4th Circuit Voids Permit for MVP Work,"

Charleston Gazette-Mail, October 3, 2018; Michael Martz, "4th Circuit Panel Vacates Two Permits for Pipeline," *Richmond Times Dispatch,* August 7, 2018.

85. Waqas Azeem, "IURC Clears Vectren to Upgrade Emissions Controls at Coal Units," *SNL Energy Finance Daily,* January 30, 2015; "PSC Approves Georgia Power Request to Remove 2 Branch Coal Units from Service," *U.S. Coal Review,* March 26, 2012.

86. Wayne Barber, "Sierra Club, Other Environmental Groups Deemed Tough Foes by Power Industry," *SNL Power Week Canada,* November 28, 2011.

11. THE TRANSFORMATION OF THE ELECTRIC POWER INDUSTRY

1. Joseph P. Tomain, *Clean Power Politics: The Democratization of Energy* (Cambridge: Cambridge University Press, 2017), 2–3; *Keeping the Lights On—Are We Doing Enough to Ensure the Reliability and Security of the U.S. Electric Grid? Hearing before the Senate Committee on Energy and Natural Resources,* 113th Cong. (April 10, 2014) (testimony of Nicholas Akins, CEO, American Electric Power); "Electric Companies Are Delivering America's Future," *Electric Perspectives,* July / August, 2016, 15 (quoting Thomas Kuhn, president, Edison Electric Institute).

2. Union of Concerned Scientists, *A Dwindling Role for Coal* (Cambridge, MA: Union of Concerned Scientists, 2017), 1; Dan Testa, "Goldman Panel on Future of Coal Sees Diminished Burn, Advantages for Newer Assets," *SNL Daily Coal Report,* August 15, 2014; "Activists See Scrapped Coal Plant Sending Signal against New Facilities," *Inside EPA Weekly Report,* December 16, 2011, 1 ("rush to coal"); Amena Saiyid, "AEP Says No to New Nuclear, Coal Generation," *Platts Coal Trader,* October 20, 2010, 3; Barry Cassell, "Report Sees Little Near-Term Impact from Recent Coal Project Cancellations," *SNL Electric Utility Report,* February 18, 2008.

3. U.S. Energy Information Administration, "Coal Production and Prices Decline in 2015," *Today in Energy,* January 8, 2016, https://www.eia.gov/todayinenergy/detail.php?id=24472; Dylan Brown, "Majority of U.S. Coal Mines Shuttered in Last Decade," *E&E News,* January 30, 2019; Dylan Brown, "Production to Rise This Year—EIA," *E&E News,* March 29, 2017; Rob Nikolewski, "State's Coal Use Drops Sharply," *Los Angeles Times,* May 12, 2016, C1 (29 percent); "The Carnage in Coal Country," editorial, *Wall Street Journal,* January 12, 2016, A12.

4. Jeffrey McDonald, "Weak Fundamentals Keep PRB Contract Coal Markets Sunk," *Platts Coal Trader,* January 8, 2016, 2; Brad Plumer, "As Coal Industry Declines, What will Happen to All Those Retired Miners," *Washington Post,* March 7, 2013.

5. Walter J. Culver and Mingguo Hong, "Coal's Decline: Driven by Policy or Technology?," *Electricity Journal,* September 2016, 50.

6. Union of Concerned Scientists, *Dwindling Role,* 2.

7. Nushin Huq, "Texas Grid Planning for Renewable Future as Coal Power Cools," *BNA Environment & Energy Report,* November 6, 2017.

8. U.S. Energy Information Administration, "Wind Adds the Most Electric Generation Capacity in 2015, Followed by Natural Gas and Solar," *Today in Energy,* March 23, 2016, http://www.eia.gov/todayinenergy/detail.php?id=25492; "Thinking Big," *Electric Perspectives,* July / August, 2016, 25 (rooftop solar statistics); Christa Marshall, "DOE Study Shows Rapid Growth, but Trouble Looms," *E&E News,* August 8, 2017 (backyard turbines); Hannah Northey and Daniel Cusick, "Wind Industry Touts Jobs, Investments in Pitch to GOP," *E&E News,* April 19, 2017; Mark Chediak, "U.S. Solar Slows after Industry Reached Record in 2016," *BNA Energy and Climate Report,* March 9, 2017; Daniel Cusick, "Wind and Solar Account for 60% of New Growth, Outpacing Gas," *E&E News,* January 11, 2017; Michael Copley, "AGA CEO: As Gas, Renewables Transform Power Sector, Future 'Impossible to Predict,'" *SNL Daily Gas Report,* February 5, 2016.

9. Joe Ryan and Christopher Martin, "California Is Plenty Green but Iowa Has the Cleanest Power," *BNA Environment & Energy Report,* November 30, 2017; Justin Gillis and Nadja Popovich, "The View from Trump Country, Where Renewable Energy Is Thriving," *New York Times,* June 8, 2017, A20; Mark Chediak, "Wind Just Blew Away Calpine's Plan to Build a Texas Gas Plant," *BNA Energy & Climate Report,* April 28, 2017; Daniel Cusick, "Midwestern States Now Get a Fifth of Their Power from Wind," *E&E News,* March 7, 2017; Benjamin Storrow, "Coal, Once King in Texas, Sees Wind as 'Real Competition,'" *E&E News,* April 14, 2017; "The Wild West of Wind," *Guardian,* March 1, 2017.

10. Natural Resources Defense Council, *A Tectonic Shift in America's Energy Landscape,* Third Annual Energy Report (2015), 4, https://www.nrdc.org/sites /default/files/energy-environment-report-2015.pdf; Joseph P. Tomain, "Traditionally-Structured Electric Utilities in a Distributed Generation World," *Nova Law Review* 38 (2014): 479.

11. U.S. Energy Information Administration, "U.S. Residential Electricity Prices Decline for the First Time in Many Years," *Today in Energy,* October 6, 2016, http://www.eia.gov/todayinenergy/detail.php?id=28252; Taylor Kuykendall, "Judicial Battlefront: Environmentalists Look to the Air to Keep Coal Underground," *SNL Electric Utility Report,* September 14, 2015 (quoting Robert Murray, Murray Energy Corp.).

12. Edward Klump, "Exelon Seeks New Path for Texas Gas Plants via Chapter 11," *E&E News,* November 8, 2017; Keith Goldberg, "Vistra Buys Dynegy to Create $20B Merchant Power Giant," *Law360,* October 30, 2017; Naureen S. Malik and Brian Eckhouse, "'Gas Apocalypse' Looms amid Power Plant Construction Boom," *BNA Energy & Climate Report,* May 23, 2017; Edward

Klump, "Change Is in the Air for Independent Power Players," *E&E News,* May 12, 2017; "FirstEnergy Corp. Sells 4 Pa. Natural Gas Units," *E&E News,* January 23, 2017.

13. Richard Martin, *Coal Wars: The Future of Energy and the Fate of the Planet* (New York: Palgrave Macmillan, 2015), 29, 41; Brad Plumer and Jim Tankersley, "Feeling Pull of Nostalgia, Trump Shapes Trade Policy," *New York Times,* June 17, 2018, A8; Tiffany Hsu and Clifford Krauss, "G.E. Says It Will Slash Jobs over Shift in Energy Market," *New York Times,* December 8, 2017, B1; Herman Wang, "Coal-Fired Power Plants Don't Deliver Promised Number of Jobs, Study Finds," *Inside Energy with Federal Lands,* April 11, 2011.

14. Martin, *Coal Wars,* 39–41; *Regional Impacts of EPA Carbon Regulations: The Case of West Virginia: Field Hearing before the Senate Committee on Environment and Public Works,* 114th Cong., 41 (March 23, 2015) (testimony of Jeremy Richardson, Union of Concerned Scientists); Taylor Kuykendall, "Narrow Band of 16 Central Appalachia Counties 'Ground Zero' in Coal Job Free Fall," *SNL Daily Coal Report,* June 18, 2015.

15. Benjamin Storrow, "Coal-Reliant Tribes Ponder a Future without Their Power Plant," *E&E News,* April 3, 2017; Dan Frosch, "A Part of Utah Built on Coal Wonders What Comes Next," *New York Times,* November 28, 2013, A19.

16. U.S. Department of Energy, *U.S. Energy and Employment Report* (Washington, DC: U.S. Department of Energy, January 2017), 8–9; Northey and Cusick, "Wind Industry Touts Jobs,"; Nadja Popovich, "Today's Energy Jobs Are in Solar, Not Coal," *New York Times,* April 25, 2017, B1; Christopher Martin, "U.S. Wind, Solar Power Focus on Rural Jobs as Trump Touts Coal," *BNA Environment Reporter,* February 3, 2017, 228.

17. U.S. Energy Information Administration, *Electric Power Annual 2016* (Washington, DC: U.S. Department of Energy, December 2017), Table 9.1; Ceres, "New Report: U.S. Power Sector Emissions Decline as the Generation Mix Continues to Evolve," press release, June 14, 2017, https://www.ceres.org/news -center/press-releases/new-report-us-power-sector-emissions-decline-generation -mix-continues; Union of Concerned Scientists, *Dwindling Role,* 2; Glen Boshart, "EEI Briefs Wall Street on Power Industry's 2016 Challenges," *SNL Electric Utility Report,* February 15, 2016.

18. U.S. Energy Information Administration, *Electric Power Annual 2016,* Table 10.8 (statistics); U.S. Department of Energy, *2014 Smart Grid System Report* (Washington, DC: U.S. Department of Energy, August 2014); Steven Ferrey, "Ring-Fencing the Power Envelope of History's Second Most Important Invention of All Time," *William & Mary Environmental Law and Policy Review* 40 (2015): 14; Trevor Houser, Jason Bordoff, and Peter Marsters, *Can Coal Make a Comeback?* (New York: Columbia / SIPA Center on Global Energy Policy, April 2017), 15; Tomain, "Traditionally-Structured Electric Utilities," 488.

19. American Council for an Energy-Efficient Economy, "State Energy Efficiency Resource Standards (EERS)," May 2016; Edison Electric Institute, "Electric Company Programs," http://www.eei.org/issuesandpolicy/efficiency/electriccompanyprograms/Pages/default.aspx; Steve Hanley, "Americans Are Using Less Electricity Today Than a Decade Ago," *CleanTechnica,* August 9, 2017.

20. U.S. Energy Information Administration, "Coal Made Up More Than 80% of Retired Electricity Generating Capacity in 2015," *Today in Energy,* March 8, 2016, http://www.eia.gov/todayinenergy/detail.php?id=25272; Lesley Fleischman et al., "Ripe for Retirement: An Economic Analysis of the U.S. Coal Fleet," *Electricity Journal,* December 2013; Bob Matyi, "FirstEnergy to Ratchet Back 2,233-MW Sammis Coal Plant in Ohio to Peaking Status in September," *Electric Utility Week,* August 27, 2012; Barry Cassell, "Interstate Power & Light Outlines Coal Unit Retirement Plans," *SNL Coal Report,* November 8, 2010.

21. Tomain, *Clean Power Politics,* 60; John M. Golden and Hannah J. Wiseman, "The Fracking Revolution: Shale Gas as a Case Study in Innovation Policy," *Emory Law Journal* 64 (2015): 964–68; U.S. Government Accountability Office, *Oil and Gas: Information on Shale Resources, Development, and Environmental and Public Health Risks* (Washington, DC, September 2012), 25–30; Chris Mooney, "It's the Same Story under Trump as under Obama: Coal Is Losing Out to Natural Gas," *Washington Post,* January 9, 2018; Robert Walton, "EIA: Gas Plant Construction Costs Fell 30% in 2015," *Utility Dive,* July 6, 2017; Bill Holland, "Study: War on Coal Is Being Driven by Shale Gas, Not EPA," *SNL Daily Coal Report,* October 12, 2016.

22. *American Energy Security and Innovation: Grid Reliability Challenges in a Shifting Energy Resource Landscape: Hearings before the Subcommittee on Energy and Power of the House Committee on Energy and Commerce,* 113th Cong. (May 9, 2013) (testimony of Daniel Weiss, Center for American Progress); Gerry Anderson, "The Transformation of Power Generation," *Electric Perspectives,* November / December, 2015, 25, 30; Culver and Hong, "Coal's Decline"; Edward Klump, "As D.C. Dawdles, CEOs Shift Power Companies to Green," *E&E News,* April 3, 2019; Naureen S. Malik, "Renewables Are Starting to Crush Aging U.S. Nukes, Coal Plants," *BNA Environment & Energy Report,* November 2, 2017.

23. Elisabeth Graffy and Steven Kihm, "Does Disruptive Competition Mean a Death Spiral for Electric Utilities?," *Energy Law Journal* 35 (2014): 7; Emily Holden, "Utilities See Demise of Climate Rule, Still Cut CO_2," *E&E News,* February 15, 2017; Diane Cardwell, "Creating Their Own Green Sources," *New York Times,* August 24, 2016, B1; "It's Easy Going Green in Minnesota," *Electric Perspectives,* January / February, 2016, 21.

24. Graffy and Kihm, "Death Spiral," 6; Alexandra B. Klass and Elizabeth J. Wilson, "Interstate Transmission Challenges for Renewable Energy: A Federalism Mismatch," *Vanderbilt Law Review* 65 (2012): 1844; Garrick B. Pursley and

Hannah J. Wiseman, "Local Energy," *Emory Law Journal* 60 (2011): 911; Holland, "War on Coal"; "Texas Town to Achieve All-Renewable Status," *Natural Gas Week,* April 6, 2015; Leticia Vasquez et al., "Dominant Role of Gas in Texas Generation Mix to Continue: Analysis," *Megawatt Daily,* June 6, 2013.

25. Travis Roach, "The Effect of the Production Tax Credit on Wind Energy Production in Deregulated Electricity Markets," *Economics Letters* 127 (2015): 86; Weiss testimony; Pursley and Wiseman, "Local Energy," 910; Christa Marshall, "Loan Guarantees Remain a Topic of Ferocious Debate," *E&E News,* February 16, 2017; Kathleen Hart, "White House Report: Gas Production at All-Time High; Renewables Set to Double," *SNL FERC Power Report,* March 14, 2012.

26. Tomain, *Clean Power Politics,* 2; "Utility Investments 'Soar,'" *Foster Natural Gas/Oil Report,* July 1, 2016, 28.

27. Herman K. Trabish, "Is Renewable Energy Threatening Power Reliability?," *Utility Dive,* June 1, 2017; Darren Epps, "BMO: Low Plant Utilization Rates to Temper Impacts of Planned Coal Retirements," *SNL Daily Coal Report,* April 12, 2013.

28. Jonathan Crawford, "Ex-DOE Official: Success of EPA Acid Rain Rules Unlikely to Be Replicated in Carbon Standard," *SNL Generation Markets Week,* October 29, 2013.

29. U.S. Energy Information Administration, *Electric Power Annual 2016,* Table 9.2.

30. Jeff McMahon, "Nearly All U.S. Coal Plants Now Comply with the EPA Mercury Rule That Was Shot Down by Supreme Court," *Forbes,* July 10, 2016; Jason Lehmann, "With Mercury Rule in Effect, Southern Reveals Wave of Coal Retirements, Future Plans," *SNL Generation Markets Week,* May 12, 2015; Christine Cordner, "APS Aims to Retire Cholla Coal-Fired Unit 2 in 2016 under Regional Haze Proposal," *SNL Daily Coal Report,* September 15, 2014; "Turning Away from Coal," *Wall Street Journal,* September 13, 2010, R3.

31. "Natural Gas Fills the Gap as Coal Drops Out of U.S. Power Market," *Daily Oil Bulletin,* June 30, 2016.

32. U.S. Government Accountability Office, *Air Pollution: EPA Needs Better Information on New Source Review Permits* (Washington, DC, June 2012), 21.

33. Housley Carr, "Progress Adds 1,088 MW to List of Aging Coal Unit Closures; Emissions Regulation a Factor," *Electric Utility Week,* December 7, 2009; "Pa., N.J. Cut Deal on 600 MW of New PPL Gas-Fired Power," *Electricity Daily,* October 20, 2003.

34. Ted Nace, *Climate Hope: On the Front Lines of the Fight against Coal* (San Francisco: CoalSwarm, 2010), chap. 15; "US Energy Industry Seen Taking Greener Tack at Incredible Speed," *Natural Gas Week,* March 7, 2016; Kuykendall, "Judicial Battlefront."

35. Annalee Grant, Everett Wheeler, and Taylor Kuykendall, "Coal Withdrawal: Can the Nation Move Away from a Top Power Fuel Symptom-Free?," *SNL Daily Coal Report,* December 21, 2015.

36. Houser, Bordoff, and Marsters, *Can Coal Make a Comeback?,* 7, 22; Union of Concerned Scientists, *Dwindling Role,* 1; Michael Grunwald, "Trump's Love Affair with Coal," *Politico,* October 15, 2017; John Schwartz, "Some See Long-Term Decline for Battered Coal Industry," *New York Times,* December 3, 2015, A12; Amy Harder, "Whose War on Coal?," *National Journal,* July 12, 2012, 23; Jason Fordney, "EIA Agrees with Many That Coal Retirements Due to More Factors Than EPA's Regulations," *Electric Utility Week,* July 2, 2012.

12. THE TRUMP EFFECT

1. Marianne Lavelle, "Fossil Fuel Industries Pumped Millions into Trump's Inauguration, Filing Shows," *Inside Climate News,* April 20, 2017; John Schwartz, "Combative, Conflicting and Confusing," *New York Times,* March 11, 2017, A14; Ken Ward Jr., "Trump Win Won't Bring Coal Rebound, Could Block Climate Progress," *Charleston Gazette-Mail,* November 12, 2016 ("spark of hope" quote); Jessica E. Trancik, "People Are Worried Trump Will Stop Climate Progress," *Washington Post,* November 21, 2016; Coral Davenport, "Parties' Divide over Climate Change Bursts into Forefront of Campaign," *New York Times,* August 2, 2016, A10.

2. Brady Dennis and Juliet Eilperin, "Scott Pruitt Steps Down as EPA Head after Ethics, Management Scandals," *Washington Post,* July 6, 2018 (phone booth); Kevin Bogardus, "Ethics Official Urged New Investigations into Pruitt," *E&E News,* July 2, 2018 (demands on staff); Robin Bravender, "Pruitt Wants 24 / 7 Bodyguards," *E&E News,* February 20, 2017; Mike Soraghan, "Environmental Enforcement Got Short Shrift in Okla. under Pruitt," *E&E News,* January 18, 2017; Eric Lipton, "Energy and Regulators on One Team," *New York Times,* December 7, 2016, A1 (regional haze case).

3. Sean Reilly, "Records: Air Chief Meets Often with Regulated Industries," *E&E News,* June 13, 2018; Jennifer Lu, "Smaller Bites in EPA Air Chief's Second Pass at Permitting Update," *BNA Daily Environment Report,* April 17, 2018; Kevin Bogardus, "Right Thrilled as Trump Taps Former Coal Lobbyist for Top Post," *E&E News,* October 6, 2017.

4. Daniel Cusick, "Perry Oversaw a Texas Wind Miracle—But Did He Lead It?," *E&E News,* December 14, 2016; Coral Davenport, "Perry Is Chosen as Energy Chief," *New York Times,* December 14, 2016, A1 (quote); Mike Lee and Edward Klump, "Perry's Rebirth from "Oops" to Energy Shows His Political Skills," *E&E News,* December 14, 2016.

5. Executive Order 13783 §§ 2, 4(a), (b); Christa Marshall, "Overhauled Website Boosts Fossil Fuels over Climate," *E&E News,* June 5, 2017; Chris Mooney and Juliet Eilperin, "EPA Website Removes Climate Science Site from Public View after Two Decades," *Washington Post,* April 29, 2017; Hannah Northey, "Trump Rails against Energy Regs, Vows to Revive Coal," *E&E News,* February 24, 2017 (coal quote); Coral Davenport, "Climate Change References Are Purged from the White House Website," *New York Times,* January 21, 2017, A20.

6. Amanda Reilly, "Some Judges Getting Impatient with Litigation Hold," *E&E News,* June 26, 2018; "Judges Suggest High Court Path for CPP Supporters to Win Rule's Release," *Inside EPA,* August 11, 2017, 1; "EPA Cites Trump Order in Bid to Halt D.C. Circuit Utility Rule Suits," *Inside EPA,* March 31, 2017, 1.

7. Coral Davenport, "Scott Pruitt Faces Anger from Right over E.P.A. Finding He Won't Fight," *New York Times,* April 12, 2017; Competitive Enterprise Institute, "Petition of the Competitive Enterprise Institute and the Science and Environmental Policy Project for Rulemaking on the Subject of Greenhouse Gases and Their Impact on Public Health and Welfare, in Connection with EPA's 2009 Endangerment Finding," February 17, 2017; Annalee Grant, "Pruitt: Endangerment Finding the Law of the Land and Must Be Respected," *SNL Power Policy Week,* January 25, 2017.

8. Robin Bravender, "Contenders for Pruitt's 'Red Team' Say It Would Be 'A Hoot,'" *E&E News,* July 25, 2017; Scott Waldman, "'Red Teams' Gain Prominence to Question Climate Science," *E&E News,* June 29, 2017; Lisa Friedman and Coral Davenport, "Pruitt's Plan for Climate Change Debates: Ask Conservative Think Tanks," *New York Times,* May 8, 2018; Steven Koonin, "A 'Red Team' Exercise Would Strengthen Climate Science," *Wall Street Journal,* April 20, 2017.

9. Lisa Friedman and Julie Hirschfeld Davis, "The E.P.A. Chief Wanted a Climate Science Debate; Trump's Chief of Staff Stopped Him," *New York Times,* March 9, 2018; Scott Waldman, "Picking 'Red-Team' Roster Presents Minefield for Pruitt," *E&E News,* October 26, 2017; Scott Waldman, "EPA Asked Heartland for Experts Who Question Climate Science," *E&E News,* September 21, 2017; Jason Samenow, "EPA's Scott Pruitt Wants to Set Up Opposing Teams to Debate Climate Change Science," *Washington Post,* June 7, 2017.

10. U.S. Environmental Protection Agency, "Repeal of Carbon Pollution Emission Guidelines for Existing Stationary Sources: Electric Utility Generating Units; Proposed Rule," 82 Fed. Reg. 48035 (October 16, 2017); Juliet Eilperin, "EPA's Pruitt Signs Proposed Rule to Unravel Clean Power Plan," *Washington Post,* October 10, 2017 (Pruitt regulatory state quote); Juliet Eilperin and Brady Dennis, "EPA Chief Scott Pruitt Tells Coal Miners He Will Repeal Power Plant Rule Tuesday," *Washington Post,* October 9, 2017 (Pruitt war quote).

11. Arianna Skibell, "233 Mayors Protest Pruitt's Rule Rollback," *E&E News,* February 20, 2018; "Little Policy Advocacy at EPA Hearings on Clean Power Plan

Repeal," *Inside EPA*, December 1, 2017 (quote); Brady Dennis, "In the Heart of Coal Country, EPA Gets an Earful about Clean Power Plan's Fate," *Washington Post*, November 28, 2017; Arianna Skibell, "Religious Leaders Condemn Trump Repeal," *E&E News*, October 11, 2017; Jennifer A. Dlouhy, "Trump Seen Replacing Obama Power Plant Overhaul with a Tuneup," *BNA Energy & Climate Report*, October 6, 2017.

12. U.S. Environmental Protection Agency, "Emissions Guidelines for Greenhouse Gas Emissions from Existing Electric Utility Generating Units; Proposed Rule," 83 Fed. Reg. 44746 (August 31, 2018).

13. U.S. Environmental Protection Agency, "National Emission Standards for Hazardous Air Pollutants: Coal-and Oil-Fired Electric Utility Steam Generating Units—Reconsideration of Supplemental Finding and Residual Risk and Technology Review," 84 Fed. Reg. 2670, 2678 (February 7, 2019); Amena H. Saiyid, "Power Plants Won't Remove Mercury Controls, Wheeler Says," *Bloomberg Law*, January 16, 2019; Lisa Friedman, "E.P.A. Puts Costs ahead of Health Gains," *New York Times*, December 29, 2018, A1; Amena H. Saiyid, "Let Mercury Air Pollution Limits Be, Power Sector Tells White House," *BNA Environment & Energy Report*, November 15, 2018; Sean Reilly, "Enviros Urge White House to Abandon Mercury Rule Overhaul," *E&E News*, November 6, 2018; Amanda Reilly, "EPA 'Still Thinking About' Obama Mercury Standards—Wehrum," *E&E News*, April 19, 2018; Jeff McMahon, "Nearly All U.S. Coal Plants Now Comply with the EPA Mercury Rule That Was Shot Down by Supreme Court," *Forbes*, July 10, 2016.

14. Sean Reilly, "Texas Emissions Linked to Hundreds of Early Deaths—Study," *E&E News*, October 30, 2018; Nushin Huq, "Texas Power Plants Praise Once-Loathed Pollution Trading Plan," *BNA Environment & Energy Report*, November 6, 2017; Sean Reilly, "Greens Urge Judge to Reject Trump EPA's Texas Haze Plan," *E&E News*, October 16, 2017; Sean Reilly, "EPA Floats Texas Coal Plant Emissions Trading Program," *E&E News*, October 3, 2017 ("do nothing" quote); Nushin Huq, "EPA Opts for Pollution Trading over Emissions Controls in Texas," *BNA Energy & Climate Report*, October 3, 2017; Juan Carlos Rodriguez, "5th Circ. Sides with EPA in Battle over Texas Haze Rule," *Law360*, March 22, 2017; Sean Reilly, "Pruitt Seen as Boon for Utilities Hit by EPA Haze Rule," *E&E News*, February 14, 2017.

15. See, e.g., Sean Reilly, "EPA To Scrap 4 Obama-era Haze Plans," *E&E News*, September 10, 2018; Sean Reilly, "EPA to Strike Obama Haze Plan for Ark. Coal Plants," *E&E News*, February 9, 2018; Sean Reilly, "EPA Seeks to Freeze Utah Regional Haze Litigation," *E&E News*, July 19, 2017.

16. Stephen Lee and Amena H. Saiyid, "No Interest, Not New Air Limits, Is Reason Navajo Plant Not Selling," *BNA Environment & Energy Report*, November 8, 2018; Jan Frisch, "The End of Coal Will Haunt Navajo Power Station," *BNA Energy & Climate Report*, October 13, 2017; Benjamin Storrow, "Giant Power

Plant in the West Gets a Short Reprieve," *E&E News,* June 28, 2017; Catherine Traywick, "Time Running Out for Trump to Rescue Navajo Power Plant," *BNA Energy & Climate Report,* April 12, 2017; Benjamin Storrow, "Southwest Asks: If Coal Dies, What Comes Next," *E&E News,* April 10, 2017; David Schlissel, "Economic Picture Worsens for Navajo Generating Station," briefing note, Institute for Energy Economics and Financial Analysis, Cleveland, OH, April 2018; Benjamin Storrow, "One of West's Largest Plants in U.S. to Close, Citing Cheap Gas," *E&E News,* February 14, 2017; Ryan Randazzo, "Utilities Vote to Close Navajo Coal Plant at End of 2019," *Arizona Republic,* February 13, 2017.

17. "OAR's Wehrum Prioritizes Piecemeal NSR Reform, Narrow Utility GHG Rule," *Inside EPA,* December 15, 2017.

18. U.S. Environmental Protection Agency, "Emissions Guidelines for Greenhouse Gas Emissions from Existing Electric Utility Generating Units; Proposed Rule," 83 Fed. Reg. 44746 (August 31, 2018); Juan Carlos Rodriguez, "4 Takeaways from EPA's New Source Review Guidance," *Law360,* March 14, 2018; Eric Roston, "EPA Clears the Air for Polluters on U.S. Emissions Rules," *BNA Energy & Climate Report,* March 13, 2018; "Critics Say Pruitt NSR Relief Memo Stymies Enforcement, 'Coerces' States," *Inside EPA,* February 23, 2018, 18 (critic quotes); Sean Reilly, "Sierra Club Sues over EPA About-Face on Power Plant Permitting," *E&E News,* February 7, 2018; "In Major Policy Shift, EPA Defers to Industry on Key NSR Permitting Test," *Inside EPA,* December 15, 2017; "OAR's Wehrum Prioritizes Piecemeal NSR Reform."

19. Gavin Bade, "EPA Enforcement Shift Will Allow Coal Plants to Pollute More, Former Air Official Says," *Utility Dive,* February 27, 2019; "EPA Eyes End to NSR Compliance Priority, Adding Water and Lead Goals," *Inside EPA,* February 8, 2019.

20. U.S. Environmental Protection Agency, "Strengthening Transparency in Regulatory Science," 83 Fed. Reg. 18768, 18773–74 (April 30, 2018); Donald J. Trump, Presidential Memorandum for the Administrator of the Environmental Protection Agency, April 12, 2018, https://www.whitehouse.gov/presidential -actions/presidential-memorandum-administrator-environmental-protection -agency/.

21. Vanessa Zainzinger, "Critics Pan EPA Plan for Weighing Toxic Chemical Risks," *Science,* August 17, 2018, 631; Robinson Meyer, "Even Geologists Hate the EPA's New Science Rule, Government Executive," *Atlantic,* July 17, 2018; Scott Waldman, "Here Are 3 Studies That Might Be Hit by Pruitt's Rule," *E&E News,* April 26, 2018; Robinson Meyer, "Scott Pruitt's New Rule Could Completely Transform the EPA," *Atlantic,* April 25, 2018; Keith Goldberg, "Pruitt Floats Overhaul to EPA Scientific Review Process," *Law360,* April 24, 2018 (quoting Dan Byers, U.S. Chamber of Commerce, and Yogin Kothari, Union of Concerned Scientists).

22. E. Scott Pruitt, Administrator, U.S. Environmental Protection Agency, Memorandum to Assistant Administrators, re: Back-to-Basics Process for Reviewing National Ambient Air Quality Standards, May 9, 2018, 4–11, https://www.epa.gov/sites/production/files/2018-05/documents/image2018-05-09 -173219.pdf; Gavin Bade, "Pruitt NAAQS Memo Part of Broad Strategy to Weaken Air Regs, Lawyers Say," *Utility Dive,* May 11, 2018 (quoting John Walke, NRDC, and Sean Hecht, UCLA); Goldberg, "EPA Floats Overhaul" (quoting Jeffrey Holmstead, Bracewell LLP).

23. U.S. Environmental Protection Agency, "Hazardous and Solid Waste Management System: Disposal of Coal Combustion Residuals from Electric Utilities; Amendments to the National Minimum Criteria (Phase One, Part One); Final Rule," 83 Fed. Reg. 11584 (March 15, 2018); "Lawsuit Over EPA's Ash Rule Overhaul Aims to Halt Regulatory Rollback," *Inside EPA,* October 26, 2018, 28; Sean Reilly, "White House Weighs EPA's Rewritten Regulations," *E&E News,* June 14, 2018; "Disposal of Coal Combustion Residuals from Electric Utilities; Amendments to the National Minimum Criteria (Phase One)," Comments of Earthjustice et al., April 30, 2018; Adam Lidgett, "EPA Urged to Reconsider Coal Ash Disposal Rule," *Law360,* May 15, 2017.

24. 16 U.S.C. § 824a(c); Jeff Horwitz, Michael Biesecker, and Matthew Daly, "A Coal Country Dispute over an Alleged Trump Promise Unmet," *Washington Post,* August 22, 2017 (Murray quote).

25. Secretary of Energy Rick Perry, letter to FERC Chairman Neil Chatterjee and Commissioners Cheryl A. LaFleur and Robert Powelson, September 28, 2017, at 2–3, 7, attachment (on file with Commission) (quotes); John H. Cushman, "Inside the Coal War Games," *Inside Climate News,* October 11, 2017.

26. Chris Mooney and Steven Mufson, "Rick Perry Just Proposed Sweeping New Steps to Help Struggling Coal and Nuclear Plants," *Washington Post,* September 29, 2017 (quoting Richard Powell, ClearPath Foundation, and Paul Bailey, American Council for Clean Coal Electricity).

27. Rod Kuckro, Ellen M. Gilmer, and Jenny Mandel, "'Strange Bedfellows' Tangle with Perry's Grid Plan," *E&E News,* October 25, 2017; Catherine Traywick, "Coal Power Generators Lukewarm on Grid Plan That Aims to Help Them," *BNA Energy & Environment Report,* October 24, 2017; Robert Walton, "Sierra Club: DOE Cost Recovery Rule Could Cost Consumers Billions in Higher Bills," *Utility Dive,* October 18, 2017; Hannah Northey, "Industrialists Want Lawmakers to Block Perry Proposal," *E&E News,* October 11, 2017; Edward Klump, "Disappointment and Hope in Perry's Texas," *E&E News,* October 4, 2017.

28. Benjamin Storrow, "Perry, an Oil Guy, Angers Allies by Rushing to Coal's Aid," *E&E News,* October 12, 2017; Edward Klump, "Regulator, Enviro, Generator: 3 Views on Perry's Resilience Push," *E&E News,* October 10, 2017 ("prop up" quote); Robbie Orvis and Mike O'Boyle, "DOE Rulemaking Threatens to Destroy

Wholesale Markets with No Tangible Benefit," *Utility Drive,* October 2, 2017; Rod
Kuckro, "Perry Calls on FERC to Dismantle U.S. Energy Markets," *E&E News,*
October 2, 2017; Sam Mintz and Hannah Northey, "Perry Proposes Regulatory
Overhaul to Boost Coal, Nuclear," *E&E News,* September 29, 2017; Keith Goldberg,
"Perry's FERC Pricing Plan Would Roil Energy Markets," *Law360,* September 29,
2017; Benjamin Storrow, "Harvey Contradicts Trump Admin's Warnings on Wind
and Gas," *E&E News,* September 11, 2017.

29. 42 U.S.C. § 7173(b); Timothy Puko, "Trump Power Plan Is Rejected," *Wall
Street Journal,* January 9, 2018, A1; Rod Kuckro and Sam Mintz, "FERC Rejects
Perry's Bid for Coal-Based Grid Resilience," *E&E News,* January 9, 2018 ("examine
holistically" quote); Keith Goldberg, "FERC Nixes Perry's Plan to Pay Coal, Nuke
Plants," *Law360,* January 8, 2018.

30. Jennifer A. Dlouhy and Jennifer Jacobs, "Trump May Invoke Cold War Era
Defense Act to Boost Coal Plants," *BNA Energy & Environment Report,* April 19,
2018; Blake Sobczak and Peter Behr, "Pipeline Fears Anchor Trump's Coal, Nuclear
Bailout," *E&E News,* June 4, 2018; James Osborne, "Perry Says Economics Are
'Secondary' When It Comes to Power Grid," *Houston Chronicle,* June 25, 2018;
Hannah Northey, "Trump Touts 'Indestructible' Coal, and Gas Industry Fumes,"
E&E News, July 5, 2018.

31. Brad Plumer and Jim Tankersley, "Feeling Pull of Nostalgia, Trump Shapes
Trade Policy," *New York Times,* June 17, 2018, A8; Max Garland, "Trump Order
Could Prop Up WV Coal Plants, but Many Warn of Consumer Cost," *West
Virginia Gazette,* June 9, 2018; "Rick Perry's Obama Imitation," editorial, *Wall
Street Journal,* June 6, 2018, A16; Erin Ailworth and Russell Gold, "Weakened
Utility Finds Few Allies," *Wall Street Journal,* April 10, 2018, B2 (quote).

32. Peter Behr, "Cyberthreats Justify Federal Intervention, DOE Tells Panel,"
E&E News, June 8, 2018 ("fail-safes" quote); Gavin Bade, "PJM: FirstEnergy Nukes
Can Retire without Reliability Threat," *Utility Drive,* April 30, 2018; Steven Mufson,
"This Is a 'Test Case' for Whether Trump Is Really Serious about Saving Coal and
Nuclear Plants," *Washington Post,* April 3, 2018; Timothy Puko, "Trump's Loyalty
to Coal Is Tested," *Wall Street Journal,* April 2, 2018, A1.

33. Eric Wolff and Darius Dixon, "Rick Perry's Coal Rescue Runs Aground at
White House," *Politico,* October 17, 2018; Chris Tomlinson, "Pity Rick Perry,
President Donald Trump Wants Him to Blow Up Electricity Markets He Pio-
neered," *Houston Chronicle,* July 20, 2018; Northey, "Trump Touts 'Indestructible'
Coal"; Greg Ip, "The Losing Proposition of Propping Up Coal," *Wall Street Journal,*
June 7, 2018, A2; Sobczak and Behr, "Pipeline Fears"; Gavin Bade, "How Trump's
'Soviet-Style' Coal Directive Would Upend Power Markets," *Utility Drive,* June 4,
2018; Adam Aton, "Did Trump Just Propose the Opposite of Pricing Carbon?,"
E&E News, June 4, 2018; Dena Wiggins and Malcolm Woolf, "Trump's Energy

Department Cries 'Emergency' in Pursuit of Coal and Nuke Bailout," *Washington Examiner,* May 15, 2018 ("picking winners" quote).

34. David J. Lynch, "Trump Imposes Tariffs on Solar Panels and Washing Machines in First Major Trade Action of 2018," *Washington Post,* January 22, 2018; David Ferris, "The 'Buy America' Company That Sourced from Abroad," *E&E News,* May 15, 2017; David Ferris, "Solar Firm Prods Trump to Start a Trade War," *E&E News,* April 27, 2017.

35. Nichola Groom, "Billions in U.S. Solar Projects Shelved after Trump Panel Tariff," Reuters, June 7, 2018; David Ferris, "Suniva Started a Trade War; Now It's Being Auctioned Off," *E&E News,* May 4, 2018.

36. Benjamin Storrow, "Explaining Perry's Grid Directive, for Dummies," *E&E News,* October 11, 2017; Coral Davenport, "Coal Has Lost Its Grip on Power," *New York Times,* April 6, 2017, F11 (quoting Jeff Burleson, Southern Company); David Schultz, "Will Decline in Greenhouse Gas Emissions Survive Trump Presidency?," *BNA Environment & Energy Report,* February 17, 2017; Vaclav Smil, "Trump's Coal Policy Will Likely Do Just What Obama's Did," *Washington Post,* March 29, 2017; Jeffrey C. Peters, "Trump May Dismantle the EPA Clean Power Plan but Its Targets Look Resilient," *The Conversation,* November 15, 2016.

37. Edward Klump, "Coal Plant Closures Are Coming to Texas; Now What?," *E&E News,* October 16, 2017 (Cohan quote); Jim Polson and Emma Ockerman, "Coal Plants Are Dying in America's Most Energy-Friendly State," *BNA Energy & Climate Report,* October 13, 2107; Timothy Puko, "Power Firms Hold Their Ground," *Wall Street Journal,* October 11, 2017, B3; Daniel Cusick, "Xcel CEO Surprised by Wind's Market Power," *E&E News,* May 25, 2017; Benjamin Storrow, "Coal Plants Keep Closing on Trump's Watch," *E&E News,* February 21, 2017.

38. Brad Plumer, "A Year After Trump's Paris Pullout, U.S. Companies Are Driving a Renewables Boom," *New York Times,* June 1, 2018; Herman K, Trabish, "The Innovations Just Keep Coming in the Corporate-Utility Deal Space," *Utility Dive,* April 26, 2018; Christa Marshall, "100% Renewables—Gimmick or Game Changer?" *E&E News.* April 20, 2018; Christopher Flavelle, "Apple, Wal-Mart Stick with Climate Pledges Despite Trump's Pivot," *BNA Environment Reporter,* March 31, 2017, 602; Brian Eckhouse, "Apple's Entire Business Now Powered with Clean Energy," *BNA Daily Environment Report,* April 10, 2018; Brad Plumer, "Despite President's Paris Pullout, U.S. Companies Pursue Clean Energy," *New York Times,* June 1, 2018, A17; Stephen Badger, "My Company's Carbon Footprint Is the Size of a Small Country," *Washington Post,* October 5, 2017; Diane Cardwell and Clifford Krauss, "Coal Country's Power Plants Are Turning Away from Coal," *New York Times,* May 27, 2017, B1; Herman K. Trabish, "Green Design: Corporate Demand Pushes New Generation of Utility Green Tariffs," *Utility Dive,* May 4, 2017; Daniel Cusick, "Mega-Retailer Makes Mega Wind Investment," *E&E News,* February 6, 2017.

39. Amanda Reilly, "Greens Sue over Water Permit for Power Plant," *E&E News,* May 24, 2018; Lisa Friedman and John Schwartz, "Democratic States Sue over E.P.A. Policy, Just as the Republicans Did," *New York Times,* March 21, 2018, A16; Dylan Brown, "Bloomberg Puts Up Another $64M for 'War on Coal,'" *E&E News,* October 11, 2017; Michael Grunwald, "Environmentalists Get a Dose of Good News," *Politico,* November 18, 2016 (quoting Bruce Niles, Sierra Club); Emily Holden, "A 'Sea Change' for Power-Sector Climate Action?," *E&E News,* November 9, 2016.

40. Kevin Miller, "Mills Sets Goal to Fight Climate Change: 100% Renewable Electricity by 2050," *Portland Press Herald,* February 28, 2019; Marianne Lavelle, "California Ups Its Clean Energy Game: Gov. Brown Signs 100% Zero-Carbon Electricity Bill," *Inside Climate News,* September 10, 2018; Anne C. Mulkern, "Donald Trump Cannot Stop the Things We Are Doing," *E&E News,* November 13, 2017; Adam Lidgett, "California Legislature Passes Cap-and-Trade Renewal," *Law360,* July 18, 2017; Michael Phillis, "17 Govs. Reaffirm GHG Goals 1 Year After Trump's Paris Vow," *Law360,* June 1, 2018; Benjamin Storrow, Debra Kahn, and Scott Waldman, "Governors, Faced with Paris Withdrawal, Pledge Climate Action," *E&E News,* June 1, 2017; Jeremy Dillon, "Territorial Shift," *Congressional Quarterly,* April 24, 2017, 26.

41. Arianna Skibell, "Post-Christie N.J. Wants Back into RGGI," *E&E News,* January 30, 2018; Benjamin Storrow, "Northeast Strengthens Carbon Goals as Federal Rules Fade," *E&E News,* August 24, 2017; Emily Holden and Rod Kuckro, "McAuliffe Puts Va. on a Path to Its Own Clean Power Plan," *E&E News,* May 17, 2017; Storrow, Kahn, and Waldman, "Governors, Faced with Paris Withdrawal"; Gerald B. Silverman, "Carbon Prices Sink in Northeast Allowance Auction," *BNA Energy & Climate Report,* March 10, 2017.

42. Uma Outka, "Cities and the Low-Carbon Grid," *Environmental Law* 46 (2016): 134–35; Dan Gearino, "100% Renewable Energy: Cleveland Sets a Big Goal as It Sheds Its Fossil Fuel Past," *Inside Climate News,* September 22, 2018; Lizette Alvarez, "Mayors, Sidestepping Trump, Vow to Fill Void on Climate Change," *New York Times,* June 27, 2017, A9.

43. Union of Concerned Scientists, *A Dwindling Role for Coal* (Cambridge, MA: Union of Concerned Scientists, 2017); Jennifer A. Dlouhy, Ari Natter, and Tim Loh, "Trump Promised To Bring Back Coal—It's Declining Again," *Bloomberg Environment and Energy Report,* August 21, 2018; Zack Colman, "Trump Can't Save Coal, Experts Say," *E&E News,* June 5, 2018; Lisa Friedman, "How a Coal Baron's Wish List Became Trump's To-Do List," *New York Times,* January 10, 2018, B1; Jennifer A. Dlouhy and Mark Chediak, "Natural Gas Moves to the Naughty List," *BNA Environment Reporter,* April 21, 2017, 760; Oliver Milman, "Scott Pruitt Hails Era of Environmental Deregulation in Speech at Coal Mine," *Guardian,* April 13, 2017.

44. Union of Concerned Scientists, *Dwindling Role,* 1; Rod Kuckro, "Kansas City Utility Will Close 6 Units by 2019," *E&E News,* June 5, 2017; Benjamin Storrow, "Coal Is About to Disappear from New England," *E&E News,* May 30, 2017.

45. Brad Plumen, "Trump Wants to Help Coal, but It Might Be Past Saving," *New York Times*, August 22, 2018, A10; Robert Walton, "We Energies to Shut Pleasant Prairie Coal Plant, Increase Renewables," *Utility Dive,* November 29, 2017; "Texas Plant to Close under Pressure from Cheap Natural Gas," *E&E News,* October 9, 2017; Aldo Svaldi, "Coal's Future as a Power Source in Colorado Flickering," *Denver Post,* September 3, 2017.

46. Benjamin Storrow, "Weaker Carbon Rules Unikely to Prompt New Power Plants," *E&E News,* July 27, 2018; Benjamin Storrow, "Will the U.S. Ever Build Another Big Coal Plant?," *E&E News,* August 21, 2017.

47. Peter Behr and Rod Kuckro, "President's Bid to Recrown Coal Meets Industry Headwinds," *E&E News,* March 29, 2017.

48. Meghan Keneally, Jessica Hopper, and Evan Simon, "Wyoming Coal Miners Have High Hopes for Trump Amid National Turbulence," *ABC News,* May 18, 2017, http://abcnews.go.com/Politics/wyoming-coal-miners-high-hopes-trump-amid-national/story?id=47465051; Ward, "Trump Win Won't Bring Coal Rebound"; Campbell Robertson, "Coal Miners Hope Trump's Order Will Help," *New York Times,* March 29, 2017 ("economic plague" quote); Darryl Fears, "Trump Promised to Bring Back Coal Jobs," *Washington Post,* March 29, 2017.

49. Clifford Krauss, "Coal Country at the Crossroads," *New York Times,* January 25, 2018, B1; Fears, "Trump Promised to Bring Back Coal Jobs" (quoting IEEFA's *Coal Outlook*); Emily Holden, "Was the Climate Rule Really Bad for the Economy?," *E&E News,* March 28, 2017; Benjamin Storrow, "Have Markets Rendered KXL, Clean Power Plan Irrelevant?," *E&E News,* March 28, 2017.

50. Jenny Mandel, "Natural Gas, Oil Loom Large in Fuel Mix for Decades— EIA," *E&E News,* September 15, 2017; "Mammoth Chinese Gas Investment Deal Questioned," *E&E News,* November 13, 2017; "Natural Gas Industry Lobbies Trump," *E&E News,* July 20, 2017; Dlouhy and Chediak, "Natural Gas Moves to the Naughty List."

51. Pat Parenteau, "A Bridge Too Far: Building Off-Ramps on the Shale Gas Superhighway," *Idaho Law Review* 49 (2013): 352; Chris Mooney, "This Massive New Wind Project Is Proof Clean Energy Is Doing Fine under Trump," *Washington Post,* July 27, 2017; Daniel Cusick, "Duke to Put More Wind Power into Hoosier State," *E&E News,* June 28, 2017; Phil McKenna, "U.S. Wind Energy Installations Surge: A New Turbine Rises Every 2.4 Hours," *Inside Climate News,* May 3, 2017; Benjamin Storrow and Debra Kahn, "PacifiCorp Bets Big on Wyo. Wind," *E&E News,* April 6, 2017; "Trump Poses New Challenge for Growing Industry," *E&E News,* February 24, 2017 ("monstrous" quote); Michael Copley,

"EIA: Renewables Growth Does Not Hang on Clean Power Plan," *SNL Electric Utility Report*, June 13, 2016.

52. Jeremy Dillon, "New Solar Stalls Amid Trump Tariffs," *E&E News*, March 13, 2019; David Ferris, "Jobs Drop for Second Straight Year," *E&E News*, February 12, 2019; Krysti Shallenberger, "Will Utilities Keep Investing in Solar after Trump's Tariffs?," *Utility Dive*, January 25, 2018; Ana Swanson and Brad Plumer, "Steep Tariffs Threaten Growth of Solar Industry," *New York Times*, January 24, 2018, A1; Danielle Paquette, "Solar's Rise Lifted These Blue-Collar Workers," *Washington Post*, June 5, 2017.

53. Chris Mooney & Brady Dennis, "U.S. Greenhouse Gas Emissions Spiked in 2018—and It Couldn't Happen at a Worse Time," *Washington Post*, January 8, 2019.

54. Ledyard King, EPA's Pruitt Vows to Continue Rolling Back Rules Despite Alarming Climate Report," *USA Today*, November 8, 2017 ("extremely likely" quote); Dino Grandoni, "The Energy 202: Pruitt Plays to GOP Base by Repealing the Clean Power Plan," Washington Post PowerPost, October 10, 2017, https://www.washingtonpost.com/news/powerpost/paloma/the-energy-202/2017/10/10/the-energy-202-trump-plays-to-gop-base-by-repealing-the-clean-power-plan/59db9d5530fb0468cea81e16/?utm_term=.9c3ba4c57809; Adam Aton, "Most Americans Want Climate Policies—That Don't Cost Much," *E&E News*, October 3, 2017; Kavya Balaraman, "Most Trump Voters See Climate as an 'Alternative Fact,'" *E&E News*, February 6, 2017.

55. Patrick Ambrosio, "Power Plant Mercury Litigation Halted to Allow EPA Review," *BNA Daily Environment Report*, April 28, 2017; Amanda Reilly, "EPA 'Still Thinking.'"

56. Joe Ryan, "Trump May Have Paths to Save Coal, Hobble Clean Energy," *BNA Energy & Climate Report*, October 2, 2017.

57. Kuckro and Mintz, "FERC Rejects Perry's Bid."

58. Elvina Nawaguna, "Coal Country Smiles a Little," *Congressional Quarterly*, January 16, 2018, 26; Brian Eckhouse and Tim Loh, "A Year after Trump's Election, Coal's Still Losing to Renewables," *BNA Environment & Energy Report*, November 8, 2017; Robert Samuels, "How 'the Energy Capital of the Nation' Regained Its Optimism in the Trump Era," *Washington Post*, June 2, 2017; Dylan Brown, "Industry Back in the Black with EPA Ready to Help," *E&E News*, May 5, 2017; Steven Mufson, "Coal in the Trump Age: Industry Has a Pulse, but Prospects for Jobs Are Weak," *Washington Post*, March 17, 2017 ("Trump bump" quote).

59. "32 Utilities Were Asked; Just One Said Trump Is Saving Coal," *E&E News*, May 5, 2017; Jennifer A. Dlouhy, "Trump to Drop Climate Change from Environmental Reviews, Source Says," *Bloomberg News*, March 14, 2017; Jeff McMahon, "To Save Coal, Trump Has to Raise Your Gas Bill," *Forbes*, November 20, 2016.

13. TOWARD A SUSTAINABLE ENERGY FUTURE

1. Joseph P. Tomain, *Clean Power Politics: The Democratization of Energy* (Cambridge: Cambridge University Press, 2017), 87–88; John C. Dernbach, "Creating Legal Pathways to a Zero-Carbon Future," *Environmental Law Reporter* 46 (2016): 10782; "Quotes from Our Native Past," http://www.ilhawaii.net/~stony /quotes.html.

2. Natural Resources Defense Council, *A Tectonic Shift in America's Energy Landscape,* Third Annual Energy Report (2015), 5, https://www.nrdc.org/sites /default/files/energy-environment-report-2015.pdf; Citi GPS, *Energy Darwinism II: Why a Low Carbon Future Doesn't Have to Cost the Earth* (August 2015), 8; Jonathan Chait, "The Sunniest Climate-Change Story You've Ever Read," *New York Magazine,* September 7, 2015, 1 ("some combination" quote).

3. Naomi Klein, *This Changes Everything* (New York: Simon & Schuster, 2014), 255; Chait, "Sunniest Climate-Change Story."

4. Richard Martin, *Coal Wars: The Future of Energy and the Fate of the Planet* (New York: Palgrave Macmillan, 2015), 241; Ted Nace, *Climate Hope: On the Front Lines of the Fight against Coal* (San Francisco: CoalSwarm, 2010), 43; Natural Resources Defense Council, *A Tectonic Shift,* 3; Joel B. Eisen, "Smart Regulation and Federalism for the Smart Grid," *Harvard Environmental Law Review* 37 (2013): 1–57; Annalee Grant, "The 'Invisible Resource': Group Points to Efficiency as Climate-Change Asset," *SNL Electric Utility Report,* August 29, 2016.

5. Adam Cooper, "Delivering Savings through Energy Efficiency Programs," *Electric Perspectives,* January / February 2016, 46, 49; Kit Kennedy, "The Role of Energy Efficiency in Deep Decarbonization," *Environmental Law Reporter* 48 (2018): 10047; Joseph P. Tomain, "Traditionally-Structured Electric Utilities in a Distributed Generation World," *Nova Law Review* 38 (2014): 511; "Energy Efficiency Program Saves Billions—Report," *E&E News,* January 12, 2017; "CAP Floats Climate, Energy Recommendations for Next Administration," *Inside EPA,* September 29, 2016.

6. 42 U.S.C. § 6295(o), 6313(a)(6)(B)(iii); Executive Order 13783 §§ 5, 6; Kennedy, "Role of Energy Efficiency," 10031–32; Chris Mooney, "Obama Has Done More to Save Energy Than Any Other President," *Washington Post,* August 5, 2016.

7. U.S. Department of Energy, *2014 Smart Grid System Report* (Washington, DC: U.S. Department of Energy, August 2014); Phillip F. Schewe, *The Grid: A Journey Through the Heart of Our Electrified World* (Washington DC: Joseph Henry Press, 2007); "The Promise of Tomorrow," *Electric Perspectives,* March / April, 2016, 34, 35; Institute for Electric Innovation, "Top 10 Things You Should Know about Developing Smarter Energy Infrastructure," May 2017, http://www .edisonfoundation.net/iei/publications/Documents/IEI_Grid%20Mod%20101%20 Top%2010.pdf.

8. U.S. Department of Energy, *2014 Smart Grid System Report*, 5; Adam Cooper, *Electric Company Smart Meter Deployments: Foundation for a Smart Grid* (Washington, DC: Institute for Electric Innovation, September 2016), 6–7; Terry Bassham, "Transforming Customers into Partners," *Electric Perspectives*, March / April, 2016, 27 ("set-it-and-forget-it" quote); Marc B. Mihaly, "Recovery of a Lost Decade (or Is It Three?): Developing the Capacity in Government Necessary to Reduce Carbon Emissions and Administer Energy Markets," *Oregon Law Review* 88 (2009): 455.

9. Alexandra B. Klass and Elizabeth J. Wilson, "Remaking Energy: The Critical Role of Energy Consumption Data," *California Law Review* 104 (2016): 1110; Amy L. Stein, "Distributed Reliability," *University of Colorado Law Review* 87 (2016): 929, 951–52; Erica Gies, "Californians Are Keeping Dirty Energy Off the Grid via Text Message," *Inside Climate News*, October 30, 2017.

10. Sara C. Bronin, "Curbing Energy Sprawl with Microgrids," *Connecticut Law Review* 43 (2010): 559–60; Garrick B. Pursley and Hannah J. Wiseman, "Local Energy," *Emory Law Journal* 60 (2011): 897; Erica Gies, "Microgrids Keep These Cities Running When the Power Goes Out," *Inside Climate News*, December 4, 2017.

11. U.S. Department of Energy, *2014 Smart Grid System Report*, 4; International Energy Agency, *Digitalization & Energy* (Paris: IEA, 2017), 16; Bronin, "Curbing Energy Sprawl," 560; Jonas J. Monast and Sarah K. Adair, "A Triple Bottom Line for Electric Utility Regulation: Aligning State-Level Energy, Environmental, and Consumer Protection Goals," *Columbia Journal of Environmental Law* 38 (2013): 19; Richard L. Revesz and Burcin Unel, "Managing the Future of the Electricity Grid: Distributed Generation and Net Metering," *Harvard Environmental Law Review* 41 (2017): 45, 79–80; Shelley Welton, "Non-Transmission Alternatives," *Harvard Environmental Law Review* 39 (2015): 467–68.

12. U.S. Department of Energy, *2014 Smart Grid System Report*, 3, 4, 15; Eisen, "Smart Regulation," 15 ; Sharon B. Jacobs, "Bypassing Federalism and the Administrative Law of Negawatts," *Iowa Law Review* 100 (2015): 905; Klass and Wilson, "Remaking Energy," 1099–1101; Revesz and Unel, "Managing the Future," 81–82; Welton, "Non-Transmission Alternatives," 466; Robert Walton, "Report: Less Than 1 / 3 of Utility Customers Enroll in Energy Management Programs," *Utility Dive*, October 20, 2017; Blake Sobczak, "Congress' Inquiry into Grid Threat Turns Up Human Error," *E&E News*, February 2, 2017; "Utilities Urged to Stem Revenue Losses Due to 'Distributed' Clean Power," *Clean Energy Report*, June 10, 2013.

13. Tomain, *Clean Power Politics*, 3, 132; Mihaly, "Lost Decade," 453–54; Rod Kuckro, "Utilities, NRDC to Regulators: Clean Energy Is Inevitable," *E&E News*, February 14, 2018.

14. Tomain, *Clean Power Politics,* 148; "Promise of Tomorrow," 34, 36; Tanuj Deora, "Rethinking the Meaning of 'Reliability' and 'Resiliency' in the Wake of DOE Grid Reliability Study," *Utility Dive,* September 5, 2017.

15. Joel B. Eisen, "Distributed Energy Resources, 'Virtual Power Plants,' and the Smart Grid," *Environmental & Energy Law & Policy Journal* 7 (2012): 207 (San Antonio); Diane Cardwell, "The Home with Its Own Personal Power Grid," *New York Times,* July 30, 2017, BU1 (Green Mountain).

16. Tomain, *Clean Power Politics,* 156; William Boyd and Ann E. Carlson, "Accidents of Federalism: Ratemaking and Policy Innovation in Public Utility Law," *UCLA Law Review* 63 (2016): 814–15, 837; Jacobs, "Bypassing Federalism," 915; Ronald L. Lehr, "New Utility Business Models: Utility and Regulatory Models for the Modern Era," *Electricity Journal,* October 2013, 35.

17. Marilyn A. Brown et al., "Alternative Business Models for Energy Efficiency: Emerging Trends in the Southeast," *Electricity Journal,* May 2015, 103; Mihaly, "Lost Decade," 465; Will Nissen, "The Link between Decoupling and Success in Utility-Led Energy Efficiency," *Electricity Journal,* March 2016, 59.

18. Galen Barbose, Charles Goldman, and Jeff Schlegel, "The Shifting Landscape of Ratepayer-Funded Energy Efficiency in the U.S.," *Electricity Journal,* October 2009, 29; Boyd and Carlson, "Accidents of Federalism," 856; Brown et al., "Alternative Business Models," 103; Kay L. Gehring, "Can Yesterday's Demand-Side Management Lessons Become Tomorrow's Market Solutions?," *Electricity Journal,* June, 2002, 15; Mihaly, "Lost Decade," 453–54 (rates increase).

19. Tomain, *Clean Power Politics,* 166; Henry Lee, "Assessing the Challenges Confronting Distributive Electric Generation," *Electricity Journal,* June, 2003; Revesz and Unel, "Managing the Future," 50, 73, 75; Hannah J. Wiseman, "Moving Past Dual Federalism to Advance Electric Grid Neutrality," *Iowa Law Review Bulletin* 100 (2015): 106; Caroline Stewart, "How Blockchain Could Change the Energy Industry: Part 2," *Law360,* November 27, 2017; Herman K. Trabish, "As Solar Matures, Rate Design and Incentive Debates Grow Ever More Complex," *Utility Dive,* May 23, 2017.

20. Klass and Wilson, "Remaking Energy," 1102, 1105, 1107–8, 1117–18.

21. Pursley and Wiseman, "Local Energy," 907–8.

22. Jacobs, "Bypassing Federalism," 927–28.

23. Max Chafkin and Dune Lawrence, "The Next Big Hack Could Turn America's Lights Out," *BNA Environment & Energy Report,* November 2. 2017; Peter Behr, "Seeking a 'Sept. 12' Mindset for the Grid," *E&E News,* March 22, 2017; Blake Sobczak, "Cyber Primer Envisions Critical Role for Utility Regulators," *E&E News,* February 3, 2017.

24. Andrew Bartholomew, "The Smart Grid in Massachusetts: A Proposal for a Consumer Data Privacy Policy," *Boston College Environmental Affairs Law Review* 43 (2016): 109–10; Megan McLean, "How Smart Is Too Smart? How Privacy

Concerns Threaten Modern Energy Infrastructure," *Vanderbilt Journal of Entertainment & Technology Law* 18 (2016): 901–2.

25. Melissa Powers, "The Cost of Coal: Climate Change and the End of Coal as a Source of 'Cheap' Electricity," *University of Pennsylvania Journal of Business Law* 12 (2010): 434; Pursley and Wiseman, "Local Energy," 897.

26. U.S. Department of Energy, *Hydropower Vision* (Washington, DC, March 2017), 9; *Laboratories of Democracy: The Economic Impacts of State Energy Policies: Hearings before the Subcommittee on Energy and Power of the House Committee on Energy and Commerce*, 113th Cong. (July 24, 2014) (testimony of Steve Clemmer, Union of Concerned Scientists); Steven Ferrey, "Soft Paths, Hard Choices: Environmental Lessons in the Aftermath of California's Electric Deregulation Debacle," *Virginia Environmental Law Journal* 23 (2004): 350; Dan Gearino, "Rural Jobs: A Big Reason Midwest Should Love Clean Energy," *Inside Climate News*, December 7, 2018; Hannah Northey and Daniel Cusick, "Wind Industry Touts Jobs, Investments in Pitch to GOP," *E&E News*, April 19, 2017; Powers, "Cost of Coal," 407, 434.

27. *Energy Tax Policy in 2016 and Beyond: Hearings before the Senate Committee on Finance*, 114th Cong. (June 14, 2016) (testimony of Benjamin Zycher, American Enterprise Institute); Hari M. Osofsky and Hannah J. Wiseman, "Dynamic Energy Federalism," *Maryland Law Review* 72 (2013): 790–91; Daniel Cusick, "R Street: 'No Credible . . . Evidence' Renewables Threaten Grid," *E&E News*, May 31, 2017; Inara Scott and David Bernell, "Planning for the Future of the Electric Power Sector through Regional Collaboratives," *Electricity Journal*, January / February 2015, 83–93; Timothy Cama, "EPA Chief: I'd 'Do Away With' Wind, Solar Tax Credits," *The Hill*, October 9, 2017; Mark Hand, "Climate Conference Panelist Sees Light at End of 'Renewable Madness Tunnel,'" *SNL Electric Utility Report*, July 21, 2014.

28. See, e.g., Union Neighbors United v. Jewell, 831 F.3d 564 (D.C. Cir. 2016); Public Employees for Environmental Responsibility v. Hopper, 827 F.3d 1077 (D.C. Cir. 2016); U.S. Department of Energy, *Quadrennial Energy Review: Energy Transmission, Storage, and Distribution Infrastructure* (Washington, DC: U.S. Department of Energy, 2015), 3–8; Alexandra B. Klass, "Expanding the U.S. Electric Transmission and Distribution Grid to Meet Deep Decarbonization Goals," *Environmental Law Reporter* 47 (2017): 10753; Pursley and Wiseman, "Local Energy," 898; Brian Eckhouse, "Cape Wind Developer Terminates Project Opposed by Kennedys, Koch," *BNA Environment & Energy Report*, December 1, 2017; Juan Carlos Rodriguez, "Nev. Solar Project Gets 9th Circ. Nod over Tortoise Concerns," *Law360*, May 18, 2017; Jeffrey Tomich, "Clean Line Makes New Pitch for $2.8B Midwest Wind Power Line," *E&E News*, March 21, 2017.

29. Tomain, *Clean Power Politics*, 3; Pat Parenteau, "A Bridge Too Far: Building Off-Ramps on the Shale Gas Superhighway," *Idaho Law Review* 49 (2013):

344; Erin Ailworth, "Wind Powers Rural U.S.," *Wall Street Journal,* September 7, 2017, B2.

30. Lyndsey Gilpin, "2 Reasons Solar Is Booming in Trump Country: Price and Energy Independence," *Inside Climate News,* November 6, 2017; Brian Baskin, "Why It's Mainly Big Companies Buying Green Power," *Wall Street Journal,* May 22, 2017, R3; Ed Felker, "It's Not Easy Goin' Green," *Congressional Quarterly,* June 8, 2015, 38.

31. *American Energy Security and Innovation: Grid Reliability Challenges in a Shifting Energy Resource Landscape, Hearings before the Subcommittee on Energy and Power of the House Committee on Energy and Commerce,* 112th Cong. (May 9, 2013) (testimony of Daniel Weiss, Center for American Progress) (ELI study); Joel B. Eisen, "Residential Renewable Energy: By Whom?," *Utah Environmental Law Review* 31 (2011): 345 (subsidies quote); Monast and Adair, "Triple Bottom Line," 18–19; Chris Martin and Jim Efstathiou, Jr., "America's Wind Farms Are Now Ready to Stand Without Subsidies," *Bloomberg Environment and Energy Report,* November 15, 2018; Jessica Shankleman, "Utilities Line Up to Profit from 'Slowest Trainwreck' in History," *BNA Energy & Climate Report,* September 18, 2017.

32. Jeffrey Tomich, "Renewable Energy Set to Expand in Bellwether Midwest Market," *E&E News,* July 13, 2017; Lizette Alvarez, "Mayors, Sidestepping Trump, Vow to Fill Void on Climate Change," *New York Times,* June 27, 2017, A9; Robert Walton, "New York Will Sink $1.5B into Renewable Energy Projects to Spur Clean Energy Jobs," *Utility Dive,* June 5, 2017.

33. Klein, *This Changes Everything,* 99; Uma Outka, "Cities and the Low-Carbon Grid," *Environmental Law* 46 (2016):144–45; "Clear the Air, Lessons from Boulder's Takeover of Energy," January 22, 2018, https://clearair.us/post/lessons-from-boulders-takeover-of-energy; Herman K. Trabish, "Join or Die: How Utilities Are Coping with 100% Renewable Energy Goals," *Utility Dive,* December 13, 2017; Alex Burness, "Boulder Municipalization Persists after Surprise Election Comeback," *Boulder Daily Camera,* November 9, 2017.

34. Lehr, "New Utility Business Models," 35.

35. Robert Walton, "NRG Will Sell 6 GW of Generation, Shed Renewable Assets in Restructuring Plan," *Utility Dive,* July 12, 2017; David Gelles, "The Messy Business of Clean Power," *New York Times,* August 14, 2016, B1.

36. Mark C. Trexler, "Not as Simple as It Sounds," *Environmental Forum,* January / February 2018, 27; Gelles, "Messy Business" ("very hard" quote); David Ferris, Edward Klump, and Debra Kahn, "Why NRG's Green Crusade Faltered, *Energy Wire,* March 7, 2016, http://www.eenews.net/stories/1060033507.

37. Sean Shimamoto and Frank Shaw, "Utilities' Build-Transfer Deals Spur Renewable Projects," *Law 360,* October 24, 2018; Chris Martin, Jim Polson, and Mark Chediak, "Growth-Starved Utilities Have Found a New Way of Making Money," *BNA Daily Environment Report,* July 28, 2017.

38. 42 U.S.C. § 1221; Klass, "Expanding," 10758; Sam Mintz, "Experts Urge Lawmakers to Focus on Transmission," *E&E News*, May 11, 2018; Sam Mintz, "Greens, Clean Power Boosters Clash over New England Project," *E&E News*, February 8, 2018.

39. Klass, "Expanding," 10764–65; Osofsky and Wiseman, "Dynamic Energy Federalism," 790; Russell Gold, "Texas Takes a Shine to Solar Power," *Wall Street Journal,* August 22, 2015, B4.

40. Amy L. Stein, "Reconsidering Regulatory Uncertainty: Making a Case for Energy Storage," *Florida State Law Review* 41 (2014): 713–14; Stein, "Distributed Reliability," 917–18; Lorne Stockman, *A Bridge Too Far: How Appalachian Basin Gas Pipeline Expansion Will Undermine U.S. Climate Goals* (Washington, DC: Oil Change International July 2016), 30; Jonathan Tirone, Google's Schmidt Sees Promise in Battery Inventor's Latest Device," *BNA Energy & Climate Report,* March 14, 2017; Diane Cardwell and Clifford Krauss, "A Big Test for Big Batteries," *New York Times,* January 15, 2017, B1.

41. Michael J. Allen, "Energy Storage: The Emerging Legal Framework (And Why It Makes a Difference)," *Natural Resources & Environment,* Spring 2016, 23–26; Daniel Hagan and Jane Rueger, "Outdated Rules Are Holding Back Energy Storage," *Law360,* March 15, 2017; Diane Cardwell, "Moving beyond Cars, Tesla Uses Batteries to Bolster Power Grid," *New York Times,* January 31, 2017, B1.

42. Tomain, *Clean Power Politics,* 40; Lincoln Davies, "Power Forward: The Argument for a National RPS," *Connecticut Law Review* 42 (2010): 1358; Donald Bryson and Jeff Glendinging, "States Are Unplugging Their Renewable-Energy Mandates," *Wall Street Journal,* July 11, 2015, A9.

43. J. R. DeShazo, Julien Gattaciecca, and Kelly Trumbull, *The Promises and Challenges of Community Choice Aggregation in California* (Los Angeles: UCLA Luskin Center for Innovation, 2017); Courtney Humphries, "Cities Cracking Down on Climate Law-Breakers," *Inside Climate News,* September 22, 2017; Keith Goldberg, "Red-State Cities Face Uphill Battle to Keep Paris Pledge," *Law360,* June 6, 2017; Benjamin Storrow, "Mayors Promise to Act on Climate," *E&E News,* July 6, 2014.

44. Jeremy Knee, "Rational Electricity Regulation: Environmental Impacts and the 'Public Interest,'" *West Virginia Law Review* 113 (2011): 754–55 756–67; Uma Outka, "Environmental Law and Fossil Fuels: Barriers to Renewable Energy," *Vanderbilt Law Review* 65 (2012): 1695; Benjamin Storrow, "Mid-Atlantic Enters 'Offshore Game' in a Big Way," *E&E News,* May 12, 2107.

45. See, e.g., *Union Neighbors United,* 831 F.3d 564; *Public Employees for Environmental Responsibility,* 827 F.3d 1077; Myrick v. Peck Electric Co., 2017 WL 129041 (Vt. 2017).

46. Michael B. Gerrard, "Legal Pathways for a Massive Increase in Utility-Scale Renewable Generation Capacity," *Environmental Law Reporter* 47 (July 2017): 10591; Justin Tschoepe and William Wood, "The Current State of Renewable Energy Disputes: Part 4," *Law360,* May 3, 2017 (two states).

47. U.S. Department of Energy, *Hydropower Vision,* 7, 13, 16, 20; Luis Berga, "The Role of Hydropower in Climate Change Mitigation and Adaptation: A Review," *Engineering* 2 (2016): 313, 316; Bobby Magill, "Tiny Turbine Developers Energized by Hydro Law," *BNA Environment & Energy Report,* November 8, 2018; Bobby Magill, "Hydropower to Become Unsustainable as Climate Changes, Study Says," *BNA Environment & Energy Report,* November 5, 2018; Rebecca Kern, "Permit Delays Dam Up Hydro Projects," *BNA Environment & Energy Report,* October 30, 2018.

48. Eduardo Porter, "Concern Grows as Senate Allies Argue Biomass is 'Neutral' Fuel," *New York Times,* October 5, 2016, B1; "EPA's Biomass Policy Draws Strong Criticism for Allowing GHG Increase," *InsideEPA/climate,* December 3, 2014.

49. M. Fischedick et al., "Mitigation Potential and Costs," in *IPCC Special Report on Renewable Energy Sources and Climate Change Mitigation,* ed. O. Edenhofer et al. (Cambridge: Cambridge University Press, 2011), 793–864; Mark Z. Jacobson et al., "Low-Cost Solution to the Grid Reliability Problem with 100 Percent Penetration of Intermittent Wind, Water, and Solar for All Purposes," *Proceedings of the National Academy of Sciences* 112 (2015): 15060–65; Trabish, "Join or Die"; Umair Irfan, "Can the World Run on Clean Energy?," *E&E News,* June 20, 2017 ("riddled" quote); Chris Mooney, "A Bitter Scientific Debate Just Erupted over the Future of America's Power Grid," *Washington Post,* June 19, 2017; "Is 100% Renewable Energy Feasible? New Paper Argues for a Different Target," *Inside Climate News,* June 19, 2017.

50. Union of Concerned Scientists, *A Dwindling Role for Coal* (Cambridge, MA: Union of Concerned Scientists, 2017); Clifford Krauss, "Rising Coal Exports Lift Ailing Industry, but Turnaround May Be Temporary," *New York Times,* December 16, 2017, B3.

51. *American Energy Security and Innovation: The Role of a Diverse Electricity Generation Portfolio: Hearings before the Subcommittee on Energy and Power of the House Committee on Energy and Commerce,* 112th Cong. (March 5, 2013) (testimony of Mark McCullough, American Electric Power); Betsy Monseu, "Coal's Place in the Energy and Electricity Space," *Electricity Journal,* July 2015, 104; "Natural Gas Boom Highlights Infrastructure Gap, Environmental Concerns," *Clean Energy Report,* January 7, 2013; "NERC Sees Potential Reliability Risks with Greater Use of Natural Gas," *Clean Energy Report,* December 3, 2012; Christine Cordner, "GAO Report Sees Continued Key Role for Coal in US Power Generation

Sector," *SNL Daily Coal Report,* November 29, 2012; Jason Fordney, "MISO Executives Warn of Dependence on Natural Gas," *Megawatt Daily,* June 21, 2012.

52. Erik Reece, *Lost Mountain: A Year in the Vanishing Wilderness—Radical Strip Mining and the Devastation of Appalachia* (New York: Riverhead Books, 2006); Joe Ryan, "Almost Half of U.S. Coal Plants Operated with Net Losses in 2017," *Bloomberg Energy and Climate Report,* March 26, 2018; Mike Soraghan, "Coal v. Gas? Not Even Close on Toxicity, Study Says," *E&E News,* October 24, 2017; Steven Mufson, "The 'Bomb Cyclone' Is Contradicting Rick Perry's Argument for Coal," *Washington Post,* January 5, 2017; Ryan Collins, "In Heart of U.S. Coal Demand, Study Shows Gas and Wind Cheaper," *BNA Daily Environment Report,* December 16, 2016; Taylor Kuykendall, "Replacing Coal, Part 3: 'The Clean Power Plan Is the Battle, Not the War,'" *SNL Daily Coal Report,* April 15, 2016 (quoting Adele Morris, Brookings Institution); Lauren Bellero, "Green Groups: Planned Coal Capacity Equals Nearly $1 Trillion In 'Wasted Capital,'" *SNL Generation Markets Week,* April 5, 2016; Matthew Bandyk, "Economic Pressure on Coal-Fired Plants Gives Opponents More Fuel," *SNL Daily Coal Report,* October 9, 2012.

53. Walter J. Culver and Mingguo Hong, "Coal's Decline: Driven by Policy or Technology?," *Electricity Journal,* September 2016, 50; Susan F. Tierney, *The U.S. Coal Industry: Challenging Transitions in the 21st Century* (September 26, 2016), 28; Peter Maloney, "A Complicated Calculus Keeps the Remaining Coal Fleet Alive," *Utility Dive,* March 19, 2018; "Officials Say Gas Could Be 'Bridge' to Deep GHG Cuts, but Debate Its Length," *Inside EPA,* September 19, 2016; Todd Woody, "Most Coal-Fired Power Plants in the US Are Nearing Retirement Age," *Quartz,* March 12, 2013; Jonathan Crawford, "NRDC's Dave Hawkins: New Coal Plants Not Dead," *SNL Daily Coal Report,* April 30, 2012.

54. "Officials Say Gas Could Be 'Bridge'"; Darren Sweeney, "Duke Energy Utilities to Ramp Up Natural Gas, Renewable Investments," *SNL Electric Utility Report,* September 12, 2016; "MISO and Spectra Energy Representatives Speak Out on Natural Gas Infrastructure's Increasing Role, and Challenges, in Support of Electric Generation," *Foster Natural Gas / Oil Report,* April 15, 2016, 14.

55. Mathew Carr and Anna Shiryaevskaya, "Natural Gas Risks Becoming Too Pricey in Climate Change Battle," *BNA Environment & Energy Report,* December 6, 2017; Ryan Collins and Emma Ockerman, "In a Twist, Texas Wind Power Can Actually Help Gas Generators," *BNA Daily Environment Report,* September 18, 2017; Maxine Joselow, "Southern Sets Sights on Flexible Generation," *E&E News,* July 12, 2017.

56. Parenteau, "A Bridge Too Far," 336–38 (methane leakage); *Rhetoric vs. Reality, Part II: Assessing the Impact of New Federal Red Tape on Hydraulic Fracturing and American Energy Independence: Hearings before the Subcommittee on Technology, Information Policy, Intergovernmental Relations and Procurement*

Reform of the House Committee on Oversight and Government Reform, 112th Cong. (May 31, 2012) (testimony of Robert Howarth, Cornell University); U.S. Government Accountability Office, *Information on Shale Resources, Development, and Environmental and Public Health Risks* (Washington, DC, September 2012), 33–53; Jeffrey Tomich, "FirstEnergy CEO: Gas Reliance 'Sets Up Security Risks,'" *E&E News,* July 31, 2017; "EIA Sees CO_2 from Gas Exceeding Coal, Intensifying Infrastructure Fights," *Inside EPA,* August 18, 2016, 1; Sarah Smith, "Gas Bridge to Renewables Already Built, Clean Energy Advocate Says," *SNL Daily Gas Report,* April 6, 2016; "Methane Emitted from Natural Gas Activities Might Offset Climate Benefits, Group Says," *BNA Environment Reporter,* June 7, 2013, 1674; "Natural Gas Boom Highlights Infrastructure Gap, Environmental Concerns," *Clean Energy Report,* January 7, 2013; Crawford, "NRDC's Dave Hawkins."

57. Clean Air Council v. Pruitt, 862 F.3d 1 (D.C. Cir. 2017); U.S. Environmental Protection Agency, "Oil and Natural Gas Sector: Emission Standards for New, Reconstructed, and Modified Sources Reconsideration," 83 Fed. Reg. 52056 (October 15, 2018); U.S. Environmental Protection Agency, "Oil and Natural Gas Sector: Emission Standards for New, Reconstructed, and Modified Sources; Stay of Certain Requirements," 82 Fed. Reg. 27645 (June 16, 2017); Parenteau, "A Bridge Too Far," 339–40; Ellen M. Gilmer, "FERC Splits on Climate Review, Reapproves Sabal Trail," *E&E News,* March 15, 2018.

58. Jonas J. Monast and Sarah K. Adair, "Completing the Energy Innovation Cycle: The View from the Public Utility Commission," *Hastings Law Journal* 65 (2014): 1369; Christa Marshall, Hannah Northey, and Sam Mintz, "White House Requests More Funding for Some Programs," *E&E News,* February 12, 2018; Jennifer A. Dlouhy and Mark Chediak, "Natural Gas Moves to the Naughty List," *BNA Environment Reporter,* April 21, 2017, 760; Hannah Hess, "To Save Coal, You Must Regulate CO_2—Bush Advisor," *E&E News,* March 28, 2017; Kathleen Hart, "AEP Puts Carbon Capture Project on Hold, Citing Weak Economy, Uncertain Climate Policy," *SNL Renewable Energy Weekly,* July 15, 2011; NRG Energy, "Petra Nova," http://www.nrg.com/generation/projects/petra-nova/.

59. Jeff Goodell, *Big Coal: The Dirty Secret behind America's Energy Future* (Boston: Houghton Mifflin, 2006), 224; Nace, *Climate Hope,* 29; Tomain, *Clean Power Politics,* 52; David E. Adelman and Ian J. Duncan, "The Limits of Liability in Promoting Safe Geologic Sequestration of CO_2," *Duke Environmental Law & Policy Forum* 22 (2011): 8; Jonathan Crawford, "Ex-DOE Official: Success of EPA Acid Rain Rules Unlikely to Be Replicated in Carbon Standard," *SNL Generation Markets Week,* October 29, 2013; Brian Hansen, "Industry Gets Most of What It Sought in EPA's New CO_2 Sequestration Rule," *Inside Energy with Federal Lands,* November 29, 2010, 7.

60. Parenteau, "A Bridge Too Far," 328; Rod Kuckro, "New England Grid Faces Challenges without More Gas," *E&E News,* January 31, 2017; Jeffrey C. Peters,

"Trump May Dismantle the EPA Clean Power Plan but Its Targets Look Resiliant," *The Conversation,* November 15, 2016; Matthew L. Wald, "Massachusetts Regulators Approve a Gas-Fired Power Plant with an Expiration Date," *New York Times,* February 21, 2014, B2.

61. U.S. Energy Information Administration, "First New U.S. Nuclear Reactor in Almost Two Decades Set to Begin Operating," *Today in Energy,* June 14, 2016, http://www.eia.gov/todayinenergy/detail.php?id=26652; Boyd and Carlson, "Accidents of Federalism," 845; Doug Vine, *Solutions for Maintaining the Existing Nuclear Fleet* (Arlington, VA: Center for Climate and Energy Solutions, May 2018), 1; World Nuclear Association, *Nuclear Power in the USA* (London: World Nuclear Association, October 2018); "Majority of U.S. Plants Losing Money," *E&E News,* June 16, 2017; Rebecca Smith, "Nuclear Power Firms Feel Squeeze," *Wall Street Journal,* March 6, 2015, B3.

62. Hannah Northey and Jeremy Dillon, "Trump's Nuclear Revival? It's a 'Black Box,'" *E&E News,* January 15, 2019; Kristi E. Swartz, "Vogtle Expansion Still on Schedule, Executives Say," *E&E News,* November 16, 2018; Kristi E. Swartz, "Vogtle Lives On, with More Risks for Georgia Power," *E&E News,* September 27, 2018; "Utility Reaches $2B Settlement over Failed S.C. Plant," *E&E News,* November 26, 2018; Brian Eckhouse, "Who Wants to Buy a Pair of Half-Built Nuclear Reactors?," *BNA Environment & Energy Report,* July 23, 2018; Brad Plumer, "U.S. Nuclear Comeback Stalls as 2 South Carolina Reactors Are Abandoned," *New York Times,* August 1, 2017; Diane Cardwell and Jonathan Soble, "Bankruptcy Rocks Nuclear Industry," *New York Times,* March 29, 2017, B1.

63. Vine, *Solutions,* 1, 5–7; "Plant Closures Would Be Bad News for Emissions—Report," *E&E News,* April 19, 2018; Rebecca Kern, "Paris Exit Doesn't End U.S. Nuclear Industry's Carbon-Free Argument," *BNA Energy & Climate Report,* June 14, 2017; "Industry Downplays California Nuclear Plant's Closure but Pressure Remains," *Inside EPA,* June 22, 2016; Glen Boshart, "EEI Panelists: Something Must Be Done to Save the Nukes," *SNL Energy Finance Daily,* June 15, 2016; Richard K. Lester, "A Roadmap for U.S. Nuclear Energy Innovation," *Issues in Science and Technology,* January 1, 2016, 32.

64. *The Future of Nuclear Power: Hearings before the Subcommittee on Energy and Water Development of the Senate Appropriations Committee,* 114th Cong. (September 14, 2016) (testimony of Ernest Moniz, secretary of energy); Lester, "Roadmap"; Powers, "Cost of Coal," 434; David A. Repka and Tyson R. Smith, "Deep Decarbonization and Nuclear Energy," *Environmental Law Reporter* 48 (2018): 10,263–64; Vine, *Solutions,* 8; "Nuclear: Providing a Foundation for Sensible Policy," *Nuclear News,* January 2016, 48; Rebecca Smith, "Power Customers Face Nuclear Bill," *Wall Street Journal,* April 18, 2015, A1.

65. Repka and Smith, "Deep Decarbonization," 10246–48; *Advanced Nuclear Technology Development Act of 2016 and the Nuclear Utilization of Keynote Energy*

Policies Act: Hearings before the Subcommittee on Energy and Power of the House Committee on Energy and Commerce, 114th Cong. (April 29, 2016) (testimony of Geoffrey Fettus, NRDC); Adrian Cho, "The Little Reactors that Could," *Science,* 363 (February 22, 2019): 806; M. V. Ramana, "The Future of Nuclear Power in the US Is Bleak," *The Hill,* June 23, 2018; John Fialka, "U.S. Receives First Plans for Small Nuclear Reactors," *E&E News,* January 13, 2017.

66. Coalition for Competitive Electricity v. Zibelman, 906 F.3d 41 (2d Cir. 2018); Electric Power Supply Ass'n v. Star, 904 F.3d 518 (7th Cir. 2018); Vine, *Solutions,* 17–19; World Nuclear Association, *Nuclear Power in the USA;* Elise Young and Brian Eckhouse, "N.J. Governor Signs Nuclear Plant Bill Environmental Groups Opposed," *BNA Daily Environment Report,* May 24, 2018.

67. Steve Clemmer et al., *The Nuclear Power Dilemma: Declining Profits, Plant Closures, and the Threat of Rising Carbon Emissions* (Cambridge, MA: Union of Concerned Scientists, November 2018), 4, 7; Rod Kuckro, "Conn. Throws Lifeline to Millstone, Seabrook Plants," *E&E News,* January 2, 2019; Juan Carlos Rodriguez, "Enviros Fight to Sink NY Nuke Subsidies," *Law 360,* December 18, 2018; Hannah Northey, "Exelon Exec: 'We're Not Going to Build More Nuclear Plants,'" *E&E News,* April 13, 2018; Amy Harder, "Environmental Groups Change Tune on Nuclear Power," *Wall Street Journal,* June 16, 2016.

68. North American Electric Reliability Corp., *2017 Long-Term Reliability Assessment* (December 2017), 6 ("robust" quote); Kassia Micek, "Coal Plant Retirements of Up to 3,000 MW Not Likely to Impact Southeast Markets, Utilities Say," *Electric Utility Week,* March 25, 2013.

69. Scott and Bernell, "Planning for the Future"; Julia Pyper, "GM Restructuring Reveals a 'Disconnect' between SUV Sales Today and an EV Future," *Greentech Media,* November 26, 2018; Maxine Joselow, "Chevy Volt Cruises to Retirement," *E&E News,* November 27, 2018.

70. Judy W. Chang et al., *Advancing Past "Baseload" to a Flexible Grid* (Boston: Brattle Group, prepared for NRDC, June 26, 2017), 1–3; Burcin Unel and Avi Zevin, *Toward Resilience: Defining, Measuring, and Monetizing Resilience in the Electricity System* (New York: Institute for Policy Integrity, New York University School of Law, August 2018), 5, 13, 37; Christa Marshall, "Perry Was Wrong about Grid Reliability—NRDC," *E&E News,* June 26, 2017; Edward Klump, "Electric Evolution Calls for 'More Eyes' on the Grid," *E&E News,* June 13, 2017 (ERCOT control room); Peter Behr, "Grid Researchers Focus on Fast-Evolving U.S. Energy System," *E&E News,* February 13, 2017.

71. James Bruggers, "Coal Ash Contaminates Groundwater at 91% of U.S. Coal Plants, Tests Show," *Inside Climate News,* March 4, 2019; Catherine Morehouse, "As 67 Coal Plants in 22 States Report Coal Ash Violations, Greens Fear Prolonged Cleanup," *Utility Dive,* December 20, 2018.

72. Motor Vehicle Manufacturers Ass'n v. State Farm Mutual Automobile Ins. Co., 463 U.S. 29 (1983).

73. *New Source Review Policy, Regulations and Enforcement Activities: Hearings before the Senate Committee on Environment and Public Works and the≈Senate Committee on the Judiciary,* 107th Cong. 127 (July 16, 2002) (testimony of Jeffrey Holmstead, EPA); ibid., 26 (statement of Ande Abbott, International Brotherhood of Boilermakers); ibid., 610–11 (testimony of John Walke, NRDC).

74. "Agency's Voluntary Programs Insufficient for Cutting Emissions, Internal Report Finds," *BNA Environment Reporter,* August 1, 2008, 1532.

75. Tomain, *Clean Power Politics,* 86; Herman K. Trabish, "How Big Can New England's Regional Cap-and-Trade Program Get?" *Utility Dive,* May 1, 2018; Stephen C. Jones and Paul R. McIntyre, "Filling the Vacuum: State and Regional Climate Change Initiatives," *BNA Environment Reporter,* July 27, 2007, 1640; "RGGI States Urge EPA to Let Success of Regional Plan Guide Power Plant Rules," *BNA Environment Reporter,* December 6, 2013, 3607.

76. Garrett Hardin, "The Tragedy of the Commons," *Science* 162 (1968): 1243–48; David B. Spence, "Regulation, 'Republican Moments,' and Energy Policy Reform," *BYU Law Review* 2011 (2011): 1610–11.

77. Spence, "'Republican Moments,'" 1608–9.

78. Scott and Bernell, "Planning for the Future."

79. Harold H. Bruff, "Presidential Management of Agency Rulemaking," *George Washington Law Review* 57 (1989): 582–86; Nina A. Mendelson, "Disclosing 'Political' Oversight of Agency Decision Making," *Michigan Law Review* 108 (2010): 1149; Rena Steinzor, "The Case for Abolishing Centralized White House Regulatory Review," *Michigan Journal of Environmental & Administrative Law* 1 (2012): 263.

80. Rena Steinzor and Sidney Shapiro, *The People's Agents and the Battle to Protect the American Public* (Chicago: University of Chicago Press, 2010), chap. 3; Thomas O. McGarity, "Some Thoughts on 'Deossifying' the Rulemaking Process," *Duke Law Journal* 41 (1992): 1385–1462.

81. *H.R. ___, the Energy Tax Prevention Act of 2011: Hearings before the Subcommittee on Energy and Power of the House Committee on Energy and Commerce,* 112th Cong. 280 (2011) (testimony of James N. Goldstene, California Air Resources Board); *EPA's Greenhouse Gas Regulations and Their Effect on American Jobs: Hearings before the Subcommittee on Energy and Power of the House Committee on Energy and Commerce,* 112th Cong. 123 (2011) (testimony of Gina A. McCarthy, Assistant Administrator for Air and Radiation, EPA); Crawford, "NRDC's Dave Hawkins."

82. Repka and Smith, "Deep Decarbonization," 10251–52; Clemmer et al., *Nuclear Power Dilemma,* 7; Arianna Skibell, "Meet the 'Eco-Right' Pushing for a

Carbon Fee," *E&E News,* November 1, 2017; Benjamin Hulac, "Exxon Doubles Down on National Carbon Tax," *E&E News,* February 24, 2017; "Congressional Budget Office Touts Carbon Tax as Better Choice Than a Cap-and-Trade System in Curbing Emissions," *Foster Natural Gas Report,* February 29, 2008, 3.

83. Evan Lehmann and Emily Holden, "Trump's Last Answer to a Carbon Tax? No Way!," *E&E News,* February 8, 2017; "FERC's Moeller Dismisses National Carbon Price as Alternative to ESPS," *Inside EPA,* October 22, 2014.

84. David Littell and Kelly Speakes-Backman, "Pricing Carbon under EPA's Proposed Rules: Cost Effectiveness and State Economic Benefits," *Electricity Journal,* October 2014, 8; Sam Napolitano et al., "The U.S. Acid Rain Program: Key Insights from the Design, Operation, and Assessment of a Cap-and-Trade Program," *Electricity Journal,* August / September 2007, 47; John Schwartz, "Inconvenient Indeed," *New York Times,* July 12, 2016, D1.

85. "Allowances Only Part of Cutting Emissions," *Megawatt Daily,* December 20, 2004, 11.

86. 5 U.S.C. § 706; Napolitano et al., "U.S. Acid Rain Program"; "Study Concludes Statutory Deadlines Spur EPA, State, Industry to Fulfill Mandates," *BNA Environment Reporter,* July 26, 1985, 509.

87. Thomas O. McGarity, "Administrative Law as Blood Sport: Policy Erosion in a Highly Partisan Era," *Duke Law Journal* 61 (2012): 1758; Mendelson, "Disclosing 'Political' Oversight," 1165.

88. Thomas O. McGarity, *Reinventing Rationality: The Role of Regulatory Analysis in the Federal Bureaucracy* (Cambridge: Cambridge University Press, 1991), 137; Robert Repetto and Duncan Austin, *The Costs of Climate Protection: A Guide for the Perplexed* (Washington, DC: World Resources Institute, 1997), 2; Cary Conglianese, "It's Time to Think Strategically about Retrospective Benefit-Cost Analysis," *Regulatory Review,* April 30, 2018; Andrew Childers, "EPA, Policy Advocates Should Improve Analyses of Job Impacts, Report Says," *BNA Environment Reporter,* April 6, 2012, 900.

89. Phil McKenna, "Power Plants' Coal Ash Reports Show Toxins Leaking into Groundwater," *Inside Climate News,* February 7, 2018.

90. Daniel A. Farber, "The Conservative as Environmentalist: From Goldwater and the Early Reagan to the 21st Century," *Arizona Law Review* 59 (2017): 1005. For an extended analysis of partisan gridlock over climate disruption legislation, see Thomas O. McGarity, "Avoiding Gridlock through Unilateral Executive Action: The Obama Administration's Clean Power Plan," *Wake Forest Journal of Law & Policy* 7 (2017): 141.

91. Coral Davenport and Marjorie Connelly, "Half in G.O.P. Say They Back Climate Action," *New York Times,* January 31, 2015, A1.

92. Klein, *This Changes Everything,* 229.

93. Ibid., 126; Martin, *Coal Wars*, 35; "Citing 'Permanent' Trends, Officials Urge West Virginia to Plan Beyond Coal," *Inside EPA*, April 14, 2016; Taylor Kuykendall, "Coal Market Squeeze Hits Differently across US Coal Basins," *SNL Daily Coal Report*, December 4, 2015; Corbin Hiar, "EPA, Sierra Club Tell Labor Groups They're Fighting for the Climate and US Jobs," *SNL Renewable Energy Weekly*, February 14, 2014.

94. Scott Martelle, "Kick Coal, Save Jobs, Right Now," *Sierra*, January / February 2012, 22.

95. Martin, *Coal Wars*, 43–44, 51; Sheryl Gay Stolberg, "In Appalachia, Breaking the Coal Mine Habit," *New York Times*, August 18, 2016, A12; Coral Davenport, "As Wind Power Lifts Wyoming's Fortunes, Coal Miners Are Left in the Dust," *New York Times*, June 20, 2016, A9; Jonathan Tasini, "No One Left Behind," *Sierra*, July / August, 2015, 46; Taylor Kuykendall, "Narrow Band of 16 Central Appalachia Counties 'Ground Zero' in Coal Job Free Fall," *SNL Daily Coal Report*, June 18, 2015 (quoting Bruce Nilles, Sierra Club).

96. *The Impact of EPA Carbon Regulations in West Virginia: Hearings before the Senate Committee on Environment and Public Works*, 114th Cong. (March 23, 2015) (testimony of Jeremy Richardson, Union of Concerned Scientists); Benjamin Storrow, "Coal-Reliant Tribes Ponder a Future without Their Power Plant," *E&E News*, April 3, 2017; Stephen Lee, "Appalachian Aid Agency Bumped from Trump's Proposed Budget," *BNA Energy & Climate Report*, March 15, 2017; Christopher Coats, "Coal Country Advocates Continue to Turn Deaf Ear to Talk of Federal Assistance," *SNL Daily Coal Report*, May 11, 2016.

97. Luke Bassett and Jason Walsh, "The Trump Budget Cuts Hit Coal Communities and Workers Where It Hurts," Center for American Progress, April 2017; Adam Aton, "Climate Loses in Trump's Plan," *E&E News*, February 13, 2018; Jose A. DelReal, "Trump's Budget Targets Rural Development Programs That Provide a Quiet Lifeline," *Washington Post*, March 21, 2017.

98. Martin, *Coal Wars*, 36, 51; Richardson testimony (West Virginia); Valerie Volcovici, "Awaiting Trump's Coal Comeback, Miners Reject Retraining," *Reuters*, November 1, 2017; Trevor Houser, Jason Bordoff, and Peter Marsters, *Can Coal Make a Comeback?* (New York: Columbia / SIPA Center on Global Energy Policy, April 2017), 44–45; Elizabeth McGowan, "Rising from the Ashes, a Buffalo Suburb Ends Its Dependence on Coal," *Grist*, July 11, 2017; Arian Campo-Flores, "Kentucky Looks to Mine Alternatives," *Wall Street Journal*, May 24, 2017; Robert Pollin and Brian Callaci, "A Just Transition for U.S. Fossil Fuel Industry Workers," *American Prospect*, Summer, 2016, 86; Tasini, "No One Left Behind."

99. Dan Gearino, "Solar Is Saving Low-Income Households Money in Colorado. It Could Be a National Model," *Inside Climate News*, July 2, 2018; Krysti Shallenberger, "Is Rooftop Solar Just a Toy for the Wealthy?," *Utility Dive*, April 27,

2017 (San Antonio); Sidney A. Shapiro and Joseph P. Tomain, "Rethinking Reform of Electricity Markets," *Wake Forest Law Review* 40 (2005): 534–35.

100. Office of Community Services, Department of Health and Human Services, About LIHEAP, http://www.acf.hhs.gov/ocs/programs/liheap/about.

101. "Veterans in Energy," *Electric Perspectives,* September / October 2016, 8, 11; George Cahlink, "Lawmakers Eye Increases for Energy, Environmental Programs," *E&E News,* February 14, 2018; Christa Marshall, "Battle Looms over Trump Bid to Kill Weatherization Program," *E&E News,* March 29, 2017; Saqil Rahim, "Trump Zeros Out LIHEAP's Low-Income Energy Assistance," *E&E News,* March 17, 2017.

CONCLUSION

1. Doug Vine, *Solutions for Maintaining the Existing Nuclear Fleet* (Arlington, VA: Center for Climate and Energy Solutions, May 2018), 5; Rachael Nealer, David Reichmuth, and Don Anair, *Cleaner Cars from Cradle to Grave* (Cambridge, MA: Union of Concerned Scientists, 2015).

2. Sean Reilly, "Scientists Need to Better 'Communicate Successes'—Study," *E&E News,* March 21, 2018; Natural Resources Defense Council, *A Tectonic Shift in America's Energy Landscape,* Third Annual Energy Report (2015), 4, https://www.nrdc.org/sites/default/files/energy-environment-report-2015.pdf; Annalee Grant, "Fall of King Coal Bodes Well for US Commitment in Paris, Expert Says," *SNL Daily Coal Report,* January 14, 2016.

3. Edward Clump, "Record Wind Generation in the Queue after 2018 Boom," *E&E News,* April 9, 2019; Gavin Bade, "DER Management Takes Center Stage at DistribuTECH," *Utility Dive,* January 29, 2018; Herman K. Trabish, "How Rural Co-ops Are Shifting to a Cleaner Power Mix," *Utility Dive,* August 21, 2017; Christopher Martin, "U.S. Wind Farms in Construction or Development Rises 41 Percent," *BNA Daily Environment Report,* July 28, 2017; Christa Marshall, "Renewable Generation Tops Nuclear for First Time since 1984," *E&E News,* July 6, 2017.

4. Benjamin Storrow, "Inside the Troubling Rise of CO_2 at Power Plants," *E&E News,* January 10, 2019; Rachel Leven and Fatima Bhojani, "What Scott Pruitt's Been Doing While You Weren't Looking," Center for Public Integrity, May 15, 2018 (old power plants); World Meteorological Organization, "The State of Greenhouse Gases in the Atmosphere Based on Global Observations through 2016," WMO Greenhouse Gas Bulletin No. 13 (October 30, 2017), 2; Jessica Shankleman, "Carbon Pollution Touched 800,000-Year Record in 2016, WMO Says," *BNA Environment & Energy Report,* October 30, 2017; Chelsea Harvey, "Carbon Dioxide in the Atmosphere Is Rising at the Fastest Rate Ever Recorded," *Washington Post,* March 13, 2017, A1; "New EIA Projections Highlight Gap between GHG Policies, Long-Term Goals," *Inside EPA,* June 29, 2016; "EPA's Final

Cooling Water Rule Faces Likely Suit from Environmentalists," *Water Regulation Alert,* May 22, 2014.

5. Herman K. Trabash, "New Campaign Will Ask Coal Users to Face the 'Cold Hard Economic Case' Against Them," *Utility Dive,* October 22, 2018; Joshua Busby, "Trump Says Goodbye to the Paris Climate Agreement," *Washington Post,* June 1, 2017; Clifford Krauss and Diane Cardwell, "Policy's Promise for Coal Has Its Limits," *New York Times,* March 29, 2017, A1.

6. David B. Spence, "Regulation, 'Republican Moments,' and Energy Policy Reform," *BYU Law Review* 2011 (2011):1602–3; Dave Anderson, "Attacks on Wind and Solar Power by the Coal and Gas Industries" (Energy and Policy Institute, February 19, 2019); Annalee Grant, "As Coal Burn Declines, Clean Power Plan Fight Heats Up," *SNL Generation Markets Week,* May 24, 2016.

7. Chris Mooney and Brady Dennis, "Even in States Suing over New Climate Regulations, Coal Use Is Shrinking," Washington Post Blogs, May 3, 2016; Paul Krugman, "Coal Country Is a State of Mind," *New York Times,* March 31, 2017, A23.

8. Naomi Klein, *This Changes Everything* (New York: Simon & Schuster, 2014), 31–34; Taylor Kuykendall, "Judicial Battlefront: Environmentalists Look to the Air to Keep Coal Underground," *SNL Electric Utility Report,* September 14, 2015.

9. Thomas O. McGarity, "The Disruptive Politics of Climate Disruption," *Nova Law Review* 38 (2014): 468–69.

10. Coral Davenport, "Energy Trends Outpace Plans for the E.P.A.," *New York Times,* December 8, 2016, A1; Michael Weisskopf, "'Tall Stacks' and Acid Rain," *Washington Post,* June 5, 1989, A1.

11. Davenport, "Energy Trends"; Annalee Grant, "Top Generators Still Emitting More CO_2 Than in 1990; Other Pollutants Decline," *SNL Daily Coal Report,* July 14, 2016.

12. Chris Morris, Iceland Expects To Use More Electricity Mining Bitcoin Than Powering Homes This Year, *Fortune,* February 13, 2018, http://fortune.com /2018/02/13/iceland-bitcoin-mining-electricity/; Radoslav Danilak, Why Energy Is a Big and Rapidly Growing Problem for Data Centers, *Forbes,* December 15, 2017, https://www.forbes.com/sites/forbestechcouncil/2017/12/15/why-energy-is-a-big -and-rapidly-growing-problem-for-data-centers/#223898ab5a30.

13. Alexandra B. Klass, "Public Utilities and Transportation Electrification," *Iowa Law Review* 104 (2019): 545, 557; Nealer, Reichmuth, and Anair, *Cleaner Cars.*

14. Brad Plumer, "An Industrial City in Colorado Is One of Dozens Across the U.S. that Are Trying to Power Themselves on 100% Clean Energy . . . but Going Green Is Easier Said than Done," *Time,* July 23, 2018, 40; "Fighting Climate Change outside the Paris Accord," *New York Times,* September 21, 2017, A19; Keith Goldberg, "NC Solar Expansion Bill Mothballs Wind Projects until 2019," *Law360,*

July 28, 2017; Gerald B. Silverman, "States See Same Climate Paths after Trump Election," *BNA Environment Reporter*, November 11, 2016, 4057; Mooney and Dennis, "Even in States Suing."

15. *Implementation of the Clean Air Act of 1970—Part 1: Hearings before the Subcommittee on Air and Water Pollution of the Senate Committee on Public Works*, 92d Cong. 325 (1972) (Ruckelshaus quote); Bill Bishop, "Americans Have Lost Faith in Institutions," *Washington Post*, March 3, 2017.

16. William Boyd and Ann E. Carlson, "Accidents of Federalism: Ratemaking and Policy Innovation in Public Utility Law," *UCLA Law Review* 63 (2016): 852–53; Tyler Norris, "The Energy Endgame: We Already Have the Tools to End the Fossil Fuel Age," Greentech Media, May 17, 2017.

17. David J. Lynch, "Trump Imposes Tariffs on Solar Panels and Washing Machines in First Major Trade Action of 2018," *Washington Post*, January 22, 2018; Krysti Shellenberger, "Pruitt's Move to Repeal CPP Sets Up Prolonged Battle over Carbon Regulations," *Utility Dive*, October 11, 2017; "Evaluating Trump's Environmental Policy—Nine Months In," *BNA Daily Environment Report*, September 26, 2017; Danielle Paquette, "Solar's Rise Lifted These Blue-Collar Workers," *Washington Post*, June 5, 2017.

18. Jacob S. Hacker and Paul Pierson, *American Amnesia: How the War on Government Led Us to Forget What Made America Prosper* (New York: Simon & Schuster, 2016), 240; Jonathan Chait, "The Sunniest Climate-Change Story You've Ever Read," *New York Magazine*, September 7, 2015, 1.

19. See Thomas O. McGarity, "Avoiding Gridlock through Unilateral Executive Action: The Obama Administration's Clean Power Plan," *Wake Forest Journal of Law & Policy* 7 (2017): 141; Spence, "'Republican Moments,'" 1611.

20. Dylan Brown, "Bloomberg Puts Up Another $64M for 'War on Coal,'" *E&E News*, October 11, 2017; Michael Bloomberg, "Climate Progress, With or Without Trump," op-ed, *New York Times*, March 31, 2017, A23.

21. Ethyl Corp. v. EPA, 541 F.2d 1 (D.C. Cir. 1976); Marianne Lavelle, "Decade of Climate Evidence Strengthens Case for EPA's Endangerment Finding," *Inside Climate News*, December 14, 2018.

22. Annalee Grant, "US May Meet Clean Power Plan's National Goal 13 Years Earlier Than Planned," *SNL Daily Coal Report*, November 30, 2016.

23. Boyd and Carlson, "Accidents of Federalism," 861 (2016); Lisa Wood, "The Digital and Distributed Grid," *Electric Perspectives*, September / October 2016; Herman K. Trabish, "The New Demand Response and the Future of the Power Sector," *Utility Dive*, December 11, 2017; Jonathan Crawford, "ACEEE: EPA Rules Create 'Tipping Point' for Expanded Use of Energy Efficiency," *SNL Coal Report*, February 6, 2012.

24. *Keeping the Lights On—Are We Doing Enough to Ensure the Reliability and Security of the U.S. Electric Grid? Hearings before the Senate Committee on Energy*

and Natural Resources, 113th Cong. (April 10, 2014) (testimony of Cheryl Roberto, Environmental Defense Fund); Mara Prentiss, *Energy Revolution: The Physics and the Promise of Efficient Technology* (Cambridge, MA: Belknap Press of Harvard University Press, 2015); Gavin Bade, "Lazard: Renewables Can Challenge Existing Coal Plants on Price," *Utility Dive,* November 12, 2018; Reed Landberg and Anna Hirtenstein, "Coal Is Being Squeezed Out of Power Industry by Cheap Renewables," *Bloomberg Energy and Climate Report,* June 19, 2018; Dianne Cardwell, "Leaving a Life in the Coal Mine Behind," *New York Times,* October 1, 2017, B1; Robert Walton, "As Operators Update Grid Planning for Renewables, Transmission Remains Key Constraint," *Utility Dive,* September 18, 2017; Christa Marshall, "80% of Rooftops Can Hold Solar—Google," *E&E News,* March 23, 2017.

25. Grant, "As Coal Burn Declines"; Kristi E. Swartz, "At Southern Co., Fanning Vows a Break with the Past," *E&E News,* May 24, 2018; U.S. Energy Information Administration, "Electric Power Sector Consumption of the Fossil Fuels at Lowest Level Since 1994," May 29, 2018.

26. Nathan Hale, "Mayors Say Cities Must Take Lead on Fighting Climate Change," *Law360,* June 23, 2017; Daniel Cusick, "Mega-Retailer Makes Mega Wind Investment," *E&E News,* February 6, 2017; Hiroko Tabuchi, "U.S. Companies Urge Trump to Uphold Climate Deal," *New York Times,* November 17, 2016, B2.

27. Spence, "'Republican Moments,'" 1621; Denise A. Grab, Iliana Paul, and Kate Fritz, *Opportunities for Valuing Climate Impacts in U.S. State Electricity Policy* (New York: Institute for Policy Integrity, 2019); Warren Leon et al., *Clean Energy Champions: The Importance of State Programs and Policies* (Montpelier, VT: Clean Energy States Alliance, 2015); Benjamin Storrow, "New Best Friends: GOP Governors and Renewables," *E&E News,* June 23, 2017; Michael Grunwald, "Environmentalists Get a Dose of Good News," *Politico,* November 18, 2016.

28. Claire Weiller, "Plug-in Hybrid Electric Vehicle Impacts on Hourly Electricity Demand in the United States," *Energy Policy* 39 (2011): 3767; Nealer, Reichmuth, and Anair, *Cleaner Cars,* 1; Skip Descant, "Growth in Electric Car Ownership Presents Challenges to the Grid," *Governing,* March 12, 2018; Rebecca Kern and Stephen Lee, "White House Denies Murray Emergency Request, Despite Trump Backing," *BNA Daily Environment Report,* August 23, 2017.

29. Adam Aton, "Most Americans Want Climate Policies—That Don't Cost Much," *E&E News,* October 3, 2017 (61 percent); Nicholas Fandos, "Alarmed by Trump's Environmental Agenda, Thousands Join Climate March," *New York Times,* April 30, 2017, A20; Tim Appenzeller, "An Unprecedented March for Science," *Science,* April 24, 2017, 356; Coral Davenport and Marjorie Connelly, "Half in G.O.P. Say They Back Climate Action," *New York Times,* January 31, 2015, A1 (January 2015 poll).

30. Scott Waldman, "Atmospheric CO_2 Sets Record High," *E&E News,* May 2, 2018; Brian L. Wolff, "Our Energy Future," *Electric Perspectives,* July / August 2016,

6 (environmental group quote); Alan Neuhauser, "Boon or Bust? States, Businesses Take Sides on Clean Power Plan," USNEWS.com, December 2, 2014; Ben Geman, "Climate Battle Plan Shifts to Bottom Line," *National Journal Daily,* July 29, 2014; Mark Schapiro, "The Carbon Taxes We're Already Paying," *Los Angeles Times,* July 20, 2014, A23.

    31. Thomas O. McGarity, *Freedom to Harm: The Lasting Legacy of the Laissez Faire Revival* (New Haven, CT: Yale University Press, 2013), 11; John Quarles, *Cleaning Up America: An Insider's View of the Environmental Protection Agency* (Boston: Houghton Mifflin, 1976), 243; Spence, "'Republican Moments,'" 1599–1600.

    32. Naomi Klein, *This Changes Everything* (New York: Simon & Schuster, 2014), 26; Philip Shabecoff, *A Fierce Green Fire: The American Environmental Movement* (New York: Hill and Wang, 1993), 283; Nicholas Lemann, "When the Earth Moved," *New Yorker,* April 15, 2013, 73; Theda Skocpol, "Naming the Problem: What It Will Take to Counter Extremism and Engage Americans in the Fight Against Global Warming" (January 2013); Spence, "'Republican Moments,'" 1622.

Acknowledgments

This book is the product of many years of teaching environmental law at the University of Texas School of Law and learning from the hundreds of students who took that course. I am especially grateful to the members of the *Texas Environmental Law Journal* who put on a symposium on power plant regulation in February 2013 that brought together legal scholars, litigators, and regulators from across the country. The law school provided additional support through the Joe R. and Teresa Lozano Long Chair in Administrative Law, without which this book would have been impossible. Barbara Bridges, the Tarlton Law Library's government documents librarian, was a constant source of information. The scholars in the Center for Progressive Reform were sources of inspiration and information. My colleagues David Adelman, Kelly Haragan, and David Spence provided critical feedback on a near-final version of the manuscript. And my cousin Ben Thomas provided helpful comments and feedback near the end of the project. I am also grateful to the two reviewers for Harvard University Press who read the manuscript and provided helpful suggestions for improving the book. My faculty assistant Dottie Lee cheerfully responded to all of my sometimes unreasonable requests for help. As always, I am deeply grateful to my wife, Cathy, who, in addition to tolerating too many evenings of silence from the other end of the workspace we share, fed me reams of clippings on power plant regulation.

Index